BIOTECHNOLOGY AND BIOINFORMATICS

Editor

Shikha Singh Sahni

scitus academics

Biotechnology and Bioinformatics
Edited by **Shikha Singh Sahni**

This Book is Printed on Acid-Free Paper. ♾

ISBN: 9781681179421
Library of Congress Control Number: 2017949080

SCITUS Academics LLC,
616, Corporate Way, Suite 2, 4766,
Valley Cottage, NY 10989

www.scitusacademics.com

Notice

Printed in United States of America.

Preface

Over the past few decades, major advances in the field of molecular biology, coupled with advances in genomic technologies, have led to an explosive growth in the biological information generated by the scientific community. This surge of genomic information has, in turn, led to an absolute requirement for computerized databases to store, organize, and index the data and for specialized tools to view and analyze the data. There are many tools in bioinformatics, with many functions to suit the needs and expertise of the scientists using them. Gene and protein databases are constantly being updated with information that aid scientists all around the world, in whatever field of the life sciences they are working. Bioinformatics carries benefits for plant researchers: it can aid in plant breeding and genetic engineering, and allow plant scientists to produce better crops for the future. By knowing which plants are closely related, scientists can figure out which sexually compatible species have desirable characteristics (such as longer stalks for rice plants, or larger grains for barley, corn, or wheat). The wild relatives of today's plants may be sources of crop improvement genes. This information, in conjunction with appropriate technology, may provide predictive measures of plant health and quality and become part of future breeding decision management systems. Next-generation sequencing coupled with high-performance computing methods have revolutionized the field of plant breeding and genetics.

The volume 'Biotechnology and Bioinformatics' contains recent advances in certain biotechnological applications. It presents some of the key concepts, methods, software packages, and databases used in bioinformatics, with an emphasis on those relevant to plant science. It also cover some fundamental issues related to biological sequence analyses, transcriptome analyses, computational proteomics, computational metabolomics, bio-ontologies, and biological databases. A focus on a few emerging research topics in bioinformatics is given. This book will be very helpful to the undergraduate and postgraduate students, researchers, teachers of microbiology, biotechnology, agriculture and horticulture.

Table of Contents

CHAPTER 1

New Traits in Crops Produced by Genome Editing Techniques Based on Deletions

C. C. M. van de Wiel[1] · J. G. Schaart[1] · L. A. P. Lotz[1] · M. J. M. Smulders[1]

Wageningen University and ResearchWageningenThe Netherlands

ABSTRACT

One of the most promising New Plant Breeding Techniques is genome editing (also called gene editing) with the help of a programmable site-directed nuclease (SDN). In this review, we focus on SDN-1, which is the generation of small deletions or insertions (indels) at a precisely defined location in the genome with zinc finger nucleases (ZFN), TALENs, or CRISPR-Cas9. The programmable nuclease is used to induce a double-strand break in the DNA, while the repair is left to the plant cell itself, and mistakes are introduced, while the cell is repairing the double-strand break using the relatively error-prone NHEJ pathway. From a biological point of view, it could be considered as a form of targeted mutagenesis. We first discuss improvements and new technical variants for SDN-1, in particular employing CRISPR-Cas, and subsequently explore the effectiveness of targeted deletions that eliminate the function of a gene, as an approach to generate novel traits useful for improving agricultural sustainability, including disease resistances. We compare them with examples of deletions that resulted in novel functionality as known from crop domestication and classical mutation breeding (both using radiation and chemical mutagens). Finally, we touch upon regulatory and access and benefit sharing issues regarding the plants produced.

Keywords: Nuclease SDN SSN Precision breeding Sustainable agriculture

INTRODUCTION

New plant breeding techniques (NPBT) encompass a set of diverse techniques and concepts that all aim to improve efficiency and/or precision of plant breeding. Most make use of transgenic plant lines at some point in the breeding process, but the final product generally contains only small mutations and in specific cased no modifications at all, and is often indistinguishable from the conventional breeding products (Lusser et al. 2012; Schaart et al. 2016). One of the most promising among these, genome editing (also called gene editing or gene targeting) with the help of a programmable nuclease (SDN: Site-directed nuclease, or SSN: sequence-specific nuclease), recently has led to a deluge of creative applications with the introduction of the CRISPR-Cas9 system. While the thorough review of NPBT by Lusser et al. (2012) not yet mentioned CRISPR-Cas, 4 years later a search using CRISPR and plants as keywords provided 246 publications in Web of Science (on 31-10-2016). Similarly to zinc finger nucleases (ZFN) and TALENs, CRISPR-Cas is able to make a double-strand break (DSB) at a precisely specified location in the genome, but it is much more versatile and easy to use, because the specificity of the target sequence is achieved by a separate guide RNA (gRNA) that can be easily designed and readily synthesised rather than by the protein structure itself (ZFN, TALEN). The use in plants has recently been reviewed by Luo et al. (2016), Paul and Qi (2016), Hilscher et al. (2016), and Rani et al. (2016).

At an early stage of the development of genome editing, three types of uses based on the DSB repair mechanism have been distinguished for regulatory purposes (Lusser et al. 2011). In SDN-1, the DSB is repaired by the non-homologous end joining (NHEJ) DNA repair machinery of the cell, during which small mistakes may be introduced, mostly consisting of small indels. Plants with such mistakes can be identified by screening. Indels in genes frequently lead to gene knockout as a result of reading frame shifts causing premature translational stops. In SDN-2, an oligonucleotide is provided to assist in the repair of the DSB that is identical to the sequence in which the DSB is made, except that a desired mutation is included. Some plants may use this oligo in an alternative cellular repair mechanism, homology-directed repair (HDR). This may lead to a higher frequency of plants with the desired change, e.g., a single amino-acid substitution. In SDN-3, the DSB is repaired in the same way, but the repair template is a longer sequence that may include a complete gene or a promoter. This will lead to the insertion of native or foreign sequences at a precisely specified location in the genome. SDN-3 is thus similar to genetic transformation and it generally results in transgenic plants, but they are produced with improved precision as the inserted sequence is targeted to a specific site in the genome, or exchanged with existing sequences at that site, e.g., in promoter swaps. Thus, SDN-3 holds much promise for diverse applications, including stacking desirable (transgenic) traits, so that they are passed on to progeny as a single block during breeding, which is desirable for efficient introgression into diverse elite materials (Kumar et al. 2016). HDR is still a relatively difficult to control mechanism. An alternative technique is called ODM (oligo-directed mutagenesis). It may produce results similar to SDN-2 but does not employ a nuclease to make a DSB. Combinations of both, improving efficiency, are also explored (Sauer et al. 2016).

Although SDN-1 is precise only in the targeted site and not in the resulting type of mutation, it appears presently to be the most frequently implemented, and is, therefore, the subject of this review. We first discuss improvements and new variants of SDN-1, in particular employing CRISPR-Cas, and subsequently explore the possibilities of targeted deletions that eliminate the function or part of the function of a gene, as an approach to generate novel traits useful for improvement of agricultural sustainability. We compare them with examples of deletions that resulted in novel functionality as known from crop domestication and classical mutation breeding. Finally, we touch upon regulatory and IP issues around the plants produced, as well as access and benefit sharing.

IMPROVEMENTS AND NEW VARIANTS OF THE TECHNOLOGY

Off-target effects

CRISPR-Cas genome editing has been reported to be accompanied by off-target effects, i.e., mutations arising from repairing DSBs induced elsewhere in the genome than the targeted sequence. Several research groups have made changes to the Cas protein and/or gRNA design to enhance the sequence target specificity (Kleinstiver et al. 2016; Slaymaker et al. 2016; Doench et al. 2016). In addition, alternative endonucleases (e.g. Cpf1) could increase specificity (e.g., Ran et al. 2015; Kim et al. 2016a). Schiml et al. (2014) turned Cas9 into a nickase, so that two adjacent gRNAs were required to make a DSB with sticky ends. This improved the specificity, alone and in combination with using truncated gRNAs (Fu et al. 2014). Tolerance for mismatches in the gRNA (actually off-target effects) may also be used to specifically target related genes from gene families (Endo et al. 2015).

Various methods for genome-wide detection of off-target mutations have been developed, including BLESS (Crosetto et al. 2013), GUIDE-seq (Tsai et al. 2015), Digenome-seq (Kim et al. 2015, 2016b), and END-seq (Canela et al. 2016). For the few plant species assessed so far, little off-target effects appear to occur using CRISPR-Cas, but this would need to be studied in more species (Peterson et al. 2016). For example, off-target effects could be limited using an optimal molar ratio of Cas9 to gRNAs by Woo et al. (2015), while Peterson et al. (2016) did not detect any in a multiplex approach by re-sequencing in *Arabidopsis*, including in computationally predicted sites.

Off-target mutations are considered a problem in, e.g., applications in humans, but they are less likely to pose a problem in plant breeding, as usually multiple plants are produced that are subsequently screened and selected. In many cases, there are multiple generations between the genome editing step and the variety produced, in which off-target mutations are selected against and in seed-propagated crops, segregate out. In addition, the frequency of off-target mutations made across the genome will be much lower than in classical mutagenesis. The type and frequency of mutations in classical mutagenesis depend on the method (chemical or ionizing radiation) used, type and concentration of chemical or type of radiation, and the duration of the treatment, which are generally adjusted case by case (Suprasanna et al. 2015). Polyploid crops, in general, tolerate a higher frequency than diploid crops (Uauy et al. 2009; Shu et al. 2011; Oladosu et al. 2016). A frequency of one mutation induced every 78 kb (what Tsuda et al. 2015 used in soybean) would in soybean or tomato mean that classical mutagenesis may introduce for every desired mutation more than 10,000 other mutations.

Delivery

Commonly plants are transformed using *Agrobacterium* or biolistic systems to introduce the Cas9- and gRNAs-encoding DNA stably into the plant genome. This may work best in tissue culture systems, as exemplified by the high success rate in rice (Paul and Qi 2016). When using floral dip, germline editing was shown to be improved in *Arabidopsis* by driving the expression of Cas9 with promoters specific for egg cells (Wang et al. 2015) or for actively dividing tissues, such as meristems and embryo sacs (Hyun et al. 2015; Yan et al. 2015).

Alternatively, viral vectors for gRNA (Ali et al. 2015; Yin et al. 2015) may improve efficiency.

Clasen et al. (2016) used a temporary expression system, i.e., a TALEN transcribed from plasmids introduced into potato protoplasts. Regenerated plants with mutations were checked for the absence of plasmid sequence insertions in their genome. Woo et al. (2015) transfected protoplasts of *Arabidopsis thaliana*, tobacco, lettuce, and rice with ribonucleoprotein complexes containing Cas9 nuclease protein and appropriate gRNAs and obtained genome-edited regenerated plants. They called their approach "DNA free genome editing". It has also been used in the unicellular green alga *Chlamydomonas reinhardtii* (Baek et al. 2016) and it was also shown to work in *Petunia hybrida* protoplasts (Subburaj et al. 2016). The approach requires protocols for protoplasting and plant regeneration from protoplasts, which are not available for many crops, or, if existing, not for all varieties within a crop. The desire to use genome editing in research and in plant breeding urges for a surge of activity into transformation, protoplasting, and regeneration protocols for crops and model species that up to now have remained recalcitrant, such as wheat (Altpeter et al. 2016).

Multiplexing

CRISPR-Cas9 is particularly suitable to multiplexing because of the versatility of the gRNAs, but also TALENs have been multiplexed: four *Nicotiana benthamiana* genes involved in glycosylation were knocked out simultaneously using TALENs for the benefit of biopharmaceutical production of glycoproteins devoid of plant-specific residues (Li et al. 2016). Peterson et al. (2016) targeted 14 loci simultaneously in *A. thaliana* with CRISPR-Cas using stacked gRNA expression arrays (Peterson et al. 2016). Steinert et al. (2015) simultaneously used two alternative modified Cas proteins, and each was directed by its own gRNA.

Larger deletions

Using two gRNAs targeted to sites at some distance from each other on the same chromosome, larger deletions become possible based on annealing the distant sites through NHEJ. In this way, Zhou et al. (2014) managed to delete a cluster of up to 10 loci in the terpenoid synthesis pathway in rice. Deleting clusters of related genes is useful, e.g., for the removal of alpha- or gamma-gliadins, which are organized in gene repeats on different chromosomes, while trying to produce wheat that is safe for people with celiac disease (Jouanin et al., in preparation).

GENERATION OF NEW TRAITS FOR CROP IMPROVEMENT THROUGH DELETIONS

'Classical' genetic modification was developed to introduce genes into plants to obtain a gain of function, although it soon was also used to silence endogenous genes, e.g., using RNAi. The reason for this is that the loss of a gene or gene function may also result in a new plant phenotype that is useful for man. During crop domestication, many traits have been selected for that are inherited as recessive, so mostly loss-of-function (Lu et al. 2006; Hancock 2012; Martínez-Ainsworth and Tenaillon 2016). They often comprised knockouts of genes. Examples include loss of seed shattering, loss of seed dormancy, reduction of shoots improving the harvest index, reduction of chemical and physical defences, and loss of photoperiodicity and/or the vernalisation requirement (Nakamichi 2015). Similar mutations have been selected for independently in different crops (e.g., Cheng et al. 2016).

Since the 1930s, (knockout) mutagenesis has been used to generate new traits, boosted by increased knowledge on the effects of various sources of radiation (which mostly induce deletions) and chemical mutagens (for example, EMS induces C > T mutations). This had led to the development of more than 3000 crop varieties, and it is still popular, for instance, for generating new flower colours in ornamentals (it is a standard procedure in Chrysanthemum). Interestingly, mutation breeding also produces, at useable frequencies, gain of function traits, such as new disease resistances (Oladosu et al. 2015). Resistance to biotic and abiotic stresses is an important goal of breeding new varieties, as one of the elements to feed the world in a sustainable way.

The SDN-1 systems make it also easier to deal with the recessive nature of the mutations, which in classical mutagenesis generally means that mutations do not show a phenotype in the mutants themselves, only in their progeny. SDN-1 can mutate all alleles of the targeted locus simultaneously, which is a particularly significant improvement in mutating polyploids. Examples are deletions in the three homoeologues (six alleles) of *MLO* for powdery mildew resistance in hexaploid wheat using TALENs by Wang et al. (2014) and the four copies of GBSS for altering starch composition in tetraploid potato by Andersson et al. (2016).

Disease resistance

Increasing pathogen resistance in crops is an important way to improve agricultural sustainability. Resistance may be generated by deleting plant genes that are used ('hijacked') by pathogens and on which the pathogens depend for growth and development (Pavan et al. 2010). These are called S genes (for Susceptibility, in contrast to dominant R genes which confer Resistance through recognition of the pathogen). The classical example is *mlo*, a recessive mutant which has been effective for mildew resistance in barley for over 30 years (reviewed by Acevedo-Garcia et al. 2014).

Knocking out S genes generated disease resistance in multiple crop species, but this may be accompanied by poor plant performance. Precisely targeted deletions using SDN-1 can make it possible to generate specific pathogen resistance without fitness costs for the plant. Using TALENs, Li et al. (2012) were able to identify and delete a small part of the rice SWEET14 promoter that was targeted by an effector protein secreted by the causal agent of Bacterial Leaf Blight, *Xanthomonas oryzae* pv. oryzae (Xoo), apparently without affecting adequate expression of the gene by the plant itself. Thus, a rice line was generated that was resistant to Xoo strains that use this effector. Later on, also CRISPR-Cas9 was implemented for mutating SWEET-type S genes (Zhou et al. 2014). S genes have been identified in a range of species, including *Arabidopsis*, rice, soybean, and potato and tomato (Sun et al. 2016a, 2016b; Zheng et al. 2016).

Resistance to viruses can also be generated following an S gene approach. Disrupting the functionality of eIF4E (eukaryotic translation initiation factor 4E) effected broad potyvirus resistance in cucumber (Chandrasekaran et al. 2016). Pyott et al. (2016) likewise achieved potyvirus resistance in *Arabidopsis*.

Product quality

In many crops, including legumes and cereals, phytate in seeds interferes with phosphorous uptake. Decreasing phytate content in feed will increase net uptake of P by livestock which may reduce losses of P to the environment. Liang et al. (2014) used both TALEN and CRISPR-Cas9 successfully in maize to obtain mutations in *IPK*, which is involved in phytate production. Reducing phytate comes with trade-offs for plant performance that may be addressed more precisely through genome editing. Recently, Yamaji et al. (2017) reported an alternative approach of reducing seed P content in rice using retrotransposon-insertional knockout mutants of *SPDT*, a P distribution transporter controlling the allocation of P to the grains. In these mutant plants, yield and seed performance were apparently not affected. An alternative TALEN-based approach was explored by Wendt et al. (2013): introducing small indels into the promoter of the most important barley grain phytase gene, *HvPAPhy_a*, which could be used to test for the possibilities of changing gene expression in the grain. Mutations were achieved in the targeted promoter site, but no further testing of plants was reported.

Clasen et al. (2016) used gene editing to knock out *Vnvl* to reduce acrylamide (a potential carcinogen) levels of potato products after heating. A similar result could be obtained by silencing asparagine synthetase genes, in this case using RNAi (Zhu et al. 2016). Some other examples of traits improving product quality targeted by SDN-1 are high amylopectin ("waxy") maize (by Pioneer, https://www.aphis. usda.gov /biotechnology/downloads/reg_loi/15-352-01_air_response_signed.pdf) and potato (Andersson et al. 2016) based on knockout of *GBSS*, and high oleic acid oilseed crops soybean (Haun et al. 2014) and camelina (Jiang et al. 2016) based on knockouts of *FAD2* and/or *FAD3*. Both traits had been addressed earlier using classical mutagenesis or RNAi. Rice fragrancy was improved by knockout of *BADH2* (Shan et al. 2015). For bio-based crops, reducing lignin contents has been studied using a transgenic silencing method up to the level of field trials in poplar for improving biofuel production (Van Acker et al. 2014). Zhou et al. (2015) reported successful SDN-1 in two poplar 4CL genes, the knockout of one of which led to lower lignin levels in stem wood.

Allergens

Removing allergens through genome editing would benefit specific groups of consumers. Dubois et al. (2015) showed that silencing Mal d 1 reduced the allergenicity of apple, which, in most patients, is a mild allergy resulting from cross-reactivity of the birch pollen Bet v 1 allergens to PR-10 proteins in Rosaceous fruits, such as apple, cherry, and strawberry. PR-10 proteins are encoded by a large gene family. Apple has 31 Mal d 1 genes of which 20 may be expressed in the fruit (Pagliarani et al. 2013) and may be targeted by genome editing. Peanut allergy is a life-threatening food allergy. Dodo et al. (2008) managed to reduce the allergenicity of the immunodominant Ara h 2 protein in peanuts using RNAi. For hypoallergenic peanuts to be safe for consumption by many patients, all genes coding for allergens would need to be silenced or removed, and genome editing offers the tools to efficiently do this.

Sensitivity of individuals with coeliac disease to cereal gluten is particularly difficult to address as gluten comprises of large gene families, with several epitopes in mainly alpha-, gamma-, and omega-gliadins that can be recognised by human T cells (Van Herpen et al. 2006; Van den Broeck et al. 2009; Salentijn et al. 2012) and that all are targets for controlled deletion or modification (Smulders et al. 2015; Jouannin et al. in prep.). Barro et al. (2016) succeeded in removing highly immunogenic gliadin proteins from wheat using RNAi. Clinical trials using breads from these silenced wheat plants were planned in Spain. Silencing will in most cases not completely remove the gliadin, and a complete knockout approach by genome editing may be more effective, and possibly more easily accepted by consumers as no foreign DNA is introduced (Laursen 2016).

Hybrid varieties

Pivotal to successful production of high-quality hybrid varieties is the availability of reliably male-sterile maternal lines. Pioneer is using a male sterility system in maize involving mutants of the male fertility gene *Ms45*. Knockout of the *Ms45* gene using CRISPR-Cas9 was recently reported by Svitashev et al. (2015).

Yield

Interestingly, a complex trait par excellence, yield, was also shown to be amenable to an SDN-1 approach. Li et al. (2017) used CRISPR-Cas9 in rice to mutate the regulatory genes *Gna1, DEP1*, and *GS3* and obtained plants with increased grain numbers, dense erect panicles plus semi-dwarf phenotype, and larger grains, respectively.

REGULATORY AND IP ISSUES

Currently, there is worldwide a discussion whether genome-edited plants should fall under existing regulatory systems that have been designed for transgenic (GM or GE) plants (for an overview, see Sprink et al. 2016). In the US, no regulatory oversight was deemed necessary by USDA-APHIS (e.g., waxy maize, high oleic acid soybean mentioned above under 'Product quality' and *sweet14*-based blight resistance in rice under 'Disease resistance', see for further examples Table 1 of Hilscher et al. 2016). In Europe, there is no definitive legal analysis yet. The EFSA GMO unit (2015) considered SDN-1 a form of mutagenesis; further analysis may be needed upon technological advancement. Among EU Member States, the German BVL (Federal Office of Consumer Protection and Food Safety) had a similar judgement, including for CRISPR, as gRNAs were not seen as recombinant DNA (i.e., no novel combination of genetic material) and the Swedish Board of Agriculture (SBA) saw CRISPR-Cas as equivalent to mutagenesis provided that no "foreign" DNA was left in plants. The German BfN (Federal Agency for Nature Conservation) held a different view, as were NGOs and organic farming organizations (see

IFOAM 2015 statement): they consider the process as most relevant and the process involves molecules not occurring naturally (Sprink et al. 2016). Mutagenesis was exempted from regulation in EU Directive 2001/18/EC as having a history of safe use (see for detailed comparison with SDN-1, the off-targets section above).

The EU GM regulation is perceived as prohibitive to most applications and to small companies because of costs and uncertainties around timing and outcomes of the application procedure. When extensive GM regulation would be applicable to SDN-1 techniques, this may lead to a paradoxical result. Parties aiming at commercializing the plant products could produce a gene-edited plant with deletions to establish whether it generates the desired functionality, and then would "re-produce" a similar plant using classical mutagenesis. Companies would be able to do that also because the quickly increasing efficiency of next-generation DNA sequencing for mutation screening in recent times has stimulated them to (re)develop mutated populations in their crops (Van de Wiel et al. 2016). However, this would not be an efficient and optimal use of resources in breeding, and would give up on advantages hardly feasible by classical mutagenesis, such as the possibility of inducing recessive mutations in all alleles of a targeted gene in polyploid crops. There is a need for regulatory systems duly addressing biosafety that are balanced in terms of workability and costs, so that they do not unduly limit the benefits of producing innovative plant products with potential advantages to agricultural sustainability.

The possibilities offered by genome editing also have ramifications to IP, access, and benefit issues, even when it concerns deletions and loss-of-function mutations. It may also raise questions in the context of the use of genetic resources: can copying a mutation already existing in a genebank accession into elite material be considered use in the sense of the '*Nagoya Protocol on Access to Genetic Resources and the Fair and Equitable Sharing of Benefits Arising from their Utilization to the Convention on Biological Diversity*'? These ramifications may attract a lot of attention in the coming years when the possibilities of SDN-1 are realized in the form of improved crop varieties (or even varieties from underutilized or novel crops). These discussions are important for the realisation of the possibilities that SDN-1 using genome editing offers for increasing the sustainability of agriculture.

ACKNOWLEDGEMENTS

Writing of this review was supported by the Ministry of Economic Affairs of The Netherlands (BO-20-003-006). The funders had no role in study design, data collection and analysis, decision to publish, or preparation of the manuscript.

REFERENCES

1. Acevedo-García J, Kusch S, Panstruga R (2014) Magical mystery tour: MLO proteins in plant immunity and beyond. New Phytol 204:273–281
2. Ali Z, Abul-faraj A, Li LX, Ghosh N, Piatek M, Mahjoub A, Aouida M, Piatek A, Baltes NJ, Voytas DF, Dinesh-Kumar S, Mahfouz MM (2015) Efficient virus-mediated genome editing in plants using the CRISPR/Cas9 system. Mol Plant 8:1288–1291
3. Altpeter F, Springer NM, Bartley LE, Blechl AE, Brutnell TP, Citovsky V, Conrad LJ, Gelvin SB, Jackson DP, Kausch AP, Lemaux PG, Medford JI, Orozco-Cardenas ML, Tricoli DM, Van Eck J, Voytas DF, Walbot V, Wang K, Zhang ZYJ, Stewart CN (2016) Advancing crop transformation in the era of genome editing. Plant Cell 28:1510–1520.
4. Andersson M, Turesson H, Nicolia A, Fält AS, Samuelsson M, Hofvander P (2016) Efficient targeted multiallelic mutagenesis in tetraploid potato (Solanum tuberosum) by transient CRISPR-Cas9 expression in protoplasts. Plant Cell Rep.
5. Baek K, Kim DH, Jeong J, Sim SJ, Melis A, Kim JS, Jin E, Bae S (2016) DNA-free two-gene knockout in Chlamydomonas reinhardtii via CRISPR-Cas9 ribonucleoproteins. Sci Rep 6:30620
6. Barro F, Iehisa JCM, Giménez MJ, García-Molina MD, Ozuna CV, Comino I, Sousa C, Gil-Humanes J (2016) Targeting of prolamins by RNAi in bread wheat: effectiveness of seven silencing-fragment combinations for obtaining lines devoid of coeliac disease epitopes from highly immunogenic gliadins. Plant Biotechnol J 14:986–996
7. Canela A, Sridharan S, Sciascia N, Tubbs A, Meltzer P, Sleckman BP, Nussenzweig A (2016) DNA breaks and end resection measured genome-wide by end sequencing. Mol Cell 63:898–911
8. Chandrasekaran J, Brumin M, Wolf D, Leibman D, Klap C, Pearlsman M, Sherman A, Arazi T, Gal-On A (2016) Development of broad virus resistance in non-transgenic cucumber using CRISPR/Cas9 technology. Mol Plant Pathol 17:1140–1153
9. Cheng F, Sun R, Hou X, Zheng H, Zhang F, Zhang Y et al (2016) Subgenome parallel selection is associated with morphotype diversification and convergent crop domestication in Brassica rapa and Brassica oleracea. Nature Genet 48:1218–1224.
10. Clasen BM, Stoddard TJ, Luo S, Demorest ZL, Li J, Cedrone F, Tibebu R, Davison S, Ray EE, Daulhac A, Coffman A, Yabandith A, Retterath A, Haun W, Baltes NJ, Mathis L, Voytas DF, Zhang F (2016) Improving cold storage and processing traits in potato through targeted gene knockout. Plant Biotechnol J 14:169–176
11. Crosetto N, A Mitra, MJ Silva, M Bienko, N Dojer, Q Wang, E Karaca, R Chiarle, M Skrzypczak, K Ginalski, P Pasero, M Rowicka, I Dikic (2013) Nucleotide-resolution DNA double-strand break mapping by next-generation sequencing. Nature Methods 10:361–365.
12. Dodo HW, Konan KN, Chen FC, Egnin M, Viquez OM (2008) Alleviating peanut allergy using genetic engineering: the silencing of the immunodominant allergen Ara h 2 leads to its significant reduction and a decrease in peanut allergenicity. Plant Biotechnol J 6:135–145
13. Doench JG, Fusi N, Sullender M, Hegde M, Vaimberg EW, Donovan KF, Smith I, Tothova Z, Wilen C, Orchard R, Virgin HW, Listgarten J, Root DE (2016) Optimized sgRNA design to maximize activity and minimize off-target effects of CRISPR-Cas9. Nat Biotechnol 34:184–191
14. Dubois AEJ, Pagliarani G, Brouwer RM, Kollen BJ, Dragsted LO, Eriksen FD, Callesen O, Gilissen LJWJ, Krens FA, Visser RGF, Smulders MJM, Vlieg-Boerstra BJ, Flokstra-de

Blok BJ, van de Weg WE (2015) First successful reduction of clinical allergenicity of food by genetic modification: Mal d 1 silenced apples cause fewer allergy symptoms than the wild type cultivar. Allergy 70:1406–1412.

15. EFSA GMO UNIT (2015) EFSA response to DG SANTE request to provide technical assistance on issues related to the legal analysis of new plant breeding techniques. Mandate Number: M-2015-0183. P 4. http://registerofquestions.efsa. europa.eu/roqFrontend/ListOfQuestionsNoLogin?1
16. Endo M, Mikami M, Toki S (2015) Multigene knockout utilizing off-target mutations of the CRISPR/Cas9 system in rice. Plant Cell Physiol 56:41–47
17. Fu YF, Sander JD, Reyon D, Cascio VM, Joung JK (2014) Improving CRISPR-Cas nuclease specificity using truncated guide RNAs. Nat Biotechnol 32:279–284
18. Hancock JF (2012) Plant evolution and the origin of crop species. CABI, WallingfordGoogle Scholar
19. Haun W, Coffman A, Clasen BM, Demorest ZL, Lowy A, Ray E, Retterath A, Stoddard T, Juillerat A, Cedrone F, Mathis L, Voytas DF, Zhang F (2014) Improved soybean oil quality by targeted mutagenesis of the fatty acid desaturase 2 gene family. Plant Biotechnol J 12:934–940
20. Hilscher J, Bürstmayr H, Stoger E (2016) Targeted modification of plant genomes for precision crop breeding. Biotechnol J.
21. IFOAM EU Group (2015) IFOAM EU position paper on New Plant Breeding Techniques. IFOAM EU Group, p 5. http://www.ifoam-eu.org/sites/default /files/ifoameu_policy_npbts_position_final_20151210.pdf
22. Jiang WZ, Henry IM, Lynagh PG, Comai L, Cahoon EB, Weeks DP (2016) Significant enhancement of fatty acid composition in seeds of the allohexaploid, Camelina sativa, using CRISPR/Cas9 gene editing. Plant Biotechnol J.
23. Kim D, Bae S, Park J, Kim E, Kim S, Yu HR, Hwang J, Kim JI, Kim JS (2015) Digenome-seq: genome-wide profiling of CRISPR-Cas9 off-target effects in human cells. Nat Methods 12:237–243
24. Kim D, Kim J, Hur JK, Been KW, Yoon S-H, Kim J-S (2016a) Genome-wide analysis reveals specificities of Cpf1 endonucleases in human cells. Nat Biotechnol 34:863–868.
25. Kim D, Kim S, Kim S, Park J, Kim J-S (2016b) Genome-wide target specificities of CRISPR-Cas9 nucleases revealed by multiplex Digenome-sEq. Genome Res 26:406–415.
26. Kleinstiver BP, Pattanayak V, Prew MS, Tsai SQ, Nguyen NT, Zheng Z, Joung JK (2016) High-fidelity CRISPR-Cas9 nucleases with no detectable genome-wide off-target effects. Nature 529:490–495
27. Kumar S, Barone P, Smith M (2016) Gene targeting and transgene stacking using intra genomic homologous recombination in plants. Plant Methods 12:11
28. Laursen L (2016) Will Europe toast GM wheat for gluten sufferers? Nat Biotechnol 34:369–371.
29. Li T, Liu B, Spalding MH, Weeks DP, Yang B (2012) High-efficiency TALEN-based gene editing produces disease-resistant rice. Nat Biotechnol 30:390–392
30. Li J, Stoddard TJ, Demorest ZL, Lavoie PO, Luo S, Clasen BM, Cedrone F, Ray EE, Coffman AP, Daulhac A, Yabandith A, Retterath AJ, Mathis L, Voytas DF, D'Aoust MA, Zhang F (2016) Multiplexed, targeted gene editing in Nicotiana benthamiana for glyco-engineering and monoclonal antibody production. Plant Biotechnol J 14:533–542
31. Li J, Sun Y, Du J, Zhao Y, Xia L (2017) Generation of targeted point mutations in rice by a modified CRISPR/Cas9 system. Mol Plant.
32. Liang Z, Zhang K, Chen K, Gao C (2014) Targeted mutagenesis in Zea mays using TALENs and the CRISPR/Cas system. J Genet Genom 41:63–68
33. Lu J, Tang T, Tang H, Huang JZ, Shi SH, Wu CI (2006) The accumulation of deleterious mutations in rice genomes: a hypothesis on the cost of domestication. Trends Genet 22:126–131

34. Luo M, Gilbert B, Ayliffe M (2016) Applications of CRISPR/Cas9 technology for targeted mutagenesis, gene replacement and stacking of genes in higher plants. Plant Cell Rep 35:1439–1450
35. Lusser M, Parisi C, Plan D, Rodríguez-Cerezo E (2011) New plant breeding techniques. State-of-the-art and prospects for commercial development. European Commission, JRC-IPTS, p 218.
36. Lusser M, Parisi C, Plan D, Rodríguez-Cerezo E (2012) FEATURE—Deployment of new biotechnologies in plant breeding. Nat Biotechnol 30:231–239
37. Martinez-Ainsworth NE, Tenaillon MI (2016) Superheroes and masterminds of plant domestication. (Special issue: trajectories of genetics, 150 years after Mendel.). CR Biol 339:268–273
38. Nakamichi N (2015) Adaptation to the local environment by modifications of the photoperiod response in crops. Plant Cell Physiol 56:594–604.
39. Oladosu Y, Rafii MY, Abdullah N, Hussin G, Ramli A, Rahim HA, Miah G, Usman M (2016) Principle and application of plant mutagenesis in crop improvement: a review. Biotechnol Biotechnol Equip 30:1–16.
40. Pagliarani G, Paris R, Arens P, Tartarini S, Ricci G, Smulders MJM, van de Weg WE (2013) A qRT-PCR assay for the expression of all Mal d 1 isoallergen genes. BMC Plant Biol 13: 51.
41. Paul JW III, Qi Y (2016) CRISPR/Cas9 for plant genome editing: accomplishments, problems and prospects. Plant Cell Rep 35:1417–1427
42. Pavan S, Jacobsen E, Visser RGF, Bai Y (2010) Loss of susceptibility as a novel breeding strategy for durable and broad-spectrum resistance. Mol Breed 25:1–12
43. Peterson BA, Haak DC, Nishimura MT, Teixeira PJPL, James SR, Dangl JL et al (2016) Genome-Wide Assessment of efficiency and specificity in CRISPR/Cas9 mediated multiple site targeting in Arabidopsis. PLoS One 11:e0162169
44. Pyott DE, Sheehan E, Molnar A (2016) Engineering of CRISPR/Cas9-mediated potyvirus resistance in transgene-free Arabidopsis plants. Mol Plant Pathol 17:1276–1288
45. Ran FA, Cong L, Yan WX, Scott DA, Gootenberg JS, Kriz AJ, Zetsche B, Shalem O, Wu XB, Makarova KS, Koonin EV, Sharp PA, Zhang F (2015) In vivo genome editing using Staphylococcus aureus Cas9. Nature 520:186–191
46. Rani R, Yadav P, Barbadikar KM, Baliyan N, Malhotra EV, Singh BK, Kumar A, Singh D (2016) CRISPR/Cas9: a promising way to exploit genetic variation in plants. Biotechnol Lett 38:1991–2006
47. Salentijn EMJ, Mitea DC, Goryunova SV, Van der Meer IM, Padioleau I, Gilissen LJWJ, Koning F, Smulders MJM (2012) Celiac disease T cell epitopes from gamma-gliadins: immunoreactivity depends on the genome of origin, transcript frequency, and flanking protein variation. BMC Genom 13:277.
48. Sauer NJ, Mozoruk J, Miller RB, Warburg ZJ, Walker KA, Beetham PR, Schöpke CR, Gocal GFW (2016) Oligonucleotide-directed mutagenesis for precision gene editing. Plant Biotechnology J 14:496–502
49. Schaart JG, Van de Wiel CCM, Lotz LAP, Smulders MJM (2016) Opportunities for products of new plant breeding techniques. Trends Plant Sci 21:438–449
50. Schiml S, Fauser F, Puchta H (2014) The CRISPR/Cas system can be used as nuclease for in planta gene targeting and as paired nickases for directed mutagenesis in Arabidopsis resulting in heritable progeny. Plant J 80:1139–1150
51. Shan Q, Zhang Y, Chen K, Zhang K, Gao C (2015) Creation of fragrant rice by targeted knockout of the OsBADH2 gene using TALEN technology. Plant Biotechnol J 13:791–800
52. Shu QY, BP Forster, H Nakagawa (eds) (2011) Plant Mutation breeding and biotechnology. FAO, Rome. ISBN 978-92-5-105000-0
53. Slaymaker IM, Gao L, Zetsche B, Scott DA, Yan WX, Zhang F (2016) Rationally engineered Cas9 nucleases with improved specificity. Science 351:84–88
54. Smulders MJM, Jouanin A, Schaart J, Visser RGF, Cockram J, Leigh F, Wallington E, Boyd LA, van den Broeck HC, van der Meer IM, Gilissen LJWJ (2015) Development of

wheat varieties with reduced contents of celiac-immunogenic epitopes through conventional and GM strategies. Proceedings of the 28th meeting of the Working Group on Prolamin Analysis and Toxicity (Peter Koehler, ed.), 25–27 September 2014, Nantes, France, pp 47–56. http://www.wgpat.com/proceeding_28th.html; http://www.wgpat.com/proceeding_28th.pdf

55. Sprink T, Eriksson D, Schiemann J, Hartung F (2016) Regulatory hurdles for genome editing: process- vs. product-based approaches in different regulatory contexts. Plant Cell Rep 35:1493–1506
56. Steinert J, Schiml S, Fauser F, Puchta H (2015) Highly efficient heritable plant genome engineering using Cas9 orthologues from Streptococcus thermophilus and Staphylococcus aureus. Plant J 84:1295–1305
57. Subburaj S, Chung SJ, Lee C, Ryu S-M, Kim DH, Kim J-S, Bae S, Lee G-J (2016) Site-directed mutagenesis in Petunia hybrida protoplast system using direct delivery of purified recombinant Cas9 ribonucleoproteins. Plant Cell Rep 35:1535–1544.
58. Sun K, Wolters AMA, Loonen AEHM, Huibers RP, Van de Vlugt R, Goverse A, Jacobsen E, Visser RGF, Bai Y (2016a) Down-regulation of Arabidopsis DND1 orthologs in potato and tomato leads to broad-spectrum resistance to late blight and powdery mildew. Transgenic Res 25:123–138
59. Sun K, Wolters AMA, Vossen JH, Rouwet ME, Loonen AEHM, Jacobsen E, Visser RGF, Bai Y (2016b) Silencing of six susceptibility genes results in potato late blight resistance. Transgenic Res 25:731–742
60. Suprasanna P, SJ Mirajkar, SG Bhagwat (2015) Induced mutations and crop improvement. Bahadur INB et al. (eds.), Plant biology and biotechnology: Volume I: plant diversity, 593 Organization, Function and Improvement,
61. Svitashev S, Young JK, Schwartz C, Gao HR, Falco SC, Cigan AM (2015) Targeted mutagenesis, precise gene editing, and site-specific gene insertion in maize using Cas9 and guide RNA. Plant Physiol 169:931–945
62. Tsai SQ, Zheng Z, Nguyen NT, Liebers M, Topkar VV, Thapar V, et al. (2015) GUIDE-seq enables genome-wide profiling of off-target cleavage by CRISPR-Cas nucleases. Nat Biotechnol 33:187–197.
63. Tsuda M, Kaga A, Anai T, Shimizu T, Sayama T, Takagi K, Machita K, Watanabe S, Nishimura M, Yamada N, Mori S, Sasaki H, Kanamori H, Katayose Y, Ishimoto M (2015) Construction of a high-density mutant library in soybean and development of a mutant retrieval method using amplicon sequencing. BMC Genom 16:1014.
64. Uauy C, Paraiso F, Colasuonno P, Tran RK, Tsai H, Berardi S, Comai L, Dubcovsky J (2009) A modified TILLING approach to detect induced mutations in tetraploid and hexaploid wheat. BMC Plant Biol 9:115.
65. Van Acker R, Leple JC, Aerts D, Storme V, Goeminne G, Ivens B, Legee F, Lapierre C, Piens K, Van Montagu MCE, Santoro N, Foster CE, Ralph J, Soetaert W, Pilate G, Boerjan W (2014) Improved saccharification and ethanol yield from field-grown transgenic poplar deficient in cinnamoyl-CoA reductase. Proc Natl Acad Sci USA 111:845–850
66. Van de Wiel CCM, Smulders MJM, Visser RGF, Schaart JG (2016) New developments in green biotechnology—an inventory for RIVM. Report, Wageningen UR Plant Breeding, Wageningen, p 51.
67. Van den Broeck HC, van Herpen TWJM, Schuit C, Salentijn EMJ, Dekking L, Bosch D, Hamer RJ, Smulders MJM, Gilissen LJWJ, van der Meer IM (2009) Removing celiac disease-related gluten proteins from bread wheat while retaining technological properties: a study with Chinese Spring deletion lines. BMC Plant Biol 9:41.
68. Van Herpen TWJM, SV Goryunova, J van der Schoot, M Mitreva, E Salentijn, O Vorst, MF Schenk, PA van Veelen, F Koning, LJM van Soest, B Vosman, D Bosch, RJ Hamer, LJWJ Gilissen, MJM Smulders (2006) Alpha-gliadin genes from the A, B, and D genomes of wheat contain different sets of celiac disease epitopes. BMC Genom 7:1.
69. Wang Y, Cheng X, Shan Q, Zhang Y, Liu J, Gao C, Qiu J-L (2014) Simultaneous editing of three homoeoalleles in hexaploid bread wheat confers heritable resistance to powdery mildew. Nat Biotechnol 32:947–951

70. Wang Z, Xing H, Dong L, Zhang H, Han C, Wang X, Chen Q (2015) Egg cell-specific promoter-controlled CRISPR/Cas9 efficiently generates homozygous mutants for multiple target genes in Arabidopsis in a single generation. Genome Biol 16:144
71. Wendt T, Holm PB, Starker CG, Christian M, Voytas DF, Brinch-Pedersen H, Holme IB (2013) TAL effector nucleases induce mutations at a pre-selected location in the genome of primary barley transformants. Plant Mol Biol 83:279–285
72. Woo J, Kim J, Kwon S, Corvalán C, Cho S, Kim H, Kim S, Kim S, Choe S, Kim J (2015) DNA-free genome editing in plants with preassembled CRISPR-Cas9 ribonucleoproteins. Nat Biotechnol 33:1162–1164
73. Yamaji N, Takemoto Y, Miyaji T, Mitani-Ueno N, Yoshida KT, JF Ma (2017) Reducing phosphorus accumulation in rice grains with an impaired transporter in the node. Nature 541:92–95.
74. Yan LH, Wei SW, Wu YR, Hu RL, Li HJ, Yang WC, Xie Q (2015) High-efficiency genome editing in Arabidopsis using YAO promoter-driven CRISPR/Cas9 system. Mol Plant 8:1820–1823
75. Zheng Z, Appiano M, Pavan S, Bracuto V, Ricciardi L, Visser RGF, Wolters AM, Bai Y (2016) Genome-wide study of the tomato SlMLO gene family and its functional characterization in response to the powdery mildew fungus Oidium neolycopersici. Front Plant Sci 7:380.
76. Zhou HB, Liu B, Weeks DP, Spalding MH, Yang B (2014) Large chromosomal deletions and heritable small genetic changes induced by CRISPR/Cas9 in rice. Nucl Acids Res 42:10903–10914
77. Zhou XH, Jacobs TB, Xue LJ, Harding SA, Tsai CJ (2015) Exploiting SNPs for biallelic CRISPR mutations in the outcrossing woody perennial Populus reveals 4-coumarate: CoA ligase specificity and redundancy. New Phytol 208:298–301
78. Zhu X, Gong H, He Q, Zeng Z, Busse JS, Jin W, Bethke PC, Jiang J (2016) Silencing of vacuolar invertase and asparagine synthetase genes and its impact on acrylamide formation of fried potato products. Plant Biotechnol J 14:709–718.

CHAPTER 2

Anthocyanin Accumulation Enhanced in *Lc*-Transgenic Cotton under Light and Increased Resistance to Bollworm

Xiaoping Fan[1] • Bohong Fan[1] • Yuxiang Wang[1] • Weicai Yang[2]

[1] *Institute of Cotton Research, Academy of ShanXi Agricultural Science, Yuncheng City, Shanxi Province 044000, China*

[2] *Key Laboratory of Molecular and Developmental Biology and National Centre for Plant Gene Research (Beijing), Institute of Genetics and Developmental Biology, Chinese Academy of Sciences, Datun Road, Beijing 100101, China*

ABSTRACT

Breeding of naturally colored cotton fiber has been hampered by the limited germplasm, an alternative way is to use transgenic approach to create more germplasm for breeding. Here, we report our effort to engineer anthocyanin production in cotton. The maize *Lc* gene, under the control of the constitutive 35S promoter, was introduced into cotton through genetic transformation. Our data showed that the expression of the *Lc* gene alone is sufficient to trigger the accumulation of anthocyanin in a variety of cell types including fiber cells in cotton. However, the accumulation of colored anthocyanin in cotton fibers requires the participation of light signaling.

These data indicate that it is feasible to engineer colored fibers through transgenic approach in cotton. Furthermore, we showed that the *Lc*-transgenic cotton plants are resistant to cotton bollworm. These transgenic plants are, therefore, potentially useful for cotton breeding against cotton bollworm.

Keywords: *Lc* gene *Gossypium hirsutum* Anthocyanin Cotton fiber

INTRODUCTION

Naturally colored cotton clothing and textile products become more and more popular in people's life. They not only have the same advantages as the common white fiber, but also require no chemical dyeing and bleaching. Therefore, they are environmental friendly as well. Early in 1990s, Bird Brothers successfully bred four colored fibers: green, brown, red (a reddish brown) and mocha (similar to tan). However, so far breeding of naturally colored cotton fiber is limited by the lack of proper germplasm, which is largely limited to brown and green fiber traits. Furthermore, genetic studies on genes controlling pigment production in cotton have been hampered by its complex genome. Therefore, it is necessary to find more cotton germplasm that introduces novel color traits into cotton through transgenic approaches, for example, by modifying anthocyanin biosynthetic pathway.

Biosynthetic pathway leading to anthocyanin pigmentation has been studied extensively in *Zea mays* and genes controlling each step have been characterized (Sharma et al. 2012). Anthocyanin biosynthesis is controlled by two groups of genes. One group is the structural genes which encode the enzymes, catalyzing the biochemical reactions of the formation of anthocyanins. The second group includes genes which encode transcription regulators that control the expression of the structural genes, and determine the spatial and temporal pattern of anthocyanin accumulation. In general, the transcription regulators belong to either the MYB-type C1 family or the basic helix-loop-helix, MYC-type R family (Dooner et al. 1991; Boudet 2007; Kong et al. 2012). Both *C1* and *R* genes have been studied extensively in many plants, such as maize (Kong et al. 2012) and petunia (Quattrocchio et al. 1993; Schwinn et al. 2014). The *C1* gene of maize plays a regulatory role in the production of anthocyanin pigments in the aleurone layer of the endosperm (Cone et al. 1986; Stinard et al. 2009) and the *R* genes control the temporal and spatial distribution of anthocyanin pigments (Ludwig et al. 1989).

The *Lc* (leaf color) gene is a bHLH (basic/helix-loop-helix) anthocyanin regulator of maize, belongs to the *R* gene family, determines the timing (Ludwig and Wessler 1990), distribution, and amount of anthocyanin pigmentation in maize and is the target for plant pigment engineering (Schwinn et al. 2014). In this family, it also consists of other regulatory genes, such as *Sn* and the B locus genes (Chandler et al. 1989; Dooner et al. 1991; Consonni et al. 1993; Petroni and Tonelli 2011). The *Lc* gene as a transcription factor can regulate the anthocyanin biosynthesis pathway and use anthocyanin-regulating transcription factors to alter anthocyanin production and pigmentation patterns in several plant systems (Boase et al. 1998; Bradley et al. 1998), such as in tomato (Goldsbrough et al. 1996; Bovy et al. 2002), tobacco (Lloyd et al. 1992; Huang et al. 2012), apple (Flachowsky et al. 2010) and petunia (Bradley et al. 1998; Albert et al. 2009). In transgenic petunia plants, the maize *Lc* gene, under the control of the constitutive CaMV 35S promoter, enhanced pigmentation in both vegetative and floral tissues. Additionally, in *Lc*-transgenic alfalfa foliage, red–purple colored anthocyanin accumulated under high light conditions (Ray et al. 2003). These reports illustrated that anthocyanin production in vegetative and reproductive tissues was enhanced and induced by *Lc* transcription factors under high light conditions (Procissi et al. 1997; Albert et al. 2009). However, in several different plant species, anthocyanin biosynthesis was stimulated by *Lc* and C1 together (Han et al. 2009), or in some species *Lc* has no visible effect (Boase et al. 1998; Bradley et al. 1999). In cotton, few studies on *Lc* transformation and expression have been reported so far. Therefore our aims for this study were to determine whether the *Lc* gene alone can active anthocyanin biosynthesis in cotton and characterize the anthocyanin pigmentation response in *Lc*-transformed cotton plants.

For another, anthocyanin in leaves has been correlated with resistance to biotic and abiotic agents, including fungi, herbivores and cold (Chalker-Scott 1999; Gould et al. 2000; Huang et al. 2012). The anthocyanin biosynthetic pathway is interlinked with the biosynthesis of flavonoid compounds, which are important to fertility and defense against insects and pathogens (Dooner et al. 1991; Flachowsky et al. 2010). Red leaves are less frequently attacked than green leaves; in *Lc*-transgenic rice, red leaves exhibited increased resistance to blast infection (Gandikota et al. 2001). Therefore, expression of *Lc* provides us a tool to engineer cotton color and insect resistance. Here, we report our approach to employ *Lc* gene to engineer cotton color and insect resistance.

MATERIALS AND METHODS

Plant materials

Cotton (*Gossypium hirsutum*) Coker 312 seeds were surface sterilized by soaking for 1 min in 70 % ethanol followed by ~2 h in 10 % H_2O_2. They were then washed with sterilized distilled water 3 times. Seeds were allowed to germinate in sterilized water for 16–20 h at 28 °C. The germinated seeds were transferred to half strength MS medium (Murashige and Skoog 1962) containing 15 g/L glucose and 3 g/L Phytagel. The pH was adjusted to 5.8 with 1 M KOH prior to autoclaving. The transformation procedure was carried out as described by Li et al. (2005). Briefly, hypocotyl explants of 5- to 7-day-old seedlings were cut into 0.8–1-cm-long segments and immersed in a bacterial culture suspension with gentle agitation. Then, the co-culturing, selection, tissue culturing, differentiation, germination and rooting processes were performed as described previously (Li et al. 2005). All tissue culture processes before transplanting to pots were performed in a culture room that was maintained at 28 °C with a 16/8 h light/dark photoperiod and light supplied by warm-white fluorescent lamps at an irradiance of 26 μmol/m^2/s.

Bacterial culture and cotton genetic transformation

Agrobacteriumtumefaciens strain LBA4404, which contains a modified plasmid pBI121 harboring a CaMV 35S::Lc::Nos transgene (provided by Dr. Lim Saw Hoon, Malaysia University of Science and Technology, Malaysia), was cultured overnight in LB liquid medium with continuous shaking at 250 rpm, at 28 °C. Bacterial cultures at OD_{600} 0.6–0.8 were collected by centrifugation and resuspended in MS liquid medium to a final OD_{600} of 0.2–0.4. Acetosyringone was added to the culture medium to a final concentration of 100 μM at ~2 h before harvesting. Cotton transformation and tissue culturing were performed as described previously (Li et al. 2005).

Screening of transgenic plants using polymerase chain reaction (PCR)

The analysis of the transgenic lines of the T0 and T1 generations was carried out by amplifying a fragment of the *Lc* gene using PCR. Genomic DNA was extracted from young leaves of greenhouse-grown transgenic and nontransgenic plants using the CTAB method (Li et al. 2002). PCR screening of putative transformants was performed using primers specific to the *Lc* gene. The plasmid was used as a positive amplification control. The sequences of the forward and reverse primers are 5′ATGGCGCTTTCAGCTTCCCGAGT3′ and 5′TCACCGCTTCCCTA TAGCT TTGC3′, respectively. The PCR master mix was made according to the PCR kit's protocol. Amplification was accomplished using a Thermal Cycler (Biometro, Germany). The reaction was as follows: incubation at 94 °C for 5 min, and then a step cycle program set to denature at 94 °C for 45 s, to anneal at 58 °C for 45 s, and then extended at 72 °C for 2 min, for a total of 35 cycles, followed by a final extension at 72 °C for 10 min.

Southern blot analysis

For the Southern blot analysis, 50 µg genomic DNA was isolated from PCR-positive plants and digested with 20 units of *BamH*1 at 37 °C overnight in a total volume of 200 µl. Digested DNA was gel fractionated and transferred onto Hybond N+ nylon membrane (Amersham) according to the manufacturer's instructions. The *Lc* probe was labeled with Digoxigenin-11-dUTP using PCR amplification according to the manufacturer's instructions (Roche). Hybridization was performed overnight at 42 °C using DIG-Easy-hybridization solution containing a DIG-labeled *Lc* probe. The hybridized membrane was washed 3 times with a 0.1× SSC plus 0.1 % SDS solution at 68 °C and detection was carried out using an anti-DIG-AP conjugate according to the manufacturer's instructions.

RNA isolation and reverse-transcription (RT)-PCR analysis

The fresh young leaves, petiole and floral tissues, including petal, stigma, anther and sepal, in addition ovule, fiber at the proper development stage, were collected and frozen in liquid nitrogen. Total RNA from these materials was extracted using the hot phenol extraction method (Li et al. 2005). Then, 1 µg of each total RNA sample was reverse transcribed using the Qiagen one-step RT-PCR protocol. Briefly, the RNA template was added to the master mix and reverse transcribed at 50 °C for 30 min. Then, the reverse transcriptase was inactivated, and the cDNA template was denatured at 95 °C for 15 min. PCR amplification was performed using *Lc* gene-specific primers, and a program of 30 cycles of 94 °C for 45 s, 56 °C for 45 s, and 72 °C for 2 min, followed by a final extension at 72 °C for 10 min. The cotton *actin* gene was used as an internal standard to normalize the cDNA concentration. The same amplification was performed using the *actin* primers: Actin-F, 5′TTTGCTGGTGATG ATGCTCC3′; and Actin-R, 5′CTCCAATCCAGACACTGTACT3′.

Ovule culture

Cotton bolls from 2-day post-anthesis T1 CaMV 35S::Lc::Nos transgenic plants of two lines, L73 and L143, and wild type (Wt) were surface sterilized in 70 % ethanol for 5–10 min, and then immature ovules were gently transferred onto refined liquid medium as described by Beasley and Ting (1974). Immature ovules from half of one boll were cultured under a 16/8 h light/dark photoperiod, while the other half were cultured in complete darkness in a culture room. Same stage ovules from non-transgenic bolls were cultured under the same conditions. The comparative experiment was repeated six times.

Anthocyanin extraction and measurement

The anthocyanin extraction was performed according to Choung et al. (2001). Air-dried old or fresh leaves were ground to a powder. Then, 5 g of powder was extracted with 100 ml of a 1 % HCl and 40 % methanol mixture in the darkness at 4 °C overnight. Subsequently, the extracts were centrifuged and filtrated through a 0.45 μm filter. The anthocyanin pigment content was determined by a pH differential method commonly used in food technology (Gong et al. 1997). Briefly, 10 ml of extract was adjusted to pH 1.0 and another 10 ml to pH 4.5 with 0.025 M potassium chloride buffer and 0.4 M sodium acetate buffer, respectively. Then, the extracts were analyzed using a UV–VIS spectrophotometer and quantified (Gong et al. 1997). The anthocyanin content, expressed as cyanidine-3-glucoside, was determined according to Lambert–Beer's Law: $A = \varepsilon C\text{Len}$ (A, absorbance; C concentration; Len, path length in cm; and ε, molar extinction coefficient) using the molar extinction coefficient of 26,900 Len/mol and molar mass of 449 g/mol.

The color density, polymeric color and the percent of tannin contribution were measured using the bisulfite solution method (Somers and Evans 1974). The metabisulfite is able to bleach the anthocyanin, but not polymeric tannin pigments, and spectrophotometric assay results were calculated according to Cortell et al. (2007). The percent contribution of tannin is the ratio of polymeric color to color density.

Insect bioassay

The 4th leaf from the apical shoot of each line (L73, L74, L143, L144 and Wt) picked for the bioassay was washed thoroughly with sterile distilled water prior to feeding. Both transgenic red leaves and non-transgenic leaves at the same distance from shoot apex were placed in one petri dish and infested with three-third-instar cotton bollworm larvae (*Helicoverpa armigera*) that had been starved for 2 h. After 3 days, the feeding preference was recorded, and 13 replicate samples from each line were used.

In addition, the third-instar larvae were placed into one petri dish with either a piece of transgenic or non-transgenic leaf, observed every day and weighed after 5 days of feeding. The insect weights on different plant lines were determined using the between-column variance method. Larval mass (to an accuracy of 0.1 mg) was measured. To compare herbivore resistance-associated parameters between Wt and *Lc*-transgenic lines a one-way analysis of variance (ANOVA) was performed, and the experiment was replicated three times.

RESULTS

Generation of transgenic cotton plants

After transformation and regeneration from tissue culture, through somatic embryogenesis, six independent transgenic lines were regenerated on *kanamycin* selective medium. To confirm the presence of the *Lc* transgene in these T0 transgenic lines, genomic DNAs were digested by *BamH*I and fractionated by gel electrophoresis. The blot was probed with a DIG-labeled *Lc* probe. Six PCR positive plantlets were confirmed by Southern hybridization. Southern blot hybridization analysis determined transgene copy numbers from these lines. Lines of L82 and L28 had multiple copies insertion and the other four lines (L73, L74, L143 and L144) had single insertion, respectively. The non-transgenic control (NC) lacks the *Lc* hybridization signal (Figure 1a). Southern blot analysis was performed on the T1 generation to reveal the *Lc* gene information of integration and separation in T1 generation. In line L143 T1 generation, 5 (L143-1, 2, 5, 6 and 7) of 15 were positive (Figure 1b).

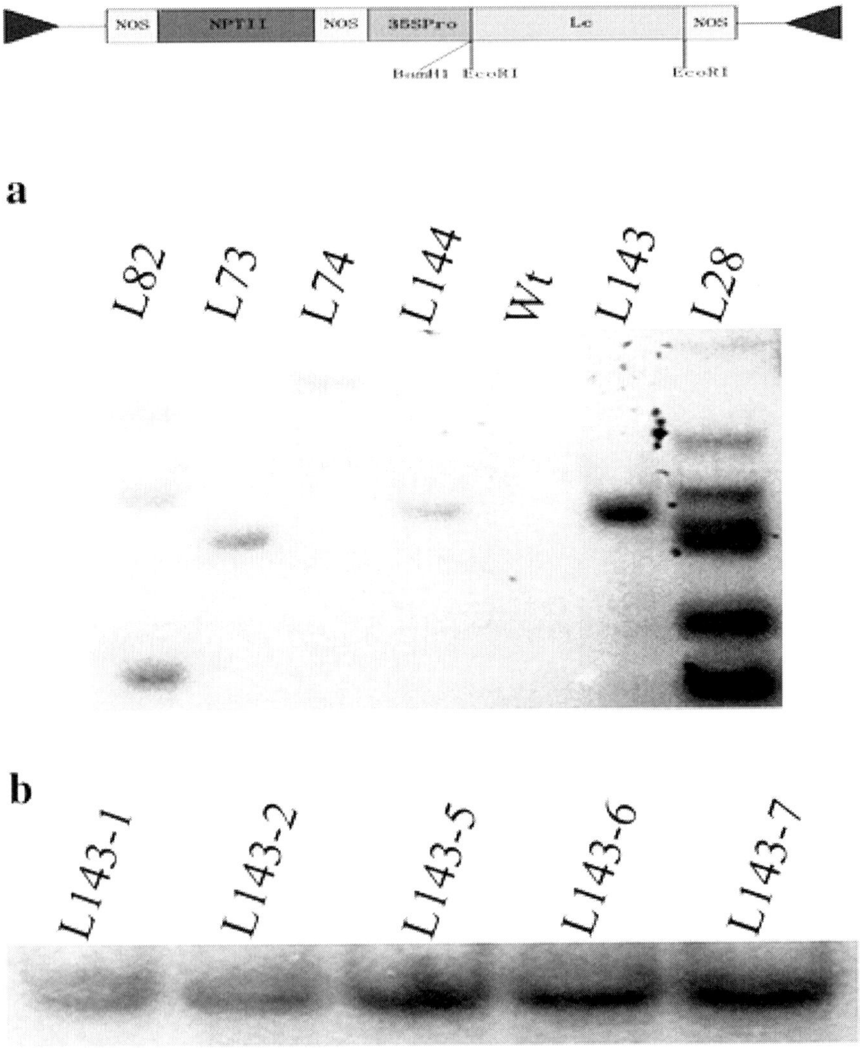

Figure 1. Southern blot analysis of the *Lc*-transgenic cotton plants. Genomic DNA from T0 transgenic (**A**) and T1 transgenic plants (**B**) was digested by *BamH*I and fractionated using gel electrophoresis. The blot was probed with a DIG-labeled *Lc* probe. **A** Lanes *1–4* and *6–7* represent six independent transgenic lines as indicated. Lane *5* is a nontransgenic control (Wt). **B** Lanes *1–5* represent five T1 siblings from L143 showing *Lc* hybridization bands

Lc gene expressed at the molecular level

To investigate whether the *Lc* transgene is expressed in the transgenic plants, RT-PCR was performed from different tissues and organs. Our results showed that *Lc* was expressed in leaves and floral tissues of the transgenic plants, and no *Lc* transcripts were detected in non-transgenic plants (data not shown). The expression levels of the *Lc* gene varied slightly among tissues and organs (Figure 2a). Leaves and petioles had relatively higher *Lc* expression levels than the floral tissues. Additionally, the *Lc* expression levels were different among flower parts. The highest expression level was in anther, while it was moderate in stigma, but no signal was detected in petals.

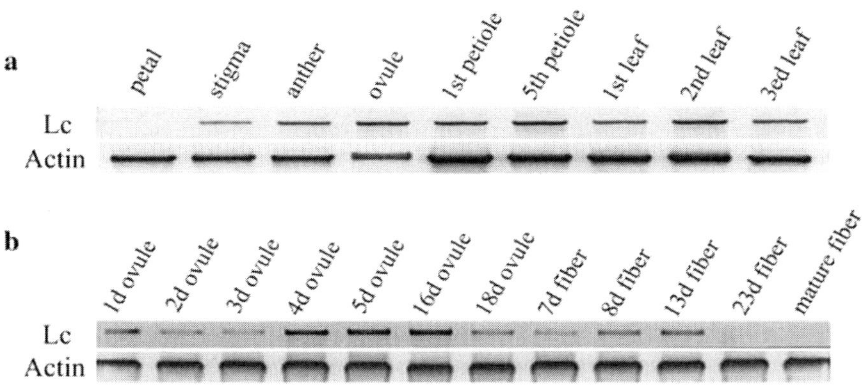

Figure 2. RT-PCR analysis of *Lc* expression patterns in transgenic cotton plants. **A***Upper panelLc* expression level in the indicated vegetative and reproductive tissues; *lower panelActin* was used as internal control indicating no DNA contamination in the samples. **B***Lc* expression in ovules and fibers at the indicated stages. *Actin* was used as an internal control

The *Lc* gene was expressed during all the ovule and fiber developmental stages in the transgenic plants. There were high expression levels in the 4–16 DPA (days post-anthesis) ovules, and moderate levels were detected in 1–3 DPA ovules and 18 DPA ovules. To determine whether the *Lc* transgene was expressed only in cotton fiber cells, we isolated RNA from fibers of 7, 8, 13 and 23 DPA ovules and mature fibers, respectively. The test results showed that in fiber, the expression level rose along with fiber development from 7 to 13 days. The expression level decreased to a weak but detectable signal in 23 DPA fibers and was weaker in mature fiber cells (Figure 2b).

Morphology of *Lc*-transgenic cotton

During the regeneration process, most of the transformants accumulated high levels of anthocyanin pigment in their vegetative and reproductive tissues. In the tissue culture process, we observed that *kanamycin*-resistant calluses and somatic embryos appeared an unusual red color (Figure 3a) that was never observed in control tissues (Figure 3b). Generally, anthocyanin pigmentation accumulated in radicles, hypocotyls and cotyledons of transgenic somatic embryos. The purple color present in cotyledons disappeared several days later, and the young root turned red gradually in a culture bottle. The red root phenotype never presented in non-transgenic plants. Juvenile leaves usually displayed red, when the transgenic plantlets were transplanted into soil. Some appeared purple in pigmentation when the leaves unfolded (Figure 3c), and Wt leaves were green under the same growth conditions (Figure 3d). With further growth of the regenerated plants, some red patches appeared on the stem's surface (Figure 3e). The sepal around the flower bud turned red (Figure 3f) when the flower opened, and the anthers and the stigma also appeared red (Figure 3g). We observed that the anthers are larger and contain less pollen compared with the Wt. However, no anthocyanin pigmentation was observed in petals at the day of anthesis. We found that the old leaves of transgenic plants turned deep red, and their air-dried leaves were more fragile when squeezed. Non-transgenic senescent leaves remained yellowish and were more flexible after air drying. The L143 and L73 lines were fertile, and their T1 and T2 generations had the same red color characteristics. The T1 generation segregated 3:1 (positive: negative). The positive seedlings had red roots (Figure 3h-r1, r2) and the negative ones were white when germinated in a bottle under the same culture conditions (Figure 3h-r3, r4).

Light-induced anthocyanin accumulation in transgenic fibers and leaves

The initiating fibers of transgenic ovules became red (Figure 3i) after being cultured in vitro under light for 7 days, and a red oil-like liquid accumulated in the large central vacuole of the single fiber cell observed under the microscope (Figure 3j). Fiber cells of non-transgenic ovules at the same developmental stage showed little color change on the fiber surface under the same light (data not show). When cultured in the dark, the other areas, the fiber no matter from transgenic or non-transgenic bolls, remained completely white (Figure 3k). Light is also an important factor that affects anthocyanin production in leaves. Freehand cross section of leaves from transgenic plants grown with exposure to normal

Figure 3. Phenotypic characterization of *Lc*-transgenic cotton. **A** Red transgenic calluses and somatic embryos (*bar* 2.5 mm). **B** Non-transgenic calluses (*bar* 2.5 mm). **C** Transgenic *purple colored* leaf, indicating anthocyanin accumulation. **D** Non-transgenic leaf. **E** Anthocyanin accumulation on stem (*red patches*). **F** Transgenic flower bud with a red sepal. **G** Transgenic flower with a red anther. **H** T1 siblings showing the segregation of *Lc*-dependent anthocyanin accumulation in roots. r1 and r2 are positive seedlings with red roots; r3 and r4 are negative seedlings with white roots. **I** Transgenic immature ovules cultured under light for 1 week (*bar* 2 mm). Note red fiber cells. **J** Magnification of J showing a single red fiber cell (*bar* 10 μm). Note red oily liquid gathered in the central big vacuole. **K** Transgenic immature ovules cultured in completely dark for 1 week (*bar* 2 mm). **L** Transgenic leaf abaxial surface toward the sun; its freehand cross section with red palisade mesophyll cells. The epidermal cells are colorless and transparent (*bar* 1 mm). **M** Anthocyanin extract from transgenic dry leaves (e1), transgenic fresh leaves (e2), Wt leaves (e3), and empty control (e4)

sunlight for 3 days exhibited anthocyanin accumulation in their palisade mesophyll cells (Figure 3l). When the upper surface of the leaf was not exposed to sunlight, for example with positioning facing down (due to abnormal growth), the spongy mesophyll cells turned red but the palisade mesophyll cells remained green (data not shown). The leaf epidermis was always maintained colorless and transparent whichever side faced toward the sun (Figure 3l).

Transgenic plants produce more anthocyanin than non-transgenic plants

We further quantified the amount of anthocyanin in transgenic plants. The leaf extract of transgenic dry or fresh leaves showed more intense red (Figure 3m-e1, e2) compared with the Wt leaves (Figure 3m-e3). Figure 3m-e4 shows the empty control.

To quantify the amount of anthocyanin, we employed the pH differential method commonly used in food technology. Measurement results showed that the anthocyanin, exemplified by cyanidin-3-glucoside, increased from 15.41 mg/100 g (Wt) to 53.60 mg/100 g (*Lc*) in the dry leaf extract and from 9.74 mg/100 g (Wt) to 12.80 mg/100 g (*Lc*) in the fresh leaf extract.

Using the bisulfite solution method, anthocyanin degradation indices data showed that the three indices, color density (total anthocyanin), and monomeric and polymeric anthocyanin (condensed tannin) (Cortell et al. 2007) increased when compared with Wt extract. Color density in *Lc*-transgenic leaf rose from 1.62 to 4.03 in the dry leaf extracts; in fresh leaf extract, it increased from 0.71 to 2.76. However, the ratio varied between monomeric anthocyanin and polymeric anthocyanin. In *Lc*-transgenic dry leaf, monomeric anthocyanin accounted for 27.2 % of the total anthocyanin relative to 17.3 % in non-transgenic dry leaf. Polymeric anthocyanin ratio from 67.2 % in Wt varied to 62.4 % of the total anthocyanin. In fresh materials, monomeric anthocyanin percentage decreased from 22.5 to 20.7 %, while the polymeric anthocyanin percentage increased from 50.7 to 51.4 % in transgenic leaf extract (Figure 4). This might be due to the formation of polymers and anthocyanin degradation that could be occurring in *Lc* leaves during different stages of growth, leading to changes in the anthocyanin ratio.

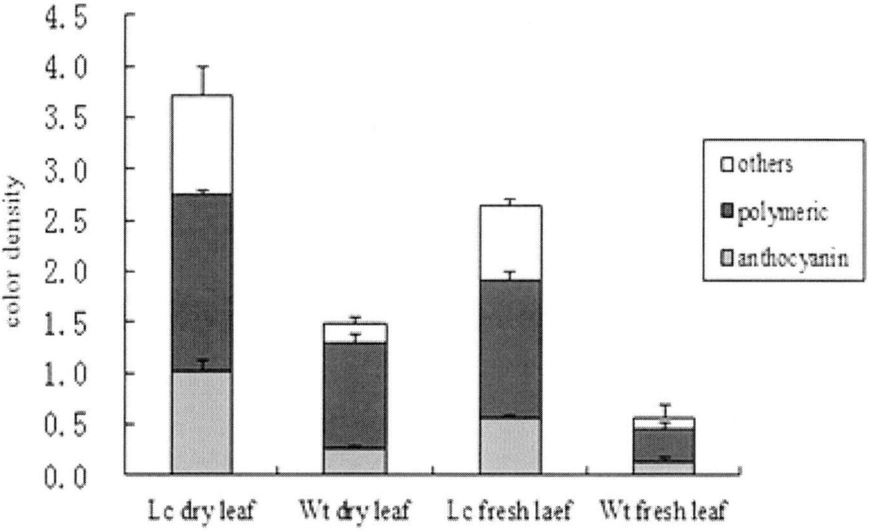

Figure 4. Anthocyanin content revealed by UV spectrometric analysis. Relative fractions of monomeric anthocyanin, polymeric anthocyanin (condensed tannins) and other compounds of the total anthocyanin calculated using the bisulfite leaching method. *Values* represent averages ± STEV, and every *bar* represents an average value from three repeated measurements

Transgenic plants are resistant to cotton bollworm

To test whether the CaMV 35S::Lc-transgenic plants were also resistant to cotton bollworm, an insect feeding experiment was performed. Our data showed that the insects fed preferably on non-transgenic leaves when both transgenic and non-transgenic leaves were present in one dish. The non-transgenic leaf was consumed more after 2 days of feeding (Figure 5a-a1), while the transgenic leaves remained more (Figure 5a-b1, c1). This trend became more obvious after 4 days of feeding (Figure 5b-a2, b2, c2). Interestingly, the green part of the leaf was preferentially eaten by larvae compared with red patches on the same leaf.

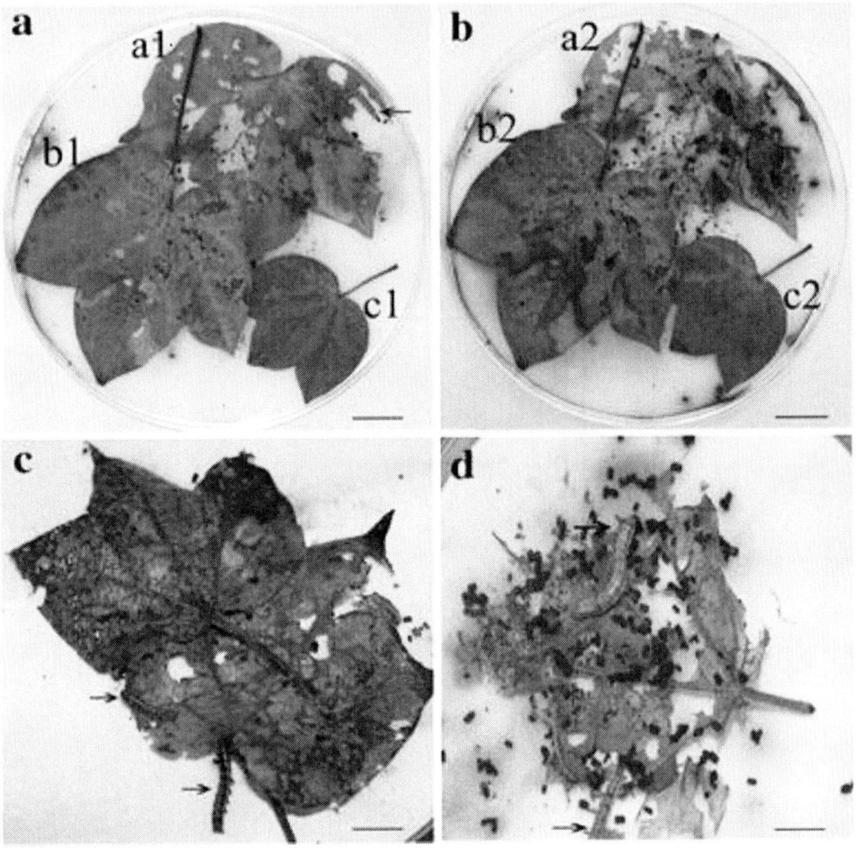

Figure 5. Bioassay for cotton bollworm feeding on *Lc* and Wt leaves. **A** Leaf images after 2 days of feeding. *b1* and *c1* are *Lc*-transgenic leaves, *a1* is a Wt leaf. The *arrow points* to worm (*bar* 2.2 cm). **B** Leaf images after 4 days of feeding. *b2* and *c2* are *Lc*-transgenic leaves, *a2* is a Wt leaf (*bar* 2.2 cm). **C** Image showing larvae after feeding 5 days on *Lc*-transgenic red leaves. The *arrows point* to worms (*bar* 2 cm). **D** Image showing larvae after feeding 5 days on Wt leaves. The *arrows point* to worms (*bar* 1.8 cm)

A force-feeding assay is shown in Figure 5; insect larvae that fed on transgenic leaves after 5 days (Figure 5c) hardly ate and were immobile, and they were visibly smaller than those fed on non-transgenic leaves those were more active(Figure 5d). To compare herbivore resistance-associated parameters between Wt and *Lc*-transgenic lines, the differences in larvae weight after feeding on Wt or *Lc*-transgenic leaves for 5 days, 13 groups data produced automatically a bar chart as Figure

6 using Excel software, and was analyzed using a one-way ANOVA with SPSS statistical software. The analysis results are presented in Table 1; the F test value is 15.792, and $P < 0.01$. This illustrated that there was a significant difference between the larvae weights which fed on transgenic *Lc* leaves or non-transgenic leaves (Table 1). The larvae weight of those fed on non-transgenic leaves was significantly heavier than those fed on *Lc*-transgenic leaves. These data showed that expression of *Lc* confirms transgenic cotton resistance to cotton bollworms.

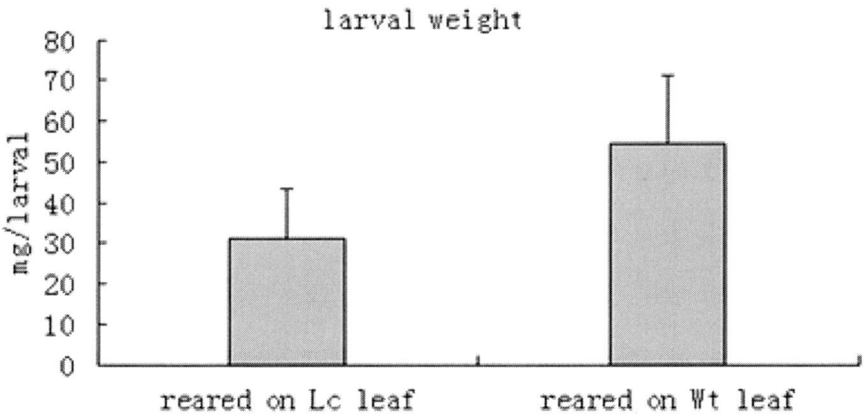

Figure 6. Statistical analysis of the larval weights after 3 days of rearing on *Lc*-transgenic and Wt leaves from Figure 5b values are from 13 samples and represent averages ± STEV. Every *bar* represents the average weight repeated three times ($n = 13$)

Table 1. Larval weights of cotton bollworms fed on *Lc*-transgenic leaves or Wt leaves were analyzed using a one-way ANOVA test by SPSS

	Sum of squares	df	Mean square	F	Sig
Between groups	3301.885	1	3301.885	15.792	0.001*
Within groups	5018.000	24	209.083		
Total	8319.885	25			

Statistical analysis of the larval weights ($n = 13$) after 3 days of rearing on *Lc* and Wt leaves. Average weight measurements were repeated three times ($n = 13$). $P < 0.01$* indicates that there was a significant difference between the two groups

DISCUSSION

Expression of maize *Lc* promotes anthocyanin production in cotton under light

Our data showed that there were 6 lines of regenerated plants obtained from 16 lines of transgenic calluses. These plants were positive and confirmed by Southern blotting using *Lc* probe. Since there is only one *BamH*1 site in the T-DNA, the number of bands of hybridization is the T-DNA copy number of insertion into plant genome (Figure 1a). From Figure 1b, we find that the T-DNA was stably transmitted to the T1 generation and segregated in a 3:1 ratio as a single gene insertion in line of L143. In the positive transgenic plants, the maize *Lc* gene under the CaMV35S promoter enhanced pigmentation in both vegetative and most floral tissues except petals. During tissue culture process, it is evident that the color of the transgenic calluses varies from pink, red, dark red to dark purple due to the amount and the localized production of anthocyanin. Similar phenomenon has been observed in rice tissues (Gandikota et al. 2001). The purple colored calluses became more intense to dark purple during culture leading to no somatic embryos formed from them. Because too much anthocyanin or toxic flavonoids in this kind of calluses was detrimental to cells (Burbulis and Winkel-Shirley 1999; Gandikota et al. 2001). And those with slight pigment, red or pink colored calluses can continuously propagate to form somatic embryos. Purple or red colored calluses never have been seen in control untransformed cultures. Therefore, *Lc* gene could be used as a color-based selection marker for genetic transformation in cotton. In transgenic embryos, color intensity varies among embryos and no consistent pattern was observed. One type of embryos with slight pink color on hypocotyl, cotyledon and radicle can develop to maturity and germinate. Another type of embryos with dark purple around the whole body or most of the embryo were often arrested before the torpedo stage. Might anthocyanin accumulation have adverse effects on hypocotyl formation? Consistent observation has reported in rice that the maize *Lc* gene up-regulated the expression levels of various anthocyanin regulatory genes leading to sterility or abnormal seed development (Li et al. 2013). Anthocyanin production and pigmentation involve multiple steps and often display cell and tissue-specific patterns. In *Lc*-transgenic leaves, anthocyanin accumulates mainly in veins and palisade mesophyll cells rather than epidermis. There was stronger expression of signal in leaves and petioles. And the expression level increased slightly from the 1st leaf to the 3rd leaf. In reproductive tissues, although pigment and signal appeared in anther, sepal and stigma at the same stage in the same cotton flower, neither red color nor transcription signal was observed in the petals. This expression pattern was mainly determined by *Lc* gene in cotton rather than the CaMV 35S promoter which stimulated higher expression levels in all floral parts, including petals in cotton (Sunilkumar et al. 2002). Therefore, it is inferred that the *Lc*-activated pathway is under developmental control, and the transgene expression is regulated both spatially and temporally (Martin and Gerats 1993); like in *Lc*-transgenic tomato fruits, the reddish-purple pigmentation disappears as fruit ripening progresses (Bovy et al. 2002). Another reasoning based on studies in petunia may be that in the petals and anthers anthocyanin synthesis is controlled by two different ternary complexes (Petroni and Tonelli 2011). In the ovule or fiber cells, no regular pattern of transgene expression was found. Although stronger expression signal presented throughout the fiber developmental phase, the fibers still appeared white when the bolls opened. Is the color due to anthocyanin in cotton fibers going to last processing or will it fade when dried? Or anthocyanins are soluble pigment molecules and generally may not be stable in tissues after drying? To resolve this problem, we open the cotton bolls in every development stage, and found that the fiber was white all the way. Therefore, the test of ovule cultured in vitro under light was done. The light induction experiment showed that the white fiber changed to red color, and leaf turned redder, suggesting that light signaling is essential for the *Lc*-activated anthocyanin biosynthesis or the conversion of colorless intermediates into colored pigments in fibers and leaves, because light can influence the accumulation of anthocyanin primarily through the activation of the transcription factors that regulate the flavonoid biosynthetic pathway (Irani and Grotewold 2005). But it is not clear whether the light signaling activates additional genes essential for the conversion of colorless anthocyanin to colored anthocyanin, or enhanced *Lc* expression in fiber cells. Therefore breeding a red fiber germplasm needs to solve this problem of having to increase anthocyanin accumulation in the fiber without having to culture them in vitro or to find another method of induction of the transgene expression.

Enhanced resistance against herbivores with anthocyanin content increased

Anthocyanins is one class of flavonoid compounds, and flavonoids are polyphenolic phytochemicals that constitute a large group of secondary metabolites in plants. Anthocyanin biosynthesis related to flavonoid biosynthesis produces important compounds for the defense against insects and pathogens (Dooner et al. 1991; Moctezuma et al. 2014). Although there have been relatively few reports concerning resistance against herbivores in *Lc*-transgenic plants, red leaves have been suggested by Hamilton and Brown (2001), to function as a visual warning to herbivores. Owing to increased amounts of defensive compounds in red leaves, the leaves containing higher concentrations of anthocyanins incur less natural herbivory. Additionally, the red leaves provide an effective chemical defense against larvae (Cooney et al. 2012) and anthocyanins in red leaf, as secondary metabolites, can function in plant defense against herbivores (Onkokesung et al. 2014). Analogous report showed that red kidney bean *Sayyad* was the most unsuitable host for feeding cotton bollworm (*H. armigera.common*) among common bean *Talash* and white kidney bean *Pak* (Namin et al. 2014). Similarly, transgenic plants that produced more anthocyanin exhibited a higher level of resistance against environmental stresses owing to their altered physiology (Flachowsky et al. 2010). Our study showed that the foreign *Lc* gene was not only expressed in cotton but also increased the anthocyanin content in transgenic leaves, and led to the green leaf changing to red or purple. In feeding experiment, cotton bollworms disfavor consuming red transgenic leaves, but ate green leaves. This is likely due to the higher anthocyanin in fresh leaves, which has been reported in the past to minimize herbivory (Cooney et al. 2012). The statistical analysis of insect weights showed that there was a significant difference between insects who fed on transgenic leaves and those fed on non-transgenic leaves ($P < 0.01$). This is most attributable to the accumulation of anthocyanin in the *Lc*-transgenic leaves. One aspect, cyanidin-3-β-glucoside is an important resistance factor in leaves that acts against the budworm and provides a potential basis for achieving insect resistance (Hedin et al. 1983). Cyanidin chloride and cyanidin-3-glucoside chloride slowed the growth of larvae when added to insect diet (Johnson et al. 2010). In another aspect, condensed tannin (polymeric anthocyanin) is also an important factor in plant–herbivore interactions (Furstenburg and van Hoven 1994; Boeckler et al. 2011). In plants, tannins serve to increase insect resistance by inhibiting several digestive enzymes, including proteases, pectinases, amylases, cellulases and lipases, as well as reducing protein digestibility. And these enzyme-inhibiting components decrease the conversion rate of food in larvae when reared on red kidney bean (Namin et al. 2014). Therefore, the resistance ability against herbivores, which was enhanced in *Lc*-transgenic leaves, was mainly attributable to the increased total anthocyanins.

In conclusion, based on our results, *Lc* alone is sufficient to trigger anthocyanin synthesis in different cell types in cotton under light conditions, and also increased resistance against bollworm. This may provide an alternative control method for cotton bollworms. Additionally, breeding a red fiber cotton variety through the regulation of anthocyanin biosynthesis in cotton fiber may be plausible in the near future.

ACKNOWLEDGMENTS

The authors thank Dr. Lim Saw Hoon at Malaysia University of Science and Technology, Malaysia, for providing the *Lc* construct. Special thanks to Dr. Yan Hong at Temasek Life Sciences Laboratory, Singapore, and Professor Nam-Hai Chua at Rockefeller University, USA, for their input. This work was supported by a grant from the Shanxi Abroad Returnee Office, Science and Research Project (2010) to Xiaoping Fan, and a grant (2006AA10A108) from the Ministry of Science and Technology of China to Weicai Yang.

REFERENCES

1. Albert NW, Lewis DH, Zhang H, Irving LJ, Jameson PE, Davies KM (2009) Light induced vegetative anthocyanin pigmentation in Petunia. J Exp Bot 60:2191–2202
2. Beasley CA, Ting IP (1974) Effects of plant growth substances on in vitro fiber development from unfertilized cotton ovules. Am J Bot 61:188–194
3. Boase MR, Bradley JM, Borst NK (1998) Genetic transformation mediated by Agrobacterium tumefaciens of florists' chrysanthemum (Dendranthema × Grandiflorum) cultivar "Peach Margaret". In Vitro Cell Dev Biol 34:46–51
4. Boeckler GA, Gershenzon J, Unsicker SB (2011) Phenolic glycosides of the salicaceae and their role as anti-herbivore defenses. Phytochemistry 72:1497–1509
5. Boudet AM (2007) Evolution and current status of research in phenolic compounds. Phytochemistry 68:2722–2735
6. Bovy A, de Vos R, Kemper M, Shijlen E, Almenar PM, Muir S, Collins G, Robinson S, Verhoeyen M, Hughes S, Santos BC, van Tunen A (2002) High-flavonol tomatoes resulting from the heterologous expression of the maize transcription factor genes Lc and C1. Plant Cell 14:2509–2526
7. Bradley JM, Davies KM, Deroles SC, Bloor SJ, Lewis DH (1998) The maize Lc regulatory gene up-regulates the flavonoid biosynthetic pathway of Petunia. Plant J 13:381–392
8. Bradley JM, Deroles SC, Boase MR, Bloor SJ, Swinny E, Davies KM (1999) Variation in the ability of the maize Lc regulatory gene to upregulate flavonoid biosynthesis in heterologous systems. Plant Sci 140:31–39
9. Burbulis IE, Winkel-Shirley B (1999) Interactions among enzymes of the Arabidopsis flavonoid biosynthetic pathway. Proc Natl Acad Sci USA 96:12929–12934
10. Chalker-Scott L (1999) Environmental significance of anthocyanin in plant stress responses. Photochem Photobiol 70:1–9
11. Chandler VL, Radicella JP, Robbins TP, Chen J, Turks D (1989) Two regulatory genes of the maize anthocyanin pathway are homologous: isolation of B utilizing R genomic sequences. Plant Cell 1:1175–1183
12. Choung MG, Baek IY, Kang ST, Han WY, Shin DC, Moon HP, Kang KH (2001) Isolation and determination of anthocyanins in seed coats of black soybean. J Agric Food Chem 49:5848–5851
13. Cone KC, Burr FR, Burr B (1986) Molecular analysis of the maize anthocyanin regulatory locus Cl. Proc Natl Acad Sci USA 83:9631–9635
14. Consonni G, Geuna F, Gavazzi G, Tonelli C (1993) Molecular homology among members of the R gene family in maize. Plant J 3:335–346

15. Cooney LJ, van Klink JW, Hughes NM, Perry NB, Schaefer HM, Menzies IJ, Gould KS (2012) Red leaf margins indicate increased polygodial content and function as visual signals to reduce herbivory in Pseudowintera colorata. New Phytol 194:488–497
16. Cortell JM, Halbleib M, Gallagher AV, Righetti TL, Kennedy JA (2007) Influence of vine vigor on grape (Vitis vinifera L. Cv. Pinot Noir) anthocyanins. 2. Anthocyanins and pigmented polymers in wine. J Agric Food Chem 55(16):6585–6595
17. Dooner HK, Robbins TP, Jorqensen RA (1991) Genetic and developmental control of anthocyanin biosynthesis. Ann Rev Genet 25:173–199
18. Flachowsky H, Szankowski Fischer TC, Richter K, Peil A, Höfer M, Dörschel C, Schmoock S, Gau AE, Halbwirth H, Hanke MV (2010) Transgenic apple plants overexpressing the Lc gene of maize show an altered growth habit and increased resistance to apple scab and fire blight. Planta 231:623–635
19. Furstenburg D, van Hoven W (1994) Condensed tannin as antidefoliate agent against browsing by giraffe (Giraffa camelopardalis) in the Kruger National Park. Comp Biochem Physiol 107:425–431
20. Gandikota M, de Kochko A, Chen L, Ithal N, Fauquet C, Reddy AR (2001) Development of transgenic rice plants expressing maize anthocyanin genes and increased blast resistance. Mol Breeding 7:73–83
21. Goldsbrough AP, Tong Y, Yode JI (1996) Lc as a non-destructive visual reporter and transposition excision maker gene for tomato. Plant J 9:927–933
22. Gong ZZ, Yamazaki M, Sugiyama M, Tanaka Y, Saito K (1997) Cloning and molecular analysis of structural genes involved in anthocyanin biosynthesis and expressed in a forma-specific manner in Perilla frutescens. Plant Mol Biol 35:915–927
23. Gould KS, Markham KR, Smith RH, Goris JJ (2000) Functional role of anthocyanin in the leaves of Quintina serrata A Cunn. J Exp Bot 51:1107–1115
24. Hamilton WD, Brown SP (2001) Autumn tree colours as a handicap signal. Proc R Soc Lond B 268:1489–1493
25. Han YJ, Kim YM, Lee JY, Kim SJ, Cho KCh, Chandrasekhar T, Song PS, Woo YM, Kim J (2009) Production of purple-colored creeping bentgrass using maize transcription factor genes Pl and Lc through Agrobacterium-mediated transformation. Plant Cell Rep 28:397–406
26. Hedin PA, Jenkins JN, Collum DH, White WH, Parrott WL, Macgown MW (1983) Cyanidin-3-ß-glucoside, a newly recognized basis for resistance in cotton to the tobacco budworm. Experientia 39:799–801
27. Huang ZA, Zhao T, Fan HJ, Wang N, Zheng SS, Ling HQ (2012) The upregulation of NtAN2 expression at low temperaturre is required for anthocyanin accumulation in juvenile leaves of Lc-transgenic tobacco (Nicotiana tabacum L.). J Genet Genomics 39:149–156
28. Irani NG, Grotewold E (2005) Light-induced morphological alteration in anthocyanin accumulating vacuoles of maize cells. BMC Plant Biol 5:1471–2229
29. Johnson ET, Berhow MA, Dowd PF (2010) Constitutive expression of the maize genes B1 and C1 in transgenic Hi II maize results in differential tissue pigmentation and generates resistance to Helicoverpa zea. J Agric Food Chem 58:2403–2409
30. Kong Q, Pattanaik S, Feller A, Werkman JR, Chai C, Wang Y, Grotewold E, Yuan L (2012) Regulatory switch enforced by basic helix-loop-helix and ACT-domain mediated dimerizations of the maize transcription factor R. Proc Natl Acad Sci USA 109:E2091–E2097
31. Li XB, Cai L, Cheng NH, Liu JW (2002) Molecular characterization of the cotton GhTUB1 gene that is preferentially expressed in fiber. Plant Physiol 130:666–674
32. Li XB, Fan XP, Wang XL, Cai L, Yang WC (2005) The cotton ACTIN1 gene is functionally expressed in fibers and participates in fiber elongation. Plant Cell 17:859–875
33. Li Y, Zhang T, Shen Z, Xu Y, Li JY (2013) Overexpression of maize anthocyanin regulatory gene Lc affects rice fertility. Biotechnol Lett 35:115–119

34. Lloyd AM, Walbot V, Davis RW (1992) Arabidopsis and Nicotiana anthocyanin production activated by maize regulators R and Cl. Science 258:1773–1775
35. Ludwig SR, Wessler SR (1990) Maize R gene family: tissue-specific helix-loop-helix proteins. Cell 62:849–851
36. Ludwig SR, Habera LF, Dellaporta SL, Wessler SR (1989) Lc, a member of the maize R gene family responsible for tissue-specific anthocyanin production, encodes a protein similar to transcriptional activators and contains the myc-homology region. Proc Natl Acad Sci USA 86:7092–7096
37. Martin C, Gerats T (1993) Control of pigment biosynthesis genes during petal development. Plant Cell 5:1253–1264
38. Moctezuma C, Hammerbacher A, Heil M, Gershenzon J, Méndez-Alonzo R, Oyama K (2014) Specific polyphenols and tannins are associated with defense against insect herbivores in the tropical oak Quercus oleoides. J Chem Ecol 40:458–467
39. Murashige T, Skoog F (1962) A revised medium for rapid growth and bioassays with tobacco tissue cultures. Physiol Plant 15:473–497
40. Namin FR, Naseri B, Razmjou J (2014) Nutritional performance and activity of some digestive enzymes of the cotton bollworm, Helicoverpa armigera, in response to seven tested bean cultivars. J Insect Sci.
41. Onkokesung N, Reichelt M, van Doorn A, Schuurink RC, van Loon JJ, Dicke M (2014) Modulation of flavonoid metabolites in Arabidopsis thaliana through overexpression of the MYB75 transcription factor: role of kaempferol-3,7-dirhamnoside in resistance to the specialist insect herbivore Pieris brassicae. J Exp Bot 65:2203–2217
42. Petroni K, Tonelli C (2011) Recent advances on the regulation of anthocyanin synthesis in reproductive organs. Plant Sci 181:219–229
43. Procissi A, Dolfini S, Ronchi A, Tonelli C (1997) Light-dependent spatial and temporal expression of pigment regulatory genes in developing maize seeds. Plant Cell 9:1547–1557
44. Quattrocchio F, Wing JF, Leppen HT, Mol J, Koes RE (1993) Regulatory genes controlling anthocyanin pigmentation are functionally conserved among plant species and have distinct sets of target genes. Plant Cell 5:1497–1512
45. Ray H, Yu M, Auser P, Blahut-Beatty L, McKersie B, Bowley S, Westcott N, Coulman B, Lloyd A, Gruber MY (2003) Expression of anthocyanins and proanthocyanidins after transformation of alfalfa with maize Lc. Plant Physiol 132:1448–1463
46. Schwinn KE, Boase MR, Bradley JM, Lewis DH, Deroles SC, Martin CR, Davies KM (2014) MYB and bHLH transcription factor transgenes increase anthocyanin pigmentation in petunia and lisianthus plants, and the petunia phenotypes are strongly enhanced under field conditions. Front Plant Sci 5:603.
47. Sharma M, Chai C, Morohashi K, Grotewold E, Snook ME, Chopra S (2012) Expression of flavonoid 3'-hydroxylase is controlled by P1, the regulator of 3-deoxyflavonoid biosynthesis in maize. BMC Plant Biol 12:196.
48. Somers TC, Evans ME (1974) Wine quality: correlations with colour density and anthocyanin equilibria in a group of young red wines. J Sci Food Agric 25:1369–1379
49. Stinard PS, Kermicle JL, Sachs MM (2009) The maize enr system of r1 haplotype-specific aleurone color enhancement factors. J Hered 100:217–228
50. Sunilkumar G, Mohr L, Lopata-Finch E, Emani C, Rathore KS (2002) Developmental and tissue-specific expression of CaMV 35S promoter in cotton as revealed by GFP. Plant Mol Biol 50:463–474

CHAPTER 3

Manipulation of Metabolic Pathways to Develop Vitamin-Enriched Crops for Human Health

Ling Jiang[1,2*], Weixuan Wang[1,2], Tong Lian[1] and Chunyi Zhang[1,2*]

[1]*Biotechnology Research Institute, Chinese Academy of Agricultural Sciences, Beijing, China*

[2]*National Key Facility for Crop Gene Resources and Genetic Improvement, Beijing, China*

ABSTRACT

Vitamin deficiencies are major forms of micronutrient deficiencies, and are associated with huge economic losses as well as severe physical and intellectual damages to humans. Much evidence has demonstrated that biofortification plays an important role in combating vitamin deficiencies due to its economical and effective delivery of nutrients to populations in need. Biofortification enables food plants to be enriched with vitamins through conventional breeding and/or biotechnology. Here, we focus on the progress in the manipulation of the vitamin metabolism, an essential part of biofortification, by the genetic modification or by the marker-assisted selection to understand mechanisms underlying metabolic improvement in food plants. We also propose to integrate new breeding technologies with metabolic pathway modification to facilitate biofortification in food plants and, thereby, to benefit human health.

Keywords: vitamins, metabolism, biofortification, crop, human health

INTRODUCTION

Vitamins are organic compounds required by human as micronutrients in trace amounts.[1,2] Adequate quantity of vitamins refers to the correct quantity of vitamins from food that is actually required by human body. People use the RDI to describe the quantity to meet the requirements of 97–98% of healthy individuals for the prevention of clinical deficiency. The RDI is a whole set of the daily intake level of a nutrient that is considered to be sufficient for people. Vitamin deficiencies are major forms of micronutrient deficiencies, due to insufficient intake of vitamins, threatening billion of people, and causing nutrition-related poor growth (FAO, 2009). For example, vitamin A deficiency results in night blindness, xerophthalmia, measles, corneal scarring and death of children; and folate deficiency in pregnant females frequently results in neural tube defects in new-borns (FAO/WHO, 2001).

Biofortification, a type of micronutrient intervention, aims to increase micronutrients in seeds, tubers, and leafy vegetables of food crops, and has the potential to reach the rural poor more effectively, the group who are often at highest risk of micronutrient deficiencies (Hotz and McClafferty, 2007). Nutritionists, biologists and breeders have, for decades, focused on promoting the bioavailability of micronutrients in edible parts of crops (Hurrell and Egli, 2010). In this review, we give a special attention to the progress in the manipulation of the vitamin metabolism for developing vitamin-enriched crops.

MANIPULATION OF VITAMIN METABOLISM

Provitamin A

Vitamin A exists in several forms known as retinoids. Human can synthesize retinal from the abundant provitamin A carotenoids present in fruits and vegetables such as oranges (*Citrus aurantium*), broccoli (*Brassica oleracea*), spinach (*Spinacia oleracea*), carrot (*Daucus carota*), squash (*Cucurbita maxima*), sweet potato (*Ipomoea batatas*), and pumpkin (*Cucurbita maxima*) (Harrison, 2005). Plants produce four types of provitamin A carotenoids from isopentenyl diphosphate (IPP), which can form the

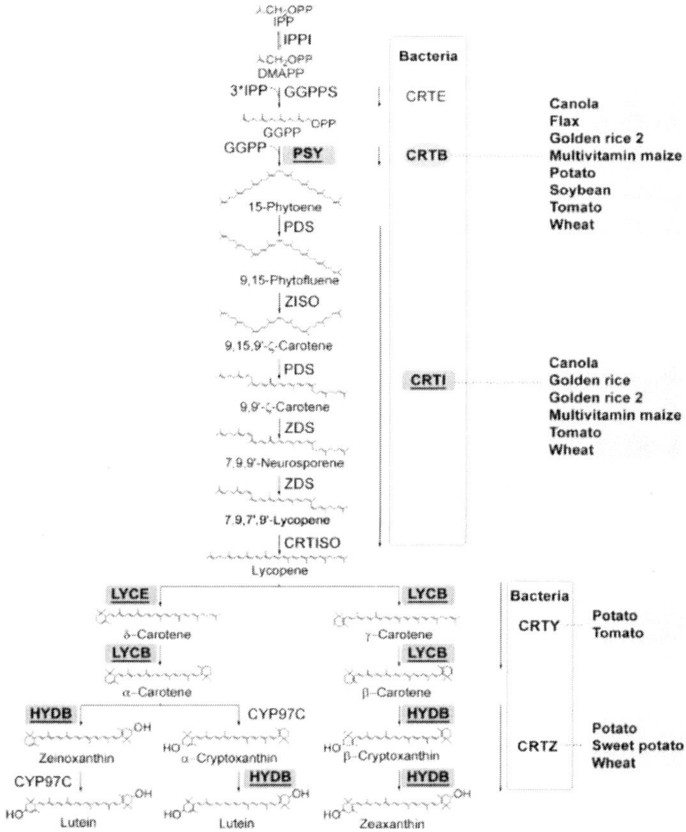

Figure 1. The carotenoid biosynthesis pathway in plants and equivalent steps in bacteria (modified from Bai et al., 2011). Enzymes: CRTB, bacterial phytoene synthase; CRTE, bacterial GGPP synthase; CRTI, bacterial phytoene desaturase/isomerase; CRTISO, carotenoid isomerase; CRTY, bacterial lycopene cyclase; CRTZ, bacterial β-carotene hydroxylase; CYP97C, carotene ε-ring hydroxylase; GGPPS, GGPP synthase; HYDB, β-carotene hydroxylase; IPPI, isopentenyl diphosphate isomerase; LYCB, lycopene β-cyclase; LYCE, lycopene ε-cyclase; PDS, phytoene desaturase; PSY, phytoene synthase; ZDS, ζ-carotene desaturase; Z-ISO, ζ-carotene isomerase. Chemicals: DMAPP, dimethylallyl diphosphate; GGPP, geranylgeranyl diphosphate; IPP, isopentenyl diphosphate. The bacterial pathway is put in a box with the hard line. The transgenic species are indicated adjacent to enzymes used in manipulation. Enzymes used only in transgenic strategies are oval and in the blue background, and enzymes used in both transgenic strategies and genetic breeding are square, underlined, and in the red background.

intermediate product phytoene. Among carotenoids, α-carotene and β-carotene accumulate in greater amounts than γ-carotene and β-cryptoxanthin, which tend to be converted rapidly into downstream products in several forms known as retinoids (reviewed by Bai et al., 2011; Figure 1).

The RDI of vitamin A is 700 μg retinol equivalents per day[2]; however, provitamin A is non-detectable in rice (*Oryza sativa*), millet (*Panicum miliaceum*), and sorghum (*Sorghum bicolor*), and low in wheat (*Triticum aestivum*), barley (*Hordeum vulgare*), and potato (*Solanum tuberosum*)[1]. In plants, β-carotene yields β-cryptoxanthin, which is further converted into zeaxanthin, whereas α-carotene yields lutein; and these chemicals are non-provitamin carotenoids (Figure 1). Both conventional and biotechnological approaches have been used to enhance the biosynthesis of provitamin A carotenoids in crops. The most commonly used gene for carotenoids biofortification is the gene encoding a phytoene synthase (PSY in plant; CRTB in bacteria; Figure 1). In Golden Rice, the introduction of a *Pantoea ananatis* phytoene desaturase (*CrtI*) and a *Narcissus pseudonarcissus* phytoene synthase (*PSY*) into rice results in an accumulation of 1.6 μg total carotenoids per gram of dry seeds (Ye et al., 2000). In Golden Rice 2, a maximum of 37 μg g^{-1} total carotenoids with β-carotene preferentially accumulated in the endosperm is achieved when a maize (*Zea mays*) *PSY1* gene is utilized (Paine et al., 2005). In these two events, the enhanced phytoene synthesis guarantees an adequate metabolic flux throughout the pathway. Similar strategies, i.e., overexpression of the phytoene synthase gene alone, phytoene desaturase gene alone, or combination of these two genes, are applied to other crops including canola (*Brassica napus*; both genes in Ravanello et al., 2003; Aluru et al., 2008; *crtB* in Shewmaker et al., 1999), flax (*Linum usitatissimum*; *crtB* in Fujisawa et al., 2008), potato (*crtB* in Ducreux et al., 2005), tomato (*Solanum lycopersicum*; *crtB* in Fraser et al., 2002; both genes in Aluru et al., 2008), maize (both genes in Naqvi et al., 2009), soybean (*Glycine max*; *crtB* in Schmidt et al., 2015), and wheat (both genes in Wang et al., 2014), among which the highest fold change of β-carotene is observed in soybean (Figure 1). A chimeric gene consisting of a chloroplast signal from pea (*Pisum sativum*) and a *crtB* from bacterium *Pantoea* is introduced into soybean via biolistics, resulting in an accumulation of 845 μg/g β-carotene, about 1500-fold increase as compared to wild type, in dry seeds (Schmidt et al., 2015).

Some other approaches are also used to enhance carotenoid biosynthesis in plants. Lycopene β-cyclase (LYCB), lycopene ε-cyclase (LYCE), and HYDB are key enzymes in the β-branch of the carotenoid biosynthesis pathway. To achieve a high level of β-carotene accumulation, suppressing the activity of LYCE by silencing *StLCY-e* in potato or the activity of HYDB by silencing *StCHY-ε* in potato is adopted to ensure the appropriate direction of the metabolic flux (Diretto et al., 2007; Van Eck et al., 2007). In wheat, simultaneous overexpression of *CrtB* and silencing of *HYDB* achieves an up to 31-fold increase of β-carotene up to 5.06 μg g^{-1} (Zeng et al., 2015). Another approach is to make use of the *Orange* (*Or*) gene from cauliflower (*Brassica oleracea* var. *botrytis*, Lu et al., 2006). The OR protein is responsible for the accumulation of carotenoids in plants, and interacts with PSY to increase the stability and activity at a post-translational level for carotenoid biosynthesis (Zhou et al., 2015). Transgenic potato tubers expressing *Or* exhibit elevated carotenoids, and the β-carotene is increased continuously during long-term cold storage (Lopez et al., 2008; Li L. et al., 2012). Similar strategy is successful in tomato (Bai et al., 2014).

However, these strategies can also have disadvantages in some cases. For example, overexpression of *PSY* in tomato increases lycopene, β-carotene, and zeaxanthin, but decreases gibberellins, which results in plant dwarfism (Fray et al., 1995). This observation indicates that manipulation of the provitamin A metabolism is still limited by our incomplete understanding of the regulation of the endogenous pathways, including rate-controlling steps and timing of expression in carotenogenic tissues (Vallabhaneni and Wurtzel, 2009). Therefore, identification of the allelic variations that control carotenoid biosynthesis by means of candidate gene-based genome-wide association studies (GWASs) to better facilitate provitamin A biofortification through allele pyramiding has become an important alternative. Much work to identify the genes or QTLs responsible for the carotenoid accumulation has been done in maize, given its marked genetic diversity (Palaisa et al., 2003; Vallabhaneni and Wurtzel, 2009; Yan et al., 2010; Azmach et al., 2013; Owens et al., 2014; Figure 1). Besides, maize lines containing such natural variations can be used as donor parents to accelerate developing provitamin A-fortified tropical maize varieties, adapted to target growth conditions and consumer preferences. For example, the β-carotene-rich hybrid maize has developed through the marker-assisted introgression of the natural alleles in β-carotene hydroxylase gene by using a crtRB1-specific DNA marker for foreground selection. (Muthusamy et al., 2014).

To identify other metabolic bottlenecks in the carotenoid pathway, the genes of *Arabidopsis* 1-deoxy-D-xylulose-5-phosphate synthase (AtDXS) and *Arabidopsis* ORANGE (AtOR) are introduced into Golden Rice 2. Upon the transformation of *AtDXS*, accumulation of the carotenoids in the endosperm is significantly enhanced, confirming that the supply of isoprenoid precursors such as geranylgeranyl diphosphate (GGPP) is a rate-limiting step (Bai et al., 2016). In Sorghum, AtDXS was coexpressed with PSY1 and CRTI, and the transgenic events accumulated higher levels of total carotenoids, and the all-trans β-carotene levels ranged from 2.5 to 9.1 μg g^{-1} DW in the mature-seed endosperm (Che et al., 2016). Upon the transformation of *AtOR*, levels of the carotenoids are also dramatically elevated, indicating that OR functions mainly through expanding the metabolic sink (Bai et al., 2016). Obviously, identification of the metabolic bottlenecks in the carotenoid pathway can help to refine the strategies for development of crops with specific carotenoids.

Folates

Folates, also called vitamin B_9, are essential water-soluble B-vitamins, including tetrahydrofolate (THF) and its derivatives. Folates play an important role as one-carbon donors and acceptors in all organisms. Folates are synthesized *de novo* in bacteria, fungi and plants. It is noteworthy that the folate biosynthesis pathway is split among cytosol, mitochondria, and chloroplasts in plants, whereas it is cytosolic in other organisms (reviewed by Storozhenko et al., 2005; Blancquaert et al., 2010, Figure 2). Usually, folates are abundant in dark-green leafy vegetables, but only tens of micrograms per 100 g weight in barley, corn, lettuce (*Lactuca sativa*), potato, rice, sweet potato, tomato, wheat, etc[1]. Therefore, folate intake is prone to inadequacy, i.e., less than the RDI of 400 μg per day for adults[2], if the dietary pattern is not optimized.

Scientists have made lots of efforts to develop various folate-fortified crops using biotechnology in recent years, including tomato, rice, maize, lettuce, potato, and Mexican common bean (*Phaseolus vulgaris*) (Diaz de la Garza et al., 2004, 2007; Storozhenko et al., 2007; Naqvi et al., 2009; Nunes et al., 2009; Blancquaert et al., 2013; Ramirez Rivera et al., 2016; Figure 2). In general, two approaches for folate biofortification have been employed. One is to overexpress dihydrofolate synthetase (DHFS; FOLE in bacteria), a rate-limiting enzyme in folate synthesis, achieving an only 2-fold increase in *folE*-overexpressing corn endosperm up to 1.94 μg g^{-1} (Naqvi et al., 2009). The other strategy, i.e., overexpression of the enzymes catalyzing the first committed step in the cytosol (CTPCHI), increases folates by 2 folds in tomato up to 2.99 nmol g^{-1}(Diaz de la Garza et al., 2004), 8.5 folds in lettuce up to 1.85 μg g^{-1} (Nunes et al., 2009), respectively. Similar strategy is used in Mexican common bean and increased pteridine by 3 folds up to 3.25 μg g^{-1} (Ramirez Rivera et al., 2016). A greater increase in tomato (25 folds, up to 8.40 μg g^{-1}) and rice (100 folds, up to 17.23 μg g^{-1}) is achieved by crossing the pteridine-overproducing traits with para-aminobenzoate (p-ABA)-overproducing traits (Diaz de la Garza et al., 2007; Storozhenko et al., 2007).

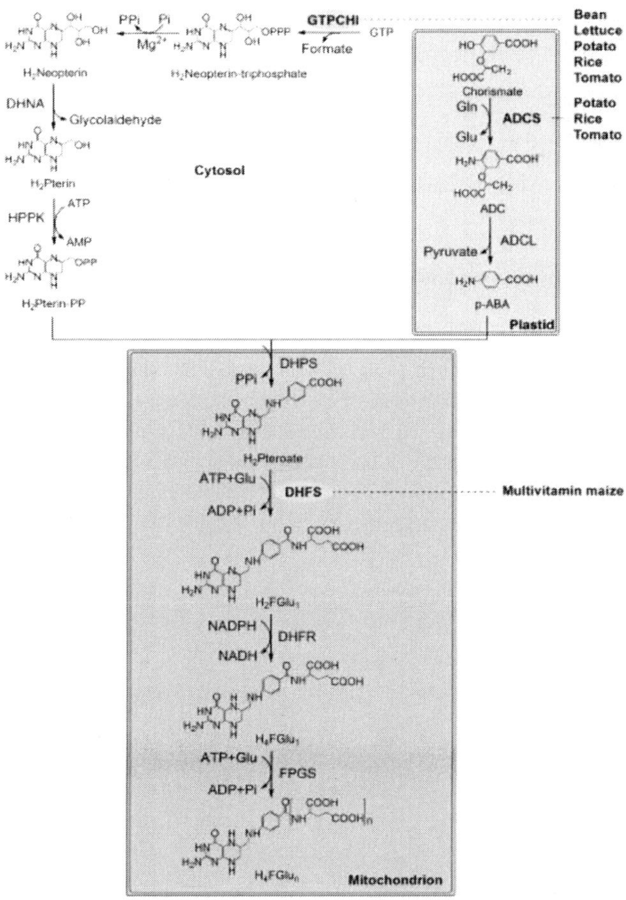

Figure 2. The folate biosynthesis pathway and its compartmentalisation in plants (modified from Storozhenko et al., 2005). p-ABA is synthesized in the plastids, pteridine in the cytosol and both are condensed to form THF in the mitochondria. Enzymes: ADCL, Aminodeoxychorismate lyase; ADCS, Aminodeoxychorismate synthase; DHFR, dihydrofolate reductase; DHFS, dihydrofolate synthetase; DHNA, dihydroneopterin aldolase; DHPS, dihydropteroate synthase; FPGS, folylpolyglutamate synthase; GTPCHI, GTP cyclohydrolase I; HPPK, dihydropterinpyrophosphokinase. Chemicals: ADC, minodeoxychorismate; H_2FGlu_1, dihydrofolate; H_2Neopterin, dihydroneopterin; H_2Pteroate, dihydropteroate; Glu, glutamate; GTP, guanosine triphosphate; H_2Pterin, dihydropterin; H_2Pterin-PP, hydroxymethyldihydropterin; p-ABA, para-aminobenzoate; H_4FGlu_1, tetrahydrofolate. The transgenic species are indicated adjacent to enzymes used in manipulation. Enzymes used in transgenic strategies are oval and in the blue background.

A similar strategy does not work in potato, regardless of the high expression of *GTPCHI* and *Aminodeoxychorismate synthase* (*ADCS*): folates in the immature potato tubers are mildly increased (~ 2 folds, up to 1.23 μg g^{-1}), and expression of the endogenous *GTPCHI* and *ADCS* does not differ between the WT and transgenic lines (Blancquaert et al., 2013). These observations are markedly different from those reported in tomato and rice (Storozhenko et al., 2007; Waller et al., 2010). Therefore, probably there is a bottleneck in the folate biosynthesis pathway in potato tubers, and a further research is needed to investigate the mechanisms of regulating folate status in these plants that are different, lacking, or of minor importance in rice seeds and tomato fruit. Besides, the combination of folylpolyglutamate synthetase (FPGS), GTPCHI, and ADCS from *Arabidopsis* with folate binding proteins (FBP) from mammals has been successful in improving folate accumulation (up to 25.30 μg g^{-1}) and stability (folate levels were stable for 4 months at 28°C) in rice (Blancquaert et al., 2015). These results indicate that a better understanding of the folate pathway is required to formulate an engineering strategy useful for the majority of other staple crops (Blancquaert et al., 2013).

Other B-class Vitamins

B-class vitamins, other than folates, include VB_1 (thiamine), VB_2 (riboflavin), VB_3 (niacin), VB_5(pantothenic acid), VB_6 (pyridoxal, pyridoxine, pyridoxamine, and their phosphorylated derivatives), VB_7(biotin), and VB_{12} (cobalamin). To our knowledge, however, only VB_6 metabolic engineering has been mainly conducted in cassava (*Manihot esculenta*; Vanderschuren et al., 2013; Li et al., 2015).

VB_6 refers to a group of six water-soluble vitamers, among which pyridoxal-5′-phosphate (PLP) is of central importance because it is required as a cofactor for over 140 chemical reactions in the cell (Vanderschuren et al., 2013; Figure 3). The RDI for vitamin VB_6 is 1.3 mg per day for adults[2]. Several groups reported overexpression of the genes either encoding pyridoxal phosphate synthase (PDX1) or encoding pyridoxal phosphate glutaminase (PDX2) in *Arabidopsis*, but the increase of VB_6 is not significant in most cases (Chen and Xiong, 2009; Leuendorf et al., 2010; Raschke et al., 2011). Later on, however, the simultaneous overexpression of *Arabidopsis PDX1.1* and *PDX2* genes in cassava achieved a 9.0-fold increase in the leaves to 54.74 μg g^{-1} by *CaMV35S* promoter, and a 15.4-fold increase in the roots to 16.21 μg g^{-1} by root-enhanced *Patatin* promoter, respectively (Li et al., 2015). Apparently, understanding of the regulatory mechanisms underlying the metabolic pathways can be improved in order for people to eventually benefit from the biofortification.

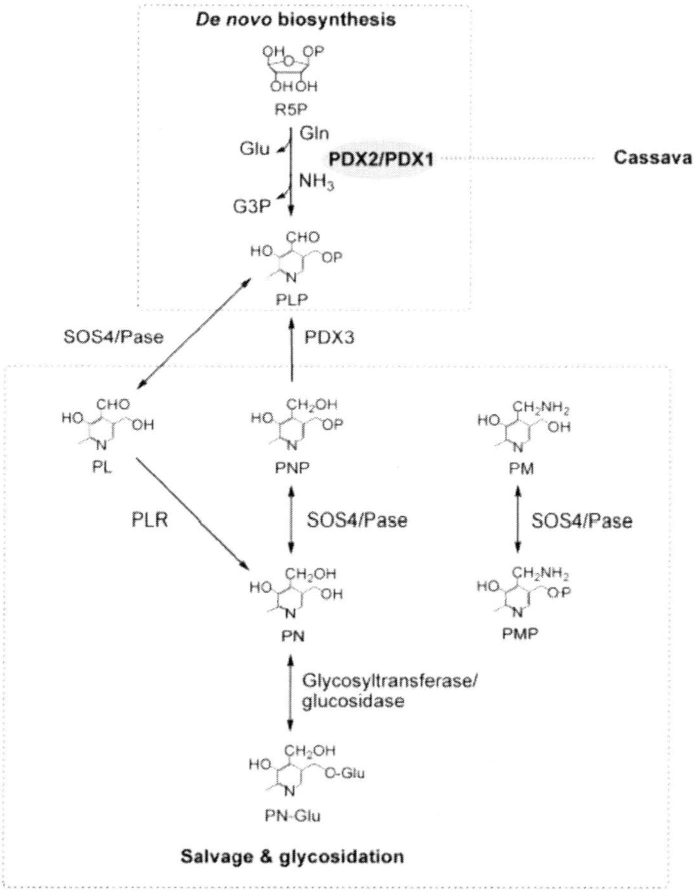

Figure 3. The VB_6 metabolism in plants (modified from Leuendorf et al., 2010). The VB_6 metabolism includes *de no* biosynthesis, salvage glycosidation. Enzymes: PDX1, pyridoxal-5′-phosphate synthase; PDX2, pyridoxal-5′-phosphate glutaminase; PDX3, pyridoxine-5′-phosphate/pyridoxamine-5′-phosphate oxidase; PLR, pyridoxal reductase; SOS4/Pase, pyridoxine/pyridoxal/pyridoxamine kinase. Chemicals: Gln, glutamine; Glu, glutamate; G3P, glyceraldehydes 3-phosphate; PA, pyridoxic acid; PN, pyridoxine; PN-Glu, PN-glycoside; PL, pyridoxal; PLP, pyridoxal-5′-phosphate; PM, pyridoxamine; PMP, pyridoxamine-5′-phosphate; PNP, pyridoxine-5′-phosphate; R5P, ribose 5-phosphate. The transgenic species are indicated adjacent to enzymes used in manipulation. Enzymes used in transgenic strategies are oval and in the blue background.

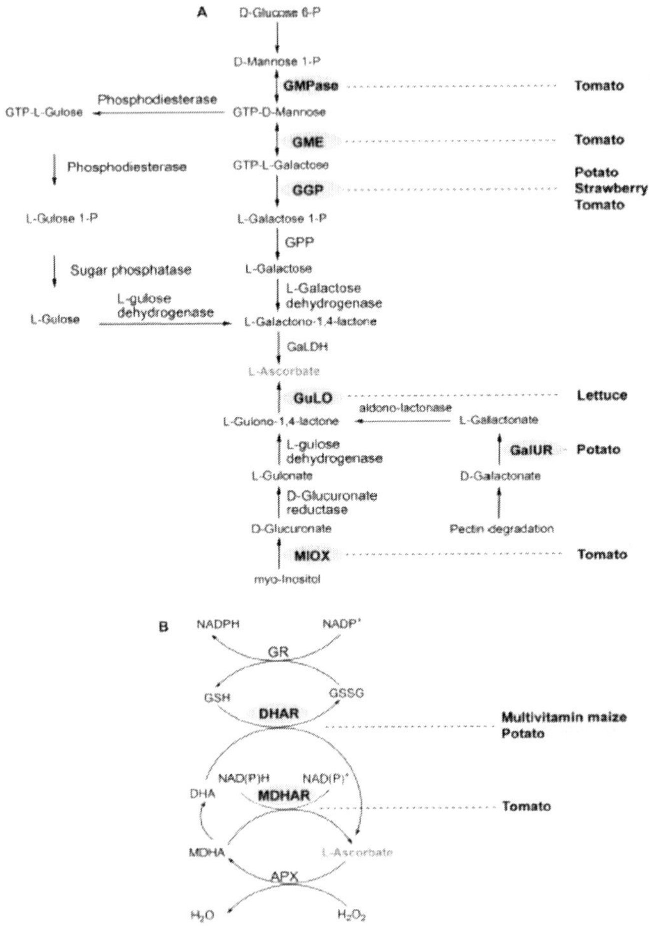

Figure 4. ASC biosynthetic routes **(A)** and Foyer-Halliwell-Asada cycle **(B)**, also known as the ASC-GSH cycle) in plants (modified from Locato et al., 2013). Enzymes: APX, ascorbateperoxidase; DHAR, dehydroascorbate reductase; GaIUR, Galacturonate reductase; GGP, GDP-Galactose phosphorylase; GME, GDP-Mannose epimerase; GMPase, GDP-Mannose pyrophosphorylase; GPP, L-Galactose 1P phosphatase; GaLDH, L-Galactono-1,4-γ-lactone dehydrogenase; GR, glutathione reductase; GuLO, L-Gulono-1,4-γ-lactone oxidase; MDHAR, monodehydroascorbate reductase; MIOX, Myo-inositol oxygenase. Chemicals: DHA, dehydroascorbate; GSH, glutathione; GSSG, oxidized glutathione; MDHA, monodehydroascorbate. The transgenic species are indicated adjacent to enzymes used in manipulation. Enzymes used in transgenic strategies are oval and in the blue background.

Vitamin C

Vitamin C, also known as ascorbate, can be synthesized via four pathways. In plants, D-glucose 6-P, D-galacturonate, and myo-inositol are the substrates of the Smirnoff-Wheeler, pectin degradation, animal, and animal-like pathways, respectively (reviewed by Locato et al., 2013; Figure 4A). The RDI for vitamin C is 75 mg per day for adults[2]. Potato and tomato contain one fourth to one fifth of the RDI for vitamin C per 100 g weight; while vitamin C is quite low and usually undetectable in the grains of cereal crops, including barley, corn, millet, rice, and wheat[1].

Vitamin C biofortification has been carried out in lettuce (Jain and Nessler, 2000), maize (Chen et al., 2003; Naqvi et al., 2009), potato (Hemavathi et al., 2009; Qin et al., 2011; Zhang et al., 2011; Bulley et al., 2012), tomato (Bulley et al., 2012; Cronje et al., 2012; Gest et al., 2013), and strawberry (*Fragaria ananassa*; Bulley et al., 2012) by either overexpressing the genes involved in biosynthesis or silencing the genes involved in ascorbate recycling (Figure 4). Most of the overexpression in *Arabidopsis*, tomato, strawberry and potato results in a 2- to 6-fold increase in ascorbate (Lorence et al., 2004; Hemavathi et al., 2009; Zhang et al., 2011; Bulley et al., 2012; Cronje et al., 2012). A 7-fold increase of ascorbate, the largest increase observed up to date, was achieved in transgenic lettuce when a rat L-Gulono-1,4-γ-lactone oxidase (GuLO) was overexpressed (Jain and Nessler, 2000). Unlike the overexpression, knockdown of the genes participating in recycling, such as monodehydroascorbate reductase (MDHAR) and dehydroascorbate reductase (DHAR), leads to less increase in tomato, maize, and potato (Chen et al., 2003; Naqvi et al., 2009; Qin et al., 2011; Gest et al., 2013; Figure 4B).

Vitamin E

Vitamin E is important for human health, and dietary or supplemental vitamin E is absorbed and delivered to the liver (Traber, 2007). Plants are the primary source of dietary vitamin E, producing the tocopherol and tocotrienol derivatives that collectively constitute vitamin E (Chen et al., 2006). The RDI for vitamin E is 15 mg per day for adults[1,2]. Tocotrienols are the major form of vitamin E in the seeds of most monocots and a limited number of dicots (Cahoon et al., 2003). Tocopherols have one saturated phytyl tail, while tocotrienols have 3-fold unsaturated side chains. Due to the presence of phenolic groups, vitamin E is easily oxidized and these derivatives are effective antioxidants. (Munne-Bosch and Falk, 2004; Figure 5). Among these derivatives, α-tocopherol has the highest biological activity for human health (Traber, 2007). Thus, the goal of vitamin E biofortification is to increase the contents of vitamin E and to convert all the other types into α-tocopherol.

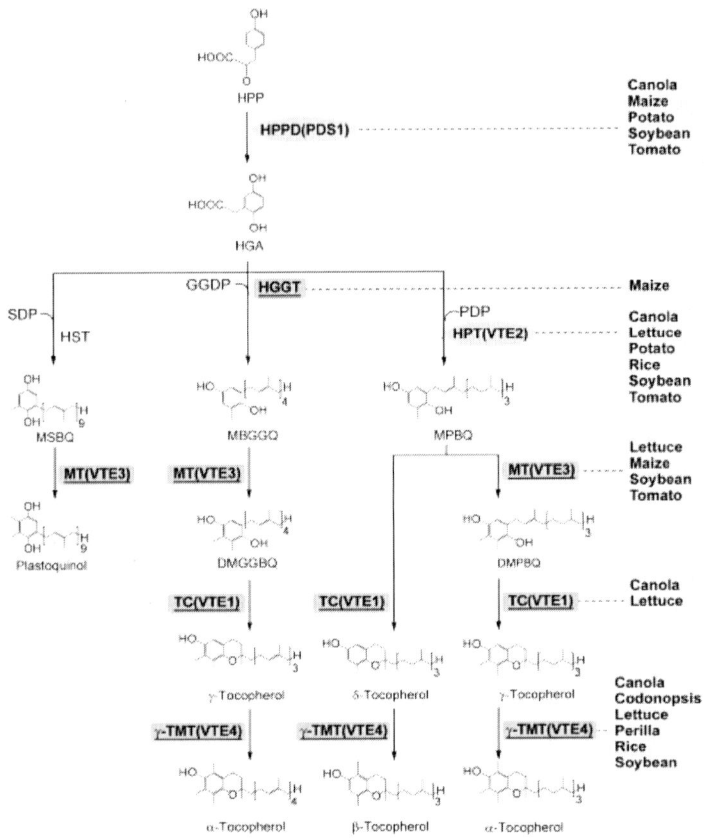

Figure 5. The enzymatic steps and metabolic products of tocopherol synthesis in plants (modified from Dörmann, 2007). Enzymes: HGGT, homogentisate geranylgeranyl transferase; HPPD/PDS1, p-Hydroxypheny lpyruvic acid dioxygenase; HPT/VTE2, homogentisic acid prenyltrans ferase; HST, homogentisate solanesyltransferase; MT/VTE3, MPBQ methyltransferase; TC/VTE1, tocopherol cyclase; γ-TMT/VTE4, γ-tocopherol methyltransferase. Chemicals: DMBPQ, 2,3-dimethyl-5-phytyl-benzoquinol; DMGGBQ, 2,3-dimethyl-5-geranylgeranylbenzoquinol; GGDP, geranylgeranyl-diphosphate; HGA, homogentisic acid; MEP, 2-*C*-methyl-D-erythritol 4-phosphate; MGGBQ, 2-methyl-6-geranylgeranyl-benzoquinol; MPBQ, 2-methyl-6-phytyl-1,4-benzoquinol; MSBQ, 2-methyl-6-solanesyl-benzoquinol; PDP, phytyldiphosphate; SDP, solanesyl-diphsophate. The transgenic species are indicated adjacent to enzymes used in manipulation. Enzymes used only in transgenic strategies are oval and in the blue background, and enzymes used in both transgenic strategies and genetic breeding are square, underlined, and in the red background.

Mostly, approaches to enhance vitamin E are to increase the activity of the enzymes in each step of the synthesis, including *p*-hydroxyphenylpyruvate dioxygenase (HPPD; Farre et al., 2012), homogentisate phytyltransferase (HPT1/VTE2; Seo et al., 2011), homogentisate geranylgeranyl transferase (HGGT; Cahoon et al., 2003), homogentisate solanesyltransferase (HST; Sadre et al., 2006), 2-methyl-6-phytyl-benzoquinol methyltransferase (MBPQ-MT/VTE3; Sattler et al., 2004; Tang et al., 2016), tocopherol cyclase (TC/VTE1; Kumar et al., 2005; Yabuta et al., 2013), and γ-tocopherol methyltransferase (γ-TMT/VTE4; Van Eenennaam et al., 2003; Cho et al., 2005; Tavva et al., 2007; Lee et al., 2008; Ghimire et al., 2011; Yabuta et al., 2013; Zhang et al., 2013; Figure 5).

Among the trials mentioned above, either a single gene or multiple genes in combination have been adopted. For example, when the barley *HGGT* gene is transformed into maize, the total tocotrienols and tocopherols increase around six folds (around 800 nmol g^{-1}) in the transgenic seeds (Cahoon et al., 2003). Barley *HGGT* gene also improves all-trans β-carotene stability in transgenic sorghum events containing *AtDXS, PSY1*, and *CRTI* (Che et al., 2016). The most commonly used single-gene approach is the overexpression of γ-TMT, which results in an increased proportion of α-tocopherol among the tocopherols, such as in canola (7.0-fold increase to over 70% of total; Van Eenennaam et al., 2003), lettuce (over 2-fold increase to 99% of total; Sattler et al., 2004; Cho et al., 2005), *Perilla frutescens* (1.8-fold increase to 99% of total with the CaMV 35S promoter, Ghimire et al., 2011; 26.6-fold increase to 75% of total with a seed-specific Vicillin promoter, Lee et al., 2008), and soybean(7.0-fold increase to 75% of total, Sattler et al., 2004; 10.4-fold increase to 88% of total,10.4 folds, Tavva et al., 2007) Compared with the single gene, the overexpression of multiple genes involved in vitamin E biosynthesis usually results in more massive accumulation of tocotrienols, and the total vitamin E activity is dramatically increased (Collakova and DellaPenna, 2003; Van Eenennaam et al., 2003; Sattler et al., 2004). For example, the combination of HPT and γ-TMT results in a 12-fold increase of vitamin E activity up to 47.4 mg α-tocopherol 100 g^{-1} tissue in canola seeds (Collakova and DellaPenna, 2003); the combination of MT and γ-TMT results in a 5-fold increase in soybean (Van Eenennaam et al., 2003; Sattler et al., 2004); and the combination of *tyrA* (*HPT* in yeast), *Arabidopsis* HPPD and HPT results in an 11-fold increase in soybean (Karunanandaa et al., 2005). Besides, modulation of the activity of TC and MT also altered the α-/γ- tocopherol ratio in plants (Van Eenennaam et al., 2003; Kumar et al., 2005). Unfortunately, similar approaches do not bring about expected effects in maize and potato. For example, overexpression of *Arabidopsis* HPPD and MT results in an only 3-fold increase in γ-tocopherol in maize kernels, and other tocopherol isomers are undetectable (Naqvi et al., 2011). Similarly, constitutive overexpression of *Arabidopsis* HPPD or HPT does not affect the tocopherol composition in potato tubers (Crowell et al., 2008).

Genome-wide association study is also used to identify the natural allelic variations controlling vitamin E. Two insertion/deletions within *VTE4*, and a single nucleotide polymorphism (SNP) located 85 kb upstream of this gene are found to be significantly associated with α-tocopherol contents in maize kernels (Li Q. et al., 2012), and three genes, *VTE1, HGGT1* and a prephenate dehydratase paralog, are also found to modestly contribute to tocotrienol variations in maize (Lipka et al., 2013). In tomato, a short interspersed nuclear elements (SINE) retrotransposon located in the promoter region of *VTE3* is identified to be responsible for vitamin E accumulation in fruits (Quadrana et al., 2014). The above findings suggest the existence of as-yet-unknown candidate genes for use in vitamin E biofortification.

CONCLUSION AND FUTURE PROSPECTS

Promotion of nutrition-sensitive agriculture and food-based strategies can solve micronutrient malnutrition (FAO, 2014). A recent study of vitamin intakes in developed countries—including Germany, the UK, the Netherlands and the United States—reveals that although inter-country differences exist, intakes of several vitamins are below the recommended levels in a significant part of the population even in these countries. Moreover, there is a gap between vitamin intake and requirements for a significant proportion of the population, despite the availability of diverse foodstuffs (Troesch et al., 2012).

Two general questions must be addressed during the design of micronutrient-enriched crops: (1) can breeding increase the micronutrient density in staple foods to reach target levels that will have a measurable and significant impact on nutritional status, and (2) will the extra nutrients bred into food crops be bioavailable and absorbed at a sufficient level to improve micronutrient status when consumed under controlled conditions (Bouis et al., 2013)? Therefore, one must take into consideration the breeding targets, target nutrients, and fortification criteria after accounting for loss during storage, milling, processing and cooking, as well as the limited bioavailability (FAO/WHO, 2001; Hotz and McClafferty, 2007). Several reports in maize, tomato and lettuce have demonstrated the potential of biofortification in alleviating vitamin deficiencies in human, by the means of ensuring an adequate uptake of nutrients (such as β-carotene and folates) from biofortified foods without processing and cooking (Tang et al., 2012; Castorena-Torres et al., 2014; Kiekens et al., 2015). For example, after cooking, the nutrients in biofortified crops retained equivalent bioavailability as synthetic compounds: the β-carotene in Golden Rice was considered as effective as that in oil in terms of vitamin A supply to children (Tang et al., 2012), and the natural folates from biofortified tomato or rice had the potential to improve folate status in human (Castorena-Torres et al., 2014; Kiekens et al., 2015).

An integrated understanding of the genetic networks and the biochemical and molecular processes that control the accumulation of target compounds in crops is required (Bhullar and Gruissem, 2013), which, to our understanding, can be achieved by employing novel technologies, such as isotope labeling-based metabolomics, metabolite-macromolecule interactions and metabolite–micromolecule interactions. Furthermore, various approaches must be used to achieve preferential accumulation in the edible parts of crops, of minerals and/or vitamins or prebiotics (non-digestible carbohydrates) and reduce the antinutrient compounds, such as phytates and polyphenolics (Hurrell and Egli, 2010).

Human nutrition research is now focused on personalized approaches. It is also conceivable that personalized plant breeding would be a future prospect to meet the nutritional needs of individuals; this will require the development of new-generation gene-sequencing technologies and large data-processing systems (Crotti et al., 2017). To this end, the following are required: (1) analysis of the whole genome of crops to understand the genetic basis of nutritional traits and to resolve the nutrition-relating haplotypes; (2) large-scale mining of genes related to nutrition and health-related phenotypes in natural germplasm resources combining metabolome, genomics, and GWAS to integrate the elite allelic variations in crops; (3) application of metabolomics to identify intermediate metabolites and the rate-limiting steps of metabolic pathways; and 4) design and creation of novel metabolic pathways that produce nutrition- and health-related nutrients. Taking all together, we can foresee that a successful manipulation of metabolism in food crops will make substantial contributions to human health by improving the nutrition and wellbeing of the world population, especially the impoverished people.

AUTHOR CONTRIBUTIONS

LJ and CZ designed and wrote the paper. WW and TL wrote the paper.

FUNDING

This work was financially supported by the National Basic Research Program of China (grant no. 2013CB127003 to CZ), the Ministry of Science and Technology of China (2016YFD0100503 to LJ), and the National Natural Science Foundation of China (grant no. 31301398 to WW).

CONFLICT OF INTEREST STATEMENT

The authors declare that the research was conducted in the absence of any commercial or financial relationships that could be construed as a potential conflict of interest.

ACKNOWLEDGMENT

We thank Dr. Lan Zhang and Ms. Ying Liu at Biotechnology Research Institute, Chinese Academy of Agricultural Sciences, for valuable discussions.

FOOTNOTES

1. ^ http://ndb.nal.usda.gov/ndb/nutrients/index
2. ^ http://www.nap.edu/catalog/11537.html

ABBREVIATIONS

γ-TMT, γ-tocopherol methyltransferase; CRTB, Bacterial phytoene synthase; CTPCHI, GTP cyclohydrolase I; FAO, Food and Agriculture Organization of the United Nations; HGGT, homogentisate geranylgeranyl transferase; HPPD, p-Hydroxyphenylpyruvate dioxygenase; HPT, homogentisate phytyltransferase; HYDB, β-carotene hydroxylase; PSY, phytoene synthase; RDI, recommended daily intake; WHO, World Health Organization.

REFERENCES

1. Aluru, M., Xu, Y., Guo, R., Wang, Z., Li, S., White, W., et al. (2008). Generation of transgenic maize with enhanced provitamin A content. *J. Exp. Bot.* 59, 3551–3562.
2. Azmach, G., Gedil, M., Menkir, A., and Spillane, C. (2013). Marker-trait association analysis of functional gene markers for provitamin A levels across diverse tropical yellow maize inbred lines. *BMC Plant Biol.* 13:227.
3. Bai, C., Capell, T., Berman, J., Medina, V., Sandmann, G., Christou, P., et al. (2016). Bottlenecks in carotenoid biosynthesis and accumulation in rice endosperm are influenced by the precursor-product balance. *Plant Biotechnol. J.* 14, 195–205.
4. Bai, C., Rivera, S. M., Medina, V., Alves, R., Vilaprinyo, E., Sorribas, A., et al. (2014). An in vitro system for the rapid functional characterization of genes involved in carotenoid biosynthesis and accumulation. *Plant J.* 77, 464–475.
5. Bai, C., Twyman, R. M., Farre, G., Sanahuja, G., Christou, P., Capell, T., et al. (2011). A golden era-pro-vitamin A enhancement in diverse crops. *In Vitro Cell. Dev. Biol. Plant* 47, 205–221.
6. Bhullar, N. K., and Gruissem, W. (2013). Nutritional enhancement of rice for human health: the contribution of biotechnology. *Biotechnol. Adv.* 31, 50–57.
7. Blancquaert, D., Storozhenko, S., Loizeau, K., De Steur, H., De Brouwer, V., Viaene, J., et al. (2010). Folates and folic acid: from fundamental research toward sustainable health. *Crit. Rev. Plant Sci.* 29, 14–35.
8. Blancquaert, D., Storozhenko, S., Van Daele, J., Stove, C., Visser, R. G., Lambert, W., et al. (2013). Enhancing pterin and para-aminobenzoate content is not sufficient to successfully biofortify potato tubers and *Arabidopsis thaliana* plants with folate. *J. Exp. Bot.* 64, 3899–3909.
9. Blancquaert, D., Van Daele, J., Strobbe, S., Kiekens, F., Storozhenko, S., De Steur, H., et al. (2015). Improving folate (vitamin B9) stability in biofortified rice through metabolic engineering. *Nat. Biotechnol.* 33, 1076–1078.
10. Bouis, H., Low, J., McEwan, M., and Tanumihardjo, S. (2013). *Biofortification: Evidence and Lessons Learned Linking Agriculture and Nutrition.* Rome: FAO/WHO.
11. Bulley, S., Wright, M., Rommens, C., Yan, H., Rassam, M., Lin-Wang, K., et al. (2012). Enhancing ascorbate in fruits and tubers through over-expression of the L-galactose pathway gene GDP-L-galactose phosphorylase. *Plant Biotechnol. J.* 10, 390–397.
12. Cahoon, E. B., Hall, S. E., Ripp, K. G., Ganzke, T. S., Hitz, W. D., and Coughlan, S. J. (2003). Metabolic redesign of vitamin E biosynthesis in plants for tocotrienol production and increased antioxidant content. *Nat. Biotechnol.* 21, 1082–1087.
13. Castorena-Torres, F., Ramos-Parra, P. A., Hernandez-Mendez, R. V., Vargas-Garcia, A., Garcia-Rivas, G., and de la Garza, R. I. (2014). Natural folates from biofortified tomato and synthetic 5-methyl-tetrahydrofolate display equivalent bioavailability in a murine model. *Plant Foods Hum. Nutr.* 69, 57–64.
14. Che, P., Zhao, Z. Y., Glassman, K., Dolde, D., Hu, T. X., Jones, T. J., et al. (2016). Elevated vitamin E content improves all-trans beta-carotene accumulation and stability in biofortified sorghum. *Proc. Natl. Acad. Sci. U.S.A.* 113, 11040–11045.
15. Chen, H., and Xiong, L. (2009). Enhancement of vitamin B(6) levels in seeds through metabolic engineering. *Plant Biotechnol. J.* 7, 673–681.
16. Chen, S., Li, H., and Liu, G. (2006). Progress of vitamin E metabolic engineering in plants. *Trans. Res.* 15, 655–665.
17. Chen, Z., Young, T. E., Ling, J., Chang, S. C., and Gallie, D. R. (2003). Increasing vitamin C content of plants through enhanced ascorbate recycling. *Proc. Natl. Acad. Sci. U.S.A.* 100, 3525–3530.
18. Cho, E. A., Lee, C. A., Kim, Y. S., Baek, S. H., de los Reyes, B. G., and Yun, S. J. (2005). Expression of gamma-tocopherol methyltransferase transgene improves tocopherol composition in lettuce (*Latuca sativa* L.). *Mol. Cells* 19, 16–22.

19. Collakova, E., and DellaPenna, D. (2003). Homogentisate phytyltransferase activity is limiting for tocopherol biosynthesis in Arabidopsis. *Plant Physiol.* 131, 632–642.
20. Cronje, C., George, G. M., Fernie, A. R., Bekker, J., Kossmann, J., and Bauer, R. (2012). Manipulation of L-ascorbic acid biosynthesis pathways in *Solanum lycopersicum*: elevated GDP-mannose pyrophosphorylase activity enhances L-ascorbate levels in red fruit. *Planta* 235, 553–564.
21. Crotti, S., Posocco, B., Marangon, E., Nitti, D., Toffoli, G., and Agostini, M. (2017). Mass spectrometry in the pharmacokinetic studies of anticancer natural products. *Mass Spectrom. Rev.* 36, 213–251.
22. Crowell, E. F., McGrath, J. M., and Douches, D. S. (2008). Accumulation of vitamin E in potato (*Solanum tuberosum*) tubers. *Transgenic Res.* 17, 205–217.
23. Diaz de la Garza, R., Quinlivan, E. P., Klaus, S. M., Basset, G. J., Gregory, J. F. III., and Hanson, A. D. (2004). Folate biofortification in tomatoes by engineering the pteridine branch of folate synthesis. *Proc. Natl. Acad. Sci. U.S.A* 101, 13720–13725.
24. Diaz de la Garza, R. I., Gregory, J. F. III, and Hanson, A. D. (2007). Folate biofortification of tomato fruit. *Proc. Natl. Acad. Sci. U.S.A.* 104, 4218–4222.
25. Diretto, G., Welsch, R., Tavazza, R., Mourgues, F., Pizzichini, D., Beyer, P., et al. (2007). Silencing of beta-carotene hydroxylase increases total carotenoid and beta-carotene levels in potato tubers. *BMC Plant Biol.* 7:11.
26. Dörmann, P. (2007). Functional diversity of tocochromanols in plants. *Planta* 225, 269–276.
27. Ducreux, L. J., Morris, W. L., Hedley, P. E., Shepherd, T., Davies, H. V., Millam, S., et al. (2005). Metabolic engineering of high carotenoid potato tubers containing enhanced levels of beta-carotene and lutein. *J. Exp. Bot.* 56, 81–89.
28. FAO (2009). *More People than Ever are Victims of Hunger.* Available at: http://www.fao.org/fileadmin/user_upload/newsroom/docs/Press%20release %20june-en.pdf
29. FAO (2014). "Rome declaration on nutrition," in *Proceedings of the Second International Conference on Nutrition Rome* (Rome: FAO).
30. FAO/WHO (2001). *Human Vitamin and Mineral Requirements.* Report of a Joint FAO/WHO Expert consultation, Bangkok, Thailand (Rome: FAO), 257.
31. Farre, G., Sudhakar, D., Naqvi, S., Sandmann, G., Christou, P., Capell, T., et al. (2012). Transgenic rice grains expressing a heterologous rho-hydroxyphenylpyruvate dioxygenase shift tocopherol synthesis from the gamma to the alpha isoform without increasing absolute tocopherol levels. *Transgenic Res.* 21, 1093–1097.
32. Fraser, P. D., Romer, S., Shipton, C. A., Mills, P. B., Kiano, J. W., Misawa, N., et al. (2002). Evaluation of transgenic tomato plants expressing an additional phytoene synthase in a fruit-specific manner. *Proc. Natl. Acad. Sci. U.S.A.* 99, 1092–1097.
33. Fray, R. G., Wallace, A., Fraser, P. D., Valero, D., Hedden, P., Bramley, P. M., et al. (1995). Constitutive expression of a fruit phytoene synthase gene in transgenic tomatoes causes dwarfism by redirecting metabolites from the gibberellin pathway. *Plant J.* 8, 693–701.
34. Fujisawa, M., Watanabe, M., Choi, S. K., Teramoto, M., Ohyama, K., and Misawa, N. (2008). Enrichment of carotenoids in flaxseed (*Linum usitatissimum*) by metabolic engineering with introduction of bacterial phytoene synthase gene *crtB*. *J. Biosci. Bioeng.* 105, 636–641.
35. Gest, N., Garchery, C., Gautier, H., Jimenez, A., and Stevens, R. (2013). Light-dependent regulation of ascorbate in tomato by a monodehydroascorbate reductase localized in peroxisomes and the cytosol. *Plant Biotechnol. J.* 11, 344–354.
36. Ghimire, B. K., Seong, E. S., Lee, C. O., Lim, J. D., Lee, J. G., Yoo, J. H., et al. (2011). Enhancement of alpha-tocopherol content in transgenic *Perilla frutescens* containing the gamma-TMT gene. *Afr. J. Biotechnol.* 10, 2430–2439.
37. Harrison, E. H. (2005). Mechanisms of digestion and absorption of dietary vitamin A. *Annu. Rev. Nutr.* 25, 87–103.
38. Hemavathi, Upadhyaya, C. P., Young, K. E., Akula, N., Kim, H. S., Heung, J. J., et al. (2009). Over-expression of strawberry D-galacturonic acid reductase in potato leads

to accumulation of vitamin C with enhanced abiotic stress tolerance. *Plant Sci.* 177, 659–667.

39. Hotz, C., and McClafferty, B. (2007). From harvest to health: challenges for developing biofortified staple foods and determining their impact on micronutrient status. *Food Nutr. Bull.* 28(2 Suppl.), S271–S279.
40. Hurrell, R., and Egli, I. (2010). Iron bioavailability and dietary reference values. *Am. J. Clin. Nutr.* 91, 1461S–1467S.
41. Jain, A. K., and Nessler, C. L. (2000). Metabolic engineering of an alternative pathway for ascorbic acid biosynthesis in plants. *Mol. Breed.* 6, 73–78.
42. Karunanandaa, B., Qi, Q., Hao, M., Baszis, S. R., Jensen, P. K., Wong, Y. H., et al. (2005). Metabolically engineered oilseed crops with enhanced seed tocopherol. *Metab. Eng.* 7, 384–400.
43. Kiekens, F., Blancquaert, D., Devisscher, L., Van Daele, J., Stove, V. V., Delanghe, J. R., et al. (2015). Folates from metabolically engineered rice: a long-term study in rats. *Mol. Nutr. Food Res.* 59, 490–500.
44. Kumar, R., Raclaru, M., Schusseler, T., Gruber, J., Sadre, R., Luhs, W., et al. (2005). Characterisation of plant tocopherol cyclases and their overexpression in transgenic *Brassica napus* seeds. *FEBS Lett.* 579, 1357–1364.
45. Lee, B. K., Kim, S. L., Kim, K. H., Yu, S. H., Lee, S. C., Zhang, Z. Y., et al. (2008). Seed specific expression of perilla gamma-tocopherol methyltransferase gene increases alpha-tocopherol content in transgenic perilla (*Perilla frutescens*). *Plant Cell Tissue Organ. Cult.* 92, 47–54.
46. Leuendorf, J. E., Osorio, S., Szewczyk, A., Fernie, A. R., and Hellmann, H. (2010). Complex assembly and metabolic profiling of *Arabidopsis thaliana* plants overexpressing vitamin B(6) biosynthesis proteins. *Mol. Plant* 3, 890–903.
47. Li, K. T., Moulin, M., Mangel, N., Albersen, M., Verhoeven-Duif, N. M., Ma, Q., et al. (2015). Increased bioavailable vitamin B6 in field-grown transgenic cassava for dietary sufficiency. *Nat. Biotechnol.* 33, 1029–1032.
48. Li, L., Yang, Y., Xu, Q., Owsiany, K., Welsch, R., Chitchumroonchokchai, C., et al. (2012). The Or gene enhances carotenoid accumulation and stability during post-harvest storage of potato tubers. *Mol. Plant* 5, 339–352.
49. Li, Q., Yang, X., Xu, S., Cai, Y., Zhang, D., Han, Y., et al. (2012). Genome-wide association studies identified three independent polymorphisms associated with alpha-tocopherol content in maize kernels. *PLoS ONE* 7:e36807.
50. Lipka, A. E., Gore, M. A., Magallanes-Lundback, M., Mesberg, A., Lin, H., Tiede, T., et al. (2013). Genome-wide association study and pathway-level analysis of tocochromanol levels in maize grain. *G3* 3, 1287–1299.
51. Locato, V., Cimini, S., and Gara, L. D. (2013). Strategies to increase vitamin C in plants: from plant defense perspective to food biofortification. *Front. Plant Sci.* 4:152.
52. Lopez, A. B., Van Eck, J., Conlin, B. J., Paolillo, D. J., O'Neill, J., and Li, L. (2008). Effect of the cauliflower Or transgene on carotenoid accumulation and chromoplast formation in transgenic potato tubers. *J. Exp. Bot.* 59, 213–223.
53. Lorence, A., Chevone, B. I., Mendes, P., and Nessler, C. L. (2004). myo-inositol oxygenase offers a possible entry point into plant ascorbate biosynthesis. *Plant Physiol.* 134, 1200–1205.
54. Lu, S., Van Eck, J., Zhou, X., Lopez, A. B., O'Halloran, D. M., Cosman, K. M., et al. (2006). The cauliflower Or gene encodes a DnaJ cysteine-rich domain-containing protein that mediates high levels of beta-carotene accumulation. *Plant Cell* 18, 3594–3605.
55. Munne-Bosch, S., and Falk, J. (2004). New insights into the function of tocopherols in plants. *Planta* 218, 323–326.
56. Muthusamy, V., Hossain, F., Thirunavukkarasu, N., Choudhary, M., Saha, S., Bhat, J. S., et al. (2014). Development of beta-carotene rich maize hybrids through marker-assisted introgression of beta-carotene hydroxylase allele. *PLoS ONE* 9:e113583.
57. Naqvi, S., Farre, G., Zhu, C., Sandmann, G., Capell, T., and Christou, P. (2011). Simultaneous expression of Arabidopsis rho-hydroxyphenylpyruvate dioxygenase

and MPBQ methyltransferase in transgenic corn kernels triples the tocopherol content. *Transgenic Res.* 20, 177–181.

58. Naqvi, S., Zhu, C., Farre, G., Ramessar, K., Bassie, L., Breitenbach, J., et al. (2009). Transgenic multivitamin corn through biofortification of endosperm with three vitamins representing three distinct metabolic pathways. *Proc. Natl. Acad. Sci. U.S.A.* 106, 7762–7767.
59. Nunes, A. C., Kalkmann, D. C., and Aragao, F. J. (2009). Folate biofortification of lettuce by expression of a codon optimized chicken GTP cyclohydrolase I gene. *Transgenic Res.* 18, 661–667.
60. Owens, B. F., Lipka, A. E., Magallanes-Lundback, M., Tiede, T., Diepenbrock, C. H., Kandianis, C. B., et al. (2014). A foundation for provitamin A biofortification of maize: genome-wide association and genomic prediction models of carotenoid levels. *Genetics* 198, 1699–1716.
61. Paine, J. A., Shipton, C. A., Chaggar, S., Howells, R. M., Kennedy, M. J., Vernon, G., et al. (2005). Improving the nutritional value of Golden Rice through increased pro-vitamin A content. *Nat. Biotechnol.* 23, 482–487.
62. Palaisa, K. A., Morgante, M., Williams, M., and Rafalski, A. (2003). Contrasting effects of selection on sequence diversity and linkage disequilibrium at two phytoene synthase loci. *Plant Cell* 15, 1795–1806.
63. Qin, A., Shi, Q., and Yu, X. (2011). Ascorbic acid contents in transgenic potato plants overexpressing two dehydroascorbate reductase genes. *Mol. Biol. Rep.* 38, 1557–1566.
64. Quadrana, L., Almeida, J., Asis, R., Duffy, T., Dominguez, P. G., Bermudez, L., et al. (2014). Natural occurring epialleles determine vitamin E accumulation in tomato fruits. *Nat. Commun.* 5:3027.
65. Ramirez Rivera, N. G., Garcia-Salinas, C., Aragao, F. J., and Diaz de la Garza, R. I. (2016). Metabolic engineering of folate and its precursors in Mexican common bean (*Phaseolus vulgaris* L.). *Plant Biotechnol. J.* 14, 2021–2032.
66. Raschke, M., Boycheva, S., Crevecoeur, M., Nunes-Nesi, A., Witt, S., Fernie, A. R., et al. (2011). Enhanced levels of vitamin B(6) increase aerial organ size and positively affect stress tolerance in Arabidopsis. *Plant J.* 66, 414–432.
67. Ravanello, M. P., Ke, D. Y., Alvarez, J., Huang, B. H., and Shewmaker, C. K. (2003). Coordinate expression of multiple bacterial carotenoid genes in canola leading to altered carotenoid production. *Metab. Eng.* 5, 255–263.
68. Sadre, R., Gruber, J., and Frentzen, M. (2006). Characterization of homogentisate prenyltransferases involved in plastoquinone-9 and tocochromanol biosynthesis. *FEBS Lett.* 580, 5357–5362.
69. Sattler, S. E., Cheng, Z., and DellaPenna, D. (2004). From Arabidopsis to agriculture: engineering improved Vitamin E content in soybean. *Trends Plant Sci.* 9, 365–367.
70. Schmidt, M. A., Parrott, W. A., Hildebrand, D. F., Berg, R. H., Cooksey, A., Pendarvis, K., et al. (2015). Transgenic soya bean seeds accumulating beta-carotene exhibit the collateral enhancements of oleate and protein content traits. *Plant Biotechnol. J.* 13, 590–600.
71. Seo, Y. S., Kim, S. J., Harn, C. H., and Kim, W. T. (2011). Ectopic expression of apple fruit homogentisate phytyltransferase gene (*MdHPT1*) increases tocopherol in transgenic tomato (*Solanum lycopersicum* cv. Micro-Tom) leaves and fruits. *Phytochemistry* 72, 321–329.
72. Shewmaker, C. K., Sheehy, J. A., Daley, M., Colburn, S., and Ke, D. Y. (1999). Seed-specific overexpression of phytoene synthase: increase in carotenoids and other metabolic effects. *Plant J.* 20, 401X–412X.
73. Storozhenko, S., De Brouwer, V., Volckaert, M., Navarrete, O., Blancquaert, D., Zhang, G. F., et al. (2007). Folate fortification of rice by metabolic engineering. *Nat. Biotechnol.* 25, 1277–1279.
74. Storozhenko, S., Ravanel, S., Zhang, G. F., Rebeille, F., Lambert, W., and Van Der Straeten, D. (2005). Folate enhancement in staple crops by metabolic engineering. *Trends Food Sci. Technol.* 16, 271–281.

75. Tang, G., Hu, Y., Yin, S. A., Wang, Y., Dallal, G. E., Grusak, M. A., et al. (2012). beta-Carotene in Golden Rice is as good as beta-carotene in oil at providing vitamin A to children. *Am. J. Clin. Nutr.* 96, 658–664.
76. Tang, Y., Fu, X., Shen, Q., and Tang, K. (2016). Roles of MPBQ-MT in promoting alpha/gamma-tocopherol production and photosynthesis under high light in lettuce. *PLoS ONE* 11:e0148490.
77. Tavva, V. S., Kim, Y. H., Kagan, I. A., Dinkins, R. D., Kim, K. H., and Collins, G. B. (2007). Increased alpha-tocopherol content in soybean seed overexpressing the *Perilla frutescens* gamma-tocopherol methyltransferase gene. *Plant Cell Rep.* 26, 61–70.
78. Traber, M. G. (2007). Vitamin E regulatory mechanisms. *Annu. Rev. Nutr.* 27, 347–362. Figure :
79. Troesch, B., Hoeft, B., McBurney, M., Eggersdorfer, M., and Weber, P. (2012). Dietary surveys indicate vitamin intakes below recommendations are common in representative Western countries. *Br. J. Nutr.* 108, 692–698.
80. Vallabhaneni, R., and Wurtzel, E. T. (2009). Timing and biosynthetic potential for carotenoid accumulation in genetically diverse germplasm of maize. *Plant Physiol.* 150, 562–572.
81. Van Eck, J., Conlin, B., Garvin, D. F., Mason, H., Navarre, D. A., and Brown, C. R. (2007). Enhancing beta-carotene content in potato by RNAi-mediated silencing of the beta-carotene hydroxylase gene. *Am. J. Potato Res.* 84, 331–342.
82. Van Eenennaam, A. L., Lincoln, K., Durrett, T. P., Valentin, H. E., Shewmaker, C. K., Thorne, G. M., et al. (2003). Engineering vitamin E content: from Arabidopsis mutant to soy oil. *Plant Cell* 15, 3007–3019.
83. Vanderschuren, H., Boycheva, S., Li, K. T., Szydlowski, N., Gruissem, W., and Fitzpatrick, T. B. (2013). Strategies for vitamin B6 biofortification of plants: a dual role as a micronutrient and a stress protectant. *Front. Plant Sci.* 4:143.
84. Waller, J. C., Akhtar, T. A., Lara-Nunez, A., Gregory, J. F. III., McQuinn, R. P., Giovannoni, J. J., et al. (2010). Developmental and feedforward control of the expression of folate biosynthesis genes in tomato fruit. *Mol. Plant* 3, 66–77.
85. Wang, C., Zeng, J., Li, Y., Hu, W., Chen, L., Miao, Y., et al. (2014). Enrichment of provitamin A content in wheat (*Triticum aestivum* L.) by introduction of the bacterial carotenoid biosynthetic genes *CrtB* and *CrtI*. *J. Exp. Bot.* 65, 2545–2556.
86. Yabuta, Y., Tanaka, H., Yoshimura, S., Suzuki, A., Tamoi, M., Maruta, T., et al. (2013). Improvement of vitamin E quality and quantity in tobacco and lettuce by chloroplast genetic engineering. *Transgenic Res.* 22, 391–402.
87. Yan, J., Kandianis, C. B., Harjes, C. E., Bai, L., Kim, E. H., Yang, X., et al. (2010). Rare genetic variation at *Zea mays* crtRB1 increases beta-carotene in maize grain. *Nat. Genet.* 42, 322–327.
88. Ye, X., Al-Babili, S., Kloti, A., Zhang, J., Lucca, P., Beyer, P., et al. (2000). Engineering the provitamin A (beta-carotene) biosynthetic pathway into (carotenoid-free) rice endosperm. *Science* 287, 303–305.
89. Zeng, J., Wang, X., Miao, Y., Wang, C., Zang, M., Chen, X., et al. (2015). Metabolic engineering of wheat provitamin A by simultaneously overexpressing *CrtB* and silencing carotenoid Hydroxylase (*TaHYD*). *J. Agric. Food Chem.* 63, 9083–9092.
90. Zhang, C., Liu, J., Zhang, Y., Cai, X., Gong, P., Zhang, J., et al. (2011). Overexpression of SlGMEs leads to ascorbate accumulation with enhanced oxidative stress, cold, and salt tolerance in tomato. *Plant Cell Rep.* 30, 389–398.
91. Zhang, G. Y., Liu, R. R., Xu, G., Zhang, P., Li, Y., Tang, K. X., et al. (2013). Increased alpha-tocotrienol content in seeds of transgenic rice overexpressing Arabidopsis gamma-tocopherol methyltransferase. *Transgenic Res.* 22, 89–99.
92. Zhou, X., Welsch, R., Yang, Y., Alvarez, D., Riediger, M., Yuan, H., et al. (2015). Arabidopsis OR proteins are the major posttranscriptional regulators of phytoene synthase in controlling carotenoid biosynthesis. *Proc. Natl. Acad. Sci. U.S.A.* 112, 3558–3563.

CHAPTER 4

The First Draft of the Pigeonpea Genome Sequence

Nagendra K. Singh[1],Deepak K. Gupta[1],Pawan K. Jayaswal[1],Ajay K. Mahato[1],Sutapa Dutta[1],Sangeeta Singh[1],Shefali Bhutani[1],Vivek Dogra[1],Bikram P. Singh[1],Giriraj Kumawat[1],Jitendra K. Pal[1],Awadhesh Pandit[1],Archana Singh[1],Hukum Rawal[1],Akhilesh Kumar[1],G. Rama Prashat[1],Ambika Khare[2],Rekha Yadav[2],Ranjit S. Raje[2],Mahendra N. Singh[3],Subhojit Datta[4],Bashasab Fakrudin[5],Keshav B. Wanjari[6],Rekha Kansal[1],Prasanta K. Dash[1],Pradeep K. Jain[1],Ramcharan Bhattacharya[1],Kishor Gaikwad[1],Trilochan Mohapatra[1],R. Srinivasan[1],Tilak R. Sharma[1]

[1].*National Research Centre on Plant BiotechnologyIndian Agricultural Research InstituteNew DelhiIndia*

[2].*Division of GeneticsIndian Agricultural Research InstituteNew DelhiIndia*

[3].*Institute of Agricultural SciencesBanaras Hindu UniversityVaranasiIndia*

[4].*Indian Institute of Pulses ResearchKanpurIndia*

[5].*University of Agricultural SciencesDharwadIndia*

[6].*Panjabrao Deshmukh Krishi VidyapeethAkolaIndia*

ABSTRACT

Pigeonpea (*Cajanus cajan*) is an important grain legume of the Indian subcontinent, South-East Asia and East Africa. More than eighty five percent of the world pigeonpea is produced and consumed in India where it is a key crop for food and nutritional security of the people. Here we present the first draft of the genome sequence of a popular pigeonpea variety 'Asha'. The genome was assembled using long sequence reads of 454 GS-FLX sequencing chemistry with mean read lengths of >550 bp and >10-fold genome coverage, resulting in 510,809,477 bp of high quality sequence. Total 47,004 protein coding genes and 12,511 transposable elements related genes were predicted. We identified 1,213 disease resistance/defense response genes and 152 abiotic stress tolerance genes in the pigeonpea genome that make it a hardy crop. In comparison to soybean, pigeonpea has relatively fewer number of genes for lipid biosynthesis and larger number of genes for cellulose synthesis. The sequence contigs were arranged in to 59,681 scaffolds, which were anchored to eleven chromosomes of pigeonpea with 347 genic-SNP markers of an intra-species reference genetic map. Eleven pigeonpea chromosomes showed low but significant synteny with the twenty chromosomes of soybean. The genome sequence was used to identify large number of hypervariable 'Arhar' simple sequence repeat (HASSR) markers, 437 of which were experimentally validated for PCR amplification and high rate of polymorphism among pigeonpea varieties. These markers will be useful for fingerprinting and diversity analysis of pigeonpea germplasm and molecular breeding applications. This is the first plant genome sequence completed entirely through a network of Indian institutions led by the Indian Council of Agricultural Research and provides a valuable resource for the pigeonpea variety improvement.

Keywords: Pigeonpea Genome sequence Disease resistance SSR markers Legumes

INTRODUCTION

Pigeonpea or Red Gram (*Cajanus cajan* (L.) Millspaugh) is an important food legume for the tropical and subtropical regions of Indian subcontinent, South-East Asia and East Africa. It is a shrub with self-compatible cleistogamous flowers, but is often cross-pollinated by bees with 10–15% out crossing. The estimated size of pigeonpea genome packed in 11 chromosomes is 858 Mbp (Greilhuber and Obermayer 1998). It plays important role in food and nutritional security because it is a rich source of proteins, minerals and vitamins. Pigeonpea seeds are consumed mainly as split pea soups such as 'Dal' and 'Sambar' but a significant proportion is also consumed as green pea vegetable and whole grain preparations. Its leaves, seed husks and pod husks are used as animal feed (Figure 1). Symbiotic bacteria (*Bradyrhizobium*) colonizing root nodules of pigeonpea fix atmospheric nitrogen up to 40 kg/ha in a cropping season and its deep root system improves soil structure and organic matter. Pigeonpea is unique among the legume crops as it is a woody shrub, therefore its stem and branches are used for firewood, fencing, thatch and making baskets by the rural population. Archeological evidence indicates that pigeonpea was domesticated in the eastern part of the Indian subcontinent along with rice and other important grain legumes, namely 'Urd' or black gram (*Vigna mungo*), 'Mung' or green gram (*Vigna radiata*) and 'Kulthi' or horse gram (*Macrotyloma uniflorum*) during prehistoric period (Fuller 2006). The world acreage of pigeonpea is 4.90 mha with annual production of about 4.22 mmt worth about 1.5 billion US dollars. India is the largest producer and consumer of pigeonpea (local names "Arhar", "Tur") with annual production of 3.07 mmt, followed by Myanmar (0.72 mmt) and Malawi (0.15 mmt) (FAOSTAT 2008).

Figure 1. The whole plant and different parts of the pigeonpea cultivar 'Asha' (ICPL 87119). **A** whole plant at fruiting stage; **B** a defoliated branch with pods; **C** a branch with heavy flowering; **D** mature seeds; **E** dehusked split seeds or 'Dal'; **F** 22 chromosomes in a root tip cell

Knowledge of the genetic basis of yield, quality and stress tolerance is important for genetic improvement of pigeonpea. Until a couple of years ago pigeonpea was considered an orphan legume crop but now substantial amount of genomic resources have been generated, largely owing to the efforts of Indo-US Agricultural Knowledge Initiative (AKI), NSF and GCP funded projects, (Varshney et al. 2009, 2010a; Dutta et al. 2011; Bohra et al. 2011). Pigeonpea cultivars have a narrow genetic base due to limited breeding efforts and poor utilization of wild pigeonpea species. Availability of genome sequence will accelerate the utilization of pigeonpea germplasm resources in breeding (Yang et al. 2006; Saxena 2008; Varshney et al. 2010b). Development of molecular markers tightly linked to the important agronomic traits is a prerequisite for undertaking molecular breeding in plants. But molecular basis of most agronomic traits in pigeonpea remains unexplored due to low level of DNA polymorphism in the primary gene pool and limited number of validated molecular markers (Ratnaparkhe et al. 1995; Yang et al. 2006; Odeny et al. 2009; Dutta et al. 2011; Bohra et al. 2011).

The aim of present study was: (a) to decode the pigeonpea genome by using next generation sequencing technologies and analyse its genes and repeat DNA contents; (b) generation of chromosome specific sequence by anchoring the sequence scaffolds to a high density reference molecular linkage map and its comparison with soybean genome; and (c) development of SSR markers for gene discovery and molecular breeding applications. Pigeonpea variety 'Asha' selected for this purpose is a popular variety with one of the highest breeder seed indents in India and is resistant to common diseases of pigeonpea, namely Fusarium wilt and sterility mosaic disease.

MATERIALS AND METHODS

Plant materials

Pigeonpea variety 'Asha' (ICPL87119) was used for genome sequencing and validation of newly designed HASSR markers. To identify informative HASSR markers, a set of 8 genotypes namely Asha, UPAS 120, HDM 04–1, Pusa Dwarf, H2004-1, Bahar, Maruti and TTB7 was screened for marker polymorphism. The seeds were obtained originally from IARI, New Delhi, ICRISAT Hyderabad, IIPR Kanpur and CCSHAU Hisar.

Genome sequence assembly and submission to NCBI GenBank

High quality genomic DNA was isolated from the leaves of a single plant of variety 'Asha' using CTAB method (Murray and Thompson 1980). Sequencing of 19 plates of whole genome shotgun libraries of short DNA fragments was carried out using GS-FLX Phase D chemistry, and 3 plates of paired end sequences from a library of 20 Kb long fragments of pigeonpea genomic DNA using GS-FLX Titanium chemistry (Margulies et al. 2005). Filtered high quality sequence reads were assembled using "Newbler GS *De Novo* assembler version 2.5.3" (Roeche Inc. Germany) with: Overlap minimum match length = 25 bp, Large genome = True, Number of CPU used = 0 (all), Exclude contigs of <500 bp. The GS Assembler is designed to compare all sequence reads in a pair wise fashion. Reads that overlap one another are joined into contigs. The consensus sequence for a contig is computed by taking an average of all aligned reads at a specific nucleotide position, the paired end reads were used for making scaffolds of sequence contigs. The large sequence contigs were quality checked and contaminating sequences were identified and removed. The quality check passed Fasta files containing 510,809,477 bp of pigeonpea genome sequence were further processed using command line software of NCBI to generate .sqn file (http://www.ncbi.nlm.nih.gov/HTGS/tbl2asninfo.html), which was submitted to GenBank as draft genome version 1 using Genomes Macro Send direct submission tool.

Gene annotation

The whole genome large sequence contigs were passed through FGENESH tool of MOLQUEST software (www.softberry.com) using *Arabidopsis thaliana* gene models as reference. From all predicted genes only those with size of >500 bp were taken for further analysis. The genes were BLAST searched against NCBI non-redundant database using optimized search parameters of gap opening penalty (G) = 4, gap extension penalty (E) = 1, mismatch score (q) = −1, match score (r) = 1, word size (W) = 11 and e-value $<e^{-20}$ (Singh et al. 2004). Low complexity regions were included in the search. The BLAST search output was processed using BLAST Parser software (http://geneproject.altervista.org/). All the hits having bit scores of >100 and e values of $<e^{-20}$ were tabulated in Microsoft Excel. Gene annotations were manually curated and categorized based on their functions. Details of pigeonpea transcriptome assemblies used for validation of predicted gene models is described earlier (Dutta et al. 2011). The predicted genes were manually curated with different

keywords/phrases using auto filters to find R-like and defense response genes and categorize them into five main classes (Hulbert et al. 2001; The Rice Chromosomes 11 and 12 sequencing consortia 2005): (a) NBS-LRR (matching with NBS-LRR, but not with LZ-NBS-LRR and LRR, CC-NBS-LRR, *Rp 1-d8*, *Lr10*, *Mla 1* and rust resistance), (b) LZ-NBS-LRR (matching with LZ-NBS-LRR, but not with NBS-LRR, CC-NBS-LRR, LRR and *RPM1*), (c) LRR-TM (matching with serine/threonine kinases and *Cf2/Cf5* resistance), (d) miscellaneous category (matching with disease resistance, viral resistance, LRR, but not with NBS-LRR, CC-NBS-LRR, LZ-NBS-LRR), (e) defense response genes (matching with glucanases, chitinases and thaumatin like proteins). Similarly, genes for abiotic stress tolerance, lipid metabolism, sugar and starch biosynthesis, cellulose synthesis and transcription factors were also identified and categorized.

Annotation of transposable elements and repeats

Both *De novo* and homology based approaches were used for the identification of repeats in the large sequence contigs of pigeonpea genome. We used Repeat Modeler software pipeline for the construction of repeat library using RECON and Repeat Scout software (Benson 1999; Bao and Eddy 2002). Repeat Masker software was used for annotation using RMBLAST as search engine (Wootton and Federhen 1993; Lander et al. 2001; Waterston et al. 2002). Same strategy was used for the identification of repeats in genetically anchored scaffolds. We developed and added two different Perl scripts (Split masker, Masked table) in the Repeat Masker to break the large data set into individual files and simultaneously run the complete file in one go. Masked table script produced results on percentage of masked elements in each scaffold and exported it in Microsoft Ecxel.

For analysis of ribosomal RNA genes we downloaded all plant rDNA data from NCBI and used BLASTN search to find 28S, 18S and 5.8S rRNA genes in the pigeonpea genome. 5S rRNA genes were searched using a pigeonpea sequence obtained by cloning of Cot1 repeat fraction. tRNAscan software (Schattner et al. 2005; Lowe and Eddy 1997) was used for prediction of transfer RNA genes. The miRNA genes were identified using BLASTN search (e $>1*10^{-5}$, top hits) of sequences present in the miRNA database, allowing no more than three mismatches (miRBAse release 17.0, Griffiths-Jones 2004; Griffiths-Jones et al. 2006, 2008; Kozomara and Griffiths-Jones 2011). Rfam database (version 10.1, May 2011, Gardner et al. 2010; Griffiths-Jones et al. 2005) was used for identification of ribosomal, small nuclear and small nucleolar RNA genes. For snRNA only those families having 100% identity and e values of

<0.001 were selected, whereas for snoRNA 80% identity and e values of <0.001 were selected.

Anchoring of sequence scaffolds to pigeonpea chromosomes

The sequence scaffolds were anchored to a high density linkage map of genic-SNP markers of an intra-species reference mapping population derived from Asha/UPAS120. The linkage map was based on two Illumina multiplex SNP assays of 1536-plex and 768-plex SNPs identified by comparing deep coverage transcriptome assemblies of the parental lines Asha and UPAS 120 (Dutta et al. 2011 and our unpublished results). The 59,681 sequence scaffolds assembled from the 454 GS-FLX sequence data were used to create a local database. Total 366 genic-SNP marker sequences genetically mapped to eleven pigeonpea chromosomes were BLASTN searched against this database at a cutoff bit score of ≥100 and e-value of $<e^{-20}$. Gene density per 50 kb of anchored scaffolds was plotted for each chromosome at respective genetic map positions (cM) using Microsoft Excel. Anchored scaffolds were also scanned for the identification and annotation of RE using Repeat Modeler and Repeat Masker software, respectively. The percentage of RE in each scaffold was plotted against the gene density. The TE related genes in the scaffolds were identified using BLAST search in the NCBI-NR database.

Comparison between pigeonpea and soybean genomes

A total of 42,094 non-TE related genes were predicted from the pseudomolecules of twenty chromosomes of soybean (*Glycine max*) using the same approach as described above and a local database was created. Genes in the anchored scaffolds of pigeonpea were searched against this database using BLASTN with optimized search parameters (Singh et al. 2004). The output was parsed using BLAST Parser software (http://geneproject.altervista.org/) and tabulated in Microsoft Excel. Chromosomal positions of both pigeonpea and soybean genes were retained in the gene headers for analysis of synteny. Numbers of hits with bit scores ≥100 for each of the eleven pigeonpea chromosomes was counted in soybean and tabulated using Microsoft Excel. Similar comparison was made using single copy pigeonpea genes against the soybean chromosomes and a circular synteny map was plotted according to Krzywinski et al. (2009). To identify single copy genes a local database of all the predicted pigeonpea genes was created using 'formatdb' script of the NCBI local BLAST (Altschul, et al. 1990). All genes in the database were searched against themselves to find their copy numbers in the genome.

In silico mining, primer design and validation of genomic-SSR markers

All assembled contigs were screened for the presence of SSRs using MISA software (http://pgrc.ipk-gatersleban.de/misa). MISA created two types of files namely, 454AllContigs.fna.misa and 454AllContigs. fna.misa. statistics. MISA files were transferred to Microsoft Excel where SSRs were classified into mono-, di-, tri-, tetra-, penta- and hexa-nucleotide and compound repeats. The minimum repeat number was set at 10 for mono-, 6 for di-, and 5 for tri-, tetra-, penta- and hexa-nucleotides. Compound SSRs were defined as those loci having ≥2 SSRs interrupted by ≤100 bp of non-repetitive sequence. Class I SSRs with repeat lengths of ≥20 bp and hypervariable SSRs with repeat lengths of ≥50 bp were extracted according to Temnykh et al. (2001) and Singh et al. (2010), respectively. Nomenclature of markers HASSR1-HASSR437 using prefix H for hypervariable A for "Arhar" (pigeonpea) followed by SSR identification number was on the same pattern as describes earlier for pigeonpea genic-ASSR markers (Dutta et al. 2011). Primer pairs flanking the repeats were designed using Primer3 software (http://frodo.wi.mit.edu/). The target amplicon size was set to 100–260 bp, annealing temperature to 60°C, primer length to 20 bp and GC content to 50%. The primers were BLAST searched against the whole genome sequence to identify those with unique binding sites. For marker validation genomic DNA of eight pigeonpea genotypes was adjusted to a final concentration of 25 ng/µl. Total 437 genomic HASSR loci were first tested for PCR amplification using genomic DNA from Asha using PTC225 Gradient Cycler (Bio-Rad). PCR was carried out in 15 µl reaction volume containing 1.5 µl of 10× reaction buffer, 0.20 µl of 10 mM dNTPs (133 µM), 1.5 µl each of forward and reverse primers (10 pmol), 2.5 µl (62.5 ng) of template genomic DNA and 0.15 µl (0.75 U) of Taq DNA polymerase (Vivantis Technologies). The PCR cycling profile was: initial denaturation at 94°C for 5 min, followed by 35 cycles of 94°C for 1 min., 55°C for 1 min., 72°C for 1 min and a final extension at 72°C for 7 min. Re-screening of primers that did not amplify at these conditions was done by sequentially decreasing the annealing temperature by 1°C; and for the primers producing multiple bands by sequentially increasing the annealing temperature by 1°C. The optimized SSR markers were then used for genotyping of eight varieties to check the level of polymorphism. PCR products were separated by electrophoresis in 4% Metaphor agarose gels (Lonza, Rockland USA) containing 0.1 µg/ml ethidium bromide in 1× TBE buffer at 130 V for 4 h, visualized and photographed in gel documentation system Fluorchem™ 5,500 (Alfa Innotech Crop., USA).

RESULTS AND DISCUSSION

Pigeonpea genome assembly

The aim of present study was to generate the first draft of pigeonpea genome sequence by making use of long sequence reads of 454 GS-FLX pyrosequencing 'Phase D' chemistry with modal read lengths of >550 bases. A total of 25,489,474 sequence reads with sequence information of 10,101,433,318 bp was generated. The primary sequence assembly included 21,102,008 sequence reads (82.79%) with 9.48 Gb sequence data, >10× coverage of the pigeonpea genome in 332,766 sequence contigs with consensus sequence of ~548 Mb. Of this, 192,089 contigs were larger than 500 bases with consensus sequence of ~511 Mb, average contig size of 2,661 bp, N50 contig size of 4,522 bp and largest contig size of 45,193 bp. After quality check (QC) 384 contig sequences were identified as bacterial contaminations and hence discarded. High quality of sequence assembly was evident from 97.9% (~500 Mb) of the consensus bases having Phred Quality scores of >40, reflecting an error rate of less than 1 in 10,000 bp (Ewing and Green 1998). Finally, 191,705 QC-passed large contigs with total 510,809,477 bp sequence. This Whole Genome Shotgun project has been deposited at DDBJ/EMBL/GenBank under the accession AFSP00000000. The version described in this paper is the first version, AFSP01000000. The contigs were arranged into 59,681 scaffolds with the help of paired end sequences of 20 kb fragment library, covering ~458 Mb of genome sequence with average scaffold size of 7,679 bp, N50 size of 13,989 bp and the largest scaffold size of 177,971 bp. Thus 83.5% of the contig sequences were arranged in the scaffolds, and 16.5% still remained as singletons. The large sequence contigs, representing about 60% of the estimated 858 Mb size of the pigeonpea genome (Greilhuber and Obermayer 1998), were used for the analysis of genes and repeat contents of the genome and mining of SSR loci. In addition, 40.7 Gb of SOLiD mate pair sequence reads and 80.4 Gb of Illumina shotgun sequence reads have been generated for improving the genome coverage and sequence quality (Figure 2). The published BAC paired end sequence data set is also available for improving the scaffolds (Bohra et al. 2011). Initial analysis showed 764.27 Mb coverage of the pigeonpea genome. However, present report describes analysis of the first draft using the 454 GS-FLX sequences only.

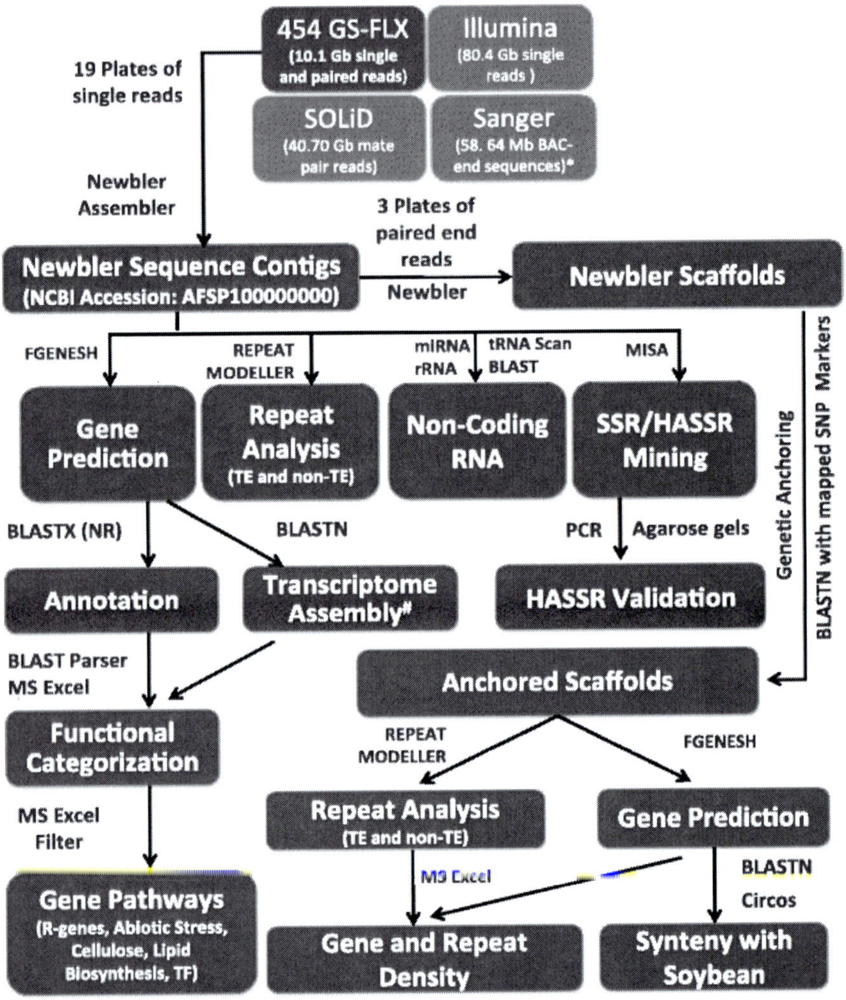

Figure 2. Flow diagram of the strategy used for the decoding of pigeonpea genome sequence

Gene content of the pigeonpea genome

The 454 GS-FLX large sequence contigs containing ~511 Mb of high quality sequence were used for gene prediction using FGENESH software. Total 59,515 genes were predicted with average gene size of 1,170 bp, largest gene size of 11,523 bp and the smallest gene size of 501 bp

(Table 1). The average exon and intron sizes were 268 bp and 288 bp, respectively, which are comparable to soybean, the species most closely related to pigeonpea, for which genome sequence is available (Schmutz et al. 2010). The predicted coding sequences of the genes were compared with a high coverage transcriptome sequence assembly database including Sanger ESTs and 454-FLX transcriptome sequence assembly (TSA) contigs (Dutta et al. 2011). Approximately 99.9% of the predicted genes showed significant matches within the pigeon pea transcriptome database. Of the 59,515 predicted genes, 42,059 showed significant matches in the NCBI-NR database with bit scores higher than 100. Total 15,558 genes showed poor hits with bit scores lower than 100 and 2,147 genes showed no hits; therefore these sequences are unique to pigeonpea. Predicted genes were classified into different functional categories (Supplementary Table S1). Total 12,511 genes (21.02%) were TE-related and 27,441 genes (46.12%) were of unknown function. The unknown category includes 2,147 genes with no matches in the NCBI–NR database, 15,558 genes showing poor BLAST hits with bit scores <100 and 9,746 genes showing significant matches in the NCBI database with hypothetical category of genes. We added these 9,746 genes to the unknown function category because they show significant matches with our pigeonpea transcriptome database and hence are real genes showing expression. Only sixteen genes belonged to hypothetical category as they did not show significant match with any transcript sequence. The remaining 19,547 genes, 41.58% of the 47,004 protein coding genes, were those with known functions. Of these 6,180 were related to physiological traits, 1,213 for disease resistance and defense response, 3,601 for cellular transportation, 216 for stress response, 139 for protein synthesis, 453 for growth and development, 751 for DNA synthesis and repair and 6,994 genes for miscellaneous functions (Figure 3).

Pigeonpea genome has a large number of 1,213 disease resistance (R-like) and defense response (DR) genes, which is 2.58% of all protein coding genes (Supplementary Table S1). These were divided into five classes based on sequence homology with the well established category of R-like and DR genes (Figure 4). Total 98 DR genes were identified which belonged to three classes, namely chitinases (31 genes), glucanases (56 genes) and thaumatin-like proteins (11 genes). Out of 1,115 R-like genes, 219 (19.6%) belonged to miscellaneous category including genes for viral resistance, verticillium wilt resistance, bacterial blight resistance and genes containing LRR motif but without NBS, CC or LZ motifs. Of the total R-like genes the largest number of 459 genes (41.1%) showed homology to LRR-TM type genes, the second largest number of 436 genes (39.1%) showed homology to NBS-LRR type genes. Only one gene belonged to LZ-NBS-LRR category. The large number of disease resistance and defense response genes makes pigeonpea a hardy crop with fewer diseases.

Table 1. Summary of gene prediction statistics in the genome sequence of pigeonpea variety 'Asha'

Description	Size/number
Size of the assembled genome sequence (bp)	510,809,477
Number of large sequence contigs	191,705
Number of protein coding genes	47,004
Number of TE-related genes	12,511
Largest gene size (bp)	11,523
Smallest gene Size (bp)	501
Average gene Size (bp)	1,170
Total number of exons	233,560
Largest exon size (bp)	6,555
Average exon size (bp)	268
Maximum number of exons in a gene	54
Total number of introns	180,000
Largest intron size (bp)	4,884
Average intron size (bp)	288

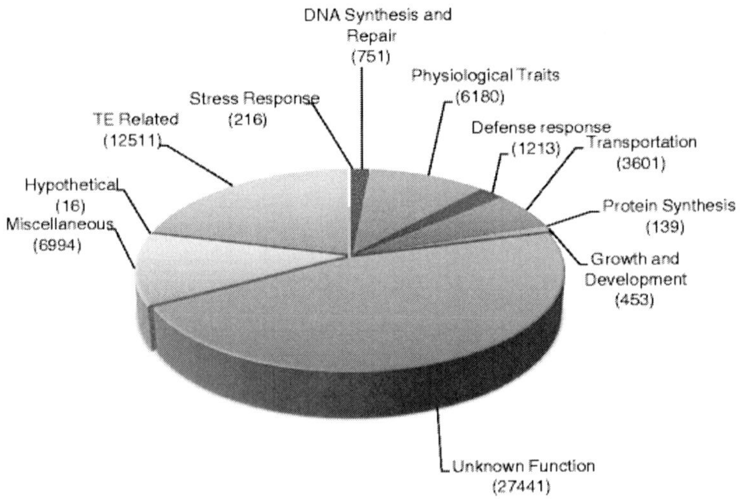

Figure 3. Frequency of different categories of genes in the 511 Mb of pigeonpea genome sequence. Unknown category includes genes unique to pigeonpea and those showing matches with hypothetical category genes of other species

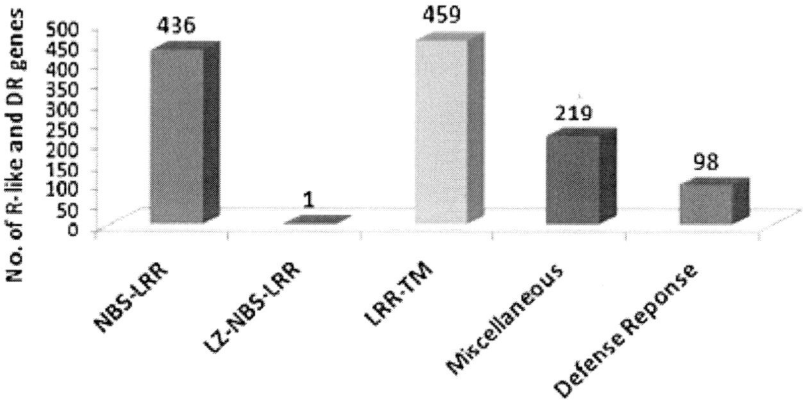

Figure 4. Frequency distribution of five main categories of resistance-like (R-like) and defense response (DR) genes predicted in the pigeonpea genome

Table 2. Frequency of some major categories of genes in the pigeonpea genome in comparison to soybean genome

Gene category	No. of genes Pigeonpea[a]	No. of genes Soybean[b]	Detailed Supplementary material
Disease resistance and defense response	1213	1174	Table S1
Abiotic stress tolerance	152	220	Table S2
Lipid metabolism	269	536	Table S3
Cellulose synthase	43	37	Table S4
Sugars and starch synthesis	108	284	Table S5
Transcription factors	1470	2300	Table S6

[a]in 511 Mb of 858 Mb genome;
[b]in 950 Mb of 1,115 Mb genome

Pigeonpea genome has 152 homologs of genes that have been implicated in abiotic stress tolerance in other plant species (Table 2). These include, 56 genes for heat shock proteins, 32 genes for glutathione-S-transferase (GST), 28 genes for trehalose-6-phosphate synthase (TPS), 8 genes for glutamine synthase (GS), 7 genes for water channel protein aquaporins and several transcription factors involved in abiotic stress response e.g. DREB, NAC and MYB genes (Supplementary Table S2).

Schmutz et al. (2010) identified 1,127 putative acyl lipid metabolism genes in the oilseed crop soybean. A similar analysis of genes for lipid metabolism in pigeonpea genome identified only 269 such genes, while soybean showed 536 genes (Table 2, Supplementary Table S3). Apart from the seed storage lipids these genes are involved in the metabolism of membrane lipids and various kinds of lipo-protein, glyco-lipid and mineral-lipid interactions. In contrast, pigeonpea genome has a higher number of 43 cellulose synthase genes as compared to only 37 genes in the soybean genome, which may be important for its woody plant architecture (Supplementary Table S4). Pigeonpea genome has 108 genes for the synthesis of various kinds of sugars, sugar transporters and starches including granule bound starch synthase, soluble starch synthase, starch branching and debranching enzymes. These have important implications for the grain yield and biomass accumulation (Supplementary Table S5). We identified 1,470 genes for different transcription factors and regulatory proteins in the pigeonpea genome (Table 2, Supplementary Table S6). These transcription factors play pivotal roles in the developmental regulation of gene expression and response of plants to various biotic and abiotic stresses. Most predominant transcription factors in the pigeonpea genome were AP2 domain-containing proteins, NAC domain containing proteins, WRKY transcription factors, Zinc finger proteins and MYB transcription factors.

Repeat elements in the pigeonpea genome

Identification and classification of repeat elements (RE) in large eukaryotic genomes is a challenging task that requires both *de novo* and homology based approaches (Lerat 2010). *De novo* analysis of RE using Repeat Modeler software revealed that pigeonpea, like other higher eukaryotic genomes, contains large proportion of repetitive DNA (Table 3). Repeat Modeler generated 1,811 different families of repeats known as the reference library. There were total 1,127,729 REs in the pigeonpea genome covering total 326,671,068 bp of sequence. Most REs (92.8%) were of interspersed type, comprising of Class I (Retro transposons), Class II (DNA transposons) or unclassified transposable elements. Simple direct repeats and low complexity repeats represented only 2.57% and 4.63% of the total RE, respectively. Homology based annotation using Repeat Masker identified REs belonging to six major categories, namely (a) LINEs including L1, R1, RTE-BovB; (b) LTR-retrotransposons including LTR, Caulimovirus, Copia, Gypsy; (c) DNA transposons including En-spm, Harbinger, hat-AC, hat-Tag1, hat-Tip100, Rc/Hiltron, MuDr, TcMAR-Pogo, RC/Hiltron; (d) Unclassified interspersed repeats; (e) Simple tandem repeats; and (f) Low complexity repeats. Similar to the other sequenced plant genomes, Gypsy and Copia type LTR-retrotransposons were the most predominant REs, constituting 16.02% and 6.10%, respectively. Interestingly, largest proportion (66.20%) of RE in the pigeonpea genome were unclassified and hence are unique to pigeonpea (Table 3). The total size of RE in the pigeonpea was 326.67 Mb which was 63.95% of the 511 Mb available genome sequence. The proportion of RE in the pigeonpea genome is higher than 23.90% in grape (Velasco et al. 2007), 25.03% in cucumber (Huang et al. 2009), 28.10% in Brachypodium (The International Brachypodium Initiative 2010), 34.79% in rice (IRGSP 2005) and 53.17% in papaya (Ming et al. 2008); similar to 61.47% in soybean (Schmutz et al. 2010), 67% in apple (Velasco et al. 2010), 64.13% in potato (The Potato Genome Sequencing Consortium 2011) and 62% in sorghum (Paterson et al. 2009); but lower than 84.20% in maize (Schnable et al. 2009) (Table 4).

Table 3. Different types of repeat elements in the 511 Mb of pigeonpea genome sequence

Repeat category	Number of elements	Sequence length (bp)	Percent of repeats
1. Interspersed repeats			
1.1 Class I (Retro transposons)	127,602	77,096,057	23.59
LINE-L1	5,239	2,270,477	0.69
LINE-R1	1,277	784,129	0.24
LINE-RTE-BovB	2,087	333,215	0.10
LTR	186	39,775	0.01
LTR-Caulimovirus	2,508	1,376,233	0.42
LTR-Copia	40,373	19,937,308	6.10
LTR-Gypsy	75,932	52,354,920	16.02
1.2 Class II (DNA transposons)	21,212	9,772,250	2.99
En-spm	5,166	2,339,643	0.71
Harbinger	1,348	467,230	0.14
hat-AC	6,572	4,059,651	1.24
hat-Tag1	1,806	586,556	0.17
hat-Tip100	934	337,903	0.10
MuDR	4,980	1,830,967	0.56
TcMAR-Pogo	153	43,654	0.01
RC/Hiltron	253	106,646	0.03
1.3 Unclassified	623,425	216,262,607	66.20
2. Simple repeats	72,522	8,405,304	2.57
3. Low complexityrepeats	282,968	15,134,850	4.63
Total	1,127,729	326,671,068	99.98

Table 4. Major classes of repeat elements (RE) in the pigeonpea genome in comparison to ten other sequenced plant genomes

Repeat Category	Pigeon pea	Soybean	Apple	Brachypodium	Cucumber	Grape	Papaya	Potato	Rice	Sorghum	Maize
Genome sequence (Mb)	511	955	742	271	227	477	271	727	370	740	2045
RE in genome (%)	63.95	61.47	67.00	28.10	25.03	23.90	53.17	64.13	34.79	62.00	84.20
1. Interspersed repeats											
1.1 Class I (Retro transposons)	23.6	68.7	55.72	83.01	48.55	85.8	82.72	50.44	55.60	87.9	89.77
Line	1.03	0.4	9.6	6.91	6.94		2.08	3.50	3.22	0.10	1.16
Copia	6.1	20.28	8.1	17.28	–	21.08	10.61	–	11.05	8.40	28.1
Gypsy	16.02	48.01	37.36	57.12	–	61.93	53.64	–	31.28	30.7	55.05
1.2 Class II (DNA transposons)	2.99	26.83	1.31	16.98	4.94	6.26	0.39	6.14	37.25	12.00	10.22
Hat super family	1.52	0.06	0.41	0.84	–	3.58	–	–	1.08	0.02	1.35
Harbinger	0.14	0.47	0	1.49	–	–	–	–	–	0.02	–
Helitron	0.03	0.86	0	0.64	–	–	–	–		1.30	2.64
1.3 Unclassified	66.2		35.35		46.49	2.99	16.44	43.40	–	1.17	
2. Simple repeats	2.57	0.75	–	–	–	–	–	–	–	–	–
3. Low complexity repeats	4.63	1.77	–	–	–	–	–	–	–	–	–

DNA transposons constituted 2.99% of the pigeonpea genome, which is higher than apple (1.31%) but much lower than rice (37.25%), soybean (26.83%) and Brachypodium (16.98%). However, these proportions might be revised upwards after all the pigeonpea REs are classified. We identified 6,572 copies of hat-AC like families which has the highest frequency among the DNA transposons, followed by En-spm (5,166 copies) and TcMAR-Pogo (153 copies). Helitrons constituted only ~0.03% of the total RE in pigeonpea while sorghum showed the highest percentage of 1.3% (Table 4). The unclassified RE sequences represented the highest copy number of 623,425, covering 216 Mb of the available genome sequence. The interspersed repeats constituted 303 Mb (92.78%) of all RE in the pigeonpea genome, which was similar to soybean (95.53%). In contrast to the interspersed transposable elements, simple repeats and low complexity repeats contributed only 2.57% and 4.63% of the pigeonpea genome, respectively. These values were higher than 0.75% and 1.77% for the soybean genome (Schmutz et al. 2010).

Non-coding RNA genes in the pigeonpea genome

Genomes of higher plants contain thousands of copies of genes for non-coding RNA including rRNA, tRNA, miRNA, snRNA and snoRNA which play important role in the cellular protein synthesis machinery and regulation of expression of protein coding genes. In the pigeonpea genome we identified 35 copies of 28S rRNA genes, 66 copies of 18S rRNA (largest match of 2,346 bp in contig number 7,811) and 77 copies of 5.8S rRNA (largest match of 2,166 bp in contig number 77,111). We identified 270 copies of 5S rRNA genes using pigeonpea specific rDNA probes. We expect more copies of rRNA genes in the finished genome. The tRNAscan-SE software identified 671 tRNA genes. Of this, twenty were pseudogenes and two have undetermined anticodon isotypes. Remaining 649 tRNA genes have 50 different anticodons, representing all the twenty amino acids (Supplementary Table S7AB). The maximum number of genes were for leucine tRNAs (49), followed by serine (47), arginine (45) and glycine (45). Thirty six of the pigeonpea tRNA genes contain introns.

MicroRNAs (miRNAs) are important regulators of several biological processes like plant growth and development. These are 20–24 nucleotides in length. The 17th release of miRBAse database contains 19,724 mature miRNA sequences including 3,423 genes of plant origin. We identified 100 miRNA genes belonging to 32 different families in the pigeonpea genome (Figure 5, Supplementary Table S8). Out of the 100 miRNA genes, 52 belong to miR829.1 family of *Arabidopsis thaliana*, which targets expression of three different proteins: (a) 3-ketoacyl-CoA reductase (b) P-glycoprotein and (c) central motor kinesin 1. There were four miRNA genes targeting ATP-sulfurylase and sulphate transporters (miR395), three genes each targeting Apetala2-like transcription factors (miR172) and F-box (miR393a) putative elements arein. There were twenty seven miRNA gene families with one or two copies of miRNA genes. Pigeonpea genome contains 226 snRNA genes showing homology in the Rfam database version 10.1 (May 2011). U6 family of snRNA showed the highest copy number of 97 genes, followed by U2 (48 genes) and U1 (34 genes) (Supplementary Table S9A). Further search with the Rfam database identified 335 sequences belonging to 90 families of snoRNA genes. The snoR71 family has the highest number of 166 genes, followed by snoRA7 and snoRD14 families having 10 genes each, the remaining families had 1–6 genes per family (Supplementary Table S9B).

Anchoring of pigeonpea sequence scaffolds to genetic map

We developed a high density intra-species reference genetic map of pigeonpea based on 366 genic-SNP markers (unpublished data). The 59,681 sequence scaffolds of pigeonpea genome were compared with the sequences of mapped genic-SNP markers and 347 (99.3%) of these showed matches with an equal number of scaffolds, covering total sequence of ~7.42 Mb. The anchored scaffolds provide genome wide nucleation points for the finishing of the pigeonpea genome and creation of large pseudomolecules for its eleven chromosomes. The 347 scaffolds were assigned to the eleven linkage groups of pigeonpea (Table 5). We predicted 1,041 genes in the anchored scaffolds, 63 of these genes were identified as TE-related genes and 26 genes did not show any hit in the database.

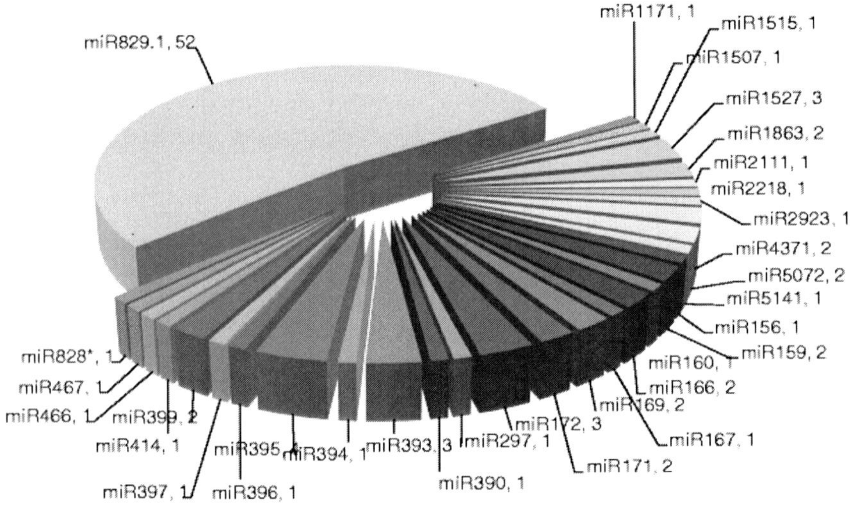

Figure 5. Distribution of 100 copies of miRNA in 32 different family in peageonpea genome

Table 5. Gene and repeat densities in the pigeonpea genome scaffolds anchored with 347 genetically mapped genic-SNP markers

Chrom. No.	No. of Markers	Size of scaffolds (bp)	No. of genes	No. of genes per 50 kb	Size of repeats (bp)	Percent repeats in scaffolds
1	40	797,775	122	7.65	182,690	22.90
2	40	763,938	103	6.74	1,74,751	22.88
3	64	1,078,018	136	6.31	264,330	24.42
4	49	1,404,117	194	6.91	274,870	19.58
5	27	431,074	60	6.96	108,700	25.22
6	19	442,848	58	6.55	108,148	24.42
7	27	768,188	101	6.57	167,903	21.86
8	18	338,877	51	7.52	85,621	25.27
9	28	640,548	99	7.73	150,894	23.56
10	23	583,427	82	7.03	152,508	26.14
11	12	175,561	35	9.97	27,072	15.42
Total	347	7,424,371	1,041	7.01	1,697,486	22.87

Out of the 7,424,371 bp of anchored scaffolds 1.697 Mb (23%) were RE which was less than half of the 63.95% RE in the whole pigeonpea genome, indicating that the anchored scaffolds represented gene-rich regions of the genome. Chromosome 10 showed the highest RE content (26.24%) while chromosome 11 showed the lowest RE content (15.43%). Anchored scaffolds represented only ~1.6% of the total ~458 Mb of assembled scaffolds, but they do provide a random sample of the genome and large number of nucleation points for the finishing of the genome. The average number of genes per 50 kb of scaffold sequence in the entire genome was 7.01 (Table 5). The gene density in the scaffolds was expected to be inversely proportional to the repeats density, which was true for many of the anchored scaffolds. For example, in chromosome 2, 3 and 5 we could clearly find this pattern for most of the scaffolds (Figure 6). There was no uniform pattern for all the chromosomes, e.g. there was higher density of repeats in the middle portion of chromosomes 1, 5, 7 and 9, in one half of the chromosomes 4 and 10 and both the telomeric ends of chromosome 3. There was no clear difference in the repeat density along the lengths of chromosomes 2, 6, 8 and 11 (Figure 6). In contrast, all twenty chromosomes of soybean have higher repeat density in the centromeric regions and higher gene density near the telomeres (Schmutz et al. 2010). The gene density in the anchored scaffolds of pigeonpea chromosomes was in the range of 6.31 to 9.97 per 50 kb. Chromosome 11 showed the highest gene density of 9.97% and lowest RE content of 15%. However, this picture may change as the size of scaffolds grows and we capture a high proportion of genome in the anchored scaffolds. We plan to merge the Illumina and Solexa data to increase the sequence coverage and BAC end sequence for increasing the size of anchored scaffolds.

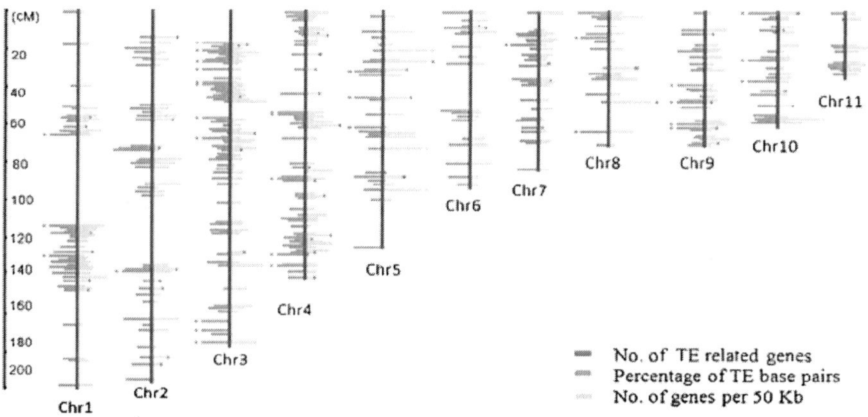

Figure 6. Density of genes and repeat elements (TE) in the 347 anchored scaffolds on the eleven chromosomes of pigeonpea. Blue bars on the left side of each chromosome represent RE percentage in the scaffold and orange bars on the right side represent gene density per 50 kb. Red segments at the end of orange bars represent number of TE-related genes in the scaffold. Discontinuous blue bars indicate RE density of >40% whereas discontinuous orange bars represent gene density in excess of >10 genes per 50 kb

Comparative analysis of pigeonpea and soybean genomes

Pigeonpea and soybean belong to the same clade Millettieae of the plant family Fabaceae (Wojciechowski 2003). Both are important crop plants but have quite different plant architecture and seed composition. Pigeonpea is a shrub grown as annual crop that has high seed protein and starch contents but minimal oil content. Soybean on the other hand is an annual herb with seeds rich in oil and protein but low in carbohydrates. Therefore, we were interested to see the difference in the genome organization and gene content of the two species. The 47,004 protein coding genes of pigeonpea were compared with 42,094 protein coding genes of soybean using BLAST search with default parameters. Total 31,937 (67.94%) of the pigeonpea genes showed matches with soybean genes at a cutoff bit score of 100, whereas 9,067 genes were unique to pigeonpea. Similarly, out of 42,094 genes predicted in soybean 40,392 showed significant matches with pigeonpea genes, whereas 1,702 genes were unique to soybean. This shows that pigeonpea has significantly higher number of unique genes that differentiate it from soybean.

Conservation of synteny between pigeonpea and soybean was analysed on the basis of 347 genetically anchored scaffolds of pigeonpea. There are total 1,041 genes in the anchored scaffolds of which 512 are single copy genes. Number of matches with all genes and single copy genes of pigeonpea in twenty chromosomes of soybean are shown in Supplementary Table S10. Genes on each of the pigeonpea chromosomes showed matches with multiple soybean chromosomes but some soybean chromosomes showed significantly higher number of matches, and therefore are likely be syntenic. Another aspect to this analysis was comparison of all genes versus single copy pigeonpea genes which are shown to have a greater conservation of synteny between rice and wheat (Singh et al. 2007). Our comparison of all genes versus single copy pigeonpea genes with soybean also showed an improved visualization of synteny with single copy genes (Supplementary Table S10AB). Therefore, we focused on comparison of homology of single copy pigeonpea genes with the protein coding genes in twenty chromosomes of soybean (Figure 7, Supplementary Table S10B). Chromosome 1 of pigeonpea showed matches with multiple soybean chromosomes even on the basis of single copy genes, but highest number of matches were found with chromosomes 8 and 5. Similarly, chromosome 2 of pigeonpea showed highest number of matches with chromosomes 19 and 10 of soybean. Chromosome 3 showed highest number of matches with chromosomes 13 and 15 of soybean. Chromosome 4 showed highest number of matches with chromosomes 12 and 13 of soybean. Chromosome 5 showed highest number of matches with chromosomes 13, 12 and 17 of soybean. Chromosome 6 showed highest number of matches with chromosomes 9 and 3 of soybean. Chromosome 7 showed highest number of matches with chromosomes 10 and 20 of soybean. Chromosome 8 of pigeonpea did not show high synteny with any specific chromosomes of soybean but it showed highest number of match with chromosomes 13 and 14. Chromosome 9 showed high number of matches with chromosomes 2, 12, 3, 11 and 16 of soybean. Chromosome 10 showed highest number of matches with chromosomes 18, 17 and 2 of soybean. Chromosome 11 of pigeonpea did not show high synteny with any specific chromosomes of soybean but highest numbers of matches were with chromosomes 14 and 18. A clear conservation of synteny was observed only with chromosomes 1, 3, 4 and 9 of pigeonpea with chromosomes 2, 5, 7, 8, 12, 13, 15 and 17 of soybean (Figure 7). Chromosomes 2, 5, 6, 7 and 10 did not show clear synteny with any soybean chromosomes. Chromosomes 8 and 11 of pigeonpea did not show more than 10 matches with any of the soybean chromosomes (Figure 7, Supplementary Table S10B). Low level of synteny between pigeonpea and soybean suggests that they might have only one genome in common and both are ancient amphipods. Their genomes have highly evolved after speciation from a common ancestral species; hence there is limited conservation of synteny between the two. This is in contrast to high conservation of macro synteny between rice and wheat, which separated about 50 mya (Singh et al. 2007).

Development and validation of hyper variable HASSR markers for pigeonpea

Pigeonpea genome was analysed to identify 1,89,895 SSR loci comprising of 100,373 mono-nucleotide, 49,325 di-nucleotide, 18,505 tri-nucleotide, 2,217 tetra-nucleotide, 512 penta-nucleotides, 815 hexa-nucleotide and 18,148 compound repeats (Table 6). Overall there is one SSR locus for every 2.88 kb of the pigeonpea genome sequence. Mononucleotide repeats are the most abundant class of SSRs in most genomes and pigeonpea was no exception to this. However, these do not serve as useful markers and excluding this category there was one SSR every 6.12 kb of the genome sequence. The frequency of SSR loci decreased successively with increasing size of the repeat unit from mono- to penta-nucleotide repeats, but frequency of hexa-nucleotide repeats was higher than penta-nucleotide repeats and compound repeats were much more abundant, comparable in frequency to the di-nucleotide repeats (Table 6). Among the two types of mono-nucleotide repeats, A/T were much more abundant than G/C (Supplementary Table S11). Among the di-nucleotide repeats, AT/AT was the most frequent while GT/AC and CG/CG were the least frequent. In the tri-nucleotide category AAT/ATT repeats were the most abundant while ACG/CGT and TAC/GTA were scarce. In the tetra-nucleotide category AAAT/ATTT was the most common motif whereas AAGG/CCTT, ACGT/ACGT, ACTA/TAGT and AGGA/TCCT were least frequent. In the penta-and hexa nucleotide categories also AT-rich repeats were more prevalent than the GC-rich repeats.

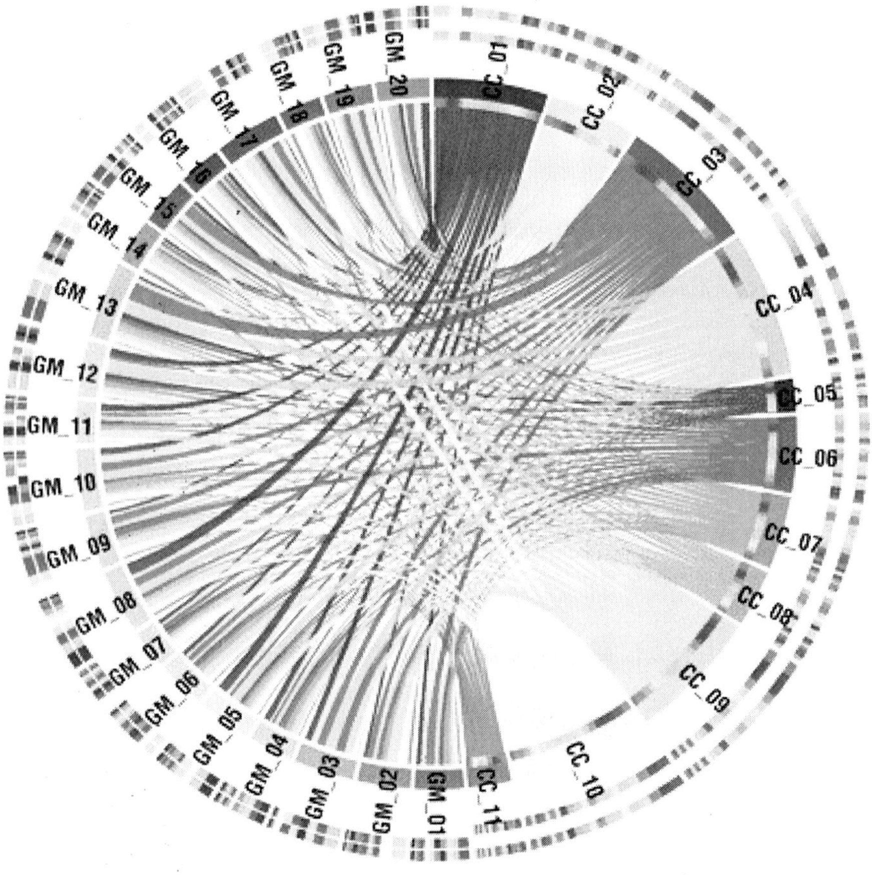

Figure 7. Circular map of syntenic relationship between 11 pairs of pigeonpea chromosomes with 20 pairs of soybean chromosomes based on 512 single copy genes in the genetically anchored scaffolds of pigeonpea genome. The outer circles depict soybean chromosome bars showing proportion of gene matches with different chromosomes of pigeonpea and vice versa

Table 6. Frequency of SSRs in the 511 Mb of pigeonpea genome sequence

Type of SSR	Total no. of SSRs	Class I SSR (n ≥20 bp)	HASSR[a] (n ≥50 bp)
Mononucleotide	1,00,373	987	0
Dinucleotide	49,325	18,000	203
Trinucleotide	18,505	5,822	515
Tetranucleotide	2,217	2,217	17
Pentanucleotide	512	512	15
Hexanucleotide	815	815	70
Compound	18,148	18,148	10,891
Total	189,895	46,501	11,711

[a] Hypervariable "Arhar" SSR

Search for class I SSRs (n ≥20 bp, Temnykh et al. 2001) and hyper variable HASSRs (n ≥50 bp, Singh et al. 2010) revealed that class I SSRs are most prevalent in the di-nucleotide category, whereas HASSRs are most abundant in the compound SSR category (Table 6). Based on the SSR length criteria 46,501 loci were classified as class I SSR and 11,711 of these were HASSR. All the SSR loci belonging to tetra-, penta-, hexa- and compound category were of class I SSR, while more than half (10,891) of the compound SSRs was of HASSR type. In contrast, mononucleotide repeats never reached a size of more than 50 bp, however this could be partly due to limitation of the 454 sequencing technology in dealing with large homopolymers. Due to their higher polymorphism longer SSR loci are more useful for routine genetic diversity analysis, fingerprinting, QTL mapping and molecular breeding applications in the laboratories lacking sophisticated fragment analysis and SNP genotyping platforms, but having simple agarose gel electrophoresis facility (Singh et al. 2010).

For wet lab validation we attempted to design PCR primers for 1,220 HASSR loci, taking 300 loci from the compound SSRs and all the loci from the remaining categories. But flanking primers could be designed successfully for amplification of only 530 of these loci, mainly due to location of the SSRs near one end of the sequence contigs. Surprisingly, no primer could be designed for the di-nucleotide category. Each of the designed primers was then compared with the whole genome sequence data to ensure that it bound to a unique position in the genome to prevent non-specific annealing. After this 93 loci were discarded due to multiple matches and primers were synthesized for 437 HASSR loci containing tri-, tetra-, penta- and hexa-nucleotide repeats as well as compound SSRs (Table 7). Details of validation results for the 437 HASSR markers, including primer sequences, Tm values, GC content and polymorphism level are shown in Supplementary Table S12. Total 358 primer pairs amplified a single PCR product of expected size and these were screened for polymorphism in a set of eight pigeonpea genotypes. We observed higher validation success rate of 81.92% for these genomic-SSR markers as compared to 72% success with genic-SSR markers described earlier (Dutta et al. 2011). HASSR markers showed 40.8% polymorphism (Table 7, Figure 8), which is three times higher than 12.9% polymorphism observed with type I genic-SSR markers on the same set of eight genotypes (Dutta et al. 2011). Among the different categories of HASSR markers, complex SSRs showed the least polymorphism of only 6.5% (Table 7). This was discouraging because most of the HASSR loci belonged to this category (Table 6). The HASSR polymorphism was much higher than the earlier reported 28.40% polymorphism for BAC-end sequence derived genomic SSR markers obtained using high resolution capillary electrophoresis (Bohra et al. 2011). This underlines the high potential utility of the HASSR markers in pigeonpea molecular breeding.

Table 7. Wet lab validation of the PCR amplification and polymorphism of 437 HASSR markers designed from pigeonpea genome sequence information

SSR category	No. of loci	Poly-morphic	Mono-morphic	Unexpected size bands	Not amplified	% Poly morphism
Trinucleotide	281	124	103	28	26	44.1
Tetranucleotide	10	5	3	2	0	50.0
Pentanucleotide	7	1	3	2	1	14.2
Hexanucleotide	16	8	6	2	0	50.0
Complex	123	8	97	6	12	6.5
Total	437	146	212	40	39	40.8

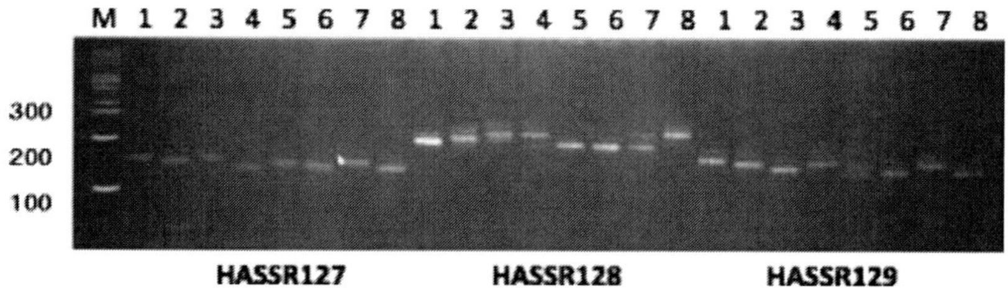

Figure 8. Agarose gel showing allelic variation in PCR product size with three different HASSR markers (HASSR27, HASSR28, HASSR27) in a set of eight pigeonpea varieties. 1 Asha, 2 UPAS 120, 3 HDM 04–1, 4 Pusa Dwarf, 5 H2004-1, 6 Bahar, 7 Maruti, 8 TTB7; M = 100 bp DNA size marker

The work presented here is the first draft of the whole genome sequence of pigeonpea and is the first report of a plant genome sequenced entirely in India. The 47,004 protein coding genes predicted in the pigeonpea genome are similar to that in soybean, potato and tomato, but significantly higher than Arabidopsis and rice. Ninety-nine point nine percent of the predicted genes were supported by RNA expression data, suggesting that these are true genes. A small proportion of genome scaffolds were genetically anchored with 347 mapped SNP markers which provide nucleation points for further finishing of the genome to large pseudomolecules of the eleven chromosomes. A comprehensive set of 46,501 Class I SSRs and 11,711 hypervariable HASSR loci were identified, and 437 HASSR markers were experimentally validated for amplification and higher rate of polymorphism. HASSR markers have high potential utility in the genetic diversity analysis, fingerprinting and molecular breeding for efficient utilization of pigeonpea germplasm resources in breeding improved varieties. The network partners under Indo-US AKI have already developed a EMS-mutagenized population and more than 24 recombinant inbred line populations for mapping of important agronomic traits including Fusarium wilt, sterility mosaic disease, flooding tolerance, seed size and number, plant type, drought tolerance and Dal (milling) quality of pigeonpea.

ACKNOWLEDGMENTS

We are grateful to the Indian Council of Agricultural Research (ICAR) New Delhi for financial support through Indo-US Agricultural Knowledge Initiative (AKI) and Network Project on Transgenics in Crops (NPTC) projects. SD is grateful to the Council of Scientific and Industrial Research,

Government of India for financial support (Grant no. 09/083/(0342)/2011/EMR-I)

REFERENCES

1. Altschul SF, Gish W, Miller W, Myers EW, Lipman DJ (1990) Basic local alignment search tool. J Mol Biol 215:403–410
2. Bao Z, Eddy SR (2002) Automated de novo identification of repeats sequence families in sequenced genomes. Genome Res 12:1269–1276
3. Benson G (1999) Tandem repeats finder: a program to analyze DNA sequences. Nucleic Acids Res 27:573–580
4. Bohra A, Dubey A, Saxena RK, Varma Penmetsa R, Poornima KN, Kumar N, Farmer AD, Srivani G, Upadhyaya HD, Gothalwal R, Ramesh S, Singh D, Saxena KB, Kavi Kishor PB, Singh NK, Town CD, May GD, Cook DR, Varshney RK (2011) Analysis of BAC-end sequences (BESs) and development of BES-SSR markers for genetic mapping and hybrid purity assessment in pigeonpea (Cajanus spp.). BMC Plant Biol 11:56
5. Dutta S, Kumawat G, Singh BP, Gupta DK, Singh S, Dogra V, Gaikwad K, Sharma TR, Raje RS, Bandhopadhya TK, Datta S, Singh MN, Fakrudin B, Kulwal P, Wanjari KB, Varshney RK, Cook DR, Singh NK (2011) Development of genic-SSR markers by deep transcriptome sequencing in pigeonpea [Cajanus cajan (L.) Millspaugh]. BMC Plant Biol 11:17
6. Ewing B, Green P (1998) Base-calling of automated sequencer traces using Phred II. Error probabilities. Genome Res 8:186–194
7. FAOSTAT 2008 [http://faostat.fao.org]
8. Fuller DQ (2006) Agricultural origins and frontiers in South Asia: a working synthesis. J World Prehist 20:1–86
9. Gardner PP, Daub J, Tate J, Moore BL, Osuch IH, Griffiths-Jones S, Finn RD, Nawrocki EP, Kolbe DL, Eddy SR, Bateman A (2010) Rfam: wikipedia, clans and the "decimal" release. Nucleic Acids Res 10:1093–1129Google Scholar
10. Greilhuber J, Obermayer R (1998) Genome size variation in Cajanus cajan Fabaceae: a reconsideration. Plant Syst Evol 212:135–141
11. Griffiths-Jones S (2004) The microRNA Registry. Nucleic Acids Res 32:D109–D111
12. Griffiths-Jones S, Moxon S, Marshall M, Khanna A, Eddy SR, Bateman A (2005) Rfam: annotating non-coding RNAs in complete genomes. Nucleic Acids Res 33:121–124
13. Griffiths-Jones S, Grocock RJ, van Dongen S, Bateman A, Enright AJ (2006) miRBase: microRNA sequences, targets and gene nomenclature. Nucleic Acids Res 34:D140–D144
14. Griffiths-Jones S, Saini HK, van Dongen S, Enright AJ (2008) miRBase: tools for microRNA genomics. Nucleic Acids Res 36:D154–D158
15. Huang S, Li R, Zhang Z, Li L, Gu X, Fan W, Lucas WJ, Wang X, Xie B, Ni P, Ren Y, Zhu H, Li J, Lin K (2009) The genome of the cucumber, Cucumis sativus L. Nat Genet 41:1275–1281
16. Hulbert SH, Webb CA, Smith SM, Sun Q (2001) Resistance gene complexes: evolution and utilization. Annu Rev Phytopathol 39:285–312
17. International Rice Genome Sequencing Project (2005) The map-based sequence of the rice genome. Nature 436:793–800
18. Kozomara A, Griffiths-Jones S (2011) miRBase: integrating microRNA annotation and deep-sequencing data. Nucleic Acids Res 39:D152–D157

19. Krzywinski M, Schein J, Birol I, Connors J, Gascoyne R, Horsman D, Jones SJ, Marra MA (2009) Circos: an information aesthetic for comparative genomics. Genome Res 19:1639–1645
20. Lander ES, Linton LM, Birren B, Nusbaum C, Zody MC, Baldwin J, Devon K, Dewar K, Doyle M, FitzHugh W, Funke R, Gage D, Harris K, Heaford A, Howland J et al (2001) Initial sequencing and analysis of the human genome. Nature 409:860–921
21. Lerat E (2010) Identifying repeats and transposable elements in sequenced genomes: how to find your way through the dense forest of programs. Heredity 104:520–533
22. Lowe TM, Eddy SR (1997) tRNAscan-SE: a program for improved detection of transfer RNA genes in genomic sequence. Nucleic Acids Res 25:955–964
23. Margulies M, Egholm M, Altman WE, Attiya S, Bader JS, Bemben LA, Berka J, Braverman MS, Chen YJ, Chen Z, Dewell SB, Du L, Fierro JM, Gomes XV, Godwin BC et al (2005) Genome sequencing in microfabricated high-density picolitre reactors. Nature 437:376–380
24. Ming R, Hou S, Feng Y, Yu Q, Dionne-Laporte A, Saw JH, Senin P, Wang W, Ly BV, Lewis KLT, Salzberg SL, Feng L, Jones MR, Skelton RL, Murray JE, Chen C, Qian W et al (2008) The draft genome of the transgenic tropical fruit tree papaya (Carica papaya L.). Nature 452:991–996
25. Murray MG, Thompson WF (1980) Rapid isolation of high molecular weight plant DNA. Nucleic Acid Res 8:4321–4325
26. Odeny DA, Jayashree B, Gebhardt C, Crouch J (2009) New microsatellite markers for pigeonpea (Cajanus cajan (L.) millsp.). BMC Res Notes 2:35
27. Paterson AH, Bowers JE, Remy B, Inna D, Jane G, Gundlach H, Georg H, Uffe H, Therese M, Alexander P, Jeremy S, Manuel S, Haibao T, Xiyin W et al (2009) The Sorghum bicolor genome and the diversification of grasses. Nature 457:551–556
28. Ratnaparkhe MB, Gupta VS, Ven Murthy MR, Ranjekar PK (1995) Genetic finger printing of pigeonpea (Cajanus cajan (L.) Millsp.) and its wild relatives using RAPD markers. Theor Appl Genet 91:893–898
29. Saxena KB (2008) Genetic improvement of pigeonpea-a review. Trop Plant Biol 1:159–178
30. Schattner P, Brooks AN, Lowe TM (2005) The tRNAscan-SE, snoscan and snoGPS web servers for the detection of tRNAs and snoRNAs. Nucleic Acids Res 33:W686–689Google Scholar
31. Schmutz J, Cannon SB, Schlueter J, Ma J, Mitros T, William N, Hyten DL, Qijian S, Thelen JJ, Jianlin C, Dong X et al (2010) Genome sequence of the palaeopolyploid Soybean. Nature 463:178–183
32. Schnable PS, Doreen W, Fulton RS, Stein JC, Fusheng W, Shiran P, Chengzhi L, Jianwei Z, Lucinda F, Graves TA, Patrick M, Reily AD, Laura C et al (2009) The B73 maize genome: complexity, diversity, and dynamics. Science 326:1112–1115
33. Singh NK, Raghuvanshi S, Srivastava SK, Gaur A, Pal AK, Dalal V, Singh A, Ghazi IA et al (2004) Sequence analysis of the long arm of rice chromosome 11 for rice-wheat synteny. Funct Integr Genomics 4:102–117
34. Singh NK, Vivek D, Kamlesh B, Singh Binay K, Chitra G, Archana S, Ghazi Irfan A, Mahavir Y, Awadhesh P, Rekha D, Singh PK, Harvinder S, Koundal Kirpa R, Kishor G, Trilochan M, Sharma Tilak R (2007) Single-copy genes define a conserved order between rice and wheat for understanding differences caused by duplication, deletion, and transposition of genes. Funct Integr Genomics 7:17–35
35. Singh H, Deshmukh RK, Singh A, Singh AK, Gaikwad K, Sharma TR, Mohapatra T, Singh NK (2010) Highly variable SSR markers suitable for rice genotyping using agarose gels. Mol Breed 25:359–364
36. Temnykh S, DeClerck G, Lukashova A, Lipovich L, Cartinhour S, McCouch S (2001) Computational and experimental analysis of microsatellites in rice (Oryza sativa L.): frequency, length variation, transposon associations, and genetic marker potential. Genome Res 11:1441–1452

37. The International Brachypodium Initiative (2010) Genome sequencing and analysis of the model grass Brachypodium distachyon. Nature 463:763–768
38. The Potato Genome Sequencing Consortium (2011) Genome sequence and analysis of the tuber crop potato. Nature 475:189–195
39. The Rice Chromosomes 11 and 12 sequencing consortia (2005) The sequence of rice chromosome 11 and 12, rich in disease resistance genes and recent gene duplication. BMC Biology 3: 1–18Google Scholar
40. Varshney RK, Close TJ, Singh NK, Hoisington DA, Cook DR (2009) Orphan legume crops enter the genomics era. Curr Opin Plant Biol 12:202–210
41. Varshney RK, Penmetsa RV, Dutta S, Kulwal PL, Saxena RK, Datta S, Sharma TR, Rosen B, Carrasquilla-Garcia N, Farmer AD, Dubey A, Saxena KB, Gao J, Fakrudin B, Singh MN, Singh BP, Wanjari KB, Yuan M, Srivastava RK, Kilian A, Upadhyaya HD, Mallikarjuna N, Town CD, Bruening GE, He G, May GD, McCombie R, Jackson SA, Singh NK, Cook DR (2010a) Pigeonpea genomics initiative (PGI): an international effort to improve crop productivity of pigeonpea (Cajanus cajan L.). Mol Breed 26:393–408
42. Varshney RK, Thudi M, May GD, Jackson SA (2010b) Legume genomics and breeding. Plant Breed Rev 33:257–304
43. Velasco R, Andrey Z, Michela T, Cartwright DA, Alessandro C, Dmitry P, Massimo P, FitzGerald LM, Silvia V, Julia R, Giulia M, Diana I, Giuseppina C, Wardell B et al (2007) A high quality draft consensus sequence of the genome of a heterozygous grapevine variety. PLoS One 12:31326Google Scholar
44. Velasco R, Zharkikh A, Jason A, Amit D, Alessandro C, Ananth K, Paolo F, Bhatnagar SK, Troggio M, Pruss D, Salvi S, Pindo M, Baldi P, Castellett S et al (2010) The genome of the domesticated apple (Malus × domestica Borkh.). Nat Genet 42:833–839
45. Waterston RH et al (2002) Initial sequencing and comparative analysis of the mouse genome. Nature 420(5):20–62Google Scholar
46. Wojciechowski MF (2003) Reconstructing the phylogeny of legumes (Leguminosae): an early 21st century perspective. In: Klitgaard BB, Bruneau A (eds) Advances in legume Systematics, Part 10, Higer level Systematics. Royal Botanic Gardens, Kew, pp 5–35Google Scholar
47. Wootton JC, Federhen S (1993) Statistics of local complexity in amino acid equences and sequence databases. Comput Chem 17:149–163
48. Yang S, Pang W, Harper J, Carling J, Wenzl P, Huttner E, Zong X, Kilian A (2006) Low level of genetic diversity in cultivated pigeonpea compared to its wild relatives is revealed by diversity arrays technology (DArT). Theor Appl Genet 113:585–595

CHAPTER 5

Overexpression of *BdMATE* Gene Improves Aluminum Tolerance in *Setaria viridis*

Ana P. Ribeiro[1,2], Wagner R. de Souza[1], Polyana K. Martins[1], Felipe Vinecky[1], Karoline E. Duarte[1], Marcos F. Basso[1], Bárbara A. D. B. da Cunha[1], Raquel B. Campanha[3], Patrícia A. de Oliveira[3], Danilo C. Centeno[4], Geraldo M. A. Cançado[5], Jurandir V. de Magalhães[6], Carlos A. F. de Sousa[1], Alan C. Andrade[2,7], Adilson K. Kobayashi[1] and Hugo B. C. Molinari[1*]

[1]*Genetics and Biotechnology Laboratory, Embrapa Agroenergy, Brasilia, Brazil*

[2]*Plant Biotechnology Program, Federal University of Lavras, Lavras, Brazil*

[3]*Biomass and Biofuels Chemistry Laboratory, Embrapa Agroenergy, Brasilia, Brazil*

[4]*Centre of Natural Sciences and Humanities, Federal University of ABC, São Bernardo do Campo, Brazil*

[5]*Center of Genetic Engineering and Molecular Biology, Embrapa GenClima, University of Campinas, Campinas, Brazil*

[6]*Applied Biology Center, Embrapa Maize and Sorghum, Sete Lagoas, Brazil*

[7]*INOVACAFÉ, Embrapa Coffee, Lavras, Brazil*

ABSTRACT

Acidic soils are distributed worldwide, predominantly in tropical and subtropical areas, reaching around 50% of the arable soil. This type of soil strongly reduces crop production, mainly because of the presence of aluminum, which has its solubility increased at low pH levels. A well-known physiological mechanism used by plants to cope with Al stress involves activation of membrane transporters responsible for organic acid anions secretion from the root apex to the rhizosphere, which chelate Al, preventing its absorption by roots. In sorghum, a membrane transporter gene belonging to multidrug and toxic compound extrusion (MATE) family was identified and characterized as an aluminum-activated citrate transporter gene responsible for Al tolerance in this crop. *Setaria viridis* is an emerging model for C4 species and it is an important model to validate some genes for further C4 crops transformation, such as sugarcane, maize, and wheat. In the present work, *Setaria viridis* was used as a model plant to overexpress a newly identified MATE gene from *Brachypodium distachyon* (*BdMATE*), closely related to *SbMATE*, for aluminum tolerance assays. Transgenic *S. viridis* plants overexpressing a *BdMATE* presented an improved Al tolerance phenotype, characterized by sustained root growth and exclusion of aluminum from the root apex in transgenic plants, as confirmed by hematoxylin assay. In addition, transgenic plants showed higher root citrate exudation into the rhizosphere, suggesting that Al tolerance improvement in these plants could be related to the chelation of the metal by the organic acid anion. These results suggest that *BdMATE* gene can be used to transform C4 crops of economic importance with improved aluminum tolerance.

Keywords: aluminum, *Setaria viridis*, *MATE*, abiotic stress, hydroponic system

INTRODUCTION

Acidic soils are widely distributed worldwide, predominantly in tropical and subtropical areas, reaching around 50% of the arable soil in the world (Von Uexküll and Mutert, 1995; Cançado, 2006; Kochian et al., 2015). This type of soil strongly reduces crop production, mainly because of the presence of aluminum therein. Aluminum (Al) is the most abundant metal on Earth, and conjugated with oxides, silicates, and hydroxides constitutes soil particles (Boff, 2006; Oliveira, 2012). In highly acidic soils (pH < 5), Al is solubilized into a phytotoxic form $Al(H_2O)_6^{3+}$, usually referred to as Al^{3+}. Aluminum is extremely rhizotoxic, affecting root growth and function and therefore limiting crop production and its toxicity, along with drought, is a primary food security concern (Von Uexküll and Mutert, 1995; Magalhães et al., 2007).

A well-known physiological mechanism used by plants to cope with Al stress involves the activation of plasma membrane transporters responsible for organic acid anions secretion from the root apex to the rhizosphere (Kochian et al., 2015). These anions form non-phytotoxic stable complexes with Al^{3+}, preventing its absorption by the roots (Boff, 2006; Kochian et al., 2015). In sorghum (*Sorghum bicolor*), a membrane transporter gene belonging to the *MATE* family was identified and characterized as an aluminum-activated citrate transporter gene responsible for the Al tolerance in this crop, codifying for a proton-dependent transporter protein (Magalhães et al., 2007). *Arabidopsis thaliana* plants overexpressing the *S. bicolor MATE* gene (*SbMATE*) demonstrated an Al tolerant phenotype and increased root citrate exudation when compared to NT plants (Magalhães et al., 2007). These results support that a member of a *MATE* family from *S. bicolor* (*Sb*MATE) is an Al-activated citrate secretion transporter conferring aluminum tolerance. In addition, it was demonstrated that MATE homologs are root citrate transporters in other monocot species such as maize (*Zm*MATE1, Maron et al., 2010, 2013), and rice (*Os*FRD1, Yokosho et al., 2011). Moreover, the expression of the citrate transporters *SbMATE* and *FDR3* in barley enhanced the Al tolerance in this sensitive cereal (Zhou et al., 2014). Other study demonstrated that a *MATE* gene cloned and isolated from *Vigna umbellata* (*VuMATE*), when expressed in tomato, was able to increase Al resistance, accompanied by enhanced citrate efflux (Yang et al., 2011). Currently, some physiological and molecular mechanisms of organic acid anions exudation are described. Several studies have implicated that plasma membrane H^+-ATPases are involved in Al tolerance mechanism by citrate exudation. For instance, in soybean and broad bean, the citrate exudation from the roots was associated with the up-regulation of the plasma membrane H^+-ATPase, indicating that this protein modulates Al-induced citrate exudation (Shen et al., 2005; Chen et al., 2013, 2015). Wang et al. (2016) demonstrated that auxin is possibly involved in Al stress tolerance in soybean, since the concentration of IAA was increased by Al and exogenous application of IAA decreased Al concentration in roots by increasing citrate exudation through upregulation of *GmMATE* and activation of the plasma membrane H^+-ATPase. In addition to the well described *SbMATE*, a *Brassica oleracea* MATE gene has also been shown to be involved in Al tolerance responses in *A. thaliana*. Transgenic Arabidopsis lines overexpressing *BoMATE* demonstrated Al tolerance and increased citrate exudation. Furthermore, these lines showed a lower K^+ efflux and higher H^+ efflux in the presence of Al, indicating that MATE is involved in the K^+ and H^+ fluxes during Al stress (Wu et al., 2014). Therefore, MATE, along with the anion channels ALMTs (Sasaki et al., 2004), are potential root organic acid anion transporters involved in Al resistance in plants and are potential targets for crop improvement under Al stress.

Setaria viridis is an emerging monocot plant model for molecular and genetic studies (Muthamilarasan and Prasad, 2015). It is a short, fast-growing, C4 plant with its genome sequence fully available (Bennetzen et al., 2012), making it a reliable model for genetic studies. In addition, *S. viridis* is amenable for genetic transformation through *Agrobacterium tumefaciens*, with well-established transformation protocol (Martins et al., 2015). In the present work, *S. viridis* was used as a model plant system to overexpress a new ortholog of *Sorghum bicolor MATE* (*SbMATE*), *Brachypodium distachyon MATE* (*BdMATE*), under the control of the maize ubiquitin constitutive promoter (*Zm*Ubi-1) for aluminum tolerance. The *BdMATE* gene was chosen because it is up-regulated under Al conditions (Contreras et al., 2014), whereas *S. viridis MATE* is not (this study, data not shown). The rationale was to use a close homolog of *SbMATE* and a patent-free gene, in order to use as proof of concept to further transform C4 crops of economic importance such as sugarcane, wheat, and maize, which are difficult and time-consuming to transform. *S. viridis* plants overexpressing *BdMATE* showed an Al tolerance phenotype, characterized by sustained root growth under {20} μM Al^{3+}, whereas NT plants showed RGI. Moreover, exclusion of aluminum from the root apex in transgenic plants was inferred based on results from hematoxylin assays. In addition, transgenic plants constitutively overexpressing *BdMATE* presented higher citrate secretion levels when compared to NT plants under Al stress, suggesting that the mechanism of organic acid anion exudation and consequent Al chelation by these compounds are involved in the Al toxicity amelioration in these plants. The expression of aluminum tolerance-related genes was evaluated, and the results obtained are discussed based on the known Al tolerance. Our results demonstrated that, in the future, the overexpression of *BdMATE* could be used to develop Al-tolerant crops closely related to *S. viridis*, such as maize, wheat, and sugarcane.

MATERIALS AND METHODS

In Silico Analysis

Searches for the *Sorghum bicolor MATE* (*Sb*MATE) gene in the Phytozome database were performed in order to find orthologs in the genome of *B. distachyon*. Based on the different sequences found, the alignment of these sequences was generated using the Clustal Omega (Sievers and Higgins, 2014). Analysis of conserved protein domains and the prediction of the molecular and cellular functions of these domains were evaluated by the online softwares FFPred and MEMSAT (Nugent and Jones, 2009; Buchan et al., 2010). The phylogenetic tree was generated based on amino acid sequences of major grasses (Contreras et al., 2014) that have been studied with *MATE* gene, using the Geneious software (Kearse et al., 2012) for alignments and MEGA7 software (Kumar et al., 2016) for building phylogenetic tree.

Tissue Culture and Plant Transformation

In the present work, we were willing to validate an orthologous of the *Sorghum bicolor MATE* (*SbMATE*) gene that confers aluminum tolerance in different species (Magalhães et al., 2007; Sivaguru et al., 2013). The gene chosen was *BdMATE*, a close homolog of *SbMATE* from *B. distachyon*, a model organism for grasses (Kellogg, 2015), also studied by our group. The gene construct containing the *B. distachyon MATE* gene (*BdMATE*), driven by *Zm*Ubi1 constitutive promoter and the *hpt* gene selection marker (which confers hygromycin resistance), was synthetized by DNA Cloning Service, Germany (Supplementary Figure S1) and introduced into *A. tumefaciens* strain EHA105 for plant transformation.

Mature seeds of *S. viridis* accession A10.1 were selected for embryogenic calli induction in CIM [consisted of MS salts (Murashige and Skoog, 1962), 1 mg/L d-biotin, 0.5 mg/L pyridoxine HCl, 0.5 mg/L nicotinic acid, 100 mg/L myo-inositol, 0.1 mg/L thiamine-HCl, 0.6 mg/L $CuSO_4$, 30 g/L sucrose, 2 mg/L 2,4-dichlorophenoxyacetic acid, 0.5 mg/L kinetin, and 4 g/L Phytagel™, pH 5.8]. Induced calli were used for *Agrobacterium*-mediated transformation, according to Martins et al. (2015). The putative transgenic calli, resistant to 30 mg/L hygromycin B, were transferred to SRM consisting of MS salts (Murashige and Skoog, 1962), 1 mg/L d-biotin, 0.5 mg/L pyridoxine HCl, 0.5 mg/L nicotinic acid, 100 mg/L myo-inositol, 0.1 mg/L thiamine HCl, 20 g/L sucrose, 2 mg/L kinetin, 150 mg/L Timentin®, 30 mg/L hygromycin B, 2 g/L Phytagel™, pH 5.8. The regenerated plants were submitted to PCR analysis in order to confirm the presence of the transgene.

Molecular Analysis and Selection of the Transgenic Events

Genomic DNA from regenerated plantlets resistant to hygromycin was extracted using a modified CTAB method (Doyle and Doyle, 1990), according to Molinari et al. (2007). The gene insertion was confirmed by PCR using specific primers designed for *BdMATE* amplification (Supplementary Table S1).

The putative transgenic events underwent gene expression analysis by real-time PCR of the target gene *BdMATE*, using the primers described in Supplementary Table S1. The expression level was calculated using endogenous genes *SiUBC* e *SiSDH2* as reference (Martins et al., 2016).

Segregation analysis was performed in T_1 seeds to estimate the number of insertions in each event. Seeds were grown in selective medium with hygromycin B 50 mg/L. After 7 days, the number of resistant and sensitive plants was counted, and the proportion was analyzed statistically using the χ-Squared test. Events with single insertion determined by 3:1 Mendelian segregation ratio were selected. The T_2 seeds of the selected events were placed in selective medium with hygromycin B 50 mg/L, for the selection of transgenic homozygous lines, which were used for aluminum stress assays.

Aluminum Treatment Assay and Root Growth Measurement

A protocol was established in order to evaluate the phenotype of *S. viridis* submitted to aluminum treatment in hydroponic system. First, NT 7-days-old plantlets were subjected to evaluation of root growth in different nutrient solutions such as Hoagland (Hoagland and Arnon, 1950), Camargo (Camargo, 1981), and Magnavaca (Magnacava et al., 1987) (Supplementary Figure S2). The growth of the plants under a solution based solely on $CaCl_2$ (calcium chloride), at different concentrations, (Supplementary Figure S3) was also tested, according to Sasaki et al. (2004). The normal plant growth verified under 500 μM $CaCl_2$solution prompted us to use it for our studies. Al^{3+} activity was estimated using the software GeoChem-EZ (Shaff et al., 2010) (Table 1 and Supplementary Figure S4). After the establishment of the stress protocol, a solution containing {20} μM of Al^{3+} free activity, pH 4.2, was used to submit *S. viridis* plants to Al^{3+} for 1, 3, and 5 days, and this protocol was followed throughout the work. The evaluation of the root growth was performed by carefully scanning the roots daily for 5 days, in order not to damage the tissue, and the total root length was measured by the software WinRhizo 2007a (Régent Instruments). Transgenic and NT plants were evaluated for liquid root growth, and the percentage of RGI in treatments in the presence or absence of aluminum was calculated as described in Ryan et al. (2009). In order to measure RNG, the length of the roots was measured before and after each day of growth in the calcium chloride solution with (+Al) and without (-Al) aluminum. The % RGI values were calculated from root growth measured over each day in +Al solution divided by root growth measured over each day in control (-Al) solution × 100, calculated individually for each plant.

Table 1. Concentration and free activity of aluminum in 500 μM $CaCl_2.2H_2O$, pH 4.2, calculated using the software GeoChem-EZ (Shaff et al., 2010).

Concentrations of $AlCl_3$	Free activity of Al^{3+}
0 μM	{0} μM
18 μM	{10} μM
35 μM	{20} μM

Values inside the brackets indicate Al^{3+} activity estimated with the speciation software, GeoChem-EZ (Shaff et al., 2010).

Real-Time qPCR Analysis

After 7-days-old plantlets had been submitted to 500 μM $CaCl_2$ solution, in the absence or the presence of {20} μM Al^{3+}, pH 4.2, in hydroponic system for 1, 3, and 5 days, total RNA was extracted from shoots using TRIzol® reagent (Thermo Fisher Scientific), according to the manufacturer's instructions. RNA from roots was extracted using a LiCl method (Chang et al., 1993). The samples were treated with RQ1 RNase-free DNase (Promega, Madison, WI, USA), according to the manufacturer's instructions. Total RNA was quantified using a NanoDrop ND-1000 Spectrophotometer (Uniscience), and RNA integrity was verified in agarose gel electrophoresis. The synthesis of first strand cDNA was accomplished using the extracted RNA as template and the RevertAid™ First Strand cDNA Synthesis kit (Thermo Fisher Scientific). These steps were all performed according to the manufacturer's instructions. The qPCR was carried out using Platinum® SYBR® Green PCR SuperMix-UDG with ROX (Invitrogen, Carlsberg, CA, USA) with synthesized single-stranded cDNA as template, using the protocol recommended by the StepOnePlusReal-Time PCR Systems (Applied Biosystems). The primers used in the qPCR are described in Supplementary Table S1.

Relative gene expression levels were calculated using the q-Gene (Muller et al., 2002). *SiEF1-α*; *SiSUI*, *SiCAC*, and *SiCUL* were used as reference genes (Martins et al., 2016) and the geometrical mean of the relative quantities (RQs) was calculated using BestKeeper software (Pfaffl et al., 2004). Individual amplification efficiencies were established with LinRegPCR v.2013.0 using a window-of-linearity (Ramakers et al., 2003). The experiment was performed using three biological replicates, with 40 plants each.

Hematoxylin Staining

The exudation of organic acid anions by the root apex will act chelating aluminum in the rhizosphere, forming a neutral complex, thus preventing the root absorption of part of the Al. Staining is described as one of the best and fastest methods to monitor the location and aluminum distribution in plants (Illéš et al., 2006). The hematoxylin method was used to evaluate Al accumulation in *S. viridis* plants under aluminum treatment, similar to treatment described above carried out for root growth measurement. The protocol was based on Tang et al. (2000). The roots were excised from the plantlets and gently shaken in 2 mL of distilled water for 60 min. The water was replaced by 2 mL of aqueous hematoxylin solution (0.2% hematoxylin and 0.02% potassium iodide, w/v) and agitated for 15 min. Finally, the solution was replaced one more time by 2 mL distilled water, thereby repeating the first step. After staining, the roots were photographed under the stereomicroscopic Leica Model S8APO.

Determination of Root Citrate Exudation

The transgenic and the NT plants were grown hydroponically as described previously, and root exudates were collected after 3 days of growth on nutrient solution containing {0} or {20} μM Al^{3+}. After the exposure period, the roots were washed with 20 mL of 500 μM $CaCl_2$ solution for 40 min, under 60 rpm in an orbital shaker. This procedure was repeated three times for each replicate. Each replicate (3 in total) was composed of 40 plantlets bulked together for extraction (120 plantlets in total). The extraction was normalized to the root biomass for each line. The solutions were lyophilized for organic acids analysis on a gas chromatography–mass spectrometry (GC-MS) system (Varian GC 3800 and MS 2000). The lyophilized material were washed in pure methanol and dried under vacuum. To the dried samples were added 100 μL of pyridine and 50 μL of BSTFA (N,O-bis(trimethylsilyl)trifluoroacetamide) + 1% TMCS (trimethylchlorosilane). The analysis of GC/MS was performed according to Centeno et al. (2016) on a 30 m HP5 column with 0.25 mm of diameter and 0.25 μM film thickness (Supelco). Helium was used as the carrier gas at a flow rate of 1 mL min^{-1}. The analysis was performed under the following temperature program: 5 min of isothermal heating at 70°C, followed by a 5°C min^{-1} oven temperature ramp to 310°C, and a final 1 min of heating at 310°C. Mass spectra were recorded at 2 scan s^{-1} with a scanning range of 50–600 m/z. Both chromatograms and mass spectra were evaluated using the OpenChrom software (Wenig and Odermatt, 2010). The peaks were identified and quantified in comparison with authentic standards and the NIST 08 Mass Spectral Library.

Statistical Analysis

Experimental data were analyzed using randomized block design (RBD) with replications for each treatment ({0} and {20} μM Al^{3+}). Samples for all analysis were collected at days 1, 3, and 5. Differences among treatments per sample were analyzed using *t*-test, considering $p < 0.05$ as significant.

RESULTS

In Silico Analysis

In order to find orthologous sequences of *S. bicolor MATE* gene, it was conducted a BLAST analysis in the Phytozome database[1]. The hits based on the lowest *E*-values, highest score values and larger alignment bars with the query sequence were chosen for alignment studies using the *MATE* genes present in different organisms. *B. distachyon* (*BdMATE)* gene showed a close similarity to the *SbMATE* gene (Supplementary Figure S5), and therefore *BdMATE* was chosen for further studies. *S. viridis MATE* gene was not chosen because the expression levels of the endogenous *SvMATE* did not change in the presence of Al under our experimental conditions (see below). The selected amino acid sequence of *B. distachyon* MATE protein (*Bd*MATE – Bradi1g69770.1) were aligned using the Geneious software (Kearse et al., 2012) with the different MATE sequences from other related organisms to generate a phylogenetic tree, using MEGA7 software (Kumar et al., 2016) (Figure 1A). The sequence Bradi1g69770.1 was chosen for the present study due to its close similarity to the *S. bicolor* MATE sequence.

In addition, *Bd*MATE and *Sb*MATE transmembrane domains analysis was performed, using the softwares FFPRED and MEMSAT-SVM (Nugent and Jones, 2009; Buchan et al., 2010). The results are shown in the Figure 1B, where it was demonstrated that the *B. distachyon* MATE possesses the same 12 transmembrane domains presented by the *S. bicolor* MATE protein. *Bd*MATE belongs to the large multifunctional transport family involved in the transport of organic solutes out of the cytoplasm. The putative molecular functions attributed to this protein are transporter activity, ion transmembrane transporter activity, receptor activity, and lipid binding; all these characteristics have shown high specificity and precision for *Bd*MATE, according to MEMSAT (McGuffin et al., 2000; Nugent and Jones, 2009).

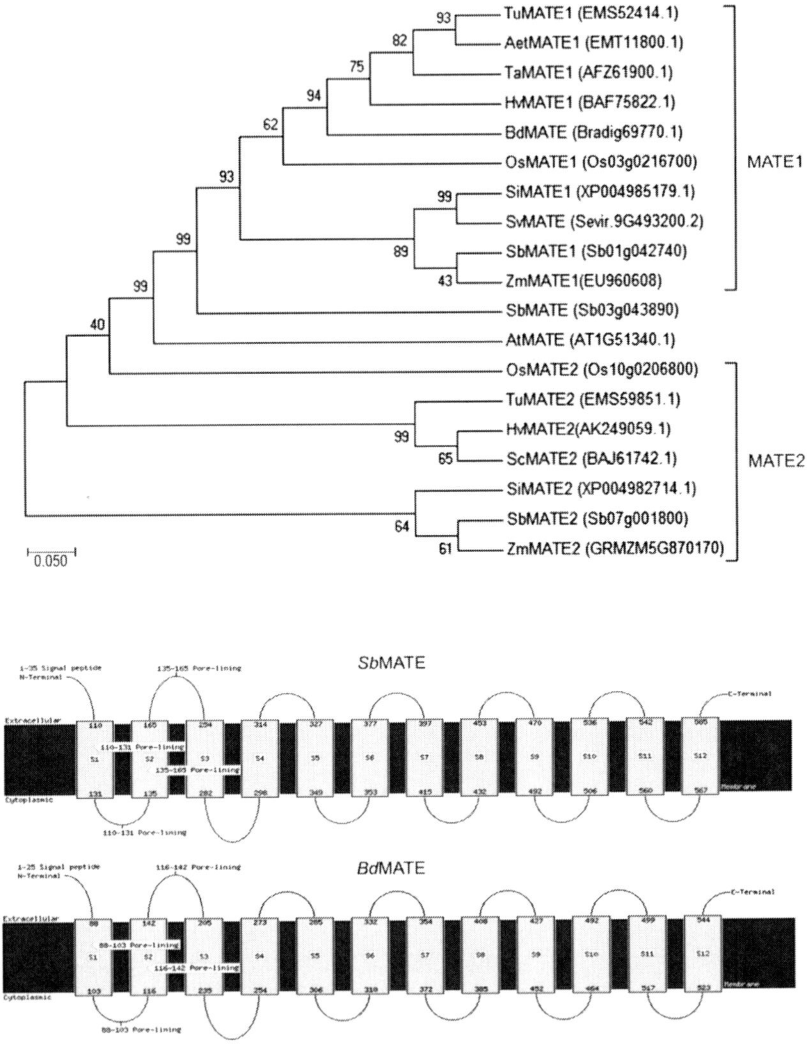

Figure 1. (A) Phylogenetic tree based on amino acid sequences. The analyses were conducted in MEGA7 software (Kumar et al., 2016) using the Maximum Likelihood method and bootstrap consensus tree inferred from 10000 replicates. The amino acids sequences of these proteins were obtained from NCBI and Phytozome. **(B)** Transmembrane domains of the *Bd*MATE and *Sb*MATE generated by FFPRED and MEMSAT-SVM (Nugent and Jones, 2009; Buchan et al., 2010).

Generation of *BdMATE* Transgenic Lines

After the similarity between *BdMATE* and *SbMATE* has been confirmed by *in silico* analysis, an expression vector containing *Zm*Ubi1::*BdMATE* was constructed in order to transform embryogenic calli of *S. viridis* accession A10.1. Transgenic calli were selected with 30 mg/L of hygromycin B and regenerated under SRM. Genomic DNA was extracted from regenerated plants and used as template for PCR using primers designed for the *BdMATE* gene amplification. The 450 bp amplification product corresponding to the specific region of the *BdMATE* gene was confirmed in the transgenic plants (Supplementary Figure S6A). The gene expression levels were analyzed by qPCR using *BdMATE* specific primers for 30 T_0 transgenic events. Representative events expressing different levels of *BdMATE* were selected for Mendelian segregation analysis. All the selected events showed a 3:1 segregation ratio, indicating a single gene insertion (Supplementary Figure S6B and Table S2).

Subsequently, the events 28 and 29, showing moderate and high expression levels of *BdMATE*, respectively, were chosen to generate homozygous lines. Two different homozygous lines at T_3 generation were obtained, 28.1 and 29.2. The expression of the *BdMATE* gene in these homozygous lines was evaluated in *S. viridis* roots using real-time quantitative PCR (Supplementary Figure S7A) and these lines were chosen for further aluminum tolerance tests.

Aluminum Tolerance Evaluation in Transgenic Plants

The most common criterion used to measure Al toxicity is the comparison of the root length of Al-affected plants with control plants grown in the absence of aluminum (Ma et al., 1997). In the present study, homozygous *S. viridis* transgenic plants overexpressing the *BdMATE* gene were evaluated by measuring the root growth along 5 days in a hydroponic system in the absence or presence of {20} μM Al^{3+} free activity, in comparison to NT plants. The results demonstrated that transgenic lines presented higher root length compared with NT plants in the presence of Al (Figure 2A). In the absence of the metal, both NT plants and the transgenic line 29.2 showed a similar root growth pattern, while the line 28.1 demonstrated a lower root growth even in control conditions (Figure ures 2A,B).

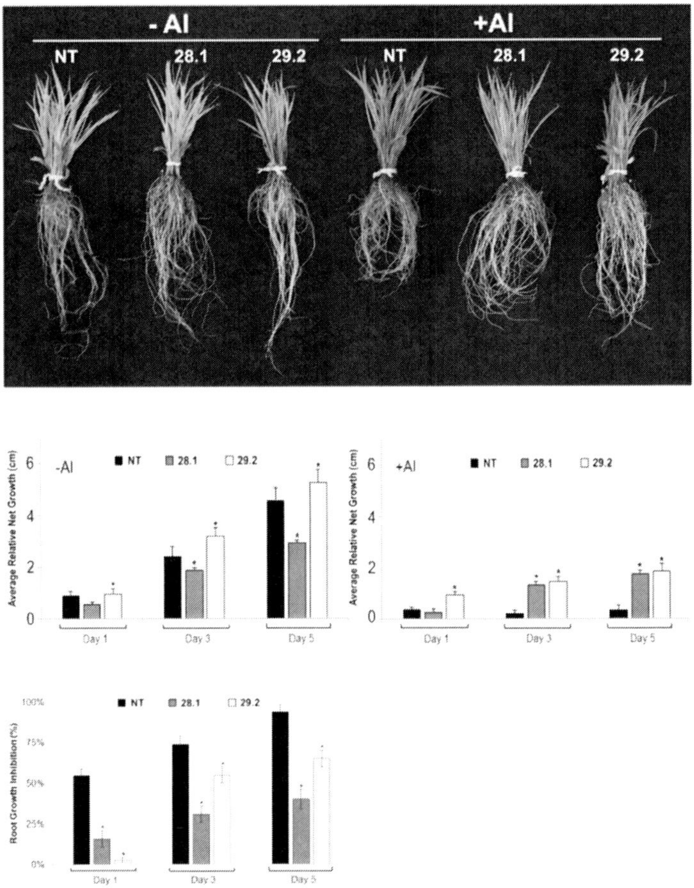

FIGURE 2. (A) *Setaria viridis* homozygous transgenic lines overexpressing *BdMATE* and NT plants in the absence (–Al; left panel) and after 5 days of exposure to {20} μM Al^{3+} (+Al; right panel). **(B,C)** Relative RNG of *S. viridis* homozygous transgenic lines overexpressing *BdMATE* and NT plants grown under the absence (–Al) or presence (+Al) of {20} μM Al^{3+} during 1, 3, and 5 days. The length of the roots was measured before and after each day of growth in the calcium chloride solution with (+Al) and without (–Al) aluminum (n = 20 plantlets). **(D)** RGI of *S. viridis* homozygous transgenic lines overexpressing *BdMATE* and NT plants grown under the absence (–Al) or presence (+Al) of {20} μM Al^{3+} during 1, 3, and 5 days. The % RGI values were calculated from root growth measured over each day in +Al solution divided by root growth measured over each day in control (–Al) solution × 100, calculated individually for each plant (n = 20 plantlets). *Significantly different at $P < 0.05$ between –Al and +Al treatments in transgenic lines.

Quantification of the relative net root growth (RNRG; Figure 3A) and the percentage of the RGI (Figure 3B) confirmed the Al tolerance phenotype demonstrated by *BdMATE* transgenic plants. It is worth to mention that the morphology of the roots was not affected from the constitutive expression of the *BdMATE* gene. A RGI above 90% was observed in NT plants after 5 days in {20} μM Al^{3+} (Figure 3B). Under the same conditions, transgenic plants overexpressing *BdMATE* demonstrated from 30 to 65% of RGI for the events 28.1 and 29.2. According to Magalhães et al. (2007), a plant can be considered tolerant to Al when its RGI during the stress is ≤70%. It is worth to mention that transgenic plants grown in the presence of Al demonstrated a small decrease of root elongation when compared to NT plants, indicating that *BdMATE* overexpression did not fully eliminate the toxic effects of Al, but ameliorates such effect.

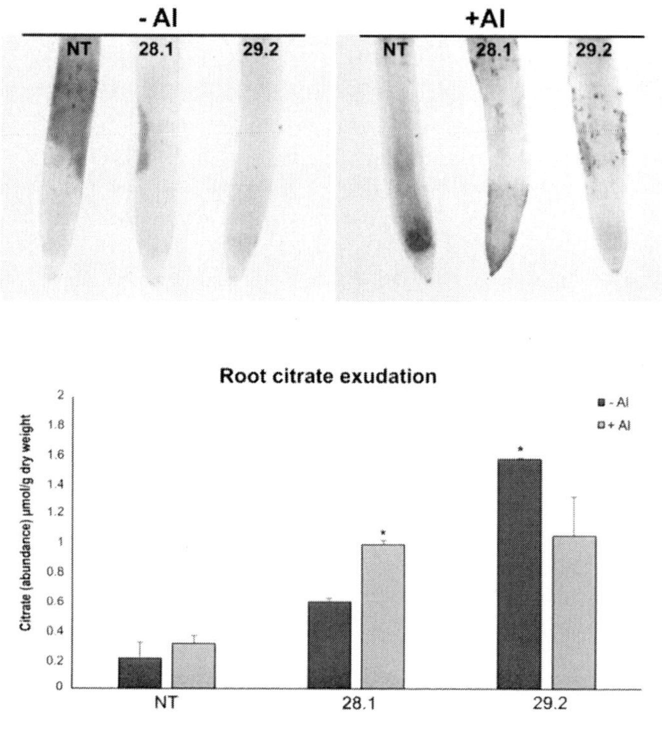

Figure 3. **(A)** Hematoxylin staining after 24 h exposure to {0} and {20} μM Al^{3+} in roots of the *S. viridis* homozygous transgenic lines overexpressing *BdMATE* and NT plant. **(B)** Citrate abundance on roots exudates in the absence and after 3 days of exposure to {20} μM Al^{3+}. Citric acid was determined by gas chromatography/mass spectrometry (GC/MS). The values are represented as the mean ± SD. *Significantly different at $P < 0.05$ between –Al and +Al treatments in transgenic lines.

Hematoxylin Staining

Hematoxylin assay was performed after 24 h of plants exposed to {0} or {20} µM Al^{3+}. This assay consists of the development of blue-purplish color in areas where aluminum is present in the root tissue. The Figure 3A shows accumulation of aluminum in a larger root portion of *S. viridis* NT plants in the presence of Al (+Al), while in the transgenic plants in the same conditions the blue-purplish color representing Al accumulation is absent. The staining of the plants is not observed in plants not submitted to Al (Figure 3A; -Al). These results suggest that NT plants are accumulating higher concentrations of Al than transgenic plants, indicating that tolerant plants are avoiding the absorption of Al by the roots.

Root Citrate Exudation

The measurements of root citrate secretion demonstrated that a new MATE homolog, *BdMATE*, when overexpressed in *S. viridis* plants, was able to increase the levels of this anion in transgenic plants when compared to NT plants (Figure 3B). In the presence of {20} µM Al^{3+}, the transgenic lines presented a threefold increase of citrate exudation in comparison to the NT plants, suggesting that *BdMATE* plants are conferring Al tolerance through the mechanism of citrate release into the rhizosphere.

Gene Expression Analysis

TCA Cycle Genes

The Figure 4 demonstrated that no difference of CS (*SiCS*) and MD (*SiMD*) gene expression levels were observed between transgenic and NT plants, in the absence or presence of Al. The signaling of Al tolerance in plants is still a matter of debate and therefore more studies are necessary in order to elucidate the relationship between the levels of citrate produced by the TCA cycle with Al tolerance in plants. The results obtained on gene expression levels will be discussed in the latter section.

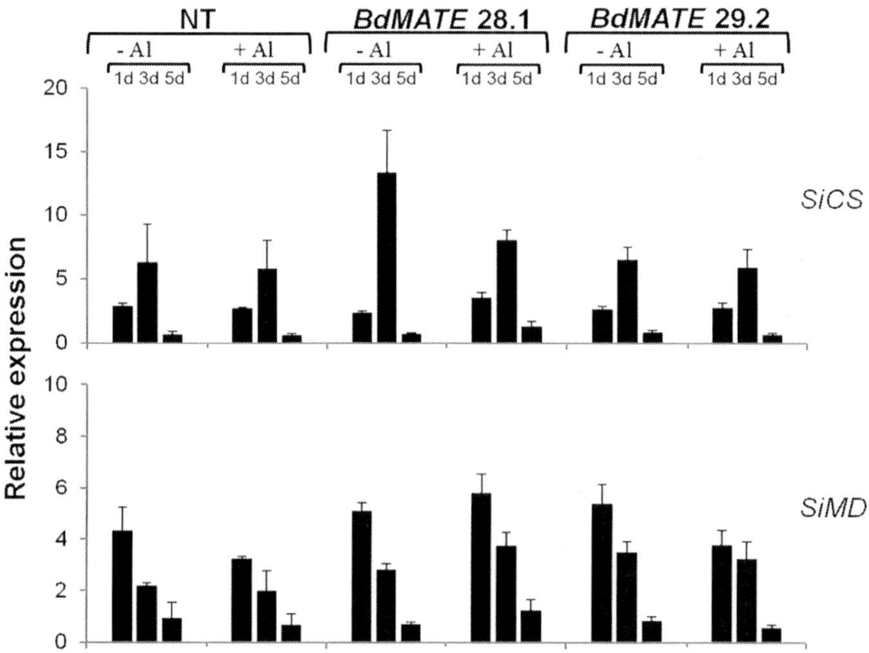

Figure 4. Relative gene expression levels of the *SiCS* (CS) and *SiMD* (MD) in the *S. viridis* homozygous transgenic lines overexpressing *BdMATE* and NT plants submitted to {0} and {20} μM Al^{3+} during 1, 3, and 5 days. No significant difference at $P < 0.05$ between –Al and +Al treatments in transgenic lines.

Expression of Endogenous *S. viridis MATE* Gene

The expression levels of endogenous *S. viridis MATE* gene (Sevir. 9g493200.2) were verified to demonstrate the involvement of *SvMATE* in Al stress responses. The Supplementary Figure S7B shows that *SvMATE* expression levels did not change in plants submitted to Al when compared to plants in the absence of the cation, indicating that the endogenous *SvMATE* does not participate in aluminum stress responses in *S. viridis*. Although the similarity of the *MATE* sequence of *S. viridis* and *B. distachyon* is 83.2% (Supplementary Figure S8).

DISCUSSION

Sugarcane is a very important economic crop for Brazil nowadays. Brazil is the largest sugarcane producer worldwide, and its production is intended mainly for ethanol and sugar production. In addition, the biomass released from the sugarcane process can be used as lignocellulosic material to be degraded by microorganisms to generate renewable fuels and added-value products. Despite of its great importance, the regions for sugarcane growth in Brazil are becoming scarce. An alternative for growing sugarcane could be the Cerrado region, characterized by its soil with acidic pH. Therefore, a great effort has been made by the scientific community in order to improve sugarcane traits for its growth in poor soils. *Setaria viridis* is an emerging model for C4 species and it is an important model to validate some genes for further sugarcane transformation (Muthamilarasan and Prasad, 2015). In the present work, a close homolog of *SbMATE* gene, the *B. distachyon MATE* gene (*BdMATE*) was constitutively overexpressed in *Setaria viridis* (Brutnell et al., 2010), in order to verify aluminum tolerance in transgenic plants. The *BdMATE* gene was chosen instead of the *S. viridis MATE* gene (*SvMATE*) because *BdMATE* is a close homolog of *SbMATE* expressed under Al conditions (Contreras et al., 2014), while *SvMATE* is not (this study, data not shown). The results obtained with the overexpression of *BdMATE* in *S. viridis* plants clearly showed an Al tolerance phenotype in these plants (Figure 2). The main symptom of aluminum toxicity is inhibition of root growth, which is observed consistently in plants sensitive to this metal (Kochian et al., 2015). In the presence of Al, the two different homozygous transgenic lines demonstrated higher root length (Figure 2A) and RNRG (Figure ures 2B,C) when compared to NT plants. On the other hand, the percentage of RGI of transgenic plants under Al stress was lower when compared to the control (Figure 2D). However, it is worth to mention that the root growth of transgenic lines was lower in the presence of Al than in the absence of this metal, indicating that *BdMATE* ameliorates but do not completely eliminates the phytotoxicity effect of this cation.

Furthermore, hematoxylin staining, a common method used to monitor the location and aluminum accumulation in plants (Illéš et al., 2006), demonstrated that *BdMATE* plants did not accumulate Al in the root apex, a phenotype demonstrated by NT plants (Figure 3B). The primary site of aluminum accumulation and damage is still a matter of debate. In many species, the root apex was identified as the primary site of perception of aluminum toxicity, and consequently the expression of tolerance (Kollmeier et al., 2000; Yang et al., 2008; Horst et al., 2010; Motoda et al., 2011; Souza et al., 2016). Sivaguru et al. (2013) demonstrated that Al affects approximately the 1–3 mm portion of the roots in sorghum lines sensitive to the metal. Although the effects of aluminum may occur in all parts of the root growing regions, ruptures above pea (Yamamoto et al., 2001), maize (Jones et al., 2006), and bean roots (Kopittke et al., 2008) occurred predominantly in regions within about 1–2 mm of the root tip, in the distal portion known by the DTZ (Sivaguru and Horst, 1998). In order to effectively confer Al resistance via Al-activated citrate exudation, it is logical to assume that the highest density of root citrate transporters should be specifically located in the root regions most affected by Al toxicity (Sivaguru et al., 2013). Our hematoxylin staining results also showed that the portion of Al accumulation in roots of *S. viridis* NT plants occurs in the DTZ (Figure 3A). Thus, it is worth to verify the gene expression profile of target genes related to Al tolerance in different portions of *S. viridis* roots in non-transformed or transgenic plants submitted to Al stress, an approach that is currently being performed by our group.

As described above, the exudation of organic acid anions by the root apex is a common mechanism to confer Al tolerance in plants, as these compounds are able to chelate this metal in the rhizosphere, avoiding root absorption. It was demonstrated that MATE homologs were root citrate transporters in different monocot species such as maize and rice (Maron et al., 2010, 2013; Yokosho et al., 2011). Here, it was demonstrated that a new MATE homolog, *BdMATE*, when overexpressed in *S. viridis* plants, was able to increase root citrate exudation in transgenic plants when compared to NT plants. We verified the exudation of citrate, the major organic acid anion secreted through the MATE transporter (Magalhães et al., 2007; Kochian et al., 2015) in *S. viridis* plants in the absence or presence of Al. Our results confirmed that transgenic plants overexpressing the *BdMATE* gene secreted more citrate than NT plants in both conditions (Figure 3B). Organic acids are metabolically active solutes with a diverse array of functions in many organisms. These acids have a potential role in the osmotic adjustment and in the balance of cation concentration of the cells, for instance. As described above, OAs anions are secreted by the root apex as an aluminum resistance mechanism in most plants. Citrate, malate, and oxalate are the main OAs anions secreted to the rhizosphere in many plant species (Kochian et al., 2015). Organic acids (OAs) are produced mainly in the mitochondria, by the TCA cycle and, in a lesser extent, in the glyoxysome (López-Bucio et al., 2000). The accumulation of OAs produced by TCA cycle is dependent upon the plant species, age of the plant and tissue type. In addition, the exudation of OAs anions by roots is dependent on the route of involvement of these OAs in specific physiological functions (Rees, 1990; López-Bucio et al., 2000). Recently, Nunes-Nesi et al. (2014) discussed the involvement of enzymes of the TCA cycle in response to the stress of aluminum, since OAs participate as key components in the mechanisms that some plants use to cope with nutritional deficiency and metal tolerance. Since citrate and malate are the main OAs anions secreted to the rhizosphere under Al stress, we aimed to investigate the expression levels of genes related to CS and MD, responsible for citrate and malate production, respectively, in the transgenic plants overexpressing *BdMATE*. It could be possible that the tolerance phenotype presented by transgenic plants under Al stress was due to an increased expression of genes responsible for the production of citrate or malate by these plants. However, we were not able to verify any difference in the gene expression levels between transgenic and NT plants, in the absence or presence of Al (Figure 4).

The relationship between increased CS activity and Al tolerance is still inconclusive. In some species such as *Triticum aestivum* L. (wheat) and *Secale cereale* L. (rye), the evaluation of CS production in aluminum stress conditions have been performed aluminum (Li et al., 2000). Plantlets of cv. King of the *S. cereale* demonstrated an increased CS activity when exposed to aluminum for more than 6 h, while in cv. Atlas 66 of the *T. aestivum* L. there was no difference in enzyme activity in plantlets subjected to aluminum (Ryan et al., 1993; Li et al., 2000). Our results demonstrated that *S. viridis* transgenic plants overexpressing *BdMATE* with Al tolerance phenotype do not increase the levels of CS gene expression. As mentioned above, the OAs that the plant uses in order to tolerate the aluminum through exudation by the root apex are metabolites originated in the TCA cycle, a key metabolic pathway to all organisms. Thus, it is plausible that the tolerance mechanism to aluminum in plants has a significant carbon cost, being highly regulated (Ryan et al., 1993; Sivaguru and Horst, 1998; Magalhães et al., 2007; Sivaguru et al., 2013). The CS gene is constantly being expressed to maintain the balance between citrate production and consumption by the cell. The product of the *CS* gene, citrate, is a molecule that is converted rapidly in the TCA cycle and/or can be degraded in other cellular compartments to ensure that the metabolic system may be constantly supplied (negative feedback). Thus, it could be very difficult to observe drastic changes in the expression levels of a gene that is involved in a fine-tune regulation of pivotal metabolic processes in the cell. Therefore, more studies are necessary in order to elucidate the relationship between the levels of citrate produced by the TCA cycle with Al tolerance in plants. We were not able to measure if the transformed plants had a carbon loss penalty compared to NT plants, especially because the experiments were performed in hydroponic conditions during a short time. We are currently establishing Al stress experiments in soil to further study the biomass of plants transformed with *BdMATE*. In addition, root-specific and/or Al-induced promoters are being currently studied by our group, to minimize the putative carbon loss caused by the constitutive expression of *BdMATE*.

Altogether, our results suggest that a C4 plant overexpressing a *B. distachyon MATE* transporter gene (*BdMATE*) was more tolerant to aluminum stress than wild plants through the exudation of citrate into the rhizosphere, with subsequent avoidance of Al absorption and accumulation in the DTZ of the roots. As consequence, the transgenic plants demonstrated sustainable root growth in the presence of Al when compared to NT plants. Our results suggested that, in the future, the overexpression of *BdMATE* could be used to develop Al tolerant crops closely related to *S. viridis* such as maize, wheat, and sugarcane.

AUTHOR CONTRIBUTIONS

HM, AK, AR, PM, WRS, and AA conceived and designed the experiments. PM and AR performed the *in silico* analysis. AR, WRS, FV, and KD performed the abiotic stress assays. AR, BC, and FV carried out the RT-qPCR assays and analyzed the data. AR, PM, WRS, and AK performed hematoxylin assay. PAO, RC, DC, WRS, MB, and HM performed and analyzed the organic acids. WRS and AR wrote the manuscript. AR, WRS, GC, JVM, CS, MB, and AA contributed to the discussion of the results. HM, AK, and GC provided intellectual input and revised the manuscript. All authors read and approved the final manuscript.

FUNDING

This work was supported by the Brazilian Agricultural Research Corporation (Embrapa), the Brazilian Federal Agency for Graduate Education (CAPES), and the São Paulo ResearchFoundation (FAPESP) grant Proc. n. 2012/23838-7.

CONFLICT OF INTEREST STATEMENT

The authors declare that the research was conducted in the absence of any commercial or financial relationships that could be construed as a potential conflict of interest.

FOOTNOTES

1. ^http://phytozome.jgi.doe.gov

REFERENCES

1. Bennetzen, J. L., Schmutz, J., Wang, H., Percifield, R., Hawkins, J., Pontaroli, A. C., et al. (2012). Reference genome sequence of the model plant *Setaria*. *Nat. Biotechnol.* 30, 555–561.
2. Boff, T. (2006). *Tolerância ao Alumínio em Trigo: Identificação e Caracterização Molecular de Genes*. Ph.D. thesis, Universidade Federal do Rio Grande do Sul, Porto Alegre.
3. Brutnell, T. P., Wang, L., Swartwood, K., Goldschmidt, A., Jackson, D., Zhu, X. G., et al. (2010). Setaria viridis: a model for C4 photosynthesis. *Plant Cell* 22, 2537–2544.
4. Buchan, D. W., Ward, S. M., Lobley, A. E., Nugent, T. C. O., Bryson, K., and Jones, D. T. (2010). Protein annotation and modelling servers at University College London. *Nucleic Acids Res.* 38, W563–W568.
5. Camargo, C. E. D. O. (1981). Melhoramento do trigo: I. Hereditariedade da tolerância à toxicidade do alumínio. *Bragantia* 40, 33–45.
6. Cançado, G. M. A. (2006). *Caracterização e Utilização de Genes Envolvidos na Tolerância ao Alumínio Tóxico em Milho*. Ph.D. thesis, Universidade Estadual de Campinas, Campinas.
7. Centeno, D. C., Hell, A. F., Braga, M. R., Del Campo, E. M., and Casano, L. M. (2016). Contrasting strategies used by lichen microalgae to cope with desiccation-rehydration stress revealed by metabolite profiling and cell wall analysis. *Environ. Microbiol.* 18, 1546–1560.
8. Chang, S., Puryear, J., and Cairney, J. (1993). A simple and efficient method for isolating RNA from pine trees. *Plant Mol. Biol. Rep.* 11, 113–116.
9. Chen, Q., Guo, C. L., Wang, P., Chen, X. Q., Wu, K. H., Li, K. Z., et al. (2013). Up-regulation and interaction of the plasma membrane H+-ATPase and the 14-3-3 protein are involved in the regulation of citrate exudation from the broad bean (*Vicia faba* L.) under Al stress. *Plant Physiol. Biochem.* 70, 504–511.
10. Chen, Q., Kan, Q., Wang, P., Yu, W., Yu, Y., Zhao, Y., et al. (2015). Phosphorylation and interaction with the 14-3-3 protein of the plasma membrane H+-ATPase are involved in the regulation of magnesium-mediated increases in aluminum-induced citrate exudation in broad bean (*Vicia faba* L). *Plant Cell Physiol.* 56, 1144–1153.
11. Contreras, R., Figure ueiras, A. M., Gallego, F. J., and Benito, C. (2014). Brachypodium distachyon: a model species for aluminium tolerance in Poaceae. *Funct. Plant Biol.* 41, 1270–1283.
12. Doyle, J. J., and Doyle, J. L. (1990). Isolation of plant DNA from fresh tissue. *Focus* 12, 13–15.
13. Hoagland, D. R., and Arnon, D. I. (1950). *The Water-Culture Method for Growing Plants Without Soil*, 2nd Edn. Berkeley, CA: University of California, 347.
14. Horst, W. J., Wang, Y., and Eticha, D. (2010). The role of the root apoplast in aluminium-induced inhibition of root elongation and in aluminium resistance of plants: a review. *Ann. Bot.* 106, 185–197.
15. Illéš, P., Schlicht, M., Pavlovkin, J., Lichtscheidl, I., Baluška, F., and Oveèka, M. (2006). Aluminium toxicity in plants: internalization of aluminium into cells of the transition zone in *Arabidopsis* root apices related to changes in plasma membrane potential, endosomal behaviour, and nitric oxide production. *J. Exp. Bot.* 57, 4201–4213.
16. Jones, D. L., Blancaflor, E. B., Kochian, L. V., and Gilroy, S. (2006). Spatial coordination of aluminium uptake, production of reactive oxygen species, callose production and wall rigidification in maize roots. *Plant Cell Environ.* 29, 1309–1318.
17. Kearse, M., Moir, R., Wilson, A., Stones-Havas, S., Cheung, M., Sturrock, S., et al. (2012). Geneious basic: an integrated and extendable desktop software platform for the organization and analysis of sequence data. *Bioinformatics* 28, 1647–1649.
18. Kellogg, E. A. (2015). Brachypodium distachyon as a genetic model system. *Annu. Rev. Genet.* 49, 1–20.

19. Kochian, L. V., Piñeros, M. A., Liu, J., and Magalhaes, J. V. (2015). Plant adaptation to acid soils: the molecular basis for crop aluminum resistance. *Annu. Rev. Plant Biol.* 66, 571–598.
20. Kollmeier, M., Felle, H. H., and Horst, W. J. (2000). Genotypical differences in aluminum resistance of maize are expressed in the distal part of the transition zone. Is reduced basipetal auxin flow involved in inhibition of root elongation by aluminum? *Plant Physiol.* 122, 945–956.
21. Kopittke, P. M., Blamey, F. P. C., and Menzies, N. W. (2008). Toxicities of soluble Al, Cu, and La include ruptures to rhizodermal and root cortical cells of cowpea. *Plant Soil* 303, 217–227.
22. Kumar, S., Stecher, G., and Tamura, K. (2016). MEGA7: molecular evolutionary genetics analysis version 7.0 for bigger datasets. *Mol. Biol. Evol.* 33, 1870–1874.
23. Li, X. F., Ma, J. F., and Matsumoto, H. (2000). Pattern of aluminum-induced secretion of organic acids differs between rye and wheat. *Plant Physiol.* 123, 1537–1544.
24. López-Bucio, J., Nieto-Jacobo, M. F., Ramírez-Rodríguez, V., and Herrera-Estrella, L. (2000). Organic acid metabolism in plants: from adaptive physiology to transgenic varieties for cultivation in extreme soils. *Plant Sci.* 160, 1–13.
25. Ma, J. F., Zheng, S. J., Li, X. F., Takeda, K., and Matsumoto, H. (1997). A rapid hydroponic screening for aluminium tolerance in barley. *Plant Soil* 191, 133–137.
26. Magalhães, J. V., Liu, J., Guimarães, C. T., Lana, U. G., Alves, V. M., Wang, Y. H., et al. (2007). A gene in the multidrug and toxic compound extrusion (MATE) family confers aluminum tolerance in sorghum. *Nat. Genet.* 39, 1156–1161.
27. Magnacava, R., Gardner, C. O., and Clark, R. B. (1987). "Evaluation of inbred maize lines for aluminum in nutrient solution," in *Genetics Aspects of Plant Mineral Nutrition*, eds H. W. Gabelman and B. C. Loughman (Dordrecht: Martinus Nijhoff Publishers), 255–265.
28. Maron, L. G., Guimarães, C. T., Kirst, M., Albert, P. S., Birchler, J. A., Bradbury, P. J., et al. (2013). Aluminum tolerance in maize is associated with higher MATE1 gene copy number. *Proc. Natl. Acad. Sci. U.S.A.* 110, 5241–5246.
29. Maron, L. G., Piñeros, M. A., Guimarães, C. T., Magalhaes, J. V., Pleiman, J. K., Mao, C., et al. (2010). Two functionally distinct members of the MATE (multi-drug and toxic compound extrusion) family of transporters potentially underlie two major aluminum tolerance QTLs in maize. *Plant J.* 61, 728–740.
30. Martins, P. K., Mafra, V., De Souza, W. R., Ribeiro, A. P., Vinecky, F., Basso, M. F., et al. (2016). Selection of reliable reference genes for RT-qPCR analysis during developmental stages and abiotic stress in *Setaria viridis*. *Sci. Rep.* 6:28348.
31. Martins, P. K., Ribeiro, A. P., da Cunha, B. A. D. B., Kobayashi, A. K., and Molinari, H. B. C. (2015). A simple and highly efficient Agrobacterium-mediated transformation protocol for *Setaria viridis*. *Biotechnol. Rep.* 6, 41–44.
32. McGuffin, L. J., Bryson, K., and Jones, D. T. (2000). The PSIPRED protein structure prediction server. *Bioinformatics* 16, 404–405.
33. Molinari, H. B. C., Marur, C. J., Daros, E., De Campos, M. K. F., De Carvalho, J. F. R. P., Pereira, L. F. P., et al. (2007). Evaluation of the stress-inducible production of proline in transgenic sugarcane (*Saccharum* spp.): osmotic adjustment, chlorophyll fluorescence and oxidative stress. *Physiol. Plant.* 130, 218–229.
34. Motoda, H., Kano, Y., Hiragami, F., Kawamura, K., and Matsumoto, H. (2011). Changes in rupture formation and zonary region stained with Evans blue during the recovery process from aluminum toxicity in the pea root apex. *Plant Signal. Behav.* 6, 98–100.
35. Muller, P. Y., Janovjak, H., Miserez, A. R., and Dobbie, Z. (2002). Processing of gene expression data generated by quantitative real-time RT-PCR. *Biotechniques* 32, 1372–1374.
36. Murashige, T., and Skoog, F. (1962). A revised medium for rapid growth and bio assays with tobacco tissue cultures. *Physiol. Plant.* 15, 473–497.
37. Muthamilarasan, M., and Prasad, M. (2015). Advances in *Setaria* genomics for genetic improvement of cereals and bioenergy grasses. *Theor. Appl. Genet.* 128, 1–14.

38. Nugent, T., and Jones, D. T. (2009). Transmembrane protein topology prediction using support vector machines. *BMC Bioinformatics* 10:159.
39. Nunes-Nesi, A., Brito, D. S., Inostroza-Blancheteau, C., Fernie, A. R., and Araújo, W. L. (2014). The complex role of mitochondrial metabolism in plant aluminum resistance. *Trends Plant Sci.* 19, 399–407.
40. Oliveira, M. D. S. (2012). *Tolerância de Variedades de Cana-De-Açúcar (Saccharum spp.) à Toxidez por Alumínio em Solução.* Master, dissertation, Universidade Federal de São Carlos, Araras.
41. Pfaffl, M. W., Tichopad, A., Prgomet, C., and Neuvians, T. P. (2004). Determination of stable housekeeping genes, differentially regulated target genes and sample integrity: BestKeeper–Excel-based tool using pair-wise correlations. *Biotechnol. Lett.* 26, 509–515.
42. Ramakers, C., Ruijter, J. M., Deprez, R. H. L., and Moorman, A. F. (2003). Assumption-free analysis of quantitative real-time polymerase chain reaction (PCR) data. *Neurosci. Lett.* 339, 62–66.
43. Rees, T. A. P. (1990). "Carbon metabolism in mitochondria," in *Plant Physiology, Biochemistry and Molecular Biology*, eds D. T. Dennis and D. H. Turpin (Essex: Longman Scientific and Technical), 106–111.
44. Ryan, P. R., Ditomaso, J. M., and Kochian, L. V. (1993). Aluminium toxicity in roots: an investigation of spatial sensitivity and the role of the root cap. *J. Exp. Bot.* 44, 437–446.
45. Ryan, P. R., Raman, H., Gupta, S., Horst, W. J., and Delhaize, E. (2009). A second mechanism for aluminum resistance in wheat relies on the constitutive efflux of citrate from roots. *Plant Physiol.* 149, 340–351.
46. Sasaki, T., Yamamoto, Y., Ezaki, B., Katsuhara, M., Ahn, S. J., Ryan, P. R., et al. (2004). A wheat gene encoding an aluminum-activated malate transporter. *Plant J.* 37, 645–653.
47. Shaff, J. E., Schultz, B. A., Craft, E. J., Clark, R. T., and Kochian, L. V. (2010). GEOCHEM-EZ: a chemical speciation program with greater power and flexibility. *Plant Soil* 330, 207–214.
48. Shen, H., He, L. F., Sasaki, T., Yamamoto, Y., Zheng, S. J., Ligaba, A., et al. (2005). Citrate secretion coupled with the modulation of soybean root tip under aluminum stress. Up-regulation of transcription, translation, and threonine-oriented phosphorylation of plasma membrane H+-ATPase. *Plant Physiol.* 138, 287–296.
49. Sievers, F., and Higgins, D. G. (2014). Clustal Omega, accurate alignment of very large numbers of sequences. *Multiple Seq. Align. Methods* 1079, 105–116.
50. Sivaguru, M., and Horst, W. J. (1998). The distal part of the transition zone is the most aluminum-sensitive apical root zone of maize. *Plant Physiol.* 116, 155–163.
51. Sivaguru, M., Liu, J., and Kochian, L. V. (2013). Targeted expression of *SbMATE* in the root distal transition zone is responsible for sorghum aluminum resistance. *Plant J.* 76, 297–307.
52. Souza, L. T. D., Cambraia, J., Ribeiro, C., Oliveira, J. A. D., and Silva, L. C. D. (2016). Effects of aluminum on the elongation and external morphology of root tips in two maize genotypes. *Bragantia* 75, 19–25.
53. Tang, Y., Sorrells, M. E., Kochian, L. V., and Garvin, D. F. (2000). Identification of RFLP markers linked to the barley aluminum tolerance gene. *Crop Sci.* 40, 778–782.
54. Von Uexküll, H. R., and Mutert, E. (1995). Global extent development and economic-impact of acid soils. *Plant Soil* 171, 115.
55. Wang, P., Yu, W., Zhang, J., Rengel, Z., Xu, J., Han, Q., et al. (2016). Auxin enhances aluminium-induced citrate exudation through upregulation of *GmMATE* and activation of the plasma membrane H+-ATPase in soybean roots. *Ann. Bot.* 118, 933–940.
56. Wenig, P., and Odermatt, J. (2010). OpenChrom: a cross-platform open source software for the mass spectrometric analysis of chromatographic data. *BMC Bioinformatics* 11:405.

57. Wu, X., Li, R., Shi, J., Wang, J., Sun, Q., Zhang, H., et al. (2014). *Brassica oleracea* MATE encodes a citrate transporter and enhances aluminum tolerance in *Arabidopsis thaliana*. *Plant Cell Physiol.* 55, 1426–1436.
58. Yamamoto, Y., Kobayashi, Y., and Matsumoto, H. (2001). Lipid peroxidation is an early symptom triggered by aluminum, but not the primary cause of elongation inhibition in pea roots. *Plant Physiol.* 125, 199–208.
59. Yang, J. L., Li, Y. Y., Zhang, Y. J., Zhang, S. S., Wu, Y. R., Wu, P., et al. (2008). Cell wall polysaccharides are specifically involved in the exclusion of aluminum from the rice root apex. *Plant Physiol.* 146, 602–611.
60. Yang, X. Y., Yang, J. L., Zhou, Y., Piñeros, M. A., Kochian, L. V., Li, G. X., et al. (2011). A de novo synthesis citrate transporter, *Vigna umbellata* multidrug and toxic compound extrusion, implicates in Al-activated citrate efflux in rice bean (*Vigna umbellata*) root apex. *Plant Cell Environ.* 34, 2138–2148.
61. Yokosho, K., Yamaji, N., and Ma, J. F. (2011). An Al-inducible MATE gene is involved in external detoxification of Al in rice. *Plant J.* 68, 1061–1069.
62. Zhou, G., Pereira, J. F., Delhaize, E., Zhou, M., Magalhães, J. V., and Ryan, P. R. (2014). Enhancing the aluminium tolerance of barley by expressing the citrate transporter genes *SbMATE* and *FRD3*. *J. Exp. Bot.* 65, 2381–2390.

CHAPTER 6

Recent Advances of Metabolomics in Plant Biotechnology

Yozo Okazaki [1]• **Kazuki Saito**[1,2]

[1].*RIKEN Plant Science CenterYokohamaJapan*

[2].*Graduate School of Pharmaceutical SciencesChiba UniversityChibaJapan*

ABSTRACT

Biotechnology, including genetic modification, is a very important approach to regulate the production of particular metabolites in plants to improve their adaptation to environmental stress, to improve food quality, and to increase crop yield. Unfortunately, these approaches do not necessarily lead to the expected results due to the highly complex mechanisms underlying metabolic regulation in plants. In this context, metabolomics plays a key role in plant molecular biotechnology, where plant cells are modified by the expression of engineered genes, because we can obtain information on the metabolic status of cells via a snapshot of their metabolome. Although metabolome analysis could be used to evaluate the effect of foreign genes and understand the metabolic state of cells, there is no single analytical method for metabolomics because of the wide range of chemicals synthesized in plants. Here, we describe the basic analytical advancements in plant metabolomics and bioinformatics and the application of metabolomics to the biological study of plants.

Keywords: Metabolomics Mass spectrometry NMR Functional genomics GMO Quantitative genetics QTL

INTRODUCTION

The modification of metabolism by biotechnological techniques is often utilized for the optimal production of plant metabolites, which directly benefit human health and plant growth. For example, "golden rice" is a transgenic line of *Oryza sativa* that was genetically engineered to biosynthesize *β*-carotene, a pro-vitamin A, in the edible parts of rice (Ye et al. 2000; Yonekura-Sakakibara and Saito 2009). The introduction of transcription factors from the snapdragon to tomato increased the production of anthocyanins, which have health protective properties (Butelli et al. 2008). However, many similar approaches do not necessarily lead to the expected results, e.g., overexpression of foreign *S*-linalool synthase in transgenic petunia did not result in the expected accumulation of free linalool, but led to the accumulation of *S*-linalyl-*β*-D-glucoside (Lucker et al. 2001). These unexpected results suggest that highly complex regulatory systems control plant metabolism and also indicate the need for more precise information on plant metabolism. In this context, metabolomics plays a key role in the field of molecular biotechnology, where plant cells are modified by the expression of engineered genes. Metabolomic analysis provides us with in-depth information on cellular metabolism via a snapshot of the metabolome, often combined with data from other "omics" (Oksman-Caldentey and Saito 2005; Saito and Matsuda 2010).

Metabolomics is one of the omics approaches that can be used to acquire comprehensive information on metabolites. It aims to grasp the global state of metabolism in measured samples. Among the omics studies used in plant sciences, genomics was the first to emerge, and uncovered the genome sequences of several organisms, including *Arabidopsis* (Arabidopsis Genome Initiative 2000) and rice (Goff et al. 2002; Sasaki et al. 2002; Yu et al. 2002; International Rice Genome Sequencing Project 2005). There should be no doubt that the development of the automated DNA sequencer has led to the current progress of genomics. Other omics studies have also been developed as a result of technical innovations. Microarrays made high throughput analysis of mRNA expression feasible and led to the emergence of transcriptomics. Two-dimensional electrophoresis and mass spectrometry (MS) significantly contributed to the development of proteomics. Similarly, MS and nuclear magnetic resonance (NMR) spectroscopy have facilitated metabolomic studies. However, metabolomics is not as advanced as the other omics because there is a critical difference between metabolites and other molecules, i.e., DNA, RNA, and proteins are linear polymers consisting of a limited numbers of monomers, and the interpretation of RNA and protein sequences can be facilitated by genome information according to the central dogma of molecular biology, whereas metabolites comprise a more heterogeneous group than these polymers in terms of their physical and chemical properties, varying widely with respect to size, polarity, quantity, and stability. In addition, there are an estimated 200,000 plant metabolites (Fiehn 2002b; Dixon and Strack 2003), and many of these metabolites remain unknown. Thus, no single method has yet been developed for plant metabolomics, and researchers who want to acquire comprehensive metabolome information have to employ several methodologies according to the chemical properties of the metabolites. In spite of these difficulties, metabolomics, metabolic profiling, and metabolic fingerprinting have been employed in many biological studies. These techniques have been applied to the functional identification of unknown genes through the metabolic profiling of plants in which some genes are up- or down-regulated (Bino et al. 2004; Hirai et al. 2005; Oksman-Caldentey and Saito 2005; Tohge et al. 2005; Watanabe et al. 2008; Yonekura-Sakakibara et al. 2008; Okazaki et al. 2009; Matsuda et al. 2010), the discovery of biomarkers associated with disease phenotypes (Soga et al. 2006; Sreekumar et al. 2009), the safety assessment of genetically modified organisms (GMOs) (Baker et al. 2006; Beale et al. 2009; Kusano et al. 2011a), and the discovery of compounds involved in plant resistance to biotic and abiotic stresses (Leiss et al. 2009; Ward et al. 2010b; Kusano et al. 2011c). When combined with genomics, transcriptomics, and/or proteomics, metabolomics can also help to interpret and understand many complex biological processes; indeed, metabolomics is now widely recognized as a cornerstone of systems biology (Quackenbush 2002; Hall 2006; Saito and Matsuda 2010). In this review, we introduce the basic analytical protocols for plant metabolomics and bioinformatics and the practical application of metabolomics to the biological study of plants.

ANALYTICAL TECHNOLOGIES FOR THE PLANT METABOLOME

Mass spectrometry

MS is the most frequently used technique in metabolic studies. MS provides mass-to-charge ratio information, which enables the structure of metabolites to be determined. The main advantage of MS is its high sensitivity. In addition, the combination of chromatographic separation with MS increases the number of compounds that can be detected by reducing the complexity of the mass spectra and the matrix effect. There are several chromatographic techniques that can be combined with MS.

Gas chromatography (GC)–MS is used for metabolite profiling. Capillary GC uses a carrier gas to move analytes through a coated, fused silica capillary. GC–MS requires the analyte to be vaporized in order for its migration through the capillary; therefore, analytes must be volatile or amenable to chemical derivatization to render them volatile. Certain types of samples (terpenoids and essential oils) are particularly well suited for GC–MS analysis. With appropriate derivatization, more polar metabolites or metabolites with highly polar functional groups can also be analyzed, e.g., amino acids, sugars, organic acids, fatty acids, and amines. This indicates that many compounds associated with primary metabolism can be analyzed by GC–MS. Electron ionization (EI) is the most commonly used GC–MS ionization technique; it is robust, highly reproducible, and considered to be less affected by the matrix effect, e.g., ion suppression. In addition, EI generates informative and characteristic mass spectra due to the relatively high degree of fragmentation reactions, which are useful for compound identification. However, molecular ions, which effectively reduce the number of candidate structures or elemental compositions, are often undetected, suggesting that GC–EI–MS can be used for the targeted analysis of known primary metabolites. Mass spectral libraries of EI–MS are commercially and non-commercially available, and these can be routinely used for peak annotation. To date, GC–MS-based metabolic profiling has been frequently used in many metabolomic studies of plants, animals, and microorganisms (Fiehn et al. 2000; Roessner et al. 2001; Fernie et al. 2004; Shellie et al. 2005; Keurentjes et al. 2006; Schauer et al. 2006, 2008; Kusano et al. 2007, 2011a, c; Ralston-Hooper et al. 2008).

Liquid chromatography (LC)–MS is also used for metabolomic studies. The major ionization method for LC–MS is atmospheric ionization (API), which includes electrospray ionization (ESI) and atmospheric chemical ionization (APCI) (Codrea et al. 2007; Dunn 2008). These methods do not require the compound to be volatile, and are thus suited for compounds with heat-labile functional groups, chemically unstable substructures, high vapor points, and high molecular weights. In plant metabolomics, LC–MS is frequently used to profile secondary metabolites (e.g., phenylpropanoids, alkaloids, and saponins) and complex lipids (e.g., glycerolipids, and sphingolipids), often with a reversed-phase column (e.g., C18) for separation (De Vos et al. 2007; Naoumkina et al. 2007; Bottcher et al. 2008; Farag et al. 2008; Matsuda et al. 2009, 2010; Okazaki et al. 2011). Many of these compounds are not amenable to volatilization by GC. Secondary metabolites are not directly involved in the normal growth or development of plants, but they have important physiological roles, e.g., as toxic compounds for predators and attractors for pollinators. In addition, they have been historically used as pharmaceutical compounds and food additives for flavor, indicating the special importance of LC–MS-based analysis in plant metabolomics. In contrast to

EI–MS, API–MS usually affords $[M+H]^+$ and $[M-H]^-$ as the main signals, which are useful for reducing the candidate structure of the detected compounds. By using collision-induced dissociation of tandem MS, i.e., MS/MS, more informative mass spectra with many fragments are obtained. Although various compounds can be analyzed by LC–MS, peak annotation is still troublesome due to the shortage of mass spectral libraries for API–MS and commercially available authentic compounds. Much effort is required for the more comprehensive identification of metabolites by the production of additional reference compounds (Nakabayashi et al. 2009, 2010) and the generation of more comprehensive databases (Kind and Fiehn 2006; Moco et al. 2006; Shinbo et al. 2006; Bocker and Rasche 2008; Horai et al. 2010). In addition to untargeted metabolomics, widely targeted metabolomic studies use a triple quadrupole mass spectrometer equipped with LC (Sawada et al. 2009; Albinsky et al. 2010b). In a targeted metabolomics strategy, predefined metabolite-specific signals (by selected reaction monitoring with tandem MS or selected ion monitoring) are often used to determine precisely and accurately the relative abundances and concentrations of a limited number of known and expected endogenous metabolites (Griffiths et al. 2010).

Capillary electrophoresis (CE)–MS is utilized to analyze a wide spectrum of ionic metabolites. Similar to LC–MS, API is the most suitable ionization method for CE–MS. In capillary zone electrophoresis, constituent ions migrate on the basis of their electrostatic force, which results from the charge and size of ions, in addition to the electro-osmotic flow derived from the capillary and the type of electrolyte used. Ionic compounds are separated at a high resolution in a narrow capillary. In metabolomic analyses using CE–MS, samples are often divided for cation and anion analyses. For cation analysis, Soga's method, using formic acid as an electrolyte, is experimentally convenient and provides excellent separation of metabolites with good reproducibility (Soga et al. 2006). In contrast, the routine analysis of anions at a high resolution is more difficult than for cations, although various approaches, including the use of coated capillaries, have been assessed (Soga et al. 2002; Harada et al. 2006). Recently, a platinum ESI electrospray needle has been developed that significantly improved the analysis of anions by CE–MS (Soga et al. 2009). Target ionic metabolites in CE–MS analyses include amino acids, organic acids, nucleotides, and sugar phosphates. The compounds detectable by CE–MS are rather similar to those detected by GC–MS, but CE–MS can analyze these compounds without derivatization. In addition, similar to LC–MS, molecular-related ions are also detectable with CE–MS. Since the metabolites suited for CE–MS analysis are physiologically important and common to all organisms, CE–MS has been used in a variety of metabolomic studies (Monton and Soga 2007; Oikawa et al. 2008; Sato et al. 2008; Ramautar et al. 2009; Urano et al. 2009; Ishikawa et al. 2010), including the identification of biomarkers for the progression of prostate cancer (Sreekumar et al. 2009), oxidative stress (Soga et al. 2006), the measurement of internal body time (Minami et al. 2009), and the functional analysis of unknown genes in *Arabidopsis* (Ohkama-Ohtsu et al. 2008, 2011; Watanabe et al. 2008).

As described above, many of the analytes in metabolomic studies are still unknown, and even the known compounds are difficult to obtain from commercial sources. Under such unfavorable circumstances, the accurate *m/z* values with ultra-high resolution generated by Fourier transform ion cyclotron resonance mass spectrometry (FT-ICR MS) are very important. FT-ICR MS is of value in metabolomic studies for the following two reasons. Firstly, different compounds with very similar molecular masses can be detected separately, even by their direct infusion without any chromatographic steps, and this can result in the rapid detection of a number of metabolites. Secondly, an accurate estimate of the chemical formula of the detected peaks can be acquired, which can facilitate annotation procedures for unknown compounds. MS/MS analysis coupled with FT-ICR MS is also helpful to estimate the chemical formula because fragment ions are also detected with high resolution and accuracy. In fact, several metabolites that accumulated in *Arabidopsis* following treatment with herbicides were identified by MS/MS analyses of FT-ICR MS (Oikawa et al. 2006). However, because of difficulties in hardware handling and the extremely large amount of data generated, the number of reports on metabolomic studies with FT-ICR MS is rather limited compared with other types of MS. The non-targeted metabolite analysis of strawberry fruit was the first published metabolomic study using FT-ICR MS (Aharoni et al. 2002). This technique has also been applied to *Arabidopsis* functional genomics (Hirai et al. 2004, 2005; Tohge et al. 2005), transgenic tobacco (Mungur et al. 2005), and the identification of metabolic biomarkers in Crohn's disease (Jansson et al. 2009). Hyphenation of FT-ICR MS and LC has increased the range of applications for FT-ICR MS in metabolomic studies of tomato (Iijima et al. 2008) and *Lotus japonicas* (Suzuki et al. 2008), although the real power of this MS approach still seems to be difficult to demonstrate due to the shortage of informatics technology. A recent report described a fine structural elucidation strategy with FT-ICR MS by using all of the isotopic peaks derived from ^{13}C-, ^{15}N-, ^{34}S-, and ^{18}O-substituted forms recorded under high magnetic field FT-ICR MS (Miura et al. 2010). In addition, the combination of FT-ICR MS and the labeling of metabolites with stable isotopes (e.g., ^{13}C and ^{15}N) has been introduced to elevate the quality of plant metabolite annotation (Hegeman et al. 2007; Giavalisco et al. 2008, 2009). According to these reports, the comparison of MS data from $^{12}C/^{13}C$ labeling experiments enabled the removal of background noise peaks and the positive identification of compounds with a true biological origin, thus eliminating the ambiguity in chemical formula assignment and resulting in the clear association of one measured mass to one chemical formula.

Since the metabolome consists of a vast array of compounds, most current metabolomic analyses using a single analytical protocol can only detect a fraction of the metabolites in a complex biological sample; thus, the combination of multiple MS systems, considering the scope of each MS system, can reveal the overall metabolic status of a sample. For example, we established a multi-MS platform for metabolomic analysis that consisted of several MS techniques (GC–MS, CE–MS, and LC–MS). This platform has been applied to the functional identification of an unknown gene (Watanabe et al. 2008) and the evaluation of a genetically modified tomato (Kusano et al. 2011a), as described later. Of course, such a multi-MS system provides a large amount of metabolite information, but the analysis of this data depends on the availability of well-structured databases that describe the assayed metabolites. Chemical identifiers are notoriously incoherent and encompass a wide range of different referencing schemes with varying scope and coverage, since chemical databases use multiple types of identifiers in parallel, but lack a common primary key for reliable database consolidation. Thus, connecting identifiers of analytes found in experimental data with the identifiers of their parent metabolites in public databases can, therefore, be very important. Recently, a tool for consolidating metabolite identifiers was developed to enable contextual and multi-platform metabolomic data analysis (Redestig et al. 2010), and we utilized this tool in practice to prepare a single consensus dataset for our multi-MS platform.

Nuclear magnetic resonance spectroscopy

NMR spectroscopy is another technology that is frequently employed in metabolomic studies since it is able to determine the atomic state of compounds and enables the identification of metabolites that are otherwise unidentifiable by MS analysis (Ward et al. 2007, 2010a; Beale et al. 2009; Kim et al. 2011). NMR spectroscopy can yield detailed information on the quantities and identities of the metabolites present in extracts or in vivo (Kikuchi et al. 2004; Lindon et al. 2004; Wang et al. 2004; Krishnan et al. 2005; Clayton et al. 2006; Ratcliffe and Shachar-Hill 2006; Sekiyama and Kikuchi 2007; Tian et al. 2007; Hagel et al. 2008; Sekiyama et al. 2010). Metabolome analysis by NMR usually does not require any pretreatment including column chromatography and derivatization, which makes standardization straightforward and points to the great potential of NMR spectroscopy in metabolomics, although attentions should be paid to pH of samples because chemical shifts could be changed depending on pH. For metabolomic analysis by NMR, buffered solutions are frequently utilized to stabilize the pH of sample solutions. Among the advantages of NMR over MS-based methods are that it is non-destructive, non-biased, highly quantitative, and enables the identification of complex unidentified metabolites. Even compounds which are bound to insoluble polymers can also be analyzed by high-resolution solid state NMR, which can afford direct and intact structural information. The major disadvantage of NMR, relative to MS, is its low sensitivity. Signal overlap from many similar molecules in biological samples is a major problem that inhibits the accurate assignment of NMR signals. However, these disadvantages, such as its lack of sensitivity and resolution, are gradually being overcome by the development of cryogenic probes (Kovacs et al. 2005), higher-strength superconducting magnets, miniaturized radiofrequency coils (Van and Veenstra 2009), and multidimensional NMR techniques (Kikuchi et al. 2004; Sekiyama and Kikuchi 2007; Chikayama et al. 2010). Dynamic nuclear polarization (DNP) enhanced NMR technique, developed by a combination of NMR and electron paramagnetic resonance (EPR) (Frydman and Blazina 2007), may provide a new approach to acquire hyper-sensitive NMR spectra for metabolomics. Labeling the samples with stables isotopes, e.g., ^{13}C and ^{15}N. is also a useful technique that selectively enhances the sensitivity of NMR and improves signal sharpness (Kikuchi and Hirayama 2007; Chikayama et al. 2010). Labeling of metabolites with isotopes receptive by NMR is also useful for metabolic flux analysis and fluxomics by tracking the selective signal enhancement of isotopologues (Mesnard and Ratcliffe 2005; Eisenreich and Bacher 2007; Sekiyama and Kikuchi 2007; Coquin et al. 2008; Bothwell and Griffin 2011; Fan and Lane 2011; Moseley et al. 2011; Ward et al. 2011). These suggest the importance of the complementary use of MS and NMR in metabolomics.

INFORMATICS FOR METABOLOMIC STUDIES

Extremely large amounts of data are generated by instrument analysis, particularly for the high-performance instruments frequently used for metabolome analysis that can detect tiny signals with high resolution. To handle the large amount of datasets generated and comprehend the metabolome data, automated software is needed that can identify peaks from raw MS or NMR data, align the peaks among samples, and identify and quantify each metabolite; therefore, informatics is an essential tool for processing large metabolomic datasets (Fukushima et al. 2009b; Tohge and Fernie 2009, 2010; Go 2010). For GC–MS data, an automated spectral deconvolution and identification system using the NIST mass spectral library is available (Ausloos et al. 1999). To process raw MS data acquired by LC–MS or CE–MS, several open source programs are available, including MetAlign (Vorst et al. 2005), MZmine2 (Katajamaa et al. 2006), XCMS (Smith et al. 2006; Benton et al. 2008), MET-IDEA (Broeckling et al. 2006), and Mass++ (http://groups.google.com/group/massplusplus). To analyze FT-ICR MS datasets, DrDmassPlus (Oikawa et al. 2006) can be used for peak picking, peak alignment, and some statistical analysis. These programs facilitate comprehensive data analysis for non-targeted metabolomic approaches.

As with any omics discipline, metabolite annotation and identification is highly dependent on the availability and quality of electronic databases. To annotate the peaks processed by the programs described above, several mass spectral databases can be used, including MassBank (Horai et al. 2010), METLIN (Smith et al. 2005), the Golm Metabolite Database (Kopka et al. 2005), FiehnLib (Kind et al. 2009), and Lipid Search (Taguchi et al. 2007). The Human Metabolome Database (Wishart et al. 2009) and the Madison Metabolomics Consortium Database (Markley et al. 2007; Cui et al. 2008) provide mass spectral and NMR data for human metabolome studies; they are also useful for annotating signals derived from primary metabolites in plants. Recently, "ReSpect," a depository of API–MS^n data from phytochemicals drawn from the literature, has been developed (http://spectra.psc.riken.jp/) as one of the web applications of PRIMe, a platform of RIKEN Metabolomics (http://prime.psc.riken.jp/; (Akiyama et al. 2008)). Users can access MS^n information at this website, and they can freely download the manually refined MS^n data. Some of these databases are integrated with software for processing raw MS data, e.g., METLIN and XCMS, and MassBank and Mass++. In addition to these spectral databases, several compound databases, including KNApSAcK (Shinbo et al. 2006), Kyoto Encyclopedia of Genes and Genomes (KEGG) (Kanehisa et al. 2008), PubChem (Wheeler et al. 2008), LipidBank (Yasugi and Watanabe 2002), and LIPIDMAPS (Fahy et al. 2007), are readily accessible. The information on compound name, chemical formula, and molecular weight that is deposited in these databases is also helpful for peak identification when authentic compounds and standard mass and NMR spectra are not available. Several databases can also be used that focus mainly on the mass spectra or metabolite information from several plant species. In MS2T (Matsuda et al. 2009), LC–MS data acquired from various plant species, including *Arabidopsis*, rice, wheat, and several non-model plant species, e.g., pear, are deposited. KOMICS (Iijima et al. 2008) and MotoDB (Moco et al. 2006) provide information for the ions that are detected in tomato, while ARMeC (http://www.armec.org/MetaboliteLibrary/) is a database of metabolites from *Arabidopsis* and potato. These data can be used to validate metabolome data among various laboratory, and as a reference source for many researchers who have to set up metabolome analytical platforms.

To comprehend the biochemical process involved in metabolism from metabolome data, metabolic pathway databases are very useful. There are several pathway databases available for plant metabolomics, including PlantCyc (http://www.plantcyc.org/), MapMan (Thimm et al. 2004), KaPPA-View (Tokimatsu et al. 2005; Sakurai et al. 2011), Arabidopsis Reactome (http://www.arabidopsisreactome.org/about.html), and KEGG (Kanehisa et al. 2008). KEGG PLANT (http://www.kegg.jp/kegg/plant/) is a new interface of KEGG that focuses on phytochemicals, especially secondary metabolites. Some of these databases import the abundance information of metabolites (and transcriptome information), and display an integrated overview of the metabolic state by reflecting the abundance of each metabolite on the putative metabolic pathway in plants. These databases also include information on the enzymatic reactions underlying these metabolic pathways and information on the proteins or genes involved in each reaction step. Although these pathway databases are helpful, it is necessary to bear in mind that many metabolic pathways in plants are divided by subcellular and intercellular compartmentalization and are highly redundant when attempting to understand the metabolic state of plants from metabolome data.

The deposition of metabolomic data in publicly available databases is also an important issue that needs to be considered. These databases enable data sharing with researchers in different laboratories and facilitate the meta-analysis of the original data. These depositories include AtMetExpress (http://prime.psc.riken.jp/lcms/AtMetExpress/) (Matsuda et al. 2010), PlantMetabolomics.org (http://plantmetabolomics.vrac.iastate.edu/) (Bais et al. 2010), and MetabolomeExpress (https://www.metabolome-express.org/) (Carroll et al. 2010).

In metabolomic studies, statistical analysis is often employed to evaluate significant differences in the metabolites detected in different samples. The methodology typically employs multivariate analysis to statistically process the massive amount of analytical chemistry data generated by high-throughput and simultaneous metabolite analyses (Fiehn 2002a; Fukusaki and Kobayashi 2005; Hall 2006; Okada et al. 2010). Various statistical methods used in conventional genetic studies are applicable to metabolomic data by considering the amount of each metabolite as a trait value. Principal component analysis (PCA), a multivariate analysis method, is commonly used in metabolomic studies. The PCA model can provide an overview of all observations or samples in a data table by projecting and clustering each sample and highlighting any holistic differences in the complex metabolic state of each sample. Many reports have described the application of PCA to metabolomic data (Catchpole et al. 2005; Takahashi et al. 2005; Tarpley et al. 2005; Tohge et al. 2005; Baker et al. 2006; Dixon et al. 2006; Oikawa et al. 2006; Kim et al. 2007; Kusano et al. 2007; Moco et al. 2007; Ku et al. 2010b). In addition, several statistical analytical methods have been used to analyze metabolomic datasets, e.g., hierarchical cluster analysis (HCA) (Grata et al. 2007; Parveen et al. 2007), partial least squares discriminant analysis (PLS-DA) (Jonsson et al. 2004; Kusano et al. 2007; Ku et al. 2010a), and batch-learning self-organizing map analysis (BL-SOM) (Hirai et al. 2004, 2005; Matsuda et al. 2010). Depending on the objective of each study, the most appropriate statistical analytical method should be exploited to evaluate the available metabolomic data.

BIOTECHNOLOGICAL APPLICATIONS

There have been many biological applications of plant metabolomics. In this section, we introduce example case studies, some of which are summarized in Table 1.

Application of metabolomics to functional genomics

The functional characterization of gene function on the genome scale is one of the most important tasks in the post-genomic era. In modern *Arabidopsis* research, loss-of-function or gain-of-function mutant lines play an important role in deciphering gene function. A combination of the metabolomic approach with these bioresources has been demonstrated to be an effective strategy to uncover the role of genes with unknown functions. Using GC–MS and CE–MS, Watanabe et al. (2008) obtained a metabolome dataset from T-DNA insertion mutants of *Arabidopsis* with immature β-substituted alanine synthase (*bsas*) genes and its corresponding wild-type. Statistical analysis revealed that one unknown metabolite was found in the wild-type plant, but did not accumulate in one of the *bsas* mutants, *bsas3;1*. This compound was eventually identified as γ-glutamyl-β-cyanoalanine, and the *bsas3;1* gene was shown to be involved in the biosynthesis of the precursor of this dipeptide. Metabolomics is also effective in high-throughput mutant screening. High-throughput GC–MS analysis of *Arabidopsis* gain-of-function mutant lines overexpressing full-length cDNAs of *Arabidopsis* led to the finding of a new transcriptional factor regulating nitrogen metabolism (Albinsky et al. 2010a). The functional characterization of unknown genes using a metabolomics approach can also be highly accelerated by combinatory use of other omics, such as transcriptomics. Several groups have used transcriptome co-expression analysis with metabolome data to find new metabolic genes using a limited number of known genes on the basis of the assumption that genes involved in the same metabolic pathway are co-expressed by a shared regulatory mechanism (Saito et al. 2008). This strategy, using the "gene-to-metabolite" correlation, successfully revealed the function of genes involved in plant primary metabolism (Okazaki et al. 2009; Kusano et al. 2011b), secondary metabolism (Hirai et al. 2005; Tohge et al. 2005; Hirai et al. 2007; Yonekura-Sakakibara et al. 2007; Farag et al. 2008, 2009; Yonekura-Sakakibara et al. 2008; Kuzina et al. 2009; Matsuda et al. 2010), and the circadian clock (Fukushima et al. 2009a).

Table 1. Application of metabolomics in plant biological studies

Application	Species	Gene	Phenotype	Analytical technology	Range	Bioinformatics	Reference
Identification of metabolic genes	*A. thaliana*	*β*-substituted alanine synthase family genes	Accumulation of *β*-cyano-Ala derivatives in the KO line	GC–MS, CE–MS	Untargeted profiling	PCA, *t* test	Watanabe et al. (2008)
	A. thaliana	Flavonoid 3-*O*-glucosyltransferase; anthocyanin 5-*O*-glucosyltransferase	Decrease of flavonol 3-glucosides; deficiency in 5-glucosylation of anthocyanins	FT-ICR-MS, LC–MS	Untargeted profiling, targeted profiling	PCA	Tohge et al. (2005)
	A. thaliana	Flavonol 3-*O*-arabinosyltransferase; UDP-rhamnose synthase	Decrease of flavonol arabinosides; decrease of flavonoid rhamnosylation	LC–MS	Untargeted profiling	PCA	Yonekura-Sakakibara et al. (2008)
	A. thaliana	Desulfo-glucosinoate sulfotransferase	–	FT-ICR-MS	Untargeted profiling	BL-SOM	Hirai et al. (2005)
	M. truncatula	Peroxidases exhibiting aurone synthase activity	Accumulation of the aurone hispidol as a major response to yeast elicitor	LC–MS	Untargeted profiling	ANOVA	Farag et al. (2009)
	A. thaliana	Transporter for Met-derived glucosinolates	Decease of Met-derived glucosinolates and increase of Met	LC–MS	Wide targeted profiling	*t* test	Sawada et al. (2009)
	A. thaliana	UDP-glucose pyrophosphorylase	Deficiency of sulfolipid in the KO line	LC–MS	Untargeted profiling	*t* test, HCA	Okazaki et al. (2009)
Evaluation of GMOs	Tomato	Miraculin	>92% of detected metabolites had an acceptable range of variation considering the variations among several tomato cultivars	GC–MS, LC–MS, CE–MS	Untargeted profiling	PCA, ANOVA, OPLS-DA	Kusano et al. (2011a)
	Wheat	Wheat gluten coding gene	Stronger influence of site and year than genotype, although a slight increase of saccharides was observed in transgenic lines	^{1}H NMR	Untargeted profiling, targeted profiling	PCA	Baker et al. (2006)
Metabolome QTL analysis	*A. thaliana*	–	Identification of QTLs for ~75% of the >2000 mass signals detected from recombinant inbred ecotype lines; discovery of mQTL hot spots	LC–MS	Untargeted profiling	ANOVA	Keurentjes et al. (2006)
	A. thaliana	–	Linkage of metabolic profile and biomass accumulation. Discovery of mQTL hot spots	GC–MS	Untargeted profiling	Pearson correlation	Lisec et al. (2008)
	Tomato	–	Association of >50% of the metabolic loci with QTLs that modify yield-associated traits	GC–MS	Untargeted profiling	ANOVA, Pearson correlation	Schauer et al. (2006)
Investigation of the stress response	*A. thaliana*	Chalcone synthase; sinapoylglucose:malate sinapoyltransferase	Reprogramming of primary and secondary metabolism against UV-B light	GC–MS, LC–MS	Untargeted profiling	Vector analysis, PCA	Kusano et al. (2011c)
	A. thaliana	9-*cis*-Epoxycarotenoid dioxygenase	Global metabolite–metabolite correlations in dehydration-increased amino acids in wild type, and reconstruction of strong correlations with raffinose in the KO line	GC–MS, CE–MS	Untargeted profiling	Pearson correlation	Urano et al. (2009)
	A. thaliana		Changes in phenolic and indolic compounds, and rapid alterations in amino acids and other nitrogenous compounds, specific classes of glucosinolates, disaccharides, and molecules that influence the prevalence of reactive oxygen species	Flow infusion ESI-MS, GC–MS, NMR	Untargeted profiling	PCA, PLS-DA, *t* test	Ward et al. (2010b)

KO knock out, *ANOVA* analysis of variance

Application of metabolomics to quantitative genetics

Natural variation provides a valuable resource to study the genetic regulation of quantitative traits. Since metabolite levels in plant tissues (m-trait) is also a quantitative trait, quantitative trait loci (QTL) analysis of m-traits is useful to enhance our understanding of the genetic architecture underlying naturally variable phenotypes with respect to metabolism. For example, QTL analysis of the levels of vitamin E in *Arabidopsis* seeds revealed the several QTLs responsible for the control of levels of this compound (Gilliland et al. 2006). QTL analysis of m-traits of *Arabidopsis* also revealed several loci regulating the levels of secondary metabolites flavonoids and glucosinolates (Keurentjes et al. 2006; Chan et al. 2010). Recently, metabolome QTL (mQTL) analysis has made a comprehensive understanding of the genetic background of m-traits possible (Keurentjes et al. 2006; Schauer et al. 2006, 2008; Wentzell et al. 2007; Lisec et al. 2008; Rowe et al. 2008; Kliebenstein 2009; Chan et al. 2010). These reports revealed that the mQTLs are unevenly distributed throughout the genome, and there are several mQTL hot-spot regions (Keurentjes et al. 2006; Lisec et al. 2008; Rowe et al. 2008), suggesting the possibility that overall metabolic state of plant cells could be controlled by modification of small genomic regions. The relationship between m-traits and other important biological traits, such as yield, taste and biomass, has been focused on since these traits are likely to closely interact with plant metabolism (Schauer et al. 2006; Meyer et al. 2007; Lisec et al. 2008). The analysis of tomato fruit traits with metabolome data indicated that there are weak correlations among these traits (Schauer et al. 2006, 2008). Regression analysis of metabolome data for *Arabidopsis* biomass traits demonstrated that the growth rate of *Arabidopsis* seedlings is, to some extent, predictable from the metabolome signature (Meyer et al. 2007). In addition, combinatory analyses of mQTL and naturally variable loci that altered circadian clock output suggested possible links between clock function and metabolism in *Arabidopsis* (Kerwin et al. 2011). These pioneering studies suggest that the interaction between metabolite composition and other traits may be predictable through future advanced metabolome analysis. The use of natural variation and mapping populations are also useful to discover new metabolic genes because transformation is still highly difficult in many plant species, except *Arabidopsis* (Morreel et al. 2006; Schilmiller et al. 2010). The accumulation of metabolome data from these lines may be useful for crop improvement through metabolomics-assisted breeding (Fernie and Schauer 2009).

Evaluation of the substantial equivalence of genetically modified organisms

To improve crop properties at the genetic level, two different strategies can be employed: traditional breeding and genetic modification (GM); however, both strategies can generate "unintended phenotypic effects" in metabolome. The public often feel insecure about the use of these strategies, particularly with GMOs, although it is not that surprising considering that we currently cannot target transgenes to specific genomic sites of the host plant, and that little is known about the mechanisms that determine the sites at which integration occurs. In addition, we do not fully understand the function of the non-coding regions of the genome. There are an increasing number of GM crops, such as barley, maize, pea, rice, soybean, and wheat (Ricroch et al. 2011); thus, we have to provide a suitable protocol to evaluate the safety of these GMOs.

A major principle and guiding tool for the food safety assessment of GMOs is the concept of "substantial equivalence" (SE), according to the principles outlined in the Organization for Economic Cooperation and Development (OECD) consensus documents (OECD 2006), and further elaborated by the Food and Agriculture Organization of the United Nations/World Health Organization. The internationally accepted SE framework proposes that risk assessment of GMOs should begin by comparing them with traditional varieties. The goal of SE evaluation is not to draw a conclusion about the novel organism's safety status because that would require the testing of all compounds, which is impossible; instead, by examining a broad set of traits, it aims to obtain a picture of the magnitude of changes to use as a screen for possible problematic changes and a starting point for further investigations. Since metabolome analysis can provide a vast amount of metabolite information, several metabolomic studies have performed an SE evaluation of GM tomato (Le Gall et al. 2003), potato (Catchpole et al. 2005), and wheat (Baker et al. 2006). Recently, we analyzed and evaluated the SE of a transgenic tomato expressing a foreign gene coding for taste-modifying miraculin using a multiple MS-based metabolomics platform (Kusano et al. 2011a). The chosen multiple-MS-based metabolomics platform (GC–MS, CE–MS, and LC–MS) detected compounds that represent 86% of the estimated chemical diversity of the metabolites listed in the LycoCyc database, and showed that >92% had an acceptable range of variation, while simultaneously indicating a reproducible transformation-related metabolic signature.

Application of metabolomics to the investigation of adaptive responses against stresses

Plants possess defense mechanisms against various stresses so they can adapt to unfavorable surroundings. Since the synthesis of specific metabolites is one of the most well-known defense reactions, the investigation of defense reactions with a metabolomics approach can help the holistic understanding of the strategies of plants against stresses. Sunlight is indispensable to all plants, except for several parasitic species, but ultra-violet (UV) light is a harmful element causing the oxidation of DNA and cellular components by the production of free radicals. Metabolome analysis of the UV-B response of wild-type *Arabidopsis* and mutants for flavonoid or sinapoyl malate revealed that the early metabolic responses only occurred at the level of primary metabolites, suggesting that these effectively prime the cell to facilitate the later production of UV-B-absorbing secondary metabolites, such as flavonoid and sinapoyl malate (Kusano et al. 2011c). In addition to abiotic stress, biotic stress is also a serious problem for plants. Ward and coworkers investigated the metabolic changes in *Arabidopsis* associated with the establishment of disease, and demonstrated clear differences in the metabolome of *Arabidopsis* leaves infected with virulent *Pseudomonas syringae* compared to uninfected leaves (Ward et al. 2010b). In addition to the confirmation of changes in phenolic and indolic compounds, they identified rapid alterations in the abundance of amino acids and other nitrogenous compounds, specific classes of glucosinolates, disaccharides, and molecules that influence the prevalence of reactive oxygen species. These reports suggest that comprehensive reprogramming of metabolic pathways is involved in the interaction of plants with stresses. Recently, many studies have examined the overall metabolic changes associated with biotic or abiotic stresses using metabolome analysis (Urano et al. 2009; Widodo et al. 2009; Chaouch et al. 2010; Consonni et al. 2010; Mhamdi et al. 2010; Simon et al. 2010).

The discovery of biomarkers is also important to understand the defense reaction of plants, and it could help to determine the disease condition at the metabolite level, leading to the development of new drugs, early disease screening, and improved pest control. Biomarker discovery by metabolomics has been actively performed in the field of human metabolomics (Soga et al. 2006; Jansson et al. 2009; Koulman et al. 2009; Sreekumar et al. 2009). In such kind of studies, the unambiguous identification or annotation of most of the metabolites is not necessarily required if the discovered biomarkers are truly useful for early disease screening. Thus, the power of the technologies used in metabolomics may directly play an effective role in this field since we still expend considerable time on processing the information derived from many unknown metabolites obtained by metabolome analysis.

OPEN ACCESS

REFERENCES

1. Aharoni A, Ric de Vos CH, Verhoeven HA, Maliepaard CA, Kruppa G, Bino R, Goodenowe DB (2002) Nontargeted metabolome analysis by use of Fourier transform ion cyclotron mass spectrometry. OMICS 6:217–234.
2. Akiyama K, Chikayama E, Yuasa H, Shimada Y, Tohge T, Shinozaki K, Hirai MY, Sakurai T, Kikuchi J, Saito K (2008) PRIMe: a Web site that assembles tools for metabolomics and transcriptomics. In Silico Biol 8:339–345
3. Albinsky D, Kusano M, Higuchi M, Hayashi N, Kobayashi M, Fukushima A, Mori M, Ichikawa T, Matsui K, Kuroda H, Horii Y, Tsumoto Y, Sakakibara H, Hirochika H, Matsui M, Saito K (2010a) Metabolomic screening applied to rice FOX *Arabidopsis* lines leads to the identification of a gene-changing nitrogen metabolism. Mol Plant 3:125–142.
4. Albinsky D, Sawada Y, Kuwahara A, Nagano M, Hirai A, Saito K, Hirai MY (2010b) Widely targeted metabolomics and coexpression analysis as tools to identify genes involved in the side-chain elongation steps of aliphatic glucosinolate biosynthesis. Amino Acids 39:1067–1075.
5. Arabidopsis Genome Initiative (2000) Analysis of the genome sequence of the flowering plant *Arabidopsis thaliana*. Nature 408:796–815.
6. Ausloos P, Clifton CL, Lias SG, Mikaya AI, Stein SE, Tchekhovskoi DV, Sparkman OD, Zaikin V, Zhu D (1999) The critical evaluation of a comprehensive mass spectral library. J Am Soc Mass Spectrom 10:287–299.
7. Bais P, Moon SM, He K, Leitao R, Dreher K, Walk T, Sucaet Y, Barkan L, Wohlgemuth G, Roth MR, Wurtele ES, Dixon P, Fiehn O, Lange BM, Shulaev V, Sumner LW, Welti R, Nikolau BJ, Rhee SY, Dickerson JA (2010) PlantMetabolomics.org: a web portal for plant metabolomics experiments. Plant Physiol 152:1807–1816.
8. Baker JM, Hawkins ND, Ward JL, Lovegrove A, Napier JA, Shewry PR, Beale MH (2006) A metabolomic study of substantial equivalence of field-grown genetically modified wheat. Plant Biotechnol J 4:381–392.
9. Beale MH, Ward JL, Baker JM (2009) Establishing substantial equivalence: metabolomics. Methods Mol Biol 478:289–303.
10. Benton HP, Wong DM, Trauger SA, Siuzdak G (2008) XCMS2: processing tandem mass spectrometry data for metabolite identification and structural characterization. Anal Chem 80:6382–6389.
11. Bino RJ, Hall RD, Fiehn O, Kopka J, Saito K, Draper J, Nikolau BJ, Mendes P, Roessner-Tunali U, Beale MH, Trethewey RN, Lange BM, Wurtele ES, Sumner LW (2004) Potential of metabolomics as a functional genomics tool. Trends Plant Sci 9:418–425.

12. Bocker S, Rasche F (2008) Towards de novo identification of metabolites by analyzing tandem mass spectra. Bioinformatics 24:i49–i55.
13. Bothwell JH, Griffin JL (2011) An introduction to biological nuclear magnetic resonance spectroscopy. Biol Rev Camb Philos Soc 86:493–510.
14. Bottcher C, von Roepenack-Lahaye E, Schmidt J, Schmotz C, Neumann S, Scheel D, Clemens S (2008) Metabolome analysis of biosynthetic mutants reveals a diversity of metabolic changes and allows identification of a large number of new compounds in *Arabidopsis*. Plant Physiol 147:2107–2120.
15. Broeckling CD, Reddy IR, Duran AL, Zhao X, Sumner LW (2006) MET-IDEA: data extraction tool for mass spectrometry-based metabolomics. Anal Chem 78:4334–4341.
16. Butelli E, Titta L, Giorgio M, Mock HP, Matros A, Peterek S, Schijlen EG, Hall RD, Bovy AG, Luo J, Martin C (2008) Enrichment of tomato fruit with health-promoting anthocyanins by expression of select transcription factors. Nat Biotechnol 26:1301–1308.
17. Carroll AJ, Badger MR, Harvey Millar A (2010) The MetabolomeExpress Project: enabling web-based processing, analysis and transparent dissemination of GC/MS metabolomics datasets. BMC Bioinform 11:376.
18. Catchpole GS, Beckmann M, Enot DP, Mondhe M, Zywicki B, Taylor J, Hardy N, Smith A, King RD, Kell DB, Fiehn O, Draper J (2005) Hierarchical metabolomics demonstrates substantial compositional similarity between genetically modified and conventional potato crops. Proc Natl Acad Sci USA 102:14458–14462.
19. Chan EK, Rowe HC, Kliebenstein DJ (2010) Understanding the evolution of defense metabolites in *Arabidopsis thaliana* using genome-wide association mapping. Genetics 185:991–1007.
20. Chaouch S, Queval G, Vanderauwera S, Mhamdi A, Vandorpe M, Langlois-Meurinne M, Van Breusegem F, Saindrenan P, Noctor G (2010) Peroxisomal hydrogen peroxide is coupled to biotic defense responses by ISOCHORISMATE SYNTHASE1 in a daylength-related manner. Plant Physiol 153:1692–1705.
21. Chikayama E, Sekiyama Y, Okamoto M, Nakanishi Y, Tsuboi Y, Akiyama K, Saito K, Shinozaki K, Kikuchi J (2010) Statistical indices for simultaneous large-scale metabolite detections for a single NMR spectrum. Anal Chem 82:1653–1658.
22. Clayton TA, Lindon JC, Cloarec O, Antti H, Charuel C, Hanton G, Provost JP, Le Net JL, Baker D, Walley RJ, Everett JR, Nicholson JK (2006) Pharmaco-metabonomic phenotyping and personalized drug treatment. Nature 440:1073–1077.
23. Codrea MC, Jimenez CR, Heringa J, Marchiori E (2007) Tools for computational processing of LC–MS datasets: a user's perspective. Comput Methods Programs Biomed 86:281–290.
24. Consonni C, Bednarek P, Humphry M, Francocci F, Ferrari S, Harzen A, Ver Loren van Themaat E, Panstruga R (2010) Tryptophan-derived metabolites are required for antifungal defense in the *Arabidopsis* mlo2 mutant. Plant Physiol 152:1544–1561.
25. Coquin L, Feala JD, McCulloch AD, Paternostro G (2008) Metabolomic and flux-balance analysis of age-related decline of hypoxia tolerance in *Drosophila* muscle tissue. Mol Syst Biol 4:233.
26. Cui Q, Lewis IA, Hegeman AD, Anderson ME, Li J, Schulte CF, Westler WM, Eghbalnia HR, Sussman MR, Markley JL (2008) Metabolite identification via the Madison Metabolomics Consortium Database. Nat Biotechnol 26:162–164.
27. De Vos RC, Moco S, Lommen A, Keurentjes JJ, Bino RJ, Hall RD (2007) Untargeted large-scale plant metabolomics using liquid chromatography coupled to mass spectrometry. Nat Protoc 2:778–791.
28. Dixon RA, Strack D (2003) Phytochemistry meets genome analysis, and beyond. Phytochemistry 62:815–816.
29. Dixon RA, Gang DR, Charlton AJ, Fiehn O, Kuiper HA, Reynolds TL, Tjeerdema RS, Jeffery EH, German JB, Ridley WP, Seiber JN (2006) Applications of metabolomics in agriculture. J Agric Food Chem 54:8984–8994.

30. Dunn WB (2008) Current trends and future requirements for the mass spectrometric investigation of microbial, mammalian and plant metabolomes. Phys Biol 5:11001.
31. Eisenreich W, Bacher A (2007) Advances of high-resolution NMR techniques in the structural and metabolic analysis of plant biochemistry. Phytochemistry 68:2799–2815.
32. Fahy E, Sud M, Cotter D, Subramaniam S (2007) LIPID MAPS online tools for lipid research. Nucleic Acids Res 35:W606–W612.
33. Fan TW, Lane AN (2011) NMR-based stable isotope resolved metabolomics in systems biochemistry. J Biomol NMR 49:267–280.
34. Farag MA, Huhman DV, Dixon RA, Sumner LW (2008) Metabolomics reveals novel pathways and differential mechanistic and elicitor-specific responses in phenylpropanoid and isoflavonoid biosynthesis in Medicago truncatula cell cultures. Plant Physiol 146:387–402.
35. Farag MA, Deavours BE, de Fatima A, Naoumkina M, Dixon RA, Sumner LW (2009) Integrated metabolite and transcript profiling identify a biosynthetic mechanism for hispidol in Medicago truncatula cell cultures. Plant Physiol 151:1096–1113.
36. Fernie AR, Schauer N (2009) Metabolomics-assisted breeding: a viable option for crop improvement? Trends Genet 25:39–48.
37. Fernie AR, Trethewey RN, Krotzky AJ, Willmitzer L (2004) Metabolite profiling: from diagnostics to systems biology. Nat Rev Mol Cell Biol 5:763–769.
38. Fiehn O (2002) Metabolomics—the link between g
39. Fiehn O, Kopka J, Dörmann P, Altmann T, Trethewey RN, Willmitzer L (2000) Metabolite profiling for plant functional genomics. Nat Biotechnol 18:1157–1161.
40. Frydman L, Blazina D (2007) Ultrafast two-dimensional nuclear magnetic resonance spectroscopy of hyperpolarized solutions. Nat Phys 3:415–419.
41. Fukusaki E, Kobayashi A (2005) Plant metabolomics: potential for practical operation. J Biosci Bioeng 100:347–354.
42. Fukushima A, Kusano M, Nakamichi N, Kobayashi M, Hayashi N, Sakakibara H, Mizuno T, Saito K (2009a) Impact of clock-associated *Arabidopsis* pseudo-response regulators in metabolic coordination. Proc Natl Acad Sci USA 106:7251–7256.
43. Fukushima A, Kusano M, Redestig H, Arita M, Saito K (2009b) Integrated omics approaches in plant systems biology. Curr Opin Chem Biol 13:532–538.
44. Giavalisco P, Hummel J, Lisec J, Inostroza AC, Catchpole G, Willmitzer L (2008) High-resolution direct infusion-based mass spectrometry in combination with whole 13C metabolome isotope labeling allows unambiguous assignment of chemical sum formulas. Anal Chem 80:9417–9425.
45. Giavalisco P, Kohl K, Hummel J, Seiwert B, Willmitzer L (2009) 13C isotope-labeled metabolomes allowing for improved compound annotation and relative quantification in liquid chromatography–mass spectrometry-based metabolomic research. Anal Chem 81:6546–6551.
46. Gilliland LU, Magallanes-Lundback M, Hemming C, Supplee A, Koornneef M, Bentsink L, Dellapenna D (2006) Genetic basis for natural variation in seed vitamin E levels in *Arabidopsis thaliana*. Proc Natl Acad Sci USA 103:18834–18841.
47. Go EP (2010) Database resources in metabolomics: an overview. J Neuroimmune Pharmacol 5:18–30.
48. Goff SA, Ricke D, Lan TH, Presting G, Wang R, Dunn M, Glazebrook J, Sessions A, Oeller P, Varma H, Hadley D, Hutchison D, Martin C, Katagiri F, Lange BM, Moughamer T, Xia Y, Budworth P, Zhong J, Miguel T, Paszkowski U, Zhang S, Colbert M, Sun WL, Chen L, Cooper B, Park S, Wood TC, Mao L, Quail P, Wing R, Dean R, Yu Y, Zharkikh A, Shen R, Sahasrabudhe S, Thomas A, Cannings R, Gutin A, Pruss D, Reid J, Tavtigian S, Mitchell J, Eldredge G, Scholl T, Miller RM, Bhatnagar S, Adey N, Rubano T, Tusneem N, Robinson R, Feldhaus J, Macalma T, Oliphant A, Briggs S (2002) A draft sequence of the rice genome (*Oryza sativa* L. ssp. *japonica*). Science 296:92–100.

49. Grata E, Boccard J, Glauser G, Carrupt PA, Farmer EE, Wolfender JL, Rudaz S (2007) Development of a two-step screening ESI–TOF–MS method for rapid determination of significant stress-induced metabolome modifications in plant leaf extracts: the wound response in *Arabidopsis thaliana* as a case study. J Sep Sci 30:2268–2278.
50. Griffiths WJ, Koal T, Wang Y, Kohl M, Enot DP, Deigner HP (2010) Targeted metabolomics for biomarker discovery. Angew Chem 49:5426–5445.
51. Hagel JM, Weljie AM, Vogel HJ, Facchini PJ (2008) Quantitative ^{1}H nuclear magnetic resonance metabolite profiling as a functional genomics platform to investigate alkaloid biosynthesis in opium poppy. Plant Physiol 147:1805–1821.
52. Hall RD (2006) Plant metabolomics: from holistic hope, to hype, to hot topic. New Phytol 169:453–468.
53. Harada K, Fukusaki E, Kobayashi A (2006) Pressure-assisted capillary electrophoresis mass spectrometry using combination of polarity reversion and electroosmotic flow for metabolomics anion analysis. J Biosci Bioeng 101:403–409.
54. Hegeman AD, Schulte CF, Cui Q, Lewis IA, Huttlin EL, Eghbalnia H, Harms AC, Ulrich EL, Markley JL, Sussman MR (2007) Stable isotope assisted assignment of elemental compositions for metabolomics. Anal Chem 79:6912–6921.
55. Hirai MY, Yano M, Goodenowe DB, Kanaya S, Kimura T, Awazuhara M, Arita M, Fujiwara T, Saito K (2004) Integration of transcriptomics and metabolomics for understanding of global responses to nutritional stresses in *Arabidopsis thaliana*. Proc Natl Acad Sci USA 101:10205–10210.
56. Hirai MY, Klein M, Fujikawa Y, Yano M, Goodenowe DB, Yamazaki Y, Kanaya S, Nakamura Y, Kitayama M, Suzuki H, Sakurai N, Shibata D, Tokuhisa J, Reichelt M, Gershenzon J, Papenbrock J, Saito K (2005) Elucidation of gene-to-gene and metabolite-to-gene networks in *Arabidopsis* by integration of metabolomics and transcriptomics. J Biol Chem 280:25590–25595.
57. Hirai MY, Sugiyama K, Sawada Y, Tohge T, Obayashi T, Suzuki A, Araki R, Sakurai N, Suzuki H, Aoki K, Goda H, Nishizawa OI, Shibata D, Saito K (2007) Omics-based identification of *Arabidopsis* Myb transcription factors regulating aliphatic glucosinolate biosynthesis. Proc Natl Acad Sci USA 104:6478–6483.
58. Horai H, Arita M, Kanaya S, Nihei Y, Ikeda T, Suwa K, Ojima Y, Tanaka K, Tanaka S, Aoshima K, Oda Y, Kakazu Y, Kusano M, Tohge T, Matsuda F, Sawada Y, Hirai MY, Nakanishi H, Ikeda K, Akimoto N, Maoka T, Takahashi H, Ara T, Sakurai N, Suzuki H, Shibata D, Neumann S, Iida T, Funatsu K, Matsuura F, Soga T, Taguchi R, Saito K, Nishioka T (2010) MassBank: a public repository for sharing mass spectral data for life sciences. J Mass Spectrom 45:703–714.
59. Iijima Y, Nakamura Y, Ogata Y, Tanaka K, Sakurai N, Suda K, Suzuki T, Suzuki H, Okazaki K, Kitayama M, Kanaya S, Aoki K, Shibata D (2008) Metabolite annotations based on the integration of mass spectral information. Plant J 54:949–962.
60. International Rice Genome Sequencing Project (2005) The map-based sequence of the rice genome. Nature 436:793–800.
61. Ishikawa T, Takahara K, Hirabayashi T, Matsumura H, Fujisawa S, Terauchi R, Uchimiya H, Kawai-Yamada M (2010) Metabolome analysis of response to oxidative stress in rice suspension cells overexpressing cell death suppressor Bax inhibitor-1. Plant Cell Physiol 51:9–20.
62. Jansson J, Willing B, Lucio M, Fekete A, Dicksved J, Halfvarson J, Tysk C, Schmitt-Kopplin P (2009) Metabolomics reveals metabolic biomarkers of Crohn's disease. PLoS ONE 4:e6386.
63. Jonsson P, Gullberg J, Nordstrom A, Kusano M, Kowalczyk M, Sjostrom M, Moritz T (2004) A strategy for identifying differences in large series of metabolomic samples analyzed by GC/MS. Anal Chem 76:1738–1745.
64. Kanehisa M, Araki M, Goto S, Hattori M, Hirakawa M, Itoh M, Katayama T, Kawashima S, Okuda S, Tokimatsu T, Yamanishi Y (2008) KEGG for linking genomes to life and the environment. Nucleic Acids Res 36:D480–D484.

65. Katajamaa M, Miettinen J, Oresic M (2006) MZmine: toolbox for processing and visualization of mass spectrometry based molecular profile data. Bioinformatics 22:634–636.
66. Kerwin RE, Jimenez-Gomez JM, Fulop D, Harmer SL, Maloof JN, Kliebenstein DJ (2011) Network quantitative trait loci mapping of circadian clock outputs identifies metabolic pathway-to-clock linkages in *Arabidopsis*. Plant Cell 23:471–485.
67. Keurentjes JJ, Fu J, de Vos CH, Lommen A, Hall RD, Bino RJ, van der Plas LH, Jansen RC, Vreugdenhil D, Koornneef M (2006) The genetics of plant metabolism. Nat Genet 38:842–849.
68. Kikuchi J, Hirayama T (2007) Practical aspects of uniform stable isotope labeling of higher plants for heteronuclear NMR-based metabolomics. Methods Mol Biol 358:273–286
69. Kikuchi J, Shinozaki K, Hirayama T (2004) Stable isotope labeling of *Arabidopsis thaliana* for an NMR-based metabolomics approach. Plant Cell Physiol 45:1099–1104.
70. Kim JK, Bamba T, Harada K, Fukusaki E, Kobayashi A (2007) Time-course metabolic profiling in *Arabidopsis thaliana* cell cultures after salt stress treatment. J Exp Bot 58:415–424.
71. Kim HK, Choi YH, Verpoorte R (2011) NMR-based plant metabolomics: where do we stand, where do we go? Trends Biotechnol.
72. Kind T, Fiehn O (2006) Metabolomic database annotations via query of elemental compositions: mass accuracy is insufficient even at less than 1 ppm. BMC Bioinform 7:234.
73. Kind T, Wohlgemuth G, Lee do Y, Lu Y, Palazoglu M, Shahbaz S, Fiehn O (2009) FiehnLib: mass spectral and retention index libraries for metabolomics based on quadrupole and time-of-flight gas chromatography/mass spectrometry. Anal Chem 81:10038–10048.
74. Kliebenstein D (2009) Advancing genetic theory and application by metabolic quantitative trait loci analysis. Plant Cell 21:1637–1646.
75. Kopka J, Schauer N, Krueger S, Birkemeyer C, Usadel B, Bergmuller E, Dörmann P, Weckwerth W, Gibon Y, Stitt M, Willmitzer L, Fernie AR, Steinhauser D (2005) GMD@CSB.DB: the Golm Metabolome Database. Bioinformatics 21:1635–1638.
76. Koulman A, Lane GA, Harrison SJ, Volmer DA (2009) From differentiating metabolites to biomarkers. Anal Bioanal Chem 394:663–670.
77. Kovacs H, Moskau D, Spraul M (2005) Cryogenically cooled probes—a leap in NMR technology. Prog Nucl Magn Reson Spectrosc 46:131–155.
78. Krishnan P, Kruger NJ, Ratcliffe RG (2005) Metabolite fingerprinting and profiling in plants using NMR. J Exp Bot 56:255–265.
79. Ku KM, Choi JN, Kim J, Kim JK, Yoo LG, Lee SJ, Hong YS, Lee CH (2010a) Metabolomics analysis reveals the compositional differences of shade grown tea (*Camellia sinensis* L.). J Agric Food Chem 58:418–426.
80. Ku KM, Kim J, Park HJ, Liu KH, Lee CH (2010b) Application of metabolomics in the analysis of manufacturing type of pu-erh tea and composition changes with different postfermentation year. J Agric Food Chem 58:345–352.
81. Kusano M, Fukushima A, Kobayashi M, Hayashi N, Jonsson P, Moritz T, Ebana K, Saito K (2007) Application of a metabolomic method combining one-dimensional and two-dimensional gas chromatography-time-of-flight/mass spectrometry to metabolic phenotyping of natural variants in rice. J Chromatogr B Analyt Technol Biomed Life Sci 855:71–79.
82. Kusano M, Redestig H, Hirai T, Oikawa A, Matsuda F, Fukushima A, Arita M, Watanabe S, Yano M, Hiwasa-Tanase K, Ezura H, Saito K (2011a) Covering chemical diversity of genetically-modified tomatoes using metabolomics for objective substantial equivalence assessment. PLoS ONE 6:e16989.
83. Kusano M, Tabuchi M, Fukushima A, Funayama K, Diaz C, Kobayashi M, Hayashi N, Tsuchiya YN, Takahashi H, Kamata A, Yamaya T, Saito K (2011b) Metabolomics data

reveal a crucial role of cytosolic glutamine synthetase 1;1 in coordinating metabolic balance in rice. Plant J 66:456–466.

84. Kusano M, Tohge T, Fukushima A, Kobayashi M, Hayashi N, Otsuki H, Kondou Y, Goto H, Kawashima M, Matsuda F, Niida R, Matsui M, Saito K, Fernie AR (2011c) Metabolomics reveals comprehensive reprogramming involving two independent metabolic responses of *Arabidopsis* to ultraviolet-B light. Plant J.

85. Kuzina V, Ekstrom CT, Andersen SB, Nielsen JK, Olsen CE, Bak S (2009) Identification of defense compounds in Barbarea vulgaris against the herbivore Phyllotreta nemorum by an ecometabolomic approach. Plant Physiol 151:1977–1990.

86. Le Gall G, Colquhoun IJ, Davis AL, Collins GJ, Verhoeyen ME (2003) Metabolite profiling of tomato (*Lycopersicon esculentum*) using ^{1}H NMR spectroscopy as a tool to detect potential unintended effects following a genetic modification. J Agric Food Chem 51:2447–2456.

87. Leiss KA, Maltese F, Choi YH, Verpoorte R, Klinkhamer PG (2009) Identification of chlorogenic acid as a resistance factor for thrips in *Chrysanthemum*. Plant Physiol 150:1567–1575.

88. Lindon JC, Holmes E, Nicholson JK (2004) Metabonomics: systems biology in pharmaceutical research and development. Curr Opin Mol Ther 6:265–272\

89. Lisec J, Meyer RC, Steinfath M, Redestig H, Becher M, Witucka-Wall H, Fiehn O, Torjek O, Selbig J, Altmann T, Willmitzer L (2008) Identification of metabolic and biomass QTL in *Arabidopsis thaliana* in a parallel analysis of RIL and IL populations. Plant J 53:960–972.

90. Lucker J, Bouwmeester HJ, Schwab W, Blaas J, van der Plas LH, Verhoeven HA (2001) Expression of Clarkia *S*-linalool synthase in transgenic petunia plants results in the accumulation of *S*-linalyl-beta-D-glucopyranoside. Plant J 27:315–324.

91. Markley JL, Anderson ME, Cui Q, Eghbalnia HR, Lewis IA, Hegeman AD, Li J, Schulte CF, Sussman MR, Westler WM, Ulrich EL, Zolnai Z (2007) New bioinformatics resources for metabolomics. Pac Symp Biocomput 12:157–168

92. Matsuda F, Yonekura-Sakakibara K, Niida R, Kuromori T, Shinozaki K, Saito K (2009) MS/MS spectral tag-based annotation of non-targeted profile of plant secondary metabolites. Plant J 57:555–577.

93. Matsuda F, Hirai MY, Sasaki E, Akiyama K, Yonekura-Sakakibara K, Provart NJ, Sakurai T, Shimada Y, Saito K (2010) AtMetExpress development: a phytochemical atlas of *Arabidopsis* development. Plant Physiol 152:566–578.

94. Mesnard F, Ratcliffe RG (2005) NMR analysis of plant nitrogen metabolism. Photosynth Res 83:163–180.

95. Meyer RC, Steinfath M, Lisec J, Becher M, Witucka-Wall H, Torjek O, Fiehn O, Eckardt A, Willmitzer L, Selbig J, Altmann T (2007) The metabolic signature related to high plant growth rate in *Arabidopsis thaliana*. Proc Natl Acad Sci USA 104:4759–4764.

96. Mhamdi A, Mauve C, Gouia H, Saindrenan P, Hodges M, Noctor G (2010) Cytosolic NADP-dependent isocitrate dehydrogenase contributes to redox homeostasis and the regulation of pathogen responses in *Arabidopsis* leaves. Plant Cell Environ 33:1112–1123.

97. Minami Y, Kasukawa T, Kakazu Y, Iigo M, Sugimoto M, Ikeda S, Yasui A, van der Horst GT, Soga T, Ueda HR (2009) Measurement of internal body time by blood metabolomics. Proc Natl Acad Sci USA 106:9890–9895.

98. Miura D, Tsuji Y, Takahashi K, Wariishi H, Saito K (2010) A strategy for the determination of the elemental composition by Fourier transform ion cyclotron resonance mass spectrometry based on isotopic peak ratios. Anal Chem 82:5887–5891.

99. Moco S, Bino RJ, Vorst O, Verhoeven HA, de Groot J, van Beek TA, Vervoort J, de Vos CH (2006) A liquid chromatography-mass spectrometry-based metabolome database for tomato. Plant Physiol 141:1205–1218.

100. Moco S, Capanoglu E, Tikunov Y, Bino RJ, Boyacioglu D, Hall RD, Vervoort J, De Vos RC (2007) Tissue specialization at the metabolite level is perceived during the development of tomato fruit. J Exp Bot 58:4131–4146.

101. Monton MR, Soga T (2007) Metabolome analysis by capillary electrophoresis-mass spectrometry. J Chromatogr A 1168:237–246.
102. Morreel K, Goeminne G, Storme V, Sterck L, Ralph J, Coppieters W, Breyne P, Steenackers M, Georges M, Messens E, Boerjan W (2006) Genetical metabolomics of flavonoid biosynthesis in *Populus*: a case study. Plant J 47:224–237.
103. Moseley HN, Lane AN, Belshoff AC, Higashi RM, Fan TW (2011) A novel deconvolution method for modeling UDP-GlcNAc biosynthetic pathways based on 13C mass isotopologue profiles under non steady-state conditions. BMC Biol 9:37.
104. Mungur R, Glass AD, Goodenow DB, Lightfoot DA (2005) Metabolite fingerprinting in transgenic *Nicotiana tabacum* altered by the *Escherichia coli* glutamate dehydrogenase gene. J Biomed Biotechnol 2005:198–214.
105. Nakabayashi R, Kusano M, Kobayashi M, Tohge T, Yonekura-Sakakibara K, Kogure N, Yamazaki M, Kitajima M, Saito K, Takayama H (2009) Metabolomics-oriented isolation and structure elucidation of 37 compounds including two anthocyanins from *Arabidopsis thaliana*. Phytochemistry 70:1017–1029.
106. Nakabayashi R, Yamazaki M, Saito K (2010) A polyhedral approach for understanding flavonoid biosynthesis in *Arabidopsis*. N Biotechnol 27:829–836.
107. Naoumkina M, Farag MA, Sumner LW, Tang Y, Liu CJ, Dixon RA (2007) Different mechanisms for phytoalexin induction by pathogen and wound signals in *Medicago truncatula*. Proc Natl Acad Sci USA 104:17909–17915.
108. OECD (2006) An Introduction to the Food/Feed Safety Consensus Documents of the Task Force. Series on the Safety of Novel Foods and Feeds, No 14. Organization for Economic Cooperation and Development, Paris, pp 7–9
109. Ohkama-Ohtsu N, Oikawa A, Zhao P, Xiang C, Saito K, Oliver DJ (2008) A gamma-glutamyl transpeptidase-independent pathway of glutathione catabolism to glutamate via 5-oxoproline in *Arabidopsis*. Plant Physiol 148:1603–1613.
110. Ohkama-Ohtsu N, Sasaki-Sekimoto Y, Oikawa A, Jikumaru Y, Shinoda S, Inoue E, Kamide Y, Yokoyama T, Hirai MY, Shirasu K, Kamiya Y, Oliver DJ, Saito K (2011) 12-oxo-phytodienoic acid–glutathione conjugate is transported into the vacuole in *Arabidopsis*. Plant Cell Physiol 52:205–209.
111. Oikawa A, Nakamura Y, Ogura T, Kimura A, Suzuki H, Sakurai N, Shinbo Y, Shibata D, Kanaya S, Ohta D (2006) Clarification of pathway-specific inhibition by Fourier transform ion cyclotron resonance/mass spectrometry-based metabolic phenotyping studies. Plant Physiol 142:398–413.
112. Oikawa A, Matsuda F, Kusano M, Okazaki Y, Saito K (2008) Rice metabolomics. Rice 1:63–71.
113. Okada T, Afendi FM, Altaf-Ul-Amin M, Takahashi H, Nakamura K, Kanaya S (2010) Metabolomics of medicinal plants: the importance of multivariate analysis of analytical chemistry data. Curr Comput Aided Drug Des 6:179–196
114. Okazaki Y, Shimojima M, Sawada Y, Toyooka K, Narisawa T, Mochida K, Tanaka H, Matsuda F, Hirai A, Hirai MY, Ohta H, Saito K (2009) A chloroplastic UDP-glucose pyrophosphorylase from *Arabidopsis* is the committed enzyme for the first step of sulfolipid biosynthesis. Plant Cell 21:892–909.
115. Okazaki Y, Kamide Y, Hirai MY, Saito K (2011) Plant lipidomics based on hydrophilic interaction chromatography coupled to ion trap time-of-flight mass spectrometry. Metabolomics (in press).
116. Oksman-Caldentey KM, Saito K (2005) Integrating genomics and metabolomics for engineering plant metabolic pathways. Curr Opin Biotechnol 16:174–179.
117. Parveen I, Moorby JM, Fraser MD, Allison GG, Kopka J (2007) Application of gas chromatography–mass spectrometry metabolite profiling techniques to the analysis of heathland plant diets of sheep. J Agric Food Chem 55:1129–1138.
118. Quackenbush J (2002) Extracting biology from high-dimensional biological data. J Exp Biol 48:1507–1517.
119. Ralston-Hooper K, Hopf A, Oh C, Zhang X, Adamec J, Sepulveda MS (2008) Development of GC×GC/TOF–MS metabolomics for use in ecotoxicological studies with invertebrates. Aquat Toxicol 88:48–52.

120. Ramautar R, Somsen GW, de Jong GJ (2009) CE-MS in metabolomics. Electrophoresis 30:276–291.
121. Ratcliffe RG, Shachar-Hill Y (2006) Measuring multiple fluxes through plant metabolic networks. Plant J 45:490–511.
122. Redestig H, Kusano M, Fukushima A, Matsuda F, Saito K, Arita M (2010) Consolidating metabolite identifiers to enable contextual and multi-platform metabolomics data analysis. BMC Bioinform 11:214.
123. Ricroch AE, Berge JB, Kuntz M (2011) Evaluation of genetically engineered crops using transcriptomic, proteomic, and metabolomic profiling techniques. Plant Physiol 155:1752–1761.
124. Roessner U, Luedemann A, Brust D, Fiehn O, Linke T, Willmitzer L, Fernie A (2001) Metabolic profiling allows comprehensive phenotyping of genetically or environmentally modified plant systems. Plant Cell 13:11–29.
125. Rowe HC, Hansen BG, Halkier BA, Kliebenstein DJ (2008) Biochemical networks and epistasis shape the *Arabidopsis thaliana* metabolome. Plant Cell 20:1199–1216.
126. Saito K, Matsuda F (2010) Metabolomics for functional genomics, systems biology, and biotechnology. Annu Rev Plant Biol 61:463–489.
127. Saito K, Hirai MY, Yonekura-Sakakibara K (2008) Decoding genes with coexpression networks and metabolomics—'majority report by precogs'. Trends Plant Sci 13:36–43.
128. Sakurai N, Ara T, Ogata Y, Sano R, Ohno T, Sugiyama K, Hiruta A, Yamazaki K, Yano K, Aoki K, Aharoni A, Hamada K, Yokoyama K, Kawamura S, Otsuka H, Tokimatsu T, Kanehisa M, Suzuki H, Saito K, Shibata D (2011) KaPPA-View4: a metabolic pathway database for representation and analysis of correlation networks of gene co-expression and metabolite co-accumulation and omics data. Nucleic Acids Res 39:D677–D684.
129. Sasaki T, Matsumoto T, Yamamoto K, Sakata K, Baba T, Katayose Y, Wu J, Niimura Y, Cheng Z, Nagamura Y, Antonio BA, Kanamori H, Hosokawa S, Masukawa M, Arikawa K, Chiden Y, Hayashi M, Okamoto M, Ando T, Aoki H, Arita K, Hamada M, Harada C, Hijishita S, Honda M, Ichikawa Y, Idonuma A, Iijima M, Ikeda M, Ikeno M, Ito S, Ito T, Ito Y, Iwabuchi A, Kamiya K, Karasawa W, Katagiri S, Kikuta A, Kobayashi N, Kono I, Machita K, Maehara T, Mizuno H, Mizubayashi T, Mukai Y, Nagasaki H, Nakashima M, Nakama Y, Nakamichi Y, Nakamura M, Namiki N, Negishi M, Ohta I, Ono N, Saji S, Sakai K, Shibata M, Shimokawa T, Shomura A, Song J, Takazaki Y, Terasawa K, Tsuji K, Waki K, Yamagata H, Yamane H, Yoshiki S, Yoshihara R, Yukawa K, Zhong H, Iwama H, Endo T, Ito H, Hahn JH, Kim HI, Eun MY, Yano M, Jiang J, Gojobori T (2002) The genome sequence and structure of rice chromosome 1. Nature 420:312–316.
130. Sato S, Arita M, Soga T, Nishioka T, Tomita M (2008) Time-resolved metabolomics reveals metabolic modulation in rice foliage. BMC Syst Biol 2:51.
131. Sawada Y, Akiyama K, Sakata A, Kuwahara A, Otsuki H, Sakurai T, Saito K, Hirai MY (2009) Widely targeted metabolomics based on large-scale MS/MS data for elucidating metabolite accumulation patterns in plants. Plant Cell Physiol 50:37–47.
132. Schauer N, Semel Y, Roessner U, Gur A, Balbo I, Carrari F, Pleban T, Perez-Melis A, Bruedigam C, Kopka J, Willmitzer L, Zamir D, Fernie AR (2006) Comprehensive metabolic profiling and phenotyping of interspecific introgression lines for tomato improvement. Nat Biotechnol 24:447–454.
133. Schauer N, Semel Y, Balbo I, Steinfath M, Repsilber D, Selbig J, Pleban T, Zamir D, Fernie AR (2008) Mode of inheritance of primary metabolic traits in tomato. Plant Cell 20:509–523.
134. Schilmiller A, Shi F, Kim J, Charbonneau AL, Holmes D, Daniel Jones A, Last RL (2010) Mass spectrometry screening reveals widespread diversity in trichome specialized metabolites of tomato chromosomal substitution lines. Plant J 62:391–403.
135. Sekiyama Y, Kikuchi J (2007) Towards dynamic metabolic network measurements by multi-dimensional NMR-based fluxomics. Phytochemistry 68:2320–2329.

136. Sekiyama Y, Chikayama E, Kikuchi J (2010) Profiling polar and semipolar plant metabolites throughout extraction processes using a combined solution-state and high-resolution magic angle spinning NMR approach. Anal Chem 82:1643–1652.

137. Shellie RA, Welthagen W, Zrostlikova J, Spranger J, Ristow M, Fiehn O, Zimmermann R (2005) Statistical methods for comparing comprehensive two-dimensional gas chromatography-time-of-flight mass spectrometry results: metabolomic analysis of mouse tissue extracts. J Chromatogr A 1086:83–90.

138. Shinbo Y, Nakamura Y, Altaf-Ul-Amin M, Asahi H, Kurokawa K, Arita M, Saito K, Ohta D, Shibata D, Kanaya S (2006) KNApSAcK: a comprehensive species-metabolite relationship database. In: Saito K, Dixon R, Willmitzer L (eds) Biotechnology in agriculture and forestry. Springer, Heidelberg, pp 165–184

139. Simon C, Langlois-Meurinne M, Bellvert F, Garmier M, Didierlaurent L, Massoud K, Chaouch S, Marie A, Bodo B, Kauffmann S, Noctor G, Saindrenan P (2010) The differential spatial distribution of secondary metabolites in *Arabidopsis* leaves reacting hypersensitively to Pseudomonas syringae pv tomato is dependent on the oxidative burst. J Exp Bot 61:3355–3370.

140. Smith CA, O'Maille G, Want EJ, Qin C, Trauger SA, Brandon TR, Custodio DE, Abagyan R, Siuzdak G (2005) METLIN: a metabolite mass spectral database. Ther Drug Monit 27:747–751

141. Smith CA, Want EJ, O'Maille G, Abagyan R, Siuzdak G (2006) XCMS: processing mass spectrometry data for metabolite profiling using nonlinear peak alignment, matching, and identification. Anal Chem 78:779–787.

142. Soga T, Ueno Y, Naraoka H, Matsuda K, Tomita M, Nishioka T (2002) Pressure-assisted capillary electrophoresis electrospray ionization mass spectrometry for analysis of multivalent anions. Anal Chem 74:6224–6229.

143. Soga T, Baran R, Suematsu M, Ueno Y, Ikeda S, Sakurakawa T, Kakazu Y, Ishikawa T, Robert M, Nishioka T, Tomita M (2006) Differential metabolomics reveals ophthalmic acid as an oxidative stress biomarker indicating hepatic glutathione consumption. J Biol Chem 281:16768–16776.

144. Soga T, Igarashi K, Ito C, Mizobuchi K, Zimmermann HP, Tomita M (2009) Metabolomic profiling of anionic metabolites by capillary electrophoresis mass spectrometry. Anal Chem 81:6165–6174.

145. Sreekumar A, Poisson LM, Rajendiran TM, Khan AP, Cao Q, Yu J, Laxman B, Mehra R, Lonigro RJ, Li Y, Nyati MK, Ahsan A, Kalyana-Sundaram S, Han B, Cao X, Byun J, Omenn GS, Ghosh D, Pennathur S, Alexander DC, Berger A, Shuster JR, Wei JT, Varambally S, Beecher C, Chinnaiyan AM (2009) Metabolomic profiles delineate potential role for sarcosine in prostate cancer progression. Nature 457:910–914.

146. Suzuki H, Sasaki R, Ogata Y, Nakamura Y, Sakurai N, Kitajima M, Takayama H, Kanaya S, Aoki K, Shibata D, Saito K (2008) Metabolic profiling of flavonoids in *Lotus japonicus* using liquid chromatography Fourier transform ion cyclotron resonance mass spectrometry. Phytochemistry 69:99–111.

147. Taguchi R, Nishijima M, Shimizu T (2007) Basic analytical systems for lipidomics by mass spectrometry in Japan. Methods Enzymol 432:185–211.

148. Takahashi H, Hotta Y, Hayashi M, Kawai-Yamada M, Komatsu S, Uchiyama H (2005) High throughput metabolome and proteome analysis of transgenic rice plants (*Oryza sativa* L.). Plant Biotechnol 22:47–50.

149. Tarpley L, Duran AL, Kebrom TH, Sumner LW (2005) Biomarker metabolites capturing the metabolite variance present in a rice plant developmental period. BMC Plant Biol 5:8.

150. Thimm O, Blasing O, Gibon Y, Nagel A, Meyer S, Kruger P, Selbig J, Muller LA, Rhee SY, Stitt M (2004) MAPMAN: a user-driven tool to display genomics data sets onto diagrams of metabolic pathways and other biological processes. Plant J 37:914–939.

151. Tian C, Chikayama E, Tsuboi Y, Kuromori T, Shinozaki K, Kikuchi J, Hirayama T (2007) Top-down phenomics of *Arabidopsis thaliana*: metabolic profiling by one- and two-dimensional nuclear magnetic resonance spectroscopy and transcriptome analysis of albino mutants. J Biol Chem 282:18532–18541.

152. Tohge T, Fernie AR (2009) Web-based resources for mass-spectrometry-based metabolomics: a user's guide. Phytochemistry 70:450–456.
153. Tohge T, Fernie AR (2010) Combining genetic diversity, informatics and metabolomics to facilitate annotation of plant gene function. Nat Protoc 5:1210–1227.
154. Tohge T, Nishiyama Y, Hirai MY, Yano M, Nakajima J, Awazuhara M, Inoue E, Takahashi H, Goodenowe DB, Kitayama M, Noji M, Yamazaki M, Saito K (2005) Functional genomics by integrated analysis of metabolome and transcriptome of *Arabidopsis* plants over-expressing an MYB transcription factor. Plant J 42:218–235.
155. Tokimatsu T, Sakurai N, Suzuki H, Ohta H, Nishitani K, Koyama T, Umezawa T, Misawa N, Saito K, Shibata D (2005) KaPPA-view: a web-based analysis tool for integration of transcript and metabolite data on plant metabolic pathway maps. Plant Physiol 138:1289–1300.
156. Urano K, Maruyama K, Ogata Y, Morishita Y, Takeda M, Sakurai N, Suzuki H, Saito K, Shibata D, Kobayashi M, Yamaguchi-Shinozaki K, Shinozaki K (2009) Characterization of the ABA-regulated global responses to dehydration in *Arabidopsis* by metabolomics. Plant J 57:1065–1078.
157. Van QN, Veenstra TD (2009) How close is the bench to the bedside? Metabolic profiling in cancer research. Genome Med 1:5.
158. Vorst O, de Vos CH, Lommen A, Staps RV, Visser RG, Bino RJ, Hall RD (2005) A non-directed approach to the differential analysis of multiple LC–MS-derived metabolic profiles. Metabolomics 1:169–180.
159. Wang Y, Holmes E, Nicholson JK, Cloarec O, Chollet J, Tanner M, Singer BH, Utzinger J (2004) Metabonomic investigations in mice infected with *Schistosoma mansoni*: an approach for biomarker identification. Proc Natl Acad Sci USA 101:12676–12681.
160. Ward JL, Baker JM, Beale MH (2007) Recent applications of NMR spectroscopy in plant metabolomics. FEBS J 274:1126–1131.
161. Ward JL, Baker JM, Miller SJ, Deborde C, Maucourt M, Biais B, Rolin D, Moing A, Moco S, Vervoort J, Lommen A, Schafer H, Humpfer E, Beale MH (2010a) An inter-laboratory comparison demonstrates that [^{1}H]-NMR metabolite fingerprinting is a robust technique for collaborative plant metabolomic data collection. Metabolomics 6:263–273.
162. Ward JL, Forcat S, Beckmann M, Bennett M, Miller SJ, Baker JM, Hawkins ND, Vermeer CP, Lu C, Lin W, Truman WM, Beale MH, Draper J, Mansfield JW, Grant M (2010b) The metabolic transition during disease following infection of *Arabidopsis thaliana* by *Pseudomonas syringae* pv. tomato. Plant J.
163. Ward JL, Baker JM, Llewellyn AM, Hawkins ND, Beale MH (2011) Metabolomic analysis of *Arabidopsis* reveals hemiterpenoid glycosides as products of a nitrate ion-regulated, carbon flux overflow. Proc Natl Acad Sci USA.
164. Watanabe M, Kusano M, Oikawa A, Fukushima A, Noji M, Saito K (2008) Physiological roles of the beta-substituted alanine synthase gene family in *Arabidopsis*. Plant Physiol 146:310–320.
165. Wentzell AM, Rowe HC, Hansen BG, Ticconi C, Halkier BA, Kliebenstein DJ (2007) Linking metabolic QTLs with network and cis-eQTLs controlling biosynthetic pathways. PLoS Genet 3:1687–1701.
166. Wheeler DL, Barrett T, Benson DA, Bryant SH, Canese K, Chetvernin V, Church DM, Dicuccio M, Edgar R, Federhen S, Feolo M, Geer LY, Helmberg W, Kapustin Y, Khovayko O, Landsman D, Lipman DJ, Madden TL, Maglott DR, Miller V, Ostell J, Pruitt KD, Schuler GD, Shumway M, Sequeira E, Sherry ST, Sirotkin K, Souvorov A, Starchenko G, Tatusov RL, Tatusova TA, Wagner L, Yaschenko E (2008) Database resources of the National Center for Biotechnology Information. Nucleic Acids Res 36:D13–D21.
167. Widodo, Patterson JH, Newbigin E, Tester M, Bacic A, Roessner U (2009) Metabolic responses to salt stress of barley (*Hordeum vulgare* L.) cultivars, Sahara and Clipper, which differ in salinity tolerance. J Exp Bot 60:4089–4103.

168. Wishart DS, Knox C, Guo AC, Eisner R, Young N, Gautam B, Hau DD, Psychogios N, Dong E, Bouatra S, Mandal R, Sinelnikov I, Xia J, Jia L, Cruz JA, Lim E, Sobsey CA, Shrivastava S, Huang P, Liu P, Fang L, Peng J, Fradette R, Cheng D, Tzur D, Clements M, Lewis A, De Souza A, Zuniga A, Dawe M, Xiong Y, Clive D, Greiner R, Nazyrova A, Shaykhutdinov R, Li L, Vogel HJ, Forsythe I (2009) HMDB: a knowledgebase for the human metabolome. Nucleic Acids Res 37:D603–D610.

169. Yasugi E, Watanabe K (2002) LIPIDBANK for Web, the newly developed lipid database. Tanpakushitsu Kakusan Koso 47:837–841

170. Ye X, Al-Babili S, Kloti A, Zhang J, Lucca P, Beyer P, Potrykus I (2000) Engineering the provitamin A (β-carotene) biosynthetic pathway into (carotenoid-free) rice endosperm. Science 287:303–305.

171. Yonekura-Sakakibara K, Saito K (2009) Functional genomics for plant natural product biosynthesis. Nat Prod Rep 26:1466–1487.

172. Yonekura-Sakakibara K, Tohge T, Niida R, Saito K (2007) Identification of a flavonol 7-*O*-rhamnosyltransferase gene determining flavonoid pattern in *Arabidopsis* by transcriptome coexpression analysis and reverse genetics. J Biol Chem 282:14932–14941.

173. Yonekura-Sakakibara K, Tohge T, Matsuda F, Nakabayashi R, Takayama H, Niida R, Watanabe-Takahashi A, Inoue E, Saito K (2008) Comprehensive flavonol profiling and transcriptome coexpression analysis leading to decoding gene-metabolite correlations in *Arabidopsis*. Plant Cell 20:2160–2176.

174. Yu J, Hu S, Wang J, Wong GK, Li S, Liu B, Deng Y, Dai L, Zhou Y, Zhang X, Cao M, Liu J, Sun J, Tang J, Chen Y, Huang X, Lin W, Ye C, Tong W, Cong L, Geng J, Han Y, Li L, Li W, Hu G, Li J, Liu Z, Qi Q, Li T, Wang X, Lu H, Wu T, Zhu M, Ni P, Han H, Dong W, Ren X, Feng X, Cui P, Li X, Wang H, Xu X, Zhai W, Xu Z, Zhang J, He S, Xu J, Zhang K, Zheng X, Dong J, Zeng W, Tao L, Ye J, Tan J, Chen X, He J, Liu D, Tian W, Tian C, Xia H, Bao Q, Li G, Gao H, Cao T, Zhao W, Li P, Chen W, Zhang Y, Hu J, Liu S, Yang J, Zhang G, Xiong Y, Li Z, Mao L, Zhou C, Zhu Z, Chen R, Hao B, Zheng W, Chen S, Guo W, Tao M, Zhu L, Yuan L, Yang H (2002) A draft sequence of the rice genome (*Oryza sativa* L. ssp. *indica*). Science 296:79–92.

CHAPTER 7

Biotechnological Advancements for Improving Floral Attributes in Ornamental Plants

Ali Noman[1,2], Muhammad Aqeel[3*], Jianming Deng[3], Noreen Khalid[4], Tayyaba Sanaullah[5] and He Shuilin[1,6*]

[1]*College of Crop Science, Fujian Agriculture and Forestry University, Fuzhou, China*

[2]*Department of Botany, Government College University, Faisalabad, Pakistan*

[3]*State Key Laboratory of Grassland Agro-Ecosystems, School of Life Science, Lanzhou University, Lanzhou, China*

[4]*Department of Botany, Government College Women University Sialkot, Sialkot, Pakistan*

[5]*Department of Botany, University of Agriculture, Faisalabad, Pakistan*

[6]*National Education Minister, Key Laboratory of Plant Genetic Improvement and Comprehensive Utilization, Fujian Agriculture and Forestry University, Fuzhou, China*

ABSTRACT

Developing new ornamental cultivars with improved floral attributes is a major goal in floriculture. Biotechnological approach together with classical breeding methods has been used to modify floral color, appearance as well as for increasing disease resistance. Transgenic strategies possess immense potential to produce novel flower phenotypes that are not found in nature. Adoption of Genetic engineering has supported the idea of floral trait modification. Ornamental plant attributes like floral color, fragrance, disease resistance, and vase life can be improved by means of genetic manipulation. Therefore, we witness transgenic plant varieties of high aesthetic and commercial value. This review focuses on biotechnological advancements in manipulating key floral traits that contribute in development of diverse ornamental plant lines. Data clearly reveals that regulation of biosynthetic pathways related to characteristics like pigment production, flower morphology and fragrance is both possible and predictable. In spite of their great significance, small number of genetically engineered varieties of ornamental plants has been field tested. Today, novel flower colors production is regarded as chief commercial benefit obtained from transgenic plants. But certain other floral traits are much more important and have high commercial potential. Other than achievements such as novel architecture, modified flower color, etc., very few reports are available regarding successful transformation of other valuable horticultural characteristics. Our review also summarized biotechnological efforts related to enhancement of fragrance and induction of early flowering along with changes in floral anatomy and morphology.

Keywords: biotechnology, commercial resource, environment, flower characteristics, horticulture, transgenic plants

INTRODUCTION

Unequivocally, horticultural industry has been revolutionized due to contribution by ornamental plants. Now a day, diverse ornamental plants are being widely used in home gardening, professional landscaping, and cut flowers as well (Dobres, 2011). Ornamental plant products are globally traded commodities. Due to rising needs, ornamental plant industry requires new plant varieties with elite traits such as improved anatomical attributes, floral color, pigments, stress tolerance, and disease resistance (Chandler and Sanchez, 2012; Azadi et al., 2016). Although we witness extensive employment of classical breeding strategies for developing new plant lines yet limitations and draw backs are also evident i.e., degree of heterozygosity (Shibata, 2008; Da Silva et al., 2011). Since inception of last decade, techniques like genetic engineering (GE), genome editing has been broadly adopted as more feasible methods to deal with intrinsic obstacles of classical techniques (Noman et al., 2016a). Global GM crop cultivation has touched its acme during the last few years (Noman et al., 2016b). Situation can be imagined from the in hand data that reveals GM crop cultivation reached 181.5 Mha in 2014 (Azadi et al., 2016).

Interest as well as contribution of private and government sector toward biotechnology and genetic engineering is increasing day by day. Over the years the main targets were food and feed along with improvements in herbicide and pesticide tolerance. Recently, scientists have also focused on improvement and enhancement of quality attributes for industry (Noman et al., 2016b, 2017; Parisi et al., 2016). The prime benefit in adopting GE is that gene from other species gene pool can be introduced in ornamental plants (Li and Pei, 2006; Chandler and Brugliera, 2011). So it is very possible to introduce genes for disease resistance and stress tolerance in ornamental plant species (Auer, 2008; Kamthan et al., 2016). Similarly, plant characteristics like floral architecture, color, fragrance, resistance to abiotic stress and post-harvest life can be addressed through GE.

This century is considered as the era of bio-economy lead by bioscience and biotechnology. This bio-economy is directly linked to sustainable developments in core areas of agriculture, environment and economy (Huang, 2011). Today, work is going on to produce GM flowers with broad color range and other attributes. Transgenic ornamental plants may become prospective benefit to growers and consumers due to their changed floral appearance, novel colors and improved fragrance (Chandler and Sanchez, 2012). Instead of their immense value, small number of GM ornamental plant varieties have been field tested and released. So far ornamental varieties released in market are mostly the color variant plant varieties e. g., rose (Tanaka and Brugliera, 2013). Therefore, in this review we have highlighted recent advancements in application of GE and biotechnology upon ornamental plants. We have tried to point out developments and requisite attention for far reaching benefits for sustainability of technology and society.

WHY FLORAL TRAITS MODIFICATIONS ARE OF SPECIAL CONCERN?

Ornamental plants generally possess extraordinarily beautiful and eye catching flowers. Other than aesthetic value, floral traits are crucial for plant survival. These characteristics like shape, shape, fragrance and color have their individual, and collaborative significance. Besides ornamental value, these flowers are utilized in pharmaceutical and other industries. For example, rose plants possess valuable secondary metabolites that are used in production of cosmetics, perfume etc. (Feng et al., 2010). Therefore, floral characteristics have unique importance for plant genetic engineers. Transgenic strategies possess prospective potential for producing innovative flower phenotypes that are not found in nature (Table 1). The genome based modifications for flowers have potential to yield utmost benefits in different aspects. Today, work is going on to produce GM flowers with multiple colors and rest of attributes. Normally, ornamental plants face problems because of troublesome sexual hybridization. This is mainly due to high heterozygosity, high chromosome number, inadequate gene pool, and elevated sterility (Van der Salm et al., 1998; Kim et al., 2006). For example, being alloploid chrysanthemum has chromosome number 36–75 rather than basic chromosome number 9. *Anthurium* sp. have life cycle of about 3 years. Therefore, developing new cultivar may require long time span of about 8–10 years (Azadi et al., 2016). Similarly, most of the carnation lines are self-fertile and unable to produce seeds (Nontaswatsri and Fukai, 2010). Huge genome size in ornamental plants e.g., lilly is a hurdle in mining genomic information (Du et al., 2015). Moreover, other than genome issues, presence or absence of certain metabolites leads to floral changes. One of the reported issues is precocious petal pigmentation and increased sepals pigmentation in *Lisianthus* sp. (Schwinn et al., 2014). Now we will evaluate role of biotechnology and GE for modifications in diverse floral characteristics.

Table 1. Successful adoption of biotechnology for modification(s) in various attributes of ornamental plants.

Species	Plant Trait(s)																	Reference(s)
	AFC	APP	DI	DR	EF	ER	EFL	GT	IFN	IRC	IFT	IF	IVL	MP	MPA	RB	RPH	
Begonia semperflorens	■																	Shaw et al., 2002
Dendrathema grandiflora	■				■									■	■		■	Zheng et al., 2001; Petty et al., 2003; Aswath et al., 2004
Dendrobium kingianum	■			■														Aida et al., 2000; Davies et al., 2003
Dianthus caryophyllus	■										■	■	■					Brugliera et al., 2000; Ahn et al., 2004
Eustoma grandiflorum	■																	Nielsen et al., 2002
Gentian triflora	■																	Nishihara and Nakatuska, 2010
Gerbera hybrida	■																	Shaw et al., 2002
Petunia hybrida	■	■		■	■		■	■										Baker et al., 2002; Shaw et al., 2002; Chang et al., 2003; Davies et al., 2003, Tsuda et al., 2004
Rosa hybrid	■		■	■		■				■						■		Li et al., 2003
Torenia fourieri	■					■			■									Aida et al., 2000
Torenia hybrid	■	■																Suzuki et al., 2002

AFC, Altered Flower Color; APP, Altered Plant Parts; DI, Disease Improved; DR, Disease resistance; EF, Early Flowering; ER, Enhanced Resistance; EFL, Extended Flower Life; GT, Glyphosate Tolerant; IFN, Increased Flower Number; IRC, Improved Rooting Characteristics; ITF, Improved Fusarium Tolerance; IF, Increased Fragrance; IVL, Increased Vase Life; MP, Modified Phenotype; MPA, Modified Plant Architecture; RB, Reduced Blackspot; RPH, Reduced Plant Height.

PLANT PIGMENTS AND FLOWER COLOR

Generally, traditional plant breeding strategies are applied to perk up attractiveness as well as effectiveness of ornamental plants. But these strategies face limitations in terms of genes pool and some other characteristics reported in sexually resembling species (Da Silva et al., 2011). During last 20 years, biotechnology has produced innovative and exclusive characters in ornamentals by adopting genes from different plant species (Li and Pei, 2006). Floriculturists and related entrepreneur are always eager to introduce innovations in flower colors. The major pigments responsible for attractiveness of flower colors are anthocyanins, flavonoids, carotenoids, and betalains. Several kinds of anthocyanins are on record (Veitch and Grayer, 2008). These pigments are primarily based upon six anthocyanidins types i.e., cyanidin, delphinidin, peonidin, petunidin, malvidin, and pelargonidin. Three of the described anthocyanidins i.e., delphinidin, cyanidin and pelargonidin are regarded as major types (Figure 1, Table 2). Blue flowers tend to have high level of delphinidin and derivatives while intense red flower color is due to pelargonidin working as anthocyanidin base (Tanaka et al., 2009).

Ornamentals like petunia and torenia are considered as more suitable plants for studying floral color modifications produced by genetic engineering. Changes in gene expression were sought out as a base line for production of altered floral colors. Initially, pelargonidin derived anthocyanin production was successfully achieved by maize *Dfr* expression in petunia deficient in *F3′5′H* and *F3′H* (Forkmann and Ruhnau, 1987). We are well aware that varied gene expression for different biosynthetic pathways is leading strategy to achieve flower color variations. For example, variations in genes for flavonoid biosynthesis may generate new colors. In 2007, Katsumoto et al. reported down regulation of *Dfr* gene in hybrid rose and over-expression of the same gene in *Iris hollandica* responsible for delphinidin accumulation in petals. Rose has been specifically emphasized for search of suitable host to achieve blue flower (Yoshida et al., 2009). It has been found that over-expression of voila gene *F3′5′H* resulted in high delphinidin accumulation and production of blue flower color (Ogata et al., 2005; Katsumoto et al., 2007). Strong role of anthocyanin related genes for rose petals color variation was suggested by Fukuchi-Mizutani et al. (2011). *RhUF3GT2* catalyze flavonol 3-glucosylation in petals which results in accumulation of anthocyanidin 3-glucoside. Zvi et al. (2012) attributed higher production of anthocyanin to introduction of *Arabidopsis* PAP1 transcription factors in transgenic rose. They also confirmed enhanced isoprenoid production in transgenic plants as compared to control plants. Similarly, Chen et al. (2012) informed that white *Chrysanthemum* cultivars e.g., Keikai and Jinba possess major genes for anthocyanins pathways i.e., *Chs, Chi,* and *F3′H*.

We can analyze consequences of cross and mutation breeding in ornamental plants. Analysis reveals ultimate consequence as an array of flower colors such as orange, yellow, red, white, and pink. These color changes are directly related to regulation of targeted genes controlling synthesis of pigment precursors. White flowers from different transgenic plants have been obtained by down regulation of genes for anthocyanin production (Tanaka and Ohmiya, 2008). For finding the most suitable promoter for chrysanthemum gene expression, *EF1α* promoter (elongation factor 1α protein) was combined with *GUS* gene for introduction into *C.*

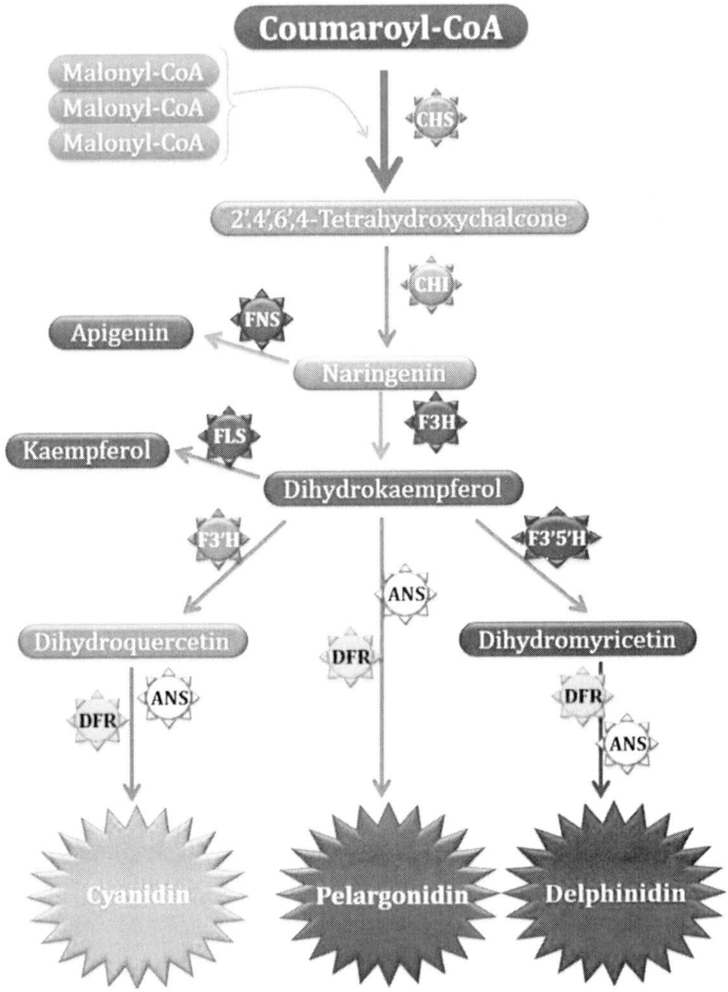

Figure 1. BIOSYNTHESIS OF ANTHOCYANIDIN**.** CHS catalyze the formation of Tetrahydroxy calchone. Later on, different enzymes such as CHI, F3H, DFR, ANS catalyze other steps of pigment production. The methyl groups are only added to anthocyanins not to anthocyanidins. The actual pigment color production is not solely dependent upon the enzyme catalyzing reactions but also depends upon other factors. CHS, chalcone synthase; F3H, flavanone 3-hydroxylase; F3′H, flavonoid 3′-hydroxylase; F3′5′H, flavonoid 3′,5′-hydroxylase; DFR, dihydroflavonol 4-reductase; ANS, anthocyanidin synthase; MT, methyltransferase, GT, glucosyltransferase; AT, acyltransferase; FNS, flavone synthase; FLS, flavonol synthase.

Table 2. Genes involved in synthesis of various enzymes for floral pigment pathways.

Sr. No.	Plant	Gene(s)	Enzyme
1	*Antirrhinum*	*ANS, CHI, CHS, DFR, F3H, F3′H, FNSII, 3GT*	*Anthocyanidin Synthase; Chalcone Flavanone Isomerase; Chalcone Synthase; Dihydroflavonol 4-reductase; Flavanone 3-hydroxylase; Flavonoid 3′ Hydroxylase; Flavone synthase II; Flavonoid-3-O-glucosyltransferase*
2	*Callistephus*	*ANS, F3H, F3′5′H, DFR*	*Anthocyanidin Synthase; Flavanone 3-hydroxylase; Flavonoid 3′,5′-hydroxylase; Dihydroflavonol 4-reductase*
3	*Chrysanthemum*	*CHS, F3H*	*Chalcone Synthase; Flavanone 3-hydroxylase*
4	*Dianthus*	*CHI, CHS, F3H, F3′H, F3′5′H, DFR*	*Chalcone Flavanone Isomerase; Chalcone Synthase; Flavanone 3-hydroxylase; Flavonoid 3′ Hydroxylase; Flavonoid 3′,5′-hydroxylase; Dihydroflavonol 4-reductase*
5	*Eustoma*	*CHI, F3′5′H*	*Chalcone Flavanone Isomerase; Flavonoid 3′,5′-hydroxylase*
6	*Gerbera*	*FNSII, DFR*	*Flavone synthase II; Dihydroflavonol 4-reductase*
7	*Orchid*	*CHS, F3H, DFR*	*Chalcone Synthase; Flavanone 3-hydroxylase; Dihydroflavonol 4-reductase*
8	*Petunia*	*ANS, CHI, FLS, F3′H, F3′5′H, DFR,*	*Anthocyanidin Synthase; Chalcone Flavanone Isomerase; Flavonol synthase; Flavonoid 3′ Hydroxylase; Dihydroflavonol 4-reductase*
9	*Rosa*	*CHS, FLS*	*Chalcone Synthase; Flavonol synthase*
10	*Gentiana*	*3GT*	*Flavonoid-3-O-glucosyltransferase*

morifolium cv. Ramat. Transgenic chrysanthemum plants exhibited high *GUS* expression and petal based transgene expression driven by the 35S CaMV promoter (Aida et al., 2005).

Diverse colors in chrysanthemum are largely resultants of carotenoids and/or red malonylated cyanidin glucosides (Kishimoto et al., 2004). A gene *CmCCD4a* is exclusively expressed in the white chrysanthemum ray petals. This is single dominant gene responsible for inhibition of formation/accumulation of carotenoids in petals (Ohmiya et al., 2009; Yoshioka et al., 2012). So, we may infer from the findings that due to suppression of *CmCCD4a* in white flowers, synthesized carotenoids break down into colorless compound (Ohmiya et al., 2006). Suppression technology like RNAi, co-suppression or antisense mediated silencing have been noticed more helpful in studying flower color variations (Tanaka et al., 2010).

Absence of delphinidin-based anthocyanins in chrysanthemum is chiefly due to deficiency of flavonoid 3′,5-hydroxylase (*F3′5′H*; Brugliera et al., 2013). That's why we do not observe violet/blue chrysanthemum flowers. Introduction of *F3′5′H* genes under different promoters control remained successful. Increase in delphinidin accumulation was very prominent under rose chalcone synthase promoter. Success was achieved by observing bluish petals in transgenic plants with higher anthocyanidin content characterized by delphinidin (Yoshida et al., 2009). Silencing of endogenous *F3′H* gene with hairpin RNA interference and over-expression of *F3′5′H* led to production of bluish petal with ~80% delphinidin derived anthocyanins (Brugliera et al., 2013). Combination of various promoters and *F3′5′H* gene proved enhanced delphinidin-based anthocyanins in transgenic plants.

Anthocyanins, carotenoids or both pigments can be noticed in Lilly sepals (Yamagishi et al., 2012). Azadi et al. (2010) confirmed numerous pigments in the transgenic lilly calli and leaves after transferring some key genes for carotenoid biosynthesis pathway under 35S CaMV. Callus and leaves presented orange color. This is due to presence of keto-carotenoids. In the variety *Lilium* oriental ₌Sorbonne', transient petal transformation by particle bombardment was conducted using three constructs (*Chs pro, HyDfr,* petunia *DifF* using 35S CaMV) possessing *Ph F3′5′H* (*Phalaenopsis F3′5′H* gene). Over-expression of *Ph F3′5′H* resulted in color change from pink to pale purple. A dark purple color was generated by synchronized expression of *Ph F3′5′H* and *HyDfr* contrary to expression alone. This propose potential role of *HyDfr* gene in facilitating delphinidin production (Qi et al., 2013, 2014). The *GMYB10* over-expression in transgenic gerbera plants considerably improved pigment accumulation and induce cyanidin biogenesis (Laitinen et al., 2008). Non-existence of delphinidin derived anthocyanins in gerbera has produced interest in blue flower production by using genetic engineering. Unfortunately we do not have successful reports for of blue gerbera production yet.

Plants express some genes in cyclical style during the day. This cyclical expression lets them to initiate photoassimilation during sunlight or release scents in the evening when pollinating agents are active. For example development of *Petunia circadia* will facilitate this internal circadian clock to control flower color leading to flower color changing over every 12 h approximately. Similarly, genome modification by means of zinc finger nucleases, CRISPR-Cas systems can help us to attain firmly changed phenotypes (Noman et al., 2016a). For redirecting pigment biosynthesis for achieving desirable color change, not only over-expression of a particular gene taking part in key enzyme synthesis is essential but a selection of appropriate host with suitable genetic background is also mandatory. This selection will be very helpful in reducing antagonism of indigenous pathways with the introduced enzyme or to permit down-regulation of another competing pathway.

INDUCTION OF EARLY FLOWERING

Together with synthetic biology, flowers with changing color characteristic have been engineered. Along with other related benefits, the color changing flower is a substantial and striking outcome of biotechnology (Figure 2). It is reflection of ability to outline our world with extensive appeal to all. Flowering time is reckoned as chief determinant for successful commercial plants (Jung and Müller, 2009). Induction of early flowering help plants to cater human needs by bearing more flowers and fruits ultimately. Hence, great economic benefit from ornamental plants can be obtained from establishment of practical floral regulation system. Fundamental studies on flowering time in model plants have described regulatory system of floral transition pathways (Blázquez, 2000). Genetic alterations by means of transforming key flowering associated genes appears to offer valuable advancements for operating floral attributes (Srikanth and Schmid, 2011).

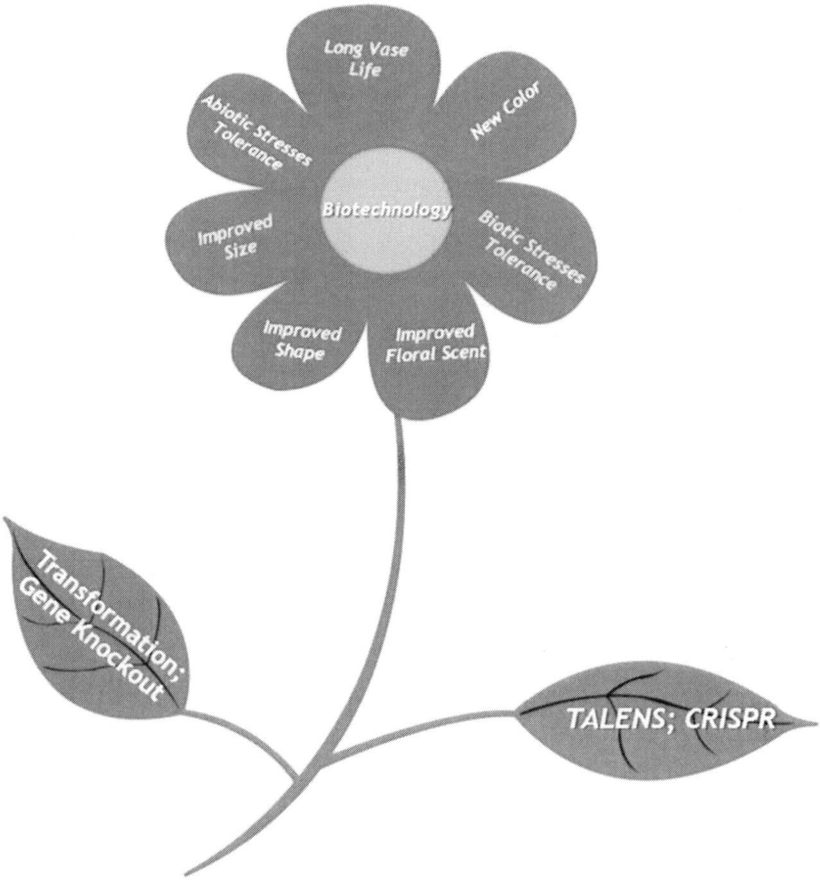

Figure 2. BENEFITS FROM FLORAL TRAITS MODIFICATION ARE SUBSTANTIAL EVIDENCE OF SUCCESSFUL APPLICATION OF BIOTECHNOLOGY. Advances in genome engineering provide innovation in ornamental plants and crops. These have the potential to circumvent the regulatory concerns raised about GMOs. The competent techniques such as ZFNs, TALENs, and CRISPR-Cas9 enable precise genome engineering by introducing DSBs and NHEJ. The techniques have ability to produce non-transgenic plants for next-generation plant breeding.

The reduction in flowering time by developing early flowering cultivars or plants able to produce flowers during long days are considerable breeding objectives in ornamental plant breeding (Shulga et al., 2011). The low cost of production make them highly feasible for growers and customers i.e., *Chrysanthemum*. Reports are in hand that comprehensively describe successful gene introduction to produce flowers in comparatively short time. One example is member of MADS box gene family e.g., *AP1* involved in flowering (Litt and Irish, 2003). MADS box genes are indispensable for floral development because they control flowering time as well as floral organ development (Thiruvengadam and Yang, 2009). In transgenic *Chrysanthemum, AP1 o*ver-expression during short-days can start bud initiation 14 days prior to non-transgenic plants. Interestingly, the inserted gene did not exert any change in plant development under long days. The differential gene over-expression during short days is seemingly linked to the plant biochemical changes. Moreover, transgenic flowers displayed prior color conFigure uration and complete inflorescence opening in comparison with non-transgenic plants (Shulga et al., 2011).

Transformation mediated by *Agrobacterium* in *Siningia* sp. supported that exogenous LFY over-expression promotes early flowering (Zheng et al., 2001). Transgenic *Gloxinia* plants with over-expressing or suppressed miRNA159, led to late or early flower appearance respectively. Meanwhile, varied miR159a expression levels resulted in *SsGAMYB* up or down-regulation during flower development. Thus, in transgenic lines transcript level of endogenous *Lfy* as well as *MADS*-box genes is changed. So it was established that miR159-mediated *GAMYB* expression also play crucial role in controlling flowering time period in ornamentals (Li et al., 2013). This will be helpful in establishing a practical regulation system that will facilitate flowering to be accelerated or delayed proportional to the demand. Fu et al. (2012) have demonstrated that rice miR156 over-expression amplify biomass and increment number of tillers in *Panicum*. Hence, it appears to be a competent strategy for boosting or curtailing target genes activity via transformation of conserved miRNA in different plant species.

In *Chrysanthemum seticuspe, CsFTL3* gene (Flowering Locus T like paralog) has been found to work as photoperiodic flowering regulator. Over-expression of this gene can induce flowering under long day conditions in *Chrysanthemum* (Oda et al., 2012). RNA sequencing has been adopted to illustrate quantitative effects of bulb vernalization on flowering and facilitates gene expression in the meristems. Investigations involving lilly bulbs subjected to cold treatment revealed many feasible candidate genes involved in vernalization (Huang et al., 2014; Villacorta-Martin et al., 2015).

In *Glebra*, transformation of *GSQUA*2 gene accelerated flowering (Ruokolainen et al., 2010). In 2011, Ruokolainen et al. detected three paralogs of *At-SOC1*. Over-expressing *Gh-SOC1* resulted in fractional loss of floral distinctiveness. This over-expression caused decrease in epidermal cell size in ray flower petals along with shape modification but flowering time remained uninfluenced. Conversely, *Gh-SOC1* down regulation failed to produce any considerable phenotypic alteration in transgenic plants (Ruokolainen et al., 2011). In *Lisianthus* sp. transformation of *OMADS1* showed significant reduction in flowering time as well as increased no. of flowers in comparison to non-transformed ones.

Of the reported genes dealing with flowering pathways, *FT* (*Flowering locus* T) is chief integrator of several flowering genes that respond to different signals such as light, temperature etc. (Fornara et al., 2010). *FT* supports flowering by up-regulating flowering genes such as *AP1, Lfy*, and *SOC1*. Over-expression of *VcFT*-Aurora in the transgenic blueberry supported early and continuous flowering in both *in vitro* shoots and greenhouse plants. Additionally, all transgenic *VcFT*-Aurora plants exhibited dwarf phenotype (Gao et al., 2016). The end result of *VcFT-OX* is differential expression of 110 pathway genes for five phytohormones. This connection between phytohormones pathway genes and *VcFT* implies a probable role of phytohormones as signals regulating plant growth and development.

Mostly, the knowledge regarding flowering time regulation is applied by means of gene over-expression or suppression of gene activity in crops (Jung and Müller, 2009). However, the research has not met full commercial execution. Due to multiple reasons, we are expecting to see wide range transgenic flowers in market. Less advancement in ornamentals can be observed in case of targeted genetic change for flowering time. The restricted access is seemingly because of relative unidentified genomic information in most of the commercial plant species. Hence, it is mandatory to search an appropriate approach for generating fresh varieties with changed flowering conduct in commercial plants. A substantial sum of try and error is certainly required to attain an agreeable final result. Analysis of in hand data pinpoints genetic modifications in ornamental plants as realistic and flourishing scheme both in scientific and commercial perspective.

FLORAL ANATOMY AND MORPHOLOGY

Customers demand for purposeful and distinctive horticultural plants urge scientists for development of novel cultivars, their large scale production and circulation as well. Ornamental plants are cultivated with rationale of beautifying, embellishing, or improving human environments irrespective of indoor or outdoor. Novel floral Figure in ornamental plants is essential for high market value. Up till now, molecular events for developing flower pattern in the meristem have remained clandestine. Primarily this is due to significant inflorescence variations among plant species. Majority of these plant types are not cross-compatible with each other. Therefore, we agree upon the decision that conventional cross-breeding strategy cannot be applicable to point out the genes responsible for flowering patterns.

The diverse gene behaviors and their differential expressions reflect diversity in plant responses. In the start of this century, tobacco *phytochrome b1* gene over-expression in chrysanthemum led to production of small sized plants but with larger branch angles (Zheng et al., 2001). Later, decrease in chrysanthemum plant height was achieved by introducing *Arabidopsis* GA insensitive gene (Petty et al., 2003). Furthermore, Aida et al. (2008) demonstrated transfer of *CAG* gene in antisense orientation into *C. morifolium*. They revealed that *CAG* gene suppression alters gynoecium and androecium to corolla-like tissues. However, the alteration pace was very small and simply changed the phenotype for flower-shape. It has been reported that *Ls*-like antisense gene expression in transgenic *Dendranthema grandiflorum* result in drop of axillary branching (Han et al., 2007). *PttKN1*ectopic expression in *D. caryophyllus* L. resulted in pleiotropic morphological wavering inclusive of phyllotaxis modifications (Meng et al., 2009).

GRAS TF restrains lateral branch production. Transformation of *DgLsL* gene with vector pCAMBIA1301-sense and antisense *DgLsL* was attributed to abundant branches in chrysanthemum. Conversely, branching was lessened in antisense transformants (Jiang et al., 2010). Collective silencing of *DmCPD* and *DmGA20ox* genes is a feasible strategy to create dwarf chrysanthemum varieties (Xie et al., 2015). It has also been proved that *D27* gene mutants (gene involved in strigolactone biosynthesis) reveal high tillering with dwarf phenotypic appearance (Lin et al., 2009). Cloned *DgD27* from *D. grandiflorum* expression in *Arabidopsis* put forward a new approach for development of chrysanthemum cultivars with reduced number of tillers (Wen et al., 2016). Similarly, *GSQUA*2 over-expressing transgenic *Gerbera* plants presented vivid vegetative alterations like elongation of vegetative axis. However, these plants were found to be infection susceptible with very limited root formation and tiny inflorescence (Ruokolainen et al., 2010). Thiruvengadam and Yang (2009) demonstrated that transgenic plants with 35S::*L MADS1-M* gene from lily produced more flowers from leafy branches in comparison to wild types. Moreover, transgenic *Lisianthus* flowers displayed a change of floral structure. Transgenic plants were characterized with conversion of petal second whorl into sepal like structures and deformation of third whorl stamens. Of note, transgenic carnations harboring *rol C* gene with promoter 35S CaMV demonstrated higher stem cutting yield and flowering stems and root organogenesis (Zuker et al., 2001; Casanova et al., 2004). So we can attribute such gene expression as solid evidence of their function for endorsing changes in floral shape and structure Conclusively, molecular breeding methods allow critical assessment of biological processes for floral changes and help scientific community as well as public to put GE of ornamental plants as deep insight regarding its effectiveness in plant breeding.

FRAGRANCE ENGINEERING

Floral scents have vital function in plants reproductive process and possess substantial economic significance. It essentially improves aesthetic characteristics of ornamental plants. Many floral scent volatiles fall into the category of terpenoid, phenylpropanoid/benzenoid, and aromatic amino acid (Oliva et al., 2015). Flowers produce a great range of specific metabolites like fragrant volatiles to attract pollinators, hormones to stimulate or repress signaling cascades and fragrant volatiles for protection against herbivores or pathogens (Baldwin, 2010; Dudareva et al., 2013). The array of particular metabolites synthesized by flowers of different plants is wide-ranging (Muhlemann et al., 2012; Zvi et al., 2012). On the other hand, different flower specific metabolites are actually produced in low amounts. Thus, their detection and characterization is cumbersome task. Therefore, enhanced production of floral specific metabolites may contribute in detection, isolation, and identification of compounds and improvement of flower properties like fragrance and pigmentation. Although the floral scent biochemistry is relatively new field yet during last decade researchers identified several scent controlling genes. Not all but a number of scent genes encode enzymes that directly catalyze the formation of volatile compounds. The manipulation of scent genes through genetic engineering has revealed success in adoption of this technology for amplifying floral scent potential.

For induction of fragrance in petals, *C. breweri BEAT* gene (benzyl alcohol acetyl transferase for benzyl acetate production) was introduced in *Lisianthus* (Aranovich et al., 2007). Recorded observations affirmed volatile compounds production including benzyl acetate among transgenic leaves and flowers upon feeding with alcoholic substrate. From the results, we can infer that alcoholic substrates are mandatory for fragrance induction by GE in lisianthus cut flowers. The innovative transcriptomic profiling offer substantial source for plant functional genomics and provide us deep insight in biological processes for petals development in *H. coronarium*. These data helps to elucidate the molecular mechanism of floral scent genesis and its regulation in monocotyledons (Yue et al., 2015).

Production of specialized metabolites for scent biosynthesis is not solely dependent upon enzymatic actions and greatly relies upon several TFs. In recent years, transcriptional regulation of fragrance biosynthesis pathways has been deeply studied (Muhlemann et al., 2012) that implies crucial roles of different transcriptional factors in controlling scent emission (Colquhoun and Clark, 2011). In spite of immense value, few TFs involved in fragrance emission regulation have been identified. From petunia petals, exclusively expressed *ODORNT1* (ODO1) has been found to regulate shikimate pathway (Verdonk et al., 2005). *ODO1* was also reportedly involved in promoter activation of plasma membrane based ABC transporter of unknown function (Van Moerkercke et al., 2012). Petunia *EOBI* (emission of benezoids 1) is R2-R3 type flower specific transcription factor that acts upstream of *ODO1* and downstream of EOBII. Silencing of this *EOBI* expression caused down regulation of several shikimate pathway and fragrance related genes (Spitzer-Rimon et al., 2012).

TFs regulating phenyl propanoid/benezoid pathways have been detected and characterized but terpenoids pathway transcriptional regulation is still obscure. Few years back, in Arabidopsis inflorescence, *MYC2* helix loop helix TF was detected for expression of two *TPS11* and *TPS21*sesquiterpene synthase genes (Hong et al., 2012). Other than identification and characterization of different TFs, master regulators which control diverse volatile compound production and different associated metabolic pathways are yet to be discovered. Recent progress in identifying the genes and their product enzymes taking part in volatile compounds biosynthesis have declared this metabolic engineering extremely practicable. Noteworthy successes are on record in improving plant defense and enhancing scent and aroma value of flowers and fruits (Table 3). The result of discussed research indicates that the GE for changing flower scents has significant potential. But, research outcomes also expose the drawbacks resulting from our insufficient knowledge of the scents metabolic processes and their regulation.

ABIOTIC AND BIOTIC STRESS TOLERANCE

Plant growth and productivity are severely affected by abiotic and biotic stresses (Ali et al., 2016; Islam et al., 2016; Noman and Aqeel, 2017). Due to inadequate number of resistance genes, breeding for stress resistance/tolerance in ornamental plants is difficult (Azadi et al., 2016). During last few years, use of biotechnological strategies for conferring resistance against abiotic and biotic stresses e.g., drought, pathogen attack have gained attention (Table 4). The critical appraisal of transgenic

Table 3. Role of different genes for improved phenotypes in gm ornamental plants.

Source	Gene	Result(s)	Reference(s)
Agrobacterium rhizogenes	*RolC*	Dwarfed Pelargoniums and Petunias	Boase et al., 2004
Agrobacterium tumefaciens	*ipT*	Increased branching and reduced internode length in Chrysanthemum	Khodakovskaya et al., 2009
Arabidopsis	*Asl38/lbd41*	Flowers turned into multiple column patterns in *Celosia cristata*	Meng et al., 2009
	Ft	Activate the floral identity genes; promotes flowering in appropriate conditions	Jung and Müller, 2009
Carnation/apple	*ACO/ACS-coding genes*	Increased vase life in carnation	Zuker et al., 2001
Chrysanthemum	*Ls*	Less branching in Chrysanthemum	Han et al., 2007, Jiang et al., 2010
	AP1	Speeds up time to flowering in Chrysanthemum	Jiang et al., 2010
	ERS1	Mutated gene slows down yellowing of leaves in *Chrysanthemum*	Narumi et al., 2005
Gentian	*Chs*	Gene silencing produces white flowers in *Gentian*	Nishihara and Nakatuska, 2011
	Ans	Gene silencing produces pale blue flowers in *Gentian*	Nakatsuka et al., 2008
Lotus	*CrtW*	Overexpression changes petal color from light yellow to deep yellow or orange in Lotus	Suzuki et al., 2002
Orchid/Lily	*MADS-Box*	Ectopic expression changes the second round of petals into calyx in orchids and lilies	Thiruvengadam and Yang, 2009
Petunia/Pansy	*F3'-5'H*	Overexpression produces blue flowers in combination with a silenced dfr gene in carnation (Petunia) and Roses (Pansy)	Katsumoto et al., 2007

ornamental plants indicates resistance to biotic stresses in comparison to the wild plants (Chandler and Sanchez, 2012; Teixeira da Silva et al., 2013). Fungal, viral, or bacterial pathogens sternly affect plants by decreasing plant growth and yields (Islam et al., 2016, 2017) inclusive of reduced quality of ornamental products. In 2002, Dohm et al., 2002 reported transgenic roses possessing antifungal genes i.e., *class II chitinase* and type I RIP (ribosome inhibiting protein). In rose, resistance to powdery mildew disease was improved by introduction of rice-chitinase gene (Pourhosseini et al., 2013). Three N-methyltransferase genes e.g., *CaXMT1, CaMXMT1,* and *CaDXMT1* were introduced in *D. grandiflorum* cv. Shinba. The observations confirmed high resistance to *Botrytis cinerea* (Kim et al., 2011). In chrysanthemum, transfer of *PGIP* gene (polygalacturonase-inhibiting protein) from *Prunus mumei* reduced disease index and occurrence of *Alternaria* leaf spot (Miao et al., 2010). Similarly, *chi*II endorsed improved resistance to septoria leaf spot disease in transgenic chrysanthemum (Sen et al., 2013). De Cáceres González et al. (2015) have produced transgenic *L. longifolium* cv. Star Gazer with high resistance against *B. cinerea*. The ectopic over-expression of the RCH10 gene (*Rice Chitinase 10*) under 35S CaMV promoter was found positively correlated with high resistance to *Botrytis*. Recently, several SNP markers have been identified for additional linkage mapping along with other transcripts that might have involvement in *Botrytis* resistance pathways (Fu et al., 2016).

Table 4. Recent biotechnological advances and milestones for improving ornamental plants.

Sr. No.	Category	Source	Transferred to	Gene	Function	References
1	Abiotic Stress Tolerance	*Arabidopsis thaliana*	*Rosa hybrida*	*PAP1* transcription factor	Enhance production of anthocyanin and also eugenol (phenylpropanoid compound) accumulation increases	Zvi et al., 2012
2		*Medicago truncatula*	*Rosa chinensis Jacq.*	*MtDREB1C*	Freezing stress tolerance enhancement	Chen et al., 2010
3		*Dendranthema grandiflora*	*Chrysanthemum morifolium* ‗ *Sei-Marine'*	*mDG-ERS1 (etr1-4)*	Reduces ethylene sensitivity	Narumi et al., 2005
4		*Arabidopsis thaliana*	*Dendranthema grandiflorum* ‗ *Fall Colour'*	*AtDREB1A*	Heat tolerance improvement	Hong et al., 2009
5		*Dendranthema vestitum*	*Chrysenthamum*	*DvDREB2A*	Enhances tolerance to heat & Water stresses	Liu et al., 2008
6		*Prunus mume*	*chrysanthemum*	*PGIP*	Reduces disease index & increases resistance to *Alternaria* leaf spot disease	Miao et al., 2010
7	Biotic Stress Resistance	*Oryza sativa*	*Chrysanthemum*	*Rice chitinase (chiII)*	Creates *Septoria* leaf spot disease Resistance	Sen et al., 2013
8		*Oryza sativa*	*Lilium oriental* ‗ *Star Gazer'*	*Rice Chitinase 10* (RCH10)	Tolerance to *Botrytis cinerea*	De Cáceres Gonzalez et al., 2015
9		*Oryza sativa*	*Rosa*	*Rice-chitinase* gene	Improves tolerance to *powdery mildew*	Pourhosseini et al., 2013
10		*Avena sativa*	*Dianthus caryophyllus*	*Oat thionin*	resistant to *Burkholderia* infections	Shirasawa-Seo et al., 2002
11		*Oryza sativa*	*Chrysenthamum*	*RCC2*	Impervious to *gray mold*	Takatsu et al., 1999
12		*Xanthomonas oryzae* cv. *oryzae*	*Chrysanthemum morifolium*	*hpaGxoo* gene	Increases resistance to *Alternaria* leaf spot disease	Xu et al., 2010
13		*Prunus mumei*	*Chrysenthamum*	*PGIP*	Increases resistance to *Alternaria* leaf spot disease	Yu et al., 2010
14		*B. thuringiensis*	*Chrysanthemum*	*cry1Ab*	Induces insect resistance	Shinoyama et al., 2008
15	Vegetative Character(s) Improvement	*Nicotiana tabacum*	*Chrysanthemum "Iridon"*	*Tobacco phytochrome b1*	Shorter plants with larger branch angles	Zheng et al., 2001
16		*Arabidopsis thaliana*	*Chrysanthemum morifolium*	*A. thaliana GA insensitive gene*	Reduction in plant height in was achieved	Petty et al., 2003
17		*Dendranthema grandiflorum*	*Chrysanthemum morifolium*	*DgLsL*	Produced profuse lateral branches	Jiang et al., 2010
18		*Cucumis melo*	*Chrysanthemum*	*CmETR1/H69A*	Significantly Reduces fertility for both male & female	Shinoyama et al., 2012
19		*Gerbera hybrida*	*Gerbera hybrida*	*Overexpression of GSQUA2*	Rapid vegetative alteration like dwarfism and vegetative axis elongation	Ruokolainen et al., 2010
20		*Clarkia breweri*	*Eustoma grandiflorum*	*Clarkia breweri* gene *(BEAT)*	Induces fragrances in petals	Aranovich et al., 2007
21		*Orchid*	*Eustoma grandiflorum*	*OMADS1*	Alteration in flowering time	Thiruvengadam and Yang, 2009
22		*Clarkia breweri*	*Dianthus caryophyllus*	*linalool synthase*	Fragrance production	Lavy et al., 2002
23		*Arabidopsis*	*chrysanthemum*	*TCP3* chimeric repressor.	Production of fringed leaves	Narumi et al., 2011
24		*Cucumus melo*	*Chrysanthemum*	*CmETR1/H69A*	Reduction in male and female sterility	Shinoyama et al., 2012
25		*Petunia*	*Lilium*	*DifF*	Production of purple colored petal cells	Azadi et al., 2016

Environmental conditions play very important role in survival of plants. Presence of pathogens and their infection dissemination is directly proportional to prevailing environmental conditions. In 2008, Clarke et al. (Clarke et al., 2008) developed GM *Euphorbia pulcherrima* with enhanced resistance to virus. Different genes from various sources have been validated for their crucial role in plant life during stress conditions. Targets of these biotic or abiotic stress-related genes also function in various cellular responses and metabolic processes e.g., transcriptional regulation, cell proliferation which indicate the variety of gene functions to tackle abiotic stress. For example, in transgenic chrysanthemum, over-expressing *hpaGxoo* gene increased resistance to *Alternaria tenuissima* (Xu et al., 2011). Agrobacterium mediated transformation of coat protein gene (cp) inducted resistance against CMV (Cucumber mosaic virus). Integration of cp gene in *C. morifolium* genome enhanced acclimatization percentage of transgenic plants displaying no or least CMV symptoms (Kumar et al., 2012). Transgenic *Grebra* plants harboring Nucleoprotein gene (*N*-gene) did not display TSWV (Tomato spotted wilt virus) symptoms (Korbin, 2006). Similar to this, transformation of glagiolus plants with sense or antisense orientation CP gene survived well under BYMV attack (bean yellow mosaic virus; Kamo et al., 2005).

Review of literature affirmed that expression of *cry1Ab* gene in chrysanthemum is strongly related to insect resistance i.e., *Spodoptera litura* (Shinoyama et al., 2012). Agrobacterium mediated transformation of LLA gene (*Lycoris Longituba agglutinin)* incremented resistance to aphids in transgenic chrysanthemum plants (He et al., 2009). Stable expression of different genes in ornamental plants like lilies has shown promising results in outdoor or green house conditions. Genes such as *bar, nptII* has been reported highly expressed in transgenic Lily plants under varying environmental conditions (Kamo, 2014). Transformation of *OcIΔD86* into *Lily* cv. Nellie White exhibited comparatively high resistant to an endoparasite, *Pratylenchus penetrans* causing RLN (root lesion disease). *OcIΔD86* expression inhibited parasitic effects of *Pratylenchus* by 75% in transgenic plants as compared to their wild relatives (Vieira et al., 2015; Table 4).

Saline soil, nutrient acidity, nutrients imbalance, water scarcity are major constraints that severely cut off plant yield, and other attributes (Noman et al., 2015; Zafar et al., 2016). Different strategies such as conventional plant breeding and genetic engineering are in use for developing abiotic stress tolerant varieties (Noman et al., 2016b). Several transcription factors e.g., ZIP, WRKY, NAC, and their products are crucial in plant response to stress conditions like drought, high temperature etc. (Vinocur and Altman, 2005). Ornamental plants also exhibit different responses upon exposure to abiotic stresses. For example, *Rosa* sp. is very susceptible to cold stress. *R. chinensis* Jacq. was transformed with *Medicago* gene *MtDREB1C*. Transgenic plants performed well under freezing temperature and hence proved improvement in tolerance to low temperature (Chen et al., 2010). It was proposed that frost tolerance in petunia may be enhanced by transforming *AtCBF3* gene (Warner, 2011). Hong et al. (2006a,b) developed low temperature resistant chrysanthemum lines by transformation of *AtDREB1A* gene. They noticed great increase in transgenic seedling and plant growth in winter. Two constructs i.e., 35S:*DREB1A* and *rd29A*:*DREB1A* were integrated into chrysanthemum genome for enhancing tolerance to salinity and water deficiency. The *rd29A*: *DREB1A* transgenic plants showed more tolerance in comparison with 35S:*DREB1A* transgenic plants (Hong et al., 2006a). Furthermore, *AtDREB1A* over-expression in *D. grandiflorum* produced sturdy heat tolerance. Upon exposure to 45°C for 36 h, wild and transgenic plants exhibited a marked difference in survival rate e.g., 20 and 70%, respectively (Hong et al., 2009). Salt stress tolerance of *chrysanthemum* cv. Jinba was enhanced with *CcSOS1*constitutive expression encoding plasma Na+/H+ antiporter (An et al., 2014; Song et al., 2014). Ascorbic acid (AsA) has very important role in abiotic stress tolerance (Noman et al., 2015). *GLDH* (L-galactono-1, 4-lactone dehydrogenase) catalyzes the oxidation of L-galactose to AsA (Agius et al., 2003). *pCAMBIA* vector harboring *GLDH* from apple was used for transformation of *L. davidii* var. unicolor. The results highlighted that *GLDH* over expression in transgenic plants, considerably amplified AsA production and endorsed resistance against abiotic stresses (Shi et al., 2012). In the light of above quoted examples, we are convinced that genetic engineering offers unprecedented arena for plant trait improvement. Particularly in ornamental plants, attributes like environmental stress tolerance, disease resistance can be retained in transgenic cultivars coupled with improved product range.

SHELF LIFE ENHANCEMENT

One of the daunting challenges for researchers is to increase shelf life of ornamental plant products coupled with maintained characteristics such as aroma, taste, etc. (Kamthan et al., 2016). Transgenic ornamental plants have potential to enhance leaf and flower longevity. Generally, cut flowers are treated with different kinds of chemicals for increasing their shelf life (Teixeira da Silva et al., 2013). With special reference to cut flowers, improved shelf life is decisive attribute and is of particular significance for breeders. To achieve the target of enhanced vase life, different biotechnological techniques have been used (Matas et al., 2009). Decreased autocatalytic ethylene synthesis by suppression of genes for ethylene production pathway such as *ACO* (ACC oxidase) or *ACS* (1-aminocyclopropane-1-carboxylicacid synthase) has been found effective to improve fruit vase life (Hamilton and Baulcombe, 1999). Transgenic carnation plants exhibiting delayed petal senescence have been developed by silencing the *ACO* gene that down regulate ethylene synthesis (Savin et al., 1995). Bovy et al. (1999) developed transgenic carnations harboring *Arabidopsis etr1-1* gene. Vase life of transgenic flowers was incremented 3 times in comparison with control plant.

Increased shelf life can be achieved by maintaining resistance to ethylene or by the inhibition of ethylene biosynthesis genes. Success has been witnessed in *Oncidium* and *Odontoglossum* by mutating ethylene receptor gene (Raffeiner et al., 2009). Different research groups have conducted experiments involving *ACS* or *ACO* genes in antisense fashion for increasing the shelf life of ornamental plant products (Bapat et al., 2010; Litz and Padilla, 2012). Transcriptional regulators such as *ERF*s (Ethylene response factors) modulate ethylene-induced fruit ripening. In transgenic *Lycopersicum esculentum*, antisense *LeERF*1 fruits exhibited more vase-life contrary to the wild-type plants (Li et al., 2007). Vase life of transgenic pot plants has been maintained by developing ethylene sensitive pot plant species with lessened ethylene sensitivity (Sanikhani et al., 2008; Milbus et al., 2009). Agrobacterium mediated transformation of *PSAG12-ipt* in Rose cultivar "Linda" augmented the *ipt* expression level in transgenic plants upon exposure to darkness and high level of exogenous ethylene. Moreover, chlorophyll amount was considerably increased (Zakizadeh et al., 2013).

Normally, leaves of cut flowers show yellowness, even before the start of senescence. On the other hand, exposure to ethylene can also speeds up yellowing. This yellowing of leaves is a feature of senescence. This destroys flower attractiveness, diminishes quality and shortens vase life. Consequently, transgenic plants with reduced ethylene sensitivity i.e., chrysanthemum is anticipated to possess improved vase life (Satoh et al., 2006). In transgenic *D. grandiflorum*, reduced leaf senescence has been proved very useful (Satoh et al., 2008). Similarly, transformation of *etr1-1* gene impeded senescence in petunia. But unfortunately this senescence reduction was coupled with curtailed rooting of cuttings that is unacceptable in market (Gubrium et al., 2000). This deficiency highlights need for a more tissue specific expression gene of genes regulating some crucial processes (Chandler and Sanchez, 2012). Additionally, the endogenous production and distribution of plant growth regulators such as IAA, iPA, and ABA during the course of flower induction and initiation may facilitate us to better plan the harvest timing for maximum flower uniformity and the quality of ornamentals (Jiang et al., 2010; Teixeira da Silva et al., 2013).

CONCLUDING REMARKS

Worldwide increase in economic worth of ornamental plants is somehow a result of promising prospects of gene transformation. Novel transgenic ornamental plants may therefore provide prospective benefits to growers and consumers by generating diverse floral appearance, novel colors, and improved fragrance. Taking aid from in hand data, we recommend that genetic modification of ornamental plants is very pragmatic and successful scheme both in scientific and commercial perspective. We should not be worried about acceptance of transgenic flowers in the market. Very luckily, consumers have shown their satisfaction over genetically modified flowers e.g., rose, carnation. Transgenic ornamental varieties have been proved successful during vegetative propagation and have not shown any undesired effect on environment or on the health of handler. Several exploitable traits of ornamental plants can be particularly associated with secondary metabolites production such as phytoceuticals. Development of molecular markers and complete genome sequences of ornamental plants will lead to discovery of new genes and related pathways. This will momentously contribute in production of new transgenic varieties of ornamental plants. Furthermore, application of novel techniques like ZFNs, TALENs, and CRISPR has substantial potential to facilitate floriculture industry by targeted genomic modifications. Despite sound scientific prospects of transgenic flowers, economic and regulatory barriers have hampered the expansion in commercialization of GM ornamentals. Time has come to address regulatory obstacles for commercial release of GM plants including flowers. In absence of internationally apt and approved process for regulation of GM products, release of ornamental products will remain very difficult due to expenditures and capability needed for commercial development. To alleviate such inconvenience the regulatory requisites for non-food plants like ornamentals ought to be decreased.

AUTHOR CONTRIBUTIONS

AN has collected research data and compiled manuscript. MA has made all Figure ures and tables. NK and TS have compiled table and corrected the references as well as DOI. JD and HS have critically read this manuscript and suggested for improvement.

CONFLICT OF INTEREST STATEMENT

The authors declare that the research was conducted in the absence of any commercial or financial relationships that could be construed as a potential conflict of interest.

REFERENCES

1. Agius, F., González-Lamonthe, R., Caballero, J. L., Muñoz-Blanco, J., Botella, M. A., and Valpuesta, V. (2003). Engineering increased vitamin C levels in plants by overexpression of a D-galacturonic acid. *Nat. Biotechnol.* 21, 177–181.
2. Ahn, B. J., Shin, H. Y., Hwang, K. H., Min, B. H., and Joung, H. Y. (2004). Transformation of carnations with jasmonate methyl transferase gene for fusarium tolerance. *In Vitro Cell. Dev. Biol.* 40:45-A.
3. Aida, R., Kishimoto, S., Tanaka, Y., and Shibata, M. (2000). Modification of flower color in torenia (*Torenia fournieri* Lind.) by genetic transformation. *Plant Sci.* 153, 33–42.
4. Aida, R., Komano, M., Saito, M., Nakase, K., and Murai, K. (2008). *Chrysanthemum* flower shape modification by suppression of chrysanthemum-AGAMOUS gene. *Plant Biotechnol.* 25, 55–59.
5. Aida, R., Nagaya, S., Yoshida, K., Kishimoto, S., Shibata, M., and Ohmiya, A. (2005). Efficient transgene expression in *Chrysanthemum morifolium* Ramat with the promoter of a gene for tobacco elongation factor 1 α protein. *JARQ* 39, 269–274.
6. Ali, Q., Haider, M. Z., Iftikhar, W., Jamil, S., Javed, M. T., Noman, A., et al. (2016). Drought tolerance potential of *Vigna mungo* L. lines as deciphered by modulated growth, antioxidant defense, and nutrient acquisition patterns. *Braz. J. Bot.* 39, 801–812.
7. An, J., Song, A., Guan, Z., Jiang, J., Chen, F., Lou, W., et al. (2014). The over-expression of *Chrysanthemum crassum CcSOS1* improves the salinity tolerance of chrysanthemum. *Mol. Biol. Rep.* 41, 4155–4162.
8. Aranovich, D., Lewinsohn, E., and Zaccai, M. (2007). Post-harvest enhancement of aroma in transgenic lisianthus (*Eustoma grandiflorum*) using the *Clarkia breweri* benzyl alcohol acetyl transferase (BEAT) gene. *Postharvest Biol. Biotechnol.* 43, 255–260.
9. Aswath, C. R., Mo, S. Y., Kim, S. H., and Kim, D. H. (2004). IbMADS4 regulates the vegetative shoot development in transgenic chrysanthemum (*Dendranthema grandiflora* (Ramat.) Kitamura). *Plant Sci.* 166, 847–854.
10. Auer, C. (2008). Ecological risk assessment and regulation for genetically-modified ornamental plants. *Crit. Rev. Plant Sci.* 27, 255–271.
11. Azadi, P., Bagheri, H., Nalousi, A. M., Nazari, F., and Chandler, S. F. (2016). Current status and biotechnological advances in genetic engineering of ornamental plants. *Biotechnol. Adv.* 34, 1073–1090.
12. Azadi, P., Otang, N. V., Chin, D. P., Nakamura, I., Fujisawa, M., Harada, H., et al. (2010). Metabolic engineering of *Lilium* × *formolongi*using multiple genes of the carotenoid biosynthesis pathway. *Plant Biotechnol. Rep.* 4, 269–280.
13. Baker, C., Zhang, H., Hall, G., Scrocki, D., Medina, A., and Dobres, M. S. (2002). The use of the GAI and CO genes to create novel ornamental plants. *In Vitro Cell. Dev. Biol.* 38:105-A.
14. Baldwin, I. T. (2010). Plant volatiles. *Curr. Biol.* 20, 392–397.
15. Bapat, V. A., Trivedi, P. K., Ghosh, A., Sane, V. A., Ganapathi, T. R., and Nath, P. (2010). Ripening of fleshy fruit: molecular insight and the role of ethylene. *Biotechnol. Adv.* 28, 94–107.
16. Blázquez, M. (2000). Flower development pathways. *J. Cell Sci.* 113, 3547–3548.
17. Boase, M. R., Winefield, C. S., Lill, T. A., and Bendall, M. J. (2004). Transgenic regal pelargoniums that express the rolC gene from Agrobacterium rhizogenes exhibit a dwarf floral and vegetative phenotype. *In Vitro Cell Dev. Biol. Plant* 40, 46–50.
18. Bovy, A. G., Angenent, G. C., Dons, H. J., and van Altvorst, A. C. (1999). Heterologous expression of the *Arabidopsis* etr1-1 allele inhibits the senescence of carnation flowers. *Mol. Breed.* 5, 301–308.

19. Brugliera, F., Kalc-Wright, G., Hyland, C., Webb, L., Herbert, S., Sheehan, B., et al. (2000). Improvement of Fusarium wilt tolerance in carnations expressing chitinase. *Int. Plant Mol. Biol. Rep.* 18, 522–529.
20. Brugliera, F., Tao, G. Q., Tems, U., Kalc, G., Mouradova, E., Price, K., et al. (2013). Violet/blue chrysanthemums metabolic engineering of the anthocyanin biosynthetic pathway results in novel petal colors. *Plant Cell Physiol.* 54, 1696–1710.
21. Casanova, E., Valdes, A. E., Zuker, A., Fernandez, B., Vainstein, A., Trillas, M. I., et al. (2004). *rolC*-transgenic carnation plants: adventitious organogenesis and levels of endogenous auxin and cytokinins. *Plant* 167, 551–560.
22. Chandler, S. F., and Brugliera, F. (2011). Genetic modification in floriculture. *Biotechnol. Lett.* 33, 207–214.
23. Chandler, S. F., and Sanchez, C. (2012). Genetic modification; the development of transgenic ornamental plant varieties. *Plant Biotech. J.* 10, 891–903.
24. Chang, H., Jones, M. L., Nanowetz, G. M., and Clark, D. G. (2003). Overproduction of cytokinins in petunia flowers transformed with PSAG12-IPT delay corolla senescence and decrease sensitivity to ethylene. *Plant Physiol.* 132, 2174–2183.
25. Chen, J. R., Lu, J. J., Liu, R., Xiong, X. Y., Wang, T. X., Chen, S. Y., et al. (2010). DREB1C from *Medicago truncatula* enhances freezing tolerance in transgenic *M. truncatula* and China Rose *(Rosa chinensis Jacq.). Plant Growth Regul.* 60, 199–211.
26. Chen, S. M., Li, C. H., Zhu, X. R., Deng, Y. M., Sun, W., Wang, L. S., et al. (2012). The identification of flavonoids and the expression of genes of anthocyanin biosynthesis in the chrysanthemum flowers. *Biol. Plant.* 56, 458–464.
27. Clarke, J. L., Spetz, C., Haugslien, S., Xing, S., Dees, M. W., Moe, R., et al. (2008). Agrobacterium tumefaciens-mediated transformation of poinsettia, *Euphorbia pulcherrima*, with virus-derived hairpin RNA constructs confers resistance to Poinsettia mosaic virus. *Plant Cell Rep.* 27, 1027–1038.
28. Colquhoun, T. A., and Clark, D. G. (2011). Unraveling the regulation of floral fragrance biosynthesis. *Plant Signal Behav.* 6, 378–381.
29. Da Silva, J. A. T., Chin, D. P., Van, P. T., and Mii, M. (2011). Transgenic orchids. *Sci. Hort.* 130, 673–680.
30. Davies, K. M., Schwinn, K. E., Deroles, S. C., Manson, D. G., Lewis, D. H., Bloor, S. J., et al. (2003). Enhancing anthocyanin production by altering competition for substrate between flavonol synthase and dihydroflavonol 4-reductase. *Euphytica* 131, 259–268.
31. De Cáceres González, F. F. N., Davey, M. R., Sanchez, E. C., and Wilson, Z. A. (2015). Conferred resistance to *Botrytis cinerea* in *Lilium* by overexpression of the *RCH10* chitinase gene. *Plant Cell Rep.* 34, 1201–1209.
32. Dobres, M. S. (2011). "Prospects for commercialisation of transgenic ornamentals," in *Transgenic Horticultural Crops Challenges and Opportunities*, eds B. Mou and R. Scorza (Boca Raton, FL: CRC Press), 305–316.
33. Dohm, A., Ludwig, C., Schilling, D., and Debener, T. (2002). Transformation of roses with genes for antifungal proteins to reduce their susceptibility to fungal diseases. *Acta Hortic.* 572, 105–111.
34. Du, F., Wu, Y., Zhang, L., Li, X. W., Zhao, X. Y., Wang, W. H., et al. (2015). *De novo* assembled transcriptome analysis and SSR marker development of a mixture of six tissues from *Lilium* Oriental hybrid-Sorbonne. *Plant Mol. Biol. Rep.* 33, 281–293.
35. Dudareva, N., Klempien, A., Muhlemann, J. K., and Kaplan, I. (2013). Biosynthesis, function and metabolic engineering of plant volatile organic compounds. *N. Phytol.* 198, 16–32.
36. Feng, L. G., Chen, C., Sheng, L. X., Liu, P., Tao, J., Su, J. L., et al. (2010). Comparative analysis of headspace volatiles of Chinese *Rosa rugosa*. *Molecules* 15, 8390–8399.
37. Forkmann, G., and Ruhnau, B. (1987). Distinct substrate specificity of dihydroflavonol 4-reductase from flowers of *Petunia Hybrida*. *Z. Naturforsch* 42, 1146–1148.

38. Fornara, F., de Montaigu, A., and Coupland, G. (2010). SnapShot: control of flowering in Arabidopsis. *Cell* 141, 550–550.

39. Fu, C., Sunkar, R., and Zhou, C. (2012). Overexpression of miR156 in switchgrass (*Panicum virgatum* L.) results in various morphological alterations and leads to improved biomass production. *Plant Biotechnol. J.* 10, 443–452.

40. Fu, Y., Esselink, G. D., Visser, R. G., Van Tuyl, J. M., and Arens, P. (2016). Transcriptome analysis of *Gerbera hybrida* including *in silico* confirmation of defense genes found. *Front. Plant Sci.* 7:247.

41. Fukuchi-Mizutani, M., Akagi, M., Ishiguro, K., Katsumoto, Y., Fukui, Y., Togami, J., et al. (2011). Biochemical and molecular characterization of anthocyanidin/flavonol 3-glucosylation pathways in *Rosa* × *hybrida*. *Plant Biotechnol.* 28, 239–244.

42. Gao, X., Walworth, A. E., Mackie, C., and Song, G. (2016). Overexpression of blueberry *FLOWERING LOCUS T* is associated with changes in the expression of phytohormone-related genes in blueberry plants. *Horti. Res.* 3, 1605–1615.

43. Gubrium, E. K., Clevenger, D. J., Clark, D. G., Barrett, J. E., and Nell, T. A. (2000). Reproduction and horticultural performance of transgenic ethylene in sensitive petunias. *J. Am. Soc. Hort Sci.* 125, 277–281.

44. Hamilton, A. J., and Baulcombe, D. C. (1999). A species of small antisense RNA in post transcriptional gene silencing in plants. *Science* 286, 950–952.

45. Han, B. H., Suh, E. J., Lee, S. Y., Shin, H. K., and Lim, Y. P. (2007). Selection of non-branching lines induced by introducing Ls-like cDNA into Chrysanthemum (*Dendranthema* × *grandiflorum* (Ramat.) Kitamura) Shuho-no-chikara. *Sci. Hortic.* 115, 70–75

46. He, J. P., Chen, F. D., Chen, S. M., Fang, W. M., Miao, H. B., and Luo, H. L. (2009). Transformation of *Lycoris longituba* agglutinin gene to cut chrysanthemum and identification of aphid resistance in the transgenic plants. *Acta Bot. Boreal. Occident. Sin.* 29, 2318–2325.

47. Hong, B., Ma, C., Yang, Y., Wang, T., Yamaguchi-Shinozaki, K., and Gao, J. (2009). Over-expression of AtDREB1A in chrysanthemum enhances tolerance to heat stress. *Plant Mol. Biol.* 70, 231–240.

48. Hong, B., Tong, Z., Li, Q. H., Ma, C., Kasuga, M., Yamaguchi-Shinozaki, K., and Gao, J. (2006b). Regeneration and transformation through somatic embryogenesis, and determination of cold stress tolerance in ground cover *Chrysanthemum* cv Fall color. *Sci. Agric. Sin.* 39, 1443–1450.

49. Hong, B., Tong, Z., Ma, C., Kasuga, M., Yamaguchi-Shinozaki, K., and Gao, J. (2006a). Heterologous expression of the AtDREB1A gene in transgenic chrysanthemum enhances tolerance to low temperature. *PLoS ONE* 49:436.

50. Hong, G. J., Xue, X. Y., Mao, Y. B., Wang, L. J., and Chen, X. Y. (2012). Arabidopsis MYC2 850 interacts with DELLA proteins in regulating sesquiterpene synthase gene 851 expression. *Plant Cell* 24, 2635–2648.

51. Huang, H. (2011). Plant diversity and conservation in China: planning a strategic bioresource for a sustainable future. *Botanical. J. Linnean Soc.* 166, 282–300.

52. Huang, J., Liu, X., Wang, J., and Lü, Y. (2014). Transcriptomic analysis of Asiatic lily in the process of vernalization *via* RNA-seq. *Mol. Biol. Rep.* 41, 3839–3852.

53. Islam, W., Awais, M., Noman, A., and Jian, W. U. (2016). Success of bio Products against bacterial leaf blight disease of rice caused by *Xanthomonas oryzae* cv. *Oryzae PSm Microbio.* 1, 50–55.

54. Islam, W., Zhang, J., Adnan, M., Noman, A., Zainab, M., and Jian, W. (2017). Plant virus ecology: a glimpse of recent accomplishments. *Appl. Econ. Environ. Res.* 15, 691–705.

55. Jiang, B., Miao, H., Chen, S., Zhang, S., Chen, F., and Fang, W. (2010). The lateral suppressor-like gene,DgLsL, alternated the axillary branching in transgenic chrysanthemum (*Chrysanthemum* × *morifolium*) by modulating IAA and GA content. *Plant Mol. Biol. Rep.* 28, 144–151.

56. Jung, C., and Müller, A. E. (2009). Flowering time control and applications in plant breeding. *Trends Plant Sci.* 14, 563–573.

57. Kamo, K. (2014). Long term, transgene expression in *Lilium longiflorum cv* Nellie white grown outdoors and in the greenhouse. *Sci. Hortic.* 167, 158–163.
58. Kamo, K., Gera, A., Cohen, J., Hammond, J., Blowers, A., Smith, F., et al. (2005). Transgenic gladiolus plants transformed with the bean yellow mosaic virus coat-protein gene in either sense or antisense orientation. *Plant Cell Rep.* 23, 654–663.
59. Kamthan, A., Chaudari, A., Kamthan, M., and Datta, A. (2016). Genetically modified (GM) crops: milestones and new advances in crop improvement. *Theo. Appl. Genet.* 29, 1639–1655.
60. Katsumoto, Y., Fukuchi-Mizutani, M., Fukui, Y., Brugliera, F., Holton, T. A., Karan, M., et al. (2007). Engineering of the rose flavonoid biosynthetic pathway successfully generated blue-hued flowers accumulating delphinidin. *Plant Cell Physiol.* 48, 1589–1600.
61. Khodakovskaya, M., Vanková, R., Malbeck, J., Li, A., Li, Y., and McAvoy, R. (2009). Enhancement of flowering and branching phenotype in chrysanthemum by expression of *ipt* under the control of a 0.821kb fragment of the LEACO1 gene promoter. *Plant Cell Rep.* 28, 1351–1362.
62. Kim, J. B., Raemakers, C. J. J. M., Jacobsen, E., and Visser, R. G. F. (2006). Efficient somatic embryogenesis in Alstroemeria. *Plant Cell Tissue Organ Cult.* 86, 233–238.
63. Kim, Y. S., Lim, S., Yoda, H., Choi, Y. E., and Sano, H. (2011). Simultaneous activation of salicylate production and fungal resistance in transgenic chrysanthemum producing caffeine. *Plant Signal. Behav.* 6, 409–412.
64. Kishimoto, S., Maoka, T., Nakayama, M., and Ohmiya, A. (2004). Carotenoid composition in petals of chrysanthemum (*Dendranthema grandiflorum* (Ramat.) Kitamura). *Phytochemistry* 65, 2781–2787.
65. Korbin, M. (2006). Assessment of gerbera plants genetically modified with TSWV nucleocapsid gene. *J. Fruit Ornam. Plant Res.* 14, 85–93.
66. Kumar, S., Raj, S. K., Sharma, A. K., and Varma, H. N. (2012). Genetic transformation and development of cucumber mosaic virus resistant transgenic plants of *Chrysanthemum morifolium* cv Kundan'. *Sci. Hortic.* 134, 40–45.
67. Laitinen, R. A., Ainasoja, M., Broholm, S. K., Teeri, T. H., and Elomaa, P. (2008). Identification of target genes for a MYB-type anthocyanin regulator in *Gerbera hybrida*. *J. Exp. Bot.* 59, 3691–3703.
68. Lavy, M., Zuker, A., Lewinsohn, E., Larkov, O., Ravid, U., Vainstein, A., et al. (2002). Linalool and linalool oxide production in transgenic carnation flowers expressing the *Clarkia breweri* linalool synthase gene. *Mol. Breed.* 9, 103–111.
69. Li, X., Bian, H., Song, D., Ma, S., Han, N., Wang, J., et al. (2013). Flowering time control in ornamental gloxinia (*Sinningia speciosa*) by manipulation of miR159 expression. *Ann. Bot.* 111, 791–799.
70. Li, X., Gasic, K., Cammue, B., Broekaert, W., and Korban, S. S. (2003). Transgenic rose lines harboring an antimicrobial protein gene, Ace-AMP1, demonstrate enhanced resistance to powdery mildew (*Sphaerotheca pannosa*). *Planta* 218, 226–232.
71. Li, Y., and Pei, Y. (2006). *Plant Biotechnology in Ornamental Horticulture*. New York, NY: Haworth Food and Agricultural Products Press.
72. Li, Y., Zhu, B., Xu, W., Zhu, H., Chen, A., Xie, Y., et al. (2007). LeERF1 positively modulated ethylene triple response on etiolated seedling, plant development and fruit ripening and softening in tomato. *Plant Cell Rep.* 26, 1999–2008.
73. Lin, H., Wang, R., Qian, Q., Yan, M., Meng, X., Fu, Z., et al. (2009). DWARF27, an iron-containing protein required for the biosynthesis of strigolactones, regulates rice tiller bud outgrowth. *Plant Cell* 21, 1512–1525.
74. Litt, A., and Irish, V. F. (2003). Duplication and diversification in the APETALA1/FRUITFULL floral homeotic gene lineage: implications for the evolution of floral development. *Genetics* 165, 821–833.
75. Litz, R. E., and Padilla, G. (2012). "Genetic transformation of fruit trees," in *Genomics of Tree Crops*, eds P. M. Priyadarshan and R. J. Schnell (Berlin: Springer), 117–153.

76. Liu, L., Zhu, K., Yang, Y., Wu, J., Chen, F., and Yu, D. (2008). Molecular cloning, expression profiling and *trans*-activation property studies of a *DREB2*-like gene from chrysanthemum (*Dendranthema vestitum*). *J. Plant Res.* 121, 215–226.
77. Matas, A. J., Gapper, N. E., Chung, M. Y., Giovannoni, J. J., and Rose, J. K. (2009). Biology and genetic engineering of fruit maturation for enhanced quality and shelf-life. *Curr. Opin. Biotechnol.* 20, 197–203.
78. Meng, L. S., Song, J. P., Sun, S. B., and Wang, C. Y. (2009). The ectopic expression of PttKN1 gene causes pleiotropic alternation of morphology in transgenic carnation (*Dianthus caryophyllus* L.). *Acta Physiol. Plant.* 31, 1155–1164.
79. Miao, H., Jiang, B., Chen, S., Zhang, S., Chen, F., Fang, W., et al. (2010). Isolation of a gibberellin 20-oxidasec DNA from and characterizationof its expression in chrysanthemum. *Plant Breed.* 129, 707–714.
80. Milbus, H., Sriskandarajah, S., and Serek, M. (2009). Genetically modified flowering potted plants with reduced ethylene sensitivity. *Acta Hortic.* 847, 75–80.
81. Muhlemann, J. K., Maeda, H., Chang, C. Y., Miguel, P. S., Baxter, I., Cooper, B., et al. (2012). Developmental changes in the metabolic network of snapdragon flowers. *PLoS ONE* 7:e40381.
82. Nakatsuka, T., Mishiba, K., Abe, Y., Kubota, A., Kakizaki, Y., Yamamura, S., et al. (2008). Flower color modification of gentian plants by RNAi-mediated gene silencing. *Plant Biotechnol.* 25, 61–68.
83. Narumi, T., Aida, R., Koyama, T., Yamaguchi, H., Sasaki, K., Shikata, M., et al. (2011). *Arabidopsis* chimeric TCP3 repressor produces novel floral traits in *Torenia fournieri* and *Chrysanthemum morifolium*. *Plant Biotechnol.* 28, 131–140.
84. Narumi, T., Aida, R., Ohmiya, A., and Satoh, S. (2005). Transformation of chrysanthemum with mutated ethylene receptor genes: mDG-ERS1 transgenes conferring reduced ethylene sensitivity and characterization of the transformants *Postharvest Biol. Technol.* 37, 101–110.
85. Nielsen, K., Deroles, S. C., Markham, K. R., Bradley, M. J., Podivinsky, E., and Manson, D. (2002). Antisense flavonol synthase alters copigmentation and flower color in lisianthus. *Mol. Breed.* 9, 217–229.
86. Nishihara, M., and Nakatsuka, T. (2010). Genetic engineering of novel flower colors in floricultural plants: recent advances via transgenic approaches. *Methods Mol. Biol.* 589, 325–347.
87. Nishihara, M., and Nakatuska, T. (2011). Genetic engineering of flavonoid pigments to modify flower color in floricultural plants. *Biotechnol. Lett.* 33, 433–441.
88. Noman, A., Ali, Q., Naheed, F., Rizwan, M., Ali, S., and Irshad, K. (2015). Foliar application of ascorbate enhances the physiological and biochemical attributes of maize (*Zea mays* L.) cultivars under drought stress. *Arch. Agron. Soil Sci.* 61, 1659–1672.
89. Noman, A., and Aqeel, M. (2017). miRNA-based heavy metal homeostasis and plant growth. *Environ. Sci. Pollut. Res.*
90. Noman, A., Aqeel, M., and He, S. (2016a). Crispr-cas9: tool for qualitative and quantitative plant genome editing. *Front. Plant Sci.* 7:1740.
91. Noman, A., Bashir, R., Aqeel, M., Anwer, S., Iftikhar, W., Zainab, M., et al. (2016b). Success of transgenic cotton (*Gossypiumhirsutum*, L.): fiction or reality? *Cogent Food Agric.* 2:1207844.
92. Noman, A., Fahad, S., Aqeel, M., Ali, U., Amanullah, Anwer, S., et al. (2017). miRNAs: major modulators for crop growth and development under abiotic stresses. *Biotechnol Lett.*
93. Nontaswatsri, C., and Fukai, S. (2010). "Genetic transformation of carnation (*Dianthus caryophylus* L.)," in *Protocols for In Vitro Propagation of Ornamental Plants*, eds S. M. Mohan and S. J. Ochatt (Amsterdam: Springer), 87–96.
94. Oda, A., Narumi, T., Li, T., Kando, T., Higuchi, Y., Sumitomo, K., et al. (2012). CsFTL3, a chrysanthemum FLOWERING LOCUS T-like gene, is a key regulator of photoperiodic flowering in chrysanthemums. *J. Exp. Bot.* 63, 1461–1477.

95. Ogata, J., Kanno, Y., Itoh, Y., Tsugawa, H., and Suzuki, M. (2005). Plant biochemistry: anthocyanin biosynthesis in roses. *Nature* 435, 757–758.

96. Ohmiya, A., Kishimoto, S., Aida, R., Yoshioka, S., and Sumitomo, K. (2006). Carotenoid cleavage dioxygenase (*CmCCD4a*) contributes to white color formation in chrysanthemum petals. *Plant Physiol.* 142, 1193–1201.

97. Ohmiya, A., Sumitomo, K., and Aida, R. (2009). "Yellow Jimba": suppression of carotenoid cleavage dioxygenase (*CmCCD4a*) expression turns white chrysanthemum petals yellow. *J. Jpn. Soc.* 78, 450–455.

98. Oliva, M., Ovadia, R., Perl, A., Bar, E., Lewinsohn, E., Galili, G., et al. (2015). Enhanced formation of aromatic amino acids increases fragrance without affecting flower longevity or pigmentation in Petunia × hybrida. *Plant Biotechnol. J.* 13, 125–136.

99. Parisi, C., Tillie, P., and Rodríguez-Cerezo, E. (2016). The global pipeline of GM crops out to 2020. *Nat. Biotechnol.* 34, 31–36.

100. Petty, L. M., Harberd, N. P., Carre, I. A., Thomas, B., and Jackson, S. D. (2003). Expression of the *Arabidopsis gai*gene under its own promoter causes a reduction in plant height in chrysanthemum by attenuation ofthe gibberellin response. *Plant Sci.* 164, 175–182.

101. Pourhosseini, L., Kermani, M. J., Habashi, A. A., and Khalighi, A. (2013). Efficiency of direct and indirect shoot organogenesis in different genotypes of *Rosa hybrida*. *Plant Cell Tissue Organ Cult.* 112, 101–108.

102. Qi, Y., Du, L., Quan, Y., Tian, F., Liu, Y., and and Wang, Y. (2014). *Agrobacterium*-mediated transformation of embryogenic cell suspension cultures and plant regeneration in *Liliumtenuifolium* oriental × trumpet cv. Robina'. *Acta Physiol.* 36, 2047–2057.

103. Qi, Y., Lou, Q., Quan, Y., Liu, Y., and Wang, Y. (2013). Flower-specific expression of the *Phalaenopsis* flavonoid 3′, 5′-hydoxylase modifies flower color pigmentation in Petunia and *Lilium*. *Plant Cell Tissue Organ Cult.* 115, 263–273.

104. Raffeiner, B., Serek, M., and Winkelmann, T. (2009). *Agrobacterium tumefaciens* mediated transformation of Oncidium and Odontoglossum orchid species with the ethylene receptor mutant gene etr1-1. *Plant Cell Tissue Organ Cult.* 98, 125–134.

105. Ruokolainen, S., Ng, Y. P., Albert, V. A., Elomaa, P., and Teeri, T. H. (2010). Large scale interaction analysis predicts that the *Gerbera hybrida* floral E function is provided both by general and specialized proteins. *BMC Plant Biol.* 10:129.

106. Ruokolainen, S., Ng, Y. P., Albert, V. A., Elomaa, P., and Teeri, T. H. (2011). Over-expression of the *Gerbera hybrida* At-SOC1-like1 gene Gh-SOC1 leads to floral organ identity deterioration. *Ann. Bot.* 107, 1491–1499.

107. Sanikhani, M., Mibus, H., Stummann, B. M., and Serek, M. (2008). Kalanchoe blossfeldiana plants expressing the Arabidopsis etr1-1 allele show reduced ethylene sensitivity. *Plant Cell Rep.* 27, 729–737.

108. Satoh, S., Kosugi, Y., Iwazaki, Y., Narumi, T., and Kinouchi, T. (2006). "Genetic engineering of ethylene production and perception in carnation and chrysanthemum," in *Floriculture, Ornamental and Plant Biotechnology: Advances and Topical Issues*, ed J. A. Teixeira da Silva (Isleworth: Global Science Books, Ltd.), 140–163.

109. Satoh, S., Watanabe, M., Chisaka, K., and Narumi, T. (2008). Suppressed leaf senescence in Chrysanthemum transformed with a mutated ethylene receptor gene mDG-ERS1 (etr1-4). *J. Plant Biol.* 51, 424–427.

110. Savin, K. W., Baudinette, S. C., Graham, M. W., Michael, M. Z., Nugent, G. D., Lu, C. Y., et al. (1995). Antisense ACC oxidase RNA delays carnation petal senescence. *HortScience* 30, 970–972.

111. Schwinn, K. E., Boase, M. R., Bradley, J. M., Lewis, D. H., Deroles, S. C., Martin, C. R., et al. (2014). MYB and bHLH transcription factor transgenes increase anthocyanin pigmentation in petunia and lisianthus plants, and the petunia phenotypes are strongly enhanced under field conditions. *Front. Plant Sci.* 5:603.

112. Sen, S., Kumar, S., Ghani, M., and Thakur, M. (2013). *Agrobacterium* mediated genetic transformation of chrysanthemum (*Dendranthema grandiflora* Tzvelev) with rice chitinase gene for improved resistance against *Septoria obesa*. *Plant Pathol. J.* 12, 1–10.
113. Shaw, J., Chen, H., Tsai, M., Kuom, C. I., and Huangm, L. C. (2002). Extended flower longevity of *Petunia hybrida* plants transformed with boers,a mutated ERS gene of *Brassica oleracea*. *Mol. Breed.* 9, 211–216.
114. Shi, S., Ma, F., Li, Y., Feng, F., and Shang, Z. (2012). Overexpression of l-galactono-1, 4-lactone dehydrogenase (GLDH) in Lanzhou lily (*Lilium davidii* var. unicolor) via particle bombardment-mediated transformation. *In Vitro Cell. Dev. Biol.* 48, 1–6.
115. Shibata, M. (2008). Importance of genetic transformation in ornamental plant breeding. *Plant Biotechnol.* 25, 3–8.
116. Shinoyama, H., Mochizuki, A., Nomura, Y., and Kamada, H. (2008). Environmental risk assessment of genetically modified chrysanthemums containing a modified cry1Ab gene from *Bacillus thuringiensis*. *Plant Biotechnol.* 25, 17–29.
117. Shinoyama, H., Sano, T., Saito, M., Ezura, H., Aida, R., Nomura, Y., et al. (2012). Induction of male sterility in transgenic chrysanthemums (*Chrysanthemum morifolium* Ramat.) by expression of a mutated ethylene receptor gene, Cm-ETR1/H69A, and the stability of this sterility at varying growth temperatures. *Mol. Breed.* 29, 285–295.
118. Shirasawa-Seo, N., Nakamura, S., Ukai, N., Honkura, R., Iwai, T., and Ohashi, Y. (2002). Ectopic expression of an oat thionin gene in carnation plants confers enhanced resistance to bacterial wilt disease. *Plant Biotechnol.* 19, 311–317.
119. Shulga, O. A., Mitiouchkina, T. Y., Shchennikova, A. V., Skryabin, K. G., and Dolgov, S. V. (2011). Overexpression of AP1-like genes from Asteraceae induces early-flowering in transgenic Chrysanthemum plants. *In Vitro Cell. Dev. Biol.* 47, 553–560.
120. Song, A., An, J., Guan, Z., Jiang, J., Chen, F., Lou, W., et al. (2014). The constitutive expression of a two transgene construct enhances the abiotic stress tolerance of *Chrysanthemum*. *Plant Physiol. Biochem.* 80, 114–120.
121. Spitzer-Rimon, B., Farhi, M., Albo, B., Cnaani, A., Ben Zvi, M. M., Masci, T., et al. (2012). The R2R3-MYB-like regulatory factor EOBI, acting downstream of EOBII, regulates scent production by activating ODO1 and structural acent-related genes in Petunia. *Plant Cell* 24, 5089–5105.
122. Srikanth, A., and Schmid, M. (2011). Regulation of flowering time: all roads lead to Rome. *Cell. Mol. Life Sci.* 68, 2013–2037.
123. Suzuki, K., Mizutani, M., Fukui, Y., Ueyama, Y., Katsumoto, Y., Miyazaki, K., et al. (2002). Flower color modification of Torenia by engineering gene expression of cytochromes P450 involved in flavonoid biosynthesis. *In Vitro Cell. Dev. Biol.* 38:109-A.
124. Takatsu, Y., Nishizawa, Y., Hibi, T., and Akutsu, K. (1999). Transgenic chrysanthemum (*Dendranthema grandiflorum* (Ramat.) Kitamura) expressing a rice chitinase gene shows enhanced resistance to gray mold (*Botrytis cinerea*). *Sci. Hortic.* 82, 113–123.
125. Tanaka, Y., and Brugliera, F. (2013). Flower colour and cytochromes P450. *Philos. Trans. R. Soc.* 368:20120432.
126. Tanaka, Y., Brugliera, F., and Chandler, S. (2009). Recent progress of flower colour modification by biotechnology. *Int. J. Mol. Sci.* 10, 5350–5369.
127. Tanaka, Y., Brugliera, F., Kalc, G., Senior, M., Dyson, B., Nakamura, N., et al. (2010). Flower color modification by engineering of the flavonoid biosynthetic pathway: practical perspectives. *Biosci. Biotechnol. Biochem.* 74, 1760–1769.
128. Tanaka, Y., and Ohmiya, A. (2008). Seeing is believing: engineering anthocyanin and carotenoid biosynthetic pathways. *Curr. Opin. Biotechnol.* 19, 190–197.
129. Teixeira da Silva, J. A., Shinoyama, H., Aida, R., Matsushita, Y., Raj, S. K., and Chen, F. (2013). Chrysanthemum biotechnology: quo vadis? *Crit. Rev. Plant Sci.* 32, 21–52.
130. Thiruvengadam, M., and Yang, C. H. (2009). Ectopic expression of two MADS box genes from orchid (*Oncidium* Gower Ramsey) and lily (*Liliumlongiflorum*) alters

flower transition and formation in *Eustoma grandiflorum*. *Plant Cell Rep.* 28, 1463–1473.

131. Tsuda, S., Fukui, Y., Makamura, N., Katsumoto, Y., Yonekura-Sakakibara, K., Fukuchi-Mizutani, M., et al. (2004). Flower color modification of *Petunia hybrida* commercial varieties by metabolic engineering. *Plant Biotechnol.* 21, 377–386.

132. Van der Salm, T. P. M., Bouwer, R. V., Van Dijk, A. J., Keizer, L. C. P., Ten Cate, C. H. H., Van Der Plas, L. H. W., et al. (1998). Stimulation of scion bud release by *rol*gene transformed rootstocks of *Rosa hybrida* L. *J. Exp. Bot.* 49, 847–852.

133. Van Moerkercke, A., Galvan-Ampudia, C. S., Verdonk, J. C., Haring, M. A., and Schuurink, R. C. (2012). Regulators of floral fragrance production and their target genes in 1122 petunia are not exclusively active in the epidermal cells of petals. *J. Exp. Bot.* 63, 3157–3171.

134. Veitch, N. C., and Grayer, R. J. (2008). Flavonoids and their glycosides, including anthocyanins. *Nat. Prod. Rep.* 25, 555–611.

135. Verdonk, J. C., Haring, M. A., van Tunen, A. J., and Schuurink, R. C. (2005). ODORANT1 regulates fragrance biosynthesis in petunia flowers. *Plant Cell* 17, 1612–1624.

136. Vieira, P., Wantoch, S., Lilley, C. J., Chitwood, D. J., Atkinson, H. J., and Kamo, K. (2015). Expression of a cystatin transgene can confer resistance to root lesion nematodes in *Lilium longiflorum* Nellie White'. *Transgenic Res.* 24, 421–432.

137. Villacorta-Martin, C., Núñez de Cáceres González, F. F., de Haan, J., Huijben, K., Passarinho, P., Hamo, M. L. B., et al. (2015). Whole transcriptome profiling of the vernalization process in *Lilium longiflorum* (cultivar White Heaven) bulbs. *BMC Genomics* 16:1.

138. Vinocur, B., and Altman, A. (2005). Recent advances in engineering plant tolerance to abiotic stress, achievements and limitations. *Curr. Opin. Biotechnol.* 16, 123–132.

139. Warner, R. (2011). *Genetic Approaches to Improve Cold Tolerance of Petunia Funding Industry Solutions Through Research and Education.* Available online at: https://endowment.org/wp-content/uploads/2014/03/306Warner.pdf

140. Wen, C., Zhao, Q., Nie, J., Liu, G., Shen, L., Cheng, C., et al. (2016). Physiological controls of *chrysanthemum DgD27* gene expression in regulation of shoot branching. *Plant Cell Rep.* 35, 1053–1070.

141. Xie, Q., Chen, G., Liu, Q., Zhu, Z., and Hu, Z. (2015). Dual silencing of DmCPD and DmGA20ox genes generates a novel miniature and delayed-flowering *Dendranthema morifolium* variety. *Mol. Breed.* 35, 1–13.

142. Xu, G., Chen, S., and Chen, F. (2010). Transgenic chrysanthemum plants expressing a harpinXoo gene demonstrate induced resistance to *Alternaria* leaf spot and accelerated development. *Russ. J. Plant Physiol.* 57, 548–553.

143. Xu, G., Liu, Y., Chen, S., and Chen, F. (2011). Potential structural and biochemical mechanisms of compositae wild species resistance to *Alternaria tenuissima*. *Russ. J. Plant Physiol.* 58, 491–497.

144. Yamagishi, M., Yoshida, Y., and Nakayama, M. (2012). The transcription factor LhMYB12 determines anthocyanin pigmentation in the tepals of Asiatic hybrid lilies (*Lilium* spp.) and regulates pigment quantity. *Mol. Breed.* 30, 913–925.

145. Yoshida, K., Mori, M., and Kondo, T. (2009). Blue flower color development by anthocyanins: from chemical structure to cell physiology. *Nat. Prod. Rep.* 26, 884–915.

146. Yoshioka, S., Aida, R., Yamamizo, C., Shibata, M., and Ohmiya, A. (2012). The carotenoid cleavage dioxygenase 4 (CmCCD4a) gene family encodes a key regulator of petal color mutation in chrysanthemum. *Euphytica* 184, 377–387.

147. Yu, M., Liu, Z. L., Chen, S. M., and Chen, F. D. (2010). Expression of *Prunus mume PGIP* gene in transgenic *Dendranthema morifolium* increased tolerance to disease resistance. *Acta Bot. Boreal. Occident. Sin.* 30, 1111–1116.

148. Yue, Y., Yu, R., and Fan, Y. (2015). Transcriptome profiling provides new insights into the formation of floral scent in *Hedychium coronarium*. *BMC Genomics* 16:470.

149. Zafar, S., Ashraf, M. Y., Anwar, S., Ali, Q., and Noman, A. (2016). Yield enhancement in wheat by soil and foliar fertilization of K and Zn under saline environment. *Soil Environ.* 35, 46–55.
150. Zakizadeh, H., Lütken, H., Sriskandarajah, S., Serek, M., and Müller, R. (2013). Transformation of miniature potted rose (*Rosa hybrida* cv Linda') with P SAG12-ipt gene delays leaf senescence and enhances resistance to exogenous ethylene. *Plant Cell Rep.* 32, 195–205.
151. Zheng, Z. L., Yang, Z., Jang, J. C., and Metzger, J. D. (2001). Modification of plant architecture in chrysanthemum by ectopic expression of the tobacco phytochrome B1 gene. *J. Am. Soc. Hortic. Sci.* 126, 19–26.
152. Zuker, A., Tzfira, T., Scovel, G., Ovadis, M., Shklarman, E., Itzhaki, H., et al. (2001). RolC transgenic carnation with improved horticultural traits: quantitative and qualitative analyses of greenhouse-grown plants. *J. Am. Soc. Hortic. Sci.* 126, 13–18.
153. Zvi, M. M. B., Shklarman, E., Masci, T., Kalev, H., Debener, T., Shafir, S., et al. (2012). PAP1 transcription factor enhances production of phenylpropanoid and terpenoid scent compounds in rose flowers. *N. Phytol.* 195, 335–345.

CHAPTER 8

Bioinformatics for Agriculture in the Next-Generation Sequencing Era

Alfonso Esposito, Chiara Colantuono, Valentino Ruggieri and Maria Luisa Chiusano*

Department of Agricultural Sciences, University of Naples Federico II, 80055 Portici (Na), Italy

ABSTRACT

The key role of bioinformatics is acquiring striking importance in the era of outstanding advances in omics technologies for its fundamental support in describing the multifaceted aspects of biological functionalities. The manifold omics efforts flourishing worldwide are also contributing fundamental novelties in many aspects of agricultural sciences and, as a consequence, bioinformatics is acquiring a crucial role also in these research fields. Indeed, the transformation of natural environment for improvement of goods from animal, plants, and microbial worlds for human nutrition and health requires the comprehension of the molecular mechanisms influencing the structure and the function of the individuals, the populations, and the communities. The expanding knowledge about the molecules and the mechanisms associated with specific phenotypic traits and specific responses to biotic or abiotic stresses, complemented with the predictive power of bioinformatics, has an impact on agriculture practices and favors innovative methods in diagnostics, monitoring, and traceability, improving human benefits at lower costs, thus supporting sustainability.

We here describe main bioinformatics approaches in the era of Next-Generation Sequencing for its impact in genomics, transcriptomics, and metagenomics efforts, describing their role in agriculture sciences. We aim to introduce common aspects, open questions and perspectives in this cutting-edge field of research.

Keywords: NGS Agriculture bioinformatics Omics technologies Biological challenges

INTRODUCTION

Bioinformatics emerged from the initial requirement of suitable informatics for biological data organization, management, and distribution [1], but soon it revealed also fundamental in providing tools for data analysis, interpretation, and modeling. Moreover, thanks to bioinformatics, it was possible to analyze and understand structure and function not only of single bio-molecules but also of larger molecular collections, derived from the so-called omics experimental approaches. These efforts permit to depict different aspects (genomics, transcriptomics, proteomics, metabolomics, etc.) of the biomolecular organization of complex biological systems, from cells to ecosystems. The fast spreading of omics techniques, with its growing power and more accessible costs, drastically increased the amount of molecular data collections from different levels of organization of an organism or an environmental sample. This favored a holistic view on systems organization and functionality, further challenging bioinformatics with data size and the need of integrative efforts [2, 3]. The recent introduction of Next-Generation Sequencing (NGS) technologies (Table 1) further revolutionized the sequencing of nucleic acids contributing to a new era in omics approaches. Indeed, on one hand these technologies introduced an incredible efficiency in terms of experimental execution time and a deeper resolution. On the other hand, they stimulated an unexpected interest from scientists due to the higher affordability in terms of experimental procedures and economical requirements. We here introduce the novelties that the advent of NGS technologies contributed in agriculture, overviewing the main bioinformatics strategies and challenges, as well as perspectives in the field.

Table 1. Main features of the most used NGS technologies in omic studies

Technology	Read length	Yield (Reads per Run)	Reference
Roche 454	700	~700 thousand	http://454.com/products/gs-flx-system/
Illumina HiSeq	300	~300 billion	http://www.illumina.com/systems/hiseq-3000-4000/specifications.html
SOLiD	100	~200 billion	https://www.thermofisher.com/it/en/home/life-science/sequencing/next-generation-sequencing/solid-next-generation-sequencing/solid-next-generation-sequencing-systems-reagents-accessories.html
Ion Torrent	200	~60 billion	http://www.thermofisher.com/it/en/home/life-science/sequencing/next-generation-sequencing/ion-torrent-next-generation-sequencing-workflow/ion-torrent-next-generation-sequencing-run-sequence/ion-proton-system-for-next-generation-sequencing/ion-proton-system-specifications.html
PacBio RS II	14,000	~47 thousand	http://files.pacb.com/pdf/PacBio_RS_II_Brochure.pdf

REVIEW

1-Single and multi species genomics for agriculture

Genomics, transcriptomics, proteomics, and metabolomics may contribute to the comprehension of the organization and the functionality of biological systems, with the possibility to also trace molecular variability during development, in different conditions, such as physiological, pathological, or influenced by environmental changes [4].

Samples for omics studies can derive from one or several individuals of a species (a population), or from multiple species (a community) [5, 6, 7]. The differences among these approaches consist mainly in the objectives of the specific studies.

In single individual approaches, the organization and the functionality of specific cells, tissues, or organs (e.g., roots, fruit, rumen) are investigated, mainly to identify factors influencing emergent properties, like the quality and the shape. This also paves the way to the characterization of even more complex traits (e.g., yield, resistance to stresses, diseases) or processes (e.g., fruit ripening, growth efficiency, senescence) [4, 8].

The description of the molecular components in a population of the same species aims to the understanding of evolutionary processes influencing genetic variability. This can also widely contribute to dissect complex quantitative traits by identifying novel and superior alleles [9, 10] or to assess the impact of genetic variation on patterns of gene expression and on phenotypic plasticity in response to environmental changes [11].

The study of the collective genetic pool deriving from communities is termed "metagenomics" [12], a term dating back to 1998. The community can derive from environmental samples, such as soil [13], seawater [14], or other [15], but also part of individuals, such as gut or roots [16, 17].

Metagenomics usually aims to describe the prokaryotic component of the community, but may be also useful to trace the different eukaryotes existing in a specific environment [18].

Nucleic acid sequencing always contributed the majority of the data in all the approaches here summarized. This is why the recent introduction of NGS technologies impressively impacted the productivity and the advancement in these research fields.

2-Impact of NGS in agriculture

The multifaceted scientific topics in agriculture sciences may be consistently supported by NGS omics for single individuals, populations, or communities [19, 20, 21].

The sequencing of whole genomes from several species permits to define their organization and provides the starting point for understanding their functionality [22, 23, 24, 25], therefore favoring human agriculture practice. Efforts addressed to the achievement of an appropriate knowledge of associated molecular information, such as the one arising from transcriptome and proteome sequencing, are also essential to better depict the gene content of a genome and its main functionalities. These efforts indeed led to major advancements in all biological sciences [4] and in agriculture as well [8, 26]. Moreover, the elucidation of the complexity of genes and their networking is also fundamental for being eventually translated into breeding practice for crops or livestock, contributing to their health, resistance, and productivity. Indeed, the contribution of genomics to agriculture spans the identification and the manipulation of genes linked to specific phenotypic traits [27] as well as genomics breeding by marker-assisted selection of variants [28, 29]. The so-called "agricultural genomics" (or agri-genomics), indeed, aims to find innovative solutions through the study of crops or livestock genomes, achieving information for protection [30, 31] and sustainable productivity for food industry, but also for alternative aspects like energy production or design of pharmaceuticals [32, 33, 34, 35].

Plant, soil, and livestock microbiome also play a key role in agriculture since it determines plant fitness [36, 37], soil biogeochemical properties [38], and affects both yield and quality traits [39, 40]. However, little knowledge is available for microbes and the communities in which they are included. As an example, it is acknowledged that soil is one of the biggest carbon reservoirs on earth, and prokaryotes constitute an important amount of the soil biomass [41]. However, culture-independent studies in the last three decades showed that, although sequencing strategies are fast evolving, the great majority of bacterial species is still unknown [42, 43]. Therefore most of the methods used for profiling microbial communities and describe their main functional features are now adopting whole DNA extraction and the use of NGS on the entire sample, with the objective of sequencing and characterizing DNA fragments of all the species included, i.e., the metagenome. The application of metagenomics in agriculture also proved to be appropriate for depicting the complex patterns of interactions occurring among microorganisms in soil [44] and in plant rhizosphere [45], as well as in specific tissues or organs [6, 46, 47]. Metagenomics recently revealed to be useful to trace the shift in taxonomic composition and functional redundancy of microbial communities in rhizosphere and in soil in connection to environmental changes associated to fertilization [48] and agricultural management [49, 50]. Metagenomics studies can also help deciphering the role of soil bacteria in plant nutrition [51, 52] or in the cycle of the elements [53]. Further applications can lead to the discovery of new genes, bio-products, plant growth promoting microorganisms consortia, useful for understanding relevant aspects such as response to stresses [36] or dysbiosis [54, 55, 56].

3-The revolution in the omics technologies and the impact on bioinformatics

The introduction of omics approaches strongly impacted bioinformatics in data collection, organization [57, 58, 59], integration, and in the implementation of suitable data mining tools [2, 60]. The support of efficient bioinformatics favored the introduction of the so-called high-throughput technologies, paving the way to the flourishing of genome sequencing efforts of key model species, such as *Homo sapiens* and *Arabidopsis thaliana* (Figure 1). The same technologies were then exploited to further push forward the genome sequencing of other model and non-model species, many of which of agriculture interest. These efforts were also preceded or accompanied by transcriptome sequencing efforts using different technologies [61, 62, 63, 64], in support of gene prediction [65], but also for depicting transcriptional processes and define cell functionality in physiological, pathological, or stress conditions. These approaches also required the design of appropriate resources to distribute the data [66] and/or dedicated collections of processed results [67, 68, 69, 70] to all the interested scientific community, enhancing the need for suitable pipelines for moving from raw to value added information and integrative data mining [71, 72] (Figure 2, Table 2).

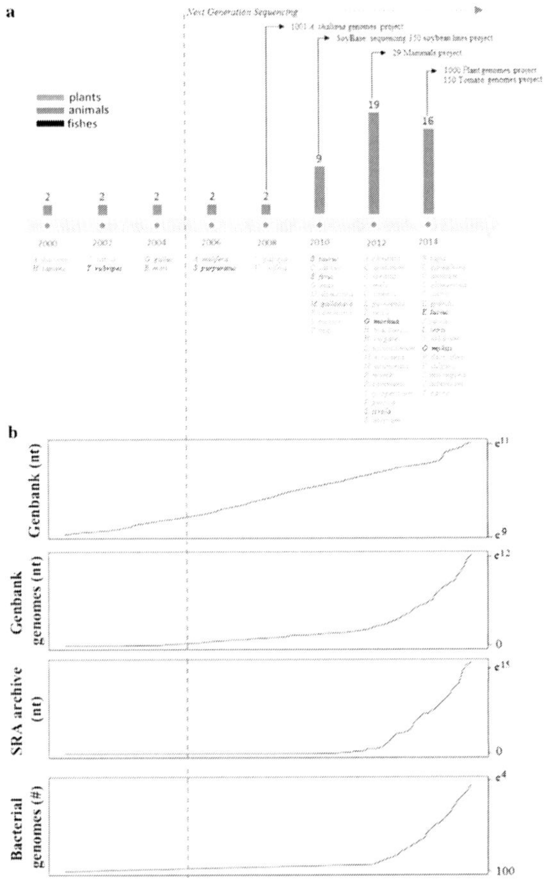

Figure 1. (**A**) Timeline from 2000 to 2014 indicating the release of the completely sequenced genomes for some of the major plants (*green*), animals (*red*), and fishes (*black*) of interest in agriculture. The *dashed line* indicates the start of Next-Generation Sequencing (NGS) era. The start of major massive sequencing projects are also indicated: the 1001 *A. thaliana* genomes project (http://1001genomes.org/), the SoyBase project (sequencing of 350 soybean lines) (http://www.soybase.org/), the 29 Mammals genomes project (https://www.broadinstitute.org/scientific-community/science/projects/mammals-models/29-mammals-project), the 1000 Plant genomes project (https://sites.google.com/ a/ualberta.ca /onekp/), and the 150 Tomato genomes ReSequencing project (http:// www.tomatogenome.net/). (**B**) *Graphs* indicate the growth as number of nucleotides (nt) of GenBank (all entries), GenBank genomes (only genome sequencing efforts), and SRA archive (all entries) and the number (#) of bacterial genomes (https://gold.jgi.doe.gov/) released in the same timeline

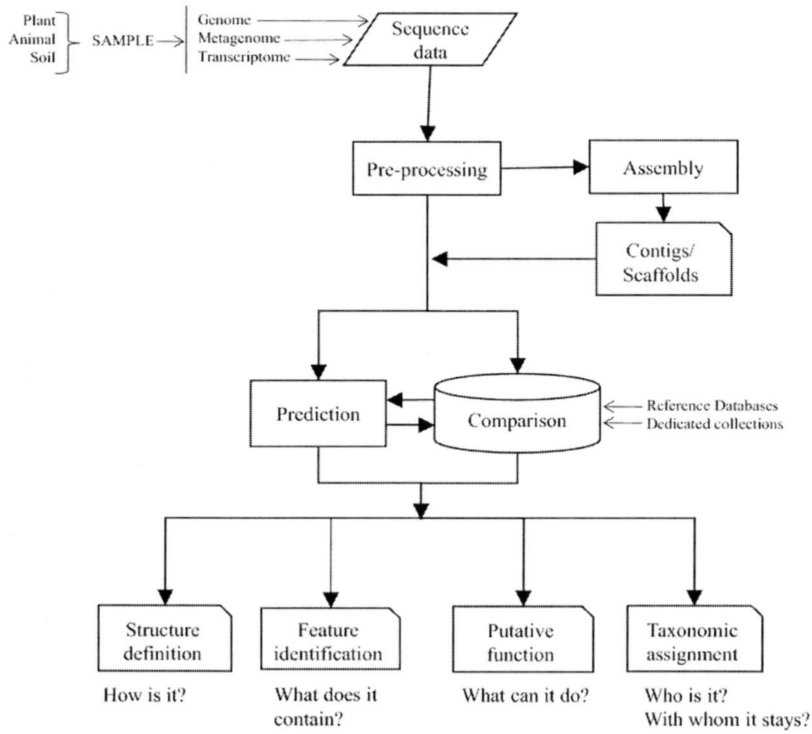

Figure 2. General description of a standard workflow in omics data analyses. Starting from the biological sample of interest and sequencing by different approaches data can be processed by assembly into contigs/scaffolds or directly undergo prediction analyses and/or comparison with reference databases. The analyses support different kinds of investigations such as those required for (i) structure definition, (ii) feature identification (e.g., genes, regulatory and repeat regions), (iii) putative functionality, and (iv) taxonomic assignment

NGS strikingly contributed to expand the number of genomes currently completely sequenced (Figure 1), as well as to the establishment of novel ambitious efforts, for instance, those focused on multi-genome sequencing [24, 25, 73] or those aiming to define global metagenomes from different environmental samples to define reference collections [74, 75]. These technologies are also exploited for the production of alternative, related collections, namely from transcriptomics, epigenomics, and metagenomics projects. The unexpected amount of raw data the new technologies are providing requires also dedicated storage for centralized data maintenance, currently solved by the SRA system [76]. Worthy to note is the size reached by the SRA archive in a short time span when compared to the entire nucleotide collection currently available (Figure 1). NGS data size represents a big challenge for bioinformatics. Indeed, main computational tasks are today focused on the optimization and adaptation of typical methodologies in bioinformatics to the magnitude of NGS collections.

Table 2. Most used open-source software and reference databases in genomic, transcriptomic, and metagenomic studies

Category	Task	Name	Aims and Scope	Usage	Reference
Software and pipelines	Reads pre-processing	FastQC	Quality check and report of NGS data	GM	http://www.bioinformatics.babraham.ac.uk/projects/fastqc/
		cutadapt	Adapter trimming algorithm	GM	[95]
		FASTX-toolkit	Toolset for manipulation of sequence data and format conversion	GM	http://hannonlab.cshl.edu/fastx_toolkit/index.html/
	Assembly	(META) VELVET/OASES	De novo genomic/transcriptomic assembly based on the de Brujin graph	GM	[96, 97]
		SOAP DE NOVO	De novo short-read assembler based on the de Brujin graph	G	[98]
		TRINITY	De novo assembly of RNA-seq data	G	[99]
	Gene prediction/annotation	Ensembl genome annotation	Gene annotation pipeline	G	http://www.ensembl.org/info/genome/genebuild/genome_annotation.html/
		Infernal	RNA secondary structure prediction based on reference multiple sequence alignments	G	[100]
		(Meta) Genemark	Gene prediction with unsupervised and semi-supervised training	GM	[101]
		(Meta) Genomethreader	Gene prediction by similarity with cDNA/EST and/or protein sequences	GM	[102]
		NCBI genome annotation	Genome annotation pipeline released by NCBI	G	http://www.ncbi.nlm.nih.gov/books/NBK169439/
		tRNAscan-SE	tRNA gene prediction	G	[103]
		Repeat masker	Similarity-based detection of DNA interspersed repeats and low complexity sequences	G	http://www.repeatmasker.org/
	Mapping	Star	RNA-seq to genome aligner	G	[104]
		Tophat/cufflinks	RNA-seq to genome aligner and quantification tools	G	[105]
	Marker-based metagenome	Mothur	Tools and software for 16S data clustering, classification, and ecological inference	M	[106]
		Qiime	Customizable pipeline for marker-gene-based metagenomics	M	[107]
		RDPipeline	RDP-based web interface for bacterial and fungal ribosomal marker gene analysis	M	[108]
	Mixed	Galaxy	Web-based platform of general purposes	GM	[109]
		transPLANT	e-infrastructure for exploring genomic data from crop and model plants	G	http://www.transplantdb.eu/
	Shotgun metagenome	Megan	Stand-alone blast output parser and mining tool for phylogenetic and functional assignment based on the lowest common ancestor algorithm	M	[110]
		Metamos	Customizable pipeline for shotgun data assembly and analysis	M	[111]
		(Mg-)Rast	Fully automated online server for analyses of shotgun data	G[a]M	[112]
	Population genomic	Metabel	Software for meta-analysis of genome-wide SNP association	G	[113]
		Metal	Tool for mining variation data and perform association studies	G	[114]
		Plink	Tools for managing genomic variation data	GM	[115]
		SVS	Genomic and phenotypic data analysis and visualization	G	http://www.goldenhelix.com
		Tassel	Tools and pipelines for genome variation studies	G	[116]
		VcfTools	Tools for genome comparisons and mining plant variation data	GM	[117]
Reference Databases	General	Genomes online database	Metadata repository for genome and metagenome sequencing projects	GM	https://gold.jgi.doe.gov/
		JGI Phytozome	Plant Comparative Genomics at the Joint Genome Institute	G	http://phytozome.jgi.doe.gov/pz/portal.html
		INSDC	DDBJ, EMBL-EBI, and NCBI, common repository	GM	http://www.insdc.org/
		PLANTGDB	Unified plant genomic database	G	http://pgdbj.jp/
	Taxonomic annotation	RDP/Silva/Greengenes	Repositories of ribosomal RNA genes	GM	[118–120]
	Functional annotation	KEGG	Integrated resources for functional annotation of genes	GM	[121]
		COG	Clusters of ortholog groups	GM	[122]
		SEED	Integrated resources for functional annotated microbial genes	G[a]M	[123]
		RFAM	RNA families collection	G	[124]
		DFAM	Repetitive DNA elements collection	G	[125]
		UNIPROT	Database of functional annotated protein sequences	G	http://www.uniprot.org/

G use in genomics and transcriptomics, *M* use in metagenomics

[a] Dedicated to microbial genomes

In Table 2, main methodologies and resources exploited for NGS data analyses are summarized. Data are usually delivered by sequencing centers in the form of raw, fragmented sequences, to be pre-processed, i.e., cleaned, from additional fragments due to the specific technology employed (Table 1), such as vectors, adaptors, barcodes, or other contaminations (Table 2). Structure definition from fragmented data usually requires an assembly step to reconstruct the most reliable original molecules, such as longer genomic sequences or transcripts. The assembly is based on sequence alignments driven by identical regions shared by the fragments (Table 2). The assembly procedure may include already available reference sequences (guided approaches), as in the case of transcript assemblies based on a genome reference, or they are based on reference-free methodologies (de novo) [77]. This procedure is a key step for many different applications, indeed high-quality longer backbones are useful to properly proceed towards successive steps in data processing, which are mainly structure and functional assignments and predictions (Figure 2). In genomics, transcriptomics, population genetics, and metagenomics, these are widespread fundamental tasks solved by different computational methodologies (Table 2), though based on similar strategies (Figure 2), mainly ab initio approaches or similarity-based ones. Ab initio-based algorithms exploit complex probabilistic models to detect expected features (genes, motives, propensities) as defined on the basis of training datasets that support the proper identification of specific features. Similarity-based methods, on the other hand, are the principal and more frequently used approaches in bioinformatics since they permit identifications, predictions, structure, and functional assignments. They rely on sequence or tridimensional similarities exploiting a typical concept in biology, which considers similarity in structure as a possible indication of similarity in role. Comparison with reference (nucleotide or amino acid general databases) or more specific collections, such as those from genomes, gene families, transcriptomes, or repeats, is fundamental to transfer information from already annotated molecules to newly defined ones. Beyond supporting structure and functional assignments by detection of common features, similarity searches also support identification of peculiarities and provide hints for evolutionary investigations [78].

The availability of an increasing number of reference genomes associated with a decreasing sequencing cost per base enabled also the analysis of genome variations based on single nucleotide polymorphisms (SNP) discovery. These studies intend to identify the variability between genomic sequences from individual genomes. By comparing the sequenced genomes, a catalog of mutations from individuals is obtained, usually defined as SNPs and/or insertion-deletion (INDELs), but also as larger rearrangements (e.g., copy number (CNV) and presence absence (PAV) variations, translocations). These features can be associated with specific phenotypes of interest. Currently, millions of polymorphisms have been discovered in plants, such as *A. thaliana* [79], rice [80, 81], soybean [82], tomato [73, 83], maize [84, 85], and in animals [86]. These resources are essential in breeding challenges for species of agriculture interest. Indeed, the possibility to exploit data from larger collections of individuals strongly increases the potential to identify more alleles useful for improved a sustainable productions, providing solutions for growing demand for better food, in a climate changing world.

As introduced, metagenomics approaches aim at the identification of the species in a sample and to the definition of their relationships. While bioinformatics approaches for what concerns pre-processing and assembly steps are similar (Figure 2), downstream analyses depend on the peculiarities of the implemented strategy. The taxonomic composition of the microbiome can be profiled using the marker-based approach, i.e., a PCR-amplification with universal primers of a taxon-specific gene, followed by the extensive sequencing of the amplicon by the preferred platform (Table 1). Sequences derived from such studies are usually compared with dedicated databases representing high-quality full-length reference tags. For example, in the case of bacteria the choice falls mainly on the *16S* gene from the ribosomal operon and the most used reference databases for comparisons and identification are listed in Table 2. Pipelines have been also implemented to aid non-experts users in a correct parsing of the metagenome-derived NGS data (Table 2). Phylogenetic relationships, obtained by sequence similarity, can be used for ecological inference using dedicated pipelines [87]. However, although widespread, the marker-based approach falls short in predicting the functionality and the activity of the microbial community. Indeed, the methodology suffers the typical PCR biases, such as (i) the misincorporation of nucleotides (which would lead to the overestimation of sequence diversity); (ii) the differential amplification of the same gene from different organisms (true for example in the case of *16S* genes whose number of copies in the genome varies among taxa [88]); and (iii) the formation of chimeric artifacts. Moreover, markers can have limits in taxonomical assignments, mainly because of lack of consistent genome information from all possible species, affecting the specificity of the identification of the components [43]. This leads to the wide use of the operative taxonomic units (OTUs) for distinguishing all the different components in a sample, since they represent groups of highly similar sequences [89]. The "shotgun" approach is alternative to the marker-based one. It consists in the high-throughput sequencing of a pool of DNA fragments that may encompass various genomic loci from all taxa represented in the sample (prokaryotic, eukaryotic, and viral genomes). Unlike target-based approaches, the shotgun technique provides more details on the genomic structure of the community, offering a wider description of its potential functionality [13, 14, 50, 55]. Data from whole metagenome shotgun consist of short DNA reads that can be assembled to obtain coding sequences or genomic contigs. Coding sequences can be identified through the comparison with specific databases (Table 2). The assembly should be carefully evaluated because most of the assemblers were developed for genome assembly and are not designed to deal with the heterogeneity of metagenomic datasets. In alternative, raw reads can be also used for direct assignment and annotation, though their short length may limit the exhaustivity of the results. Ultimately, other limits of the shotgun method are (i) the initial amount of extracted DNA for library production should be rather high (>10 ng); and (ii) in case of large and complex communities, or communities where one or few species dominate over the others, the coverage of the entire components may be limited. Indeed, the likelihood that the species poorly represented will be covered by sufficient reads that will also permit the assembly of representative contigs is rather low. Examples of main software dedicated to metagenome analyses are reported in Table 2. Ultimately, other limits of the shotgun method are (i) the initial amount of extracted DNA for library production is rather high (>10 ng); and (ii) in case of large and complex communities, or communities where one or few species dominate over the others, the likelihood that more than one read will cover a single gene is rather low; therefore, little information will be obtained about the species with low abundance, and assembly would probably result in short contigs (if any). As introduced, reference data collections are fundamental since the beginning of bioinformatics. Data sharing, by general reference or specialized databases, is precious to the majority of the bioinformatics strategies here presented. They not only support fundamental analyses for straightforward characterization of the investigated molecules, but are also essential to offer results to the whole scientific community. To this aim, the effort of setting up comprehensive collections, suitably representing the metadata derived by their mining and by the integration of different resources, demands major efforts in the NGS bioinformatics era. Indeed, NGS technologies attracted an unexpected interest from the scientific world for their accessibility and their resolution power, further challenging the stabilization of resources and data integration. As an example, the fast sequencing of complete genomes from different species, such as those from different genotypes or cultivars, faces the bottleneck caused by the need of suitable data analyses and curation. On the other hand, the fast production of collections from parallel efforts makes the update of novel release hard, though the presence of reference databases, favoring the flourishing of community specific collections, often misaligned with reference ones, affecting also the quality of the results.

CONCLUSIONS

Bioinformatics is the exclusive approach capable of exploiting and sharing the large amount of omics data the different technologies may provide. Suitable computational methods and appropriate resources are fundamental for detecting value added biological information providing novel insights into the organization of biological systems. The identification of structure and functional properties of the molecular data in a specific process allows the in-depth understanding of systems organization and behavior, supporting the design of reliable and representative models and paving the way to the comprehension of emergent properties that only holistic approaches can offer.

However, despite the introduction of highly processive experimental technologies and of innovative computational approaches in support of the molecular characterizations, only 10 % of the genome organization and associated functionalities have been today understood and an even lower percentage of metagenomics datasets have been confidently annotated [90]. This confirms that though at quite 70 years from the discovery of the DNA structure and the beginning of bioinformatics, we are still at the very early stage of the genomics era, and surely quite far from achieving the ambitious goal of the in silico simulation of complex living organisms as well as ecosystem relationships. Indeed, these efforts still demand for extensive and suitable studies of genomes, transcriptomes, and metagenome data for proper links with sample organization and functionality, considering single species analyses and community approaches.

Despite these limits, the NGS bioinformatics era is revolutionizing the experimental design in molecular biology, strikingly contributing in increasing scientific knowledge while impacting relevant applications in many different aspects of agriculture. Data from disparate research fields, such as breeding, microbiology, and environmental sciences, are favoring a common exploitation and advances in molecular knowledge from massive efforts, with bioinformatics as driving methodology for its power and multifaceted capabilities. Organizing, detecting, integrating data information content, and data sharing are contributing to multidisciplinary interactions, expanding resources and spreading common methodologies. This revolutionize agriculture practice and production, offering knowledge and tools for improved product quality and ameliorated strategies of protection against environmental stress, diseases, and parasites [40, 91]. The different applications here overviewed, beyond providing relevant scientific knowledge based on their specificities, are also fundamental for translational approaches providing contributions with technological innovation, novel products, predictive and monitoring approaches [92], also supporting innovative applications for crop and livestock management [93, 94].

The increase of omics-based studies needs education in the associated technologies and in bioinformatics for appropriate experimental design and analyses, and for properly conveying experimental and computational efforts towards an in-depth knowledge and appropriate modeling of the biological systems [126].

AUTHORS' CONTRIBUTIONS

AE introduced the metagenomics section and contributed to the organization of the manuscript; CC and VR introduced the genomics and transcriptomics topics; MLC planned, organized and supervised the entire effort. All authors contributed to the writing of the manuscript. All authors read and approved the final version.

ACKNOWLEDGEMENTS

This work was supported by the Genopom PRO and GenHORT Projects (Ministero dell'Istruzione, dell'Università e della Ricerca (MIUR), Italy) and it was in the frame of the COST ACTION (FA1106).

COMPETING INTERESTS

The authors declare that they have no competing interests.

REFERENCES

1. Dayhoff MO. Atlas of protein sequence and structure. Silver Spring: National Biomedical Research Foundation; 1965.
2. Chiusano ML, D'Agostino N, Traini A, Licciardello C, Raimondo E, Aversano M, Frusciante L, Monti L. ISOL@: an Italian SOLAnaceae genomics resource. BMC Bioinform. 2008;9(Suppl 2):S7.
3. Bostan H, Chiusano ML. NexGenEx-Tom: a gene expression platform to investigate the functionalities of the tomato genome. BMC Plant Biol. 2015;15(1):48.
4. Barh D, Zambare V, Azevedo V, Omics. Applications in biomedical, agricultural, and environmental sciences. Boca Raton: CRC Press. 2013.
5. Grice EA, Segre JA. The human microbiome: our second genome. Annu Rev Genomics Hum Genet. 2012;13:151.
6. Knief C. Analysis of plant microbe interactions in the era of next generation sequencing technologies. Front Plant Sci. 2014;5.
7. Wang J, McLenachan PA, Biggs PJ, Winder LH, Schoenfeld BI, Narayan VV, Phiri BJ, Lockhart PJ. Environmental bio-monitoring with high-throughput sequencing. Brief Bioinform. 2013;14(5):575–88.
8. Van Emon J. Omics revolution in agricultural research. J Agri Food Chem. 2015.
9. Zhu B, Pennack JA, McQuilton P, Forero MG, Mizuguchi K, Sutcliffe B, Gu CJ, Fenton JC, Hidalgo A. Drosophila neurotrophins reveal a common mechanism for nervous system formation. PLoS Biol. 2008;6(11):e284.
10. Semagn K, Bjørnstad Å, Xu Y. The genetic dissection of quantitative traits in crops. Electron J Biotechnol. 2010;13(5):16–7.
11. Ekblom R, Wolf JBW. A field guide to whole-genome sequencing, assembly and annotation. Evol Appl. 2014;7(9):1026–42.
12. Handelsman J, Rondon MR, Brady SF, Clardy J, Goodman RM. Molecular biological access to the chemistry of unknown soil microbes: a new frontier for natural products. Chem Biol. 1998;5(10):R245–9.
13. Howe AC, Jansson JK, Malfatti SA, Tringe SG, Tiedje JM, Brown CT. Tackling soil diversity with the assembly of large, complex metagenomes. Proc Natl Acad Sci. 2014;111(13):4904–9.
14. Kodzius R, Gojobori T. Marine metagenomics as a source for bioprospecting. Mar Genom. 2015.
15. Esposito A, Ahmed E, Ciccazzo S, Sikorski J, Overmann J, Holmström SJ, Brusetti L. Comparison of rock varnish bacterial communities with surrounding non-varnished rock surfaces: taxon-specific analysis and morphological description. Microb Ecol 2015; 1–10.
16. Consortium HMP. Structure, function and diversity of the healthy human microbiome. Nature. 2012;486(7402):207–14.
17. Ofek-Lalzar M, Sela N, Goldman-Voronov M, Green SJ, Hadar Y, Minz D. Niche and host-associated functional signatures of the root surface microbiome. Nat Commun. 2014; 5.
18. Epp LS, Boessenkool S, Bellemain EP, Haile J, Esposito A, Riaz T, Erseus C, Gusarov VI, Edwards ME, Johnsen A. New environmental metabarcodes for analysing soil DNA: potential for studying past and present ecosystems. Mol Ecol. 2012;21(8):1821–33.
19. Benkeblia N. Sustainable agriculture and new biotechnologies. Boca Raton: CRC Press; 2011.
20. Siol M, Wright SI, Barrett SC. The population genomics of plant adaptation. New Phytol. 2010;188(2):313–32.
21. Barabaschi D, Tondelli A, Desiderio F, Volante A, Vaccino P, Valè G, Cattivelli L. Next generation breeding. Plant Sci. 2015.

22. Morrell PL, Buckler ES, Ross-Ibarra J. Crop genomics: advances and applications. Nat Rev Genet. 2012;13(2):85–96.
23. Ellegren H. Genome sequencing and population genomics in non-model organisms. Trends Ecol Evol. 2014;29(1):51–63.
24. Weigel D, Mott R. The 1001 genomes project for Arabidopsis thaliana. Genome Biol. 2009;10(5):107.
25. Lindblad-Toh K, Garber M, Zuk O, Lin MF, Parker BJ, Washietl S, Kheradpour P, Ernst J, Jordan G, Mauceli E, Ward LD, Lowe CB, Holloway AK, Clamp M, Gnerre S, Alfoldi J, Beal K, Chang J, Clawson H, Cuff J, Di Palma F, Fitzgerald S, Flicek P, Guttman M, Hubisz MJ, Jaffe DB, Jungreis I, Kent WJ, Kostka D, Lara M, et al. A high-resolution map of human evolutionary constraint using 29 mammals. Nature. 2011;478(7370):476–82.
26. Chiusano M, D'Agostino N, Barone A, Carputo D, Frusciante L. Genome analysis of species of agricultural interest, In: Advances in modeling agricultural systems. Berlin: Springer; 2009. p. 385–402.
27. Zhang Z, Ober U, Erbe M, Zhang H, Gao N, He J, Li J, Simianer H. Improving the accuracy of whole genome prediction for complex traits using the results of genome wide association studies. PLoS One. 2014;9(3):e93017.
28. Organization EPS. European plant science: a field of opportunities. J Exp Bot. 2005;56(417):1699–709.
29. Iovene M, Barone A, Frusciante L, Monti L, Carputo D. Selection for aneuploid potato hybrids combining a low wild genome content and resistance traits from Solanum commersonii. Theor Appl Genet. 2004;109(6):1139–46.
30. van der Vlugt R, Minafra A, Olmos A, Ravnikar M, Wetzel T, Varveri C, Massart S. Application of next generation sequencing for study and diagnosis of plant viral diseases in agriculture. 2015.
31. Van Borm S, Belák S, Freimanis G, Fusaro A, Granberg F, Höper D, King DP, Monne I, Orton R, Rosseel T. Next-generation sequencing in veterinary medicine: how can the massive amount of information arising from high-throughput technologies improve diagnosis, control, and management of infectious diseases? In: Veterinary infection biology: molecular diagnostics and high-throughput strategies. Berlin: Springer; 2015. p. 415–36.
32. Blanchfield J. Genetically modified food crops and their contribution to human nutrition and food quality. J Food Science. 2004, 69(1):CRH28-CRH30.
33. Yuan JS, Tiller KH, Al-Ahmad H, Stewart NR, Stewart CN Jr. Plants to power: bioenergy to fuel the future. Trends Plant Sci. 2008;13(8):421–9.
34. Ma JKC, Drake PMW, Christou P. The production of recombinant pharmaceutical proteins in plants. Nat Rev Genet. 2003;4(10):794–805.
35. Wilson SA, Roberts SC. Metabolic engineering approaches for production of biochemicals in food and medicinal plants. Curr Opin Biotechnol. 2014;26:174–82.
36. Timmusk S, El-Daim IAA, Copolovici L, Tanilas T, Kännaste A, Behers L, Nevo E, Seisenbaeva G, Stenström E, Niinemets Ü. Drought-tolerance of wheat improved by rhizosphere bacteria from harsh environments: enhanced biomass production and reduced emissions of stress volatiles. 2014.
37. Haney CH, Samuel BS, Bush J, Ausubel FM. Associations with rhizosphere bacteria can confer an adaptive advantage to plants. Nat Plants 2015.
38. Acosta-Martínez V, Cotton J, Gardner T, Moore-Kucera J, Zak J, Wester D, Cox S. Predominant bacterial and fungal assemblages in agricultural soils during a record drought/heat wave and linkages to enzyme activities of biogeochemical cycling. Applied Soil Ecology. 2014;84:69–82.
39. Babu AN, Jogaiah S. Ito S-i, Nagaraj AK, Tran L-SP. Improvement of growth, fruit weight and early blight disease protection of tomato plants by rhizosphere bacteria is correlated with their beneficial traits and induced biosynthesis of antioxidant peroxidase and polyphenol oxidase. Plant Sci. 2015;231:62–73.

40. Deusch S, Tilocca B, Camarinha-Silva A, Seifert J. News in livestock research—use of Omics-technologies to study the microbiota in the gastrointestinal tract of farm animals. Computational and Structural Biotechnology Journal. 2015;13:55–63.
41. Whitman WB, Coleman DC, Wiebe WJ. Prokaryotes: the unseen majority. Proc Natl Acad Sci. 1998;95(12):6578–83.
42. Staley JT, Konopka A. Measurement of in situ activities of nonphotosynthetic microorganisms in aquatic and terrestrial habitats. Ann Rev Microbiol. 1985;39(1):321–46.
43. Rappé MS, Giovannoni SJ. The uncultured microbial majority. Ann Rev Microbiol. 2003;57(1):369–94.
44. Carbonetto B, Rascovan N, Álvarez R, Mentaberry A, Vázquez MP. Structure, composition and metagenomic profile of soil microbiomes associated to agricultural land use and tillage systems in Argentine Pampas. 2014.
45. Mendes LW, Kuramae EE, Navarrete AA, van Veen JA, Tsai SM. Taxonomical and functional microbial community selection in soybean rhizosphere. The ISME journal. 2014;8(8):1577–87.
46. Fouts DE, Szpakowski S, Purushe J, Torralba M, Waterman RC, MacNeil MD, Alexander LJ, Nelson KE. Next generation sequencing to define prokaryotic and fungal diversity in the bovine rumen. 2012.
47. Rastogi G, Coaker GL, Leveau JH. New insights into the structure and function of phyllosphere microbiota through high-throughput molecular approaches. FEMS Microbiol Lett. 2013;348(1):1–10.
48. Pan Y, Cassman N, de Hollander M, Mendes LW, Korevaar H, Geerts RH, van Veen JA, Kuramae EE. Impact of long-term N, P, K, and NPK fertilization on the composition and potential functions of the bacterial community in grassland soil. FEMS Microbiol Ecol. 2014;90(1):195–205.
49. Bevivino A, Paganin P, Bacci G, Florio A, Pellicer MS, Papaleo MC, Mengoni A, Ledda L, Fani R, Benedetti A. Soil Bacterial community response to differences in agricultural management along with seasonal changes in a mediterranean region. 2014.
50. Souza RC, Hungria M, Cantão ME, Vasconcelos ATR, Nogueira MA, Vicente VA. Metagenomic analysis reveals microbial functional redundancies and specificities in a soil under different tillage and crop-management regimes. Appl Soil Ecol. 2015;86:106–12.
51. Lavecchia A, Curci M, Jangid K, Whitman WB, Ricciuti P, Pascazio S, Crecchio C. Microbial 16S gene-based composition of a sorghum cropped rhizosphere soil under different fertilization managements. Biol Fertil Soils. 2015;51(6):661–72.
52. Pii Y, Borruso L, Brusetti L, Crecchio C, Cesco S, Mimmo T. The interaction between iron nutrition, plant species and soil type shapes the rhizosphere microbiome. Plant Physiol Biochem. 2016;99:39–48.
53. Stempfhuber B, Richter-Heitmann T, Regan KM, Kölbl A, Kaul P, Marhan S, Sikorski J, Overmann J, Friedrich MW, Kandeler E. Spatial interaction of archaeal ammonia-oxidizers and nitrite-oxidizing bacteria in an unfertilized grassland soil. Front Microbiol. 2015;6:1567.
54. Vayssier-Taussat M, Albina E, Citti C, Cosson J-F, Jacques M-A, Lebrun M-H, Le Loir Y, Ogliastro M, Petit M-A, Roumagnac P. Shifting the paradigm from pathogens to pathobiome: new concepts in the light of meta-omics. Front Cell Infect Microbiol. 2014; 4.
55. Handelsman J. Metagenomics: application of genomics to uncultured microorganisms. Microbiol Mol Biol Rev. 2004;68(4):669–85.
56. Urano K, Kurihara Y, Seki M, Shinozaki K. 'Omics' analyses of regulatory networks in plant abiotic stress responses. Curr Opin Plant Biol. 2010;13(2):132–8.
57. Benson DA, Karsch-Mizrachi I, Lipman DJ, Ostell J, Rapp BA, Wheeler DL. GenBank. Nucleic Acids Res. 2000;28(1):15–8.

58. Pruitt KD, Tatusova T, Maglott DR. NCBI reference sequences (RefSeq): a curated non-redundant sequence database of genomes, transcripts and proteins. Nucleic Acids Res. 2007;35(suppl 1):D61–5.
59. Hubbard T, Barker D, Birney E, Cameron G, Chen Y, Clark L, Cox T, Cuff J, Curwen V, Down T. The Ensembl genome database project. Nucleic Acids Res. 2002;30(1):38–41.
60. Dong Q, Schlueter SD, Brendel V. PlantGDB, plant genome database and analysis tools. Nucleic Acids Res. 2004;32(suppl 1):D354–9.
61. Adams MD, Kelley JM, Gocayne JD, Dubnick M, Polymeropoulos MH, Xiao H, Merril CR, Wu A, Olde B, Moreno RF, et al. Complementary DNA sequencing: expressed sequence tags and human genome project. Science. 1991;252(5013):1651–6.
62. Velculescu VE, Zhang L, Vogelstein B, Kinzler KW. Serial analysis of gene expression. Science. 1995;270(5235):484–7.
63. Kodzius R, Kojima M, Nishiyori H, Nakamura M, Fukuda S, Tagami M, Sasaki D, Imamura K, Kai C, Harbers M. CAGE: cap analysis of gene expression. Nat Methods. 2006;3(3):211–22.
64. Brenner S, Johnson M, Bridgham J, Golda G, Lloyd DH, Johnson D, Luo S, McCurdy S, Foy M, Ewan M. Gene expression analysis by massively parallel signature sequencing (MPSS) on microbead arrays. Nat Biotechnol. 2000;18(6):630–4.
65. Iquebal M, Jaiswal S, Mukhopadhyay C, Sarkar C, Rai A, Kumar D: Applications of Bioinformatics in Plant and Agriculture in PlantOmics: The Omics of Plant Science. 2015, Springer:755-89.
66. Boguski MS, Lowe TM, Tolstoshev CM. dbEST–database for "expressed sequence tags". Nat Genet. 1993;4(4):332–3.
67. Pontius JU, Wagner L, Schuler GD. 21. UniGene: A Unified View of the Transcriptome. The NCBI Handbook. Bethesda, MD: National Library of Medicine (US), NCBI 2003.
68. Pertea G, Huang X, Liang F, Antonescu V, Sultana R, Karamycheva S, Lee Y, White J, Cheung F, Parvizi B. TIGR Gene Indices clustering tools (TGICL): a software system for fast clustering of large EST datasets. Bioinformatics. 2003;19(5):651–2.
69. D'Agostino N, Traini A, Frusciante L, Chiusano ML. SolEST database: a. BMC Plant Biol. 2009;9(1):142.
70. Mita K, Morimyo M, Okano K, Koike Y, Nohata J, Kawasaki H, Kadono-Okuda K, Yamamoto K, Suzuki MG, Shimada T. The construction of an EST database for Bombyx mori and its application. Proc Natl Acad Sci. 2003;100(24):14121–6.
71. Christoffels A, van Gelder A, Greyling G, Miller R, Hide T, Hide W. STACK: Sequence Tag Alignment and Consensus Knowledgebase. Nucleic Acids Res. 2001;29(1):234–8.
72. D'Agostino N, Aversano M, Chiusano ML. ParPEST: a pipeline for EST data analysis based on parallel computing. BMC Bioinform. 2005;6(Suppl 4):S9.
73. Aflitos S, Schijlen E, Jong H, Ridder D, Smit S, Finkers R, Wang J, Zhang G, Li N, Mao L. Exploring genetic variation in the tomato (Solanum section Lycopersicon) clade by whole-genome sequencing. Plant J. 2014;80(1):136–48.
74. Vogel TM, Simonet P, Jansson JK, Hirsch PR, Tiedje JM, van Elsas JD, Bailey MJ, Nalin R, Philippot L. TerraGenome: a consortium for the sequencing of a soil metagenome. Nat Rev Microbiol. 2009;7(4):252.
75. Gilbert JA, Meyer F, Antonopoulos D, Balaji P, Brown CT, Brown CT, Desai N, Eisen JA, Evers D, Field D. Meeting report: the terabase metagenomics workshop and the vision of an Earth microbiome project. Stand Genomic Sci. 2010;3(3):243.
76. Leinonen R, Sugawara H, Shumway M. The sequence read archive. Nucleic Acids Res. 2010;gkq1019.
77. Martin JA, Wang Z. Next-generation transcriptome assembly. Nat Rev Genet. 2011;12(10):671–82.
78. Mathé C, Sagot MF, Schiex T, Rouze P. Current methods of gene prediction, their strengths and weaknesses. Nucleic Acids Res. 2002;30(19):4103–17.

79. Gan X, Stegle O, Behr J, Steffen JG, Drewe P, Hildebrand KL, Lyngsoe R, Schultheiss SJ, Osborne EJ, Sreedharan VT. Multiple reference genomes and transcriptomes for Arabidopsis thaliana. Nature. 2011;477(7365):419–23.
80. Xu X, Liu X, Ge S, Jensen JD, Hu F, Li X, Dong Y, Gutenkunst RN, Fang L, Huang L. Resequencing 50 accessions of cultivated and wild rice yields markers for identifying agronomically important genes. Nat Biotechnol. 2012;30(1):105–11.
81. Subbaiyan GK, Waters DL, Katiyar SK, Sadananda AR, Vaddadi S, Henry RJ. Genome-wide DNA polymorphisms in elite indica rice inbreds discovered by whole-genome sequencing. Plant Biotechnol J. 2012;10(6):623–34.
82. Lam H-M, Xu X, Liu X, Chen W, Yang G, Wong F-L, Li M-W, He W, Qin N, Wang B. Resequencing of 31 wild and cultivated soybean genomes identifies patterns of genetic diversity and selection. Nat Genet. 2010;42(12):1053–9.
83. Aflitos SA, Sanchez-Perez G, Ridder D, Fransz P, Schranz ME, Jong H, Peters SA. Introgression browser: high-throughput whole-genome SNP visualization. Plant J. 2015;82(1):174–82.
84. Lai J, Li R, Xu X, Jin W, Xu M, Zhao H, Xiang Z, Song W, Ying K, Zhang M. Genome-wide patterns of genetic variation among elite maize inbred lines. Nat Genet. 2010;42(11):1027–30.
85. Hufford MB, Xu X, Van Heerwaarden J, Pyhäjärvi T, Chia J-M, Cartwright RA, Elshire RJ, Glaubitz JC, Guill KE, Kaeppler SM. Comparative population genomics of maize domestication and improvement. Nat Genet. 2012;44(7):808–11.
86. Romiguier J, Gayral P, Ballenghien M, Bernard A, Cahais V, Chenuil A, Chiari Y, Dernat R, Duret L, Faivre N. Comparative population genomics in animals uncovers the determinants of genetic diversity. Nature. 2014;515(7526):261–3.
87. Lozupone C, Knight R. UniFrac: a new phylogenetic method for comparing microbial communities. Appl Environ Microbiol. 2005;71(12):8228–35.
88. Rastogi R, Wu M, DasGupta I, Fox GE. Visualization of ribosomal RNA operon copy number distribution. BMC Microbiol. 2009;9(1):208.
89. Schloss PD, Westcott SL. Assessing and improving methods used in operational taxonomic unit-based approaches for 16S rRNA gene sequence analysis. Appl Environ Microbiol. 2011;77(10):3219–26.
90. Gilbert JA, Field D, Swift P, Thomas S, Cummings D, Temperton B, Weynberg K, Huse S, Hughes M, Joint I. The taxonomic and functional diversity of microbes at a temperate coastal site: a 'multi-omic' study of seasonal and diel temporal variation. PLoS ONE. 2010;5(11):e15545.
91. Coleman-Derr D, Tringe SG. Building the crops of tomorrow: advantages of symbiont-based approaches to improving abiotic stress tolerance. Front Microbiol. 2014;5(283):1–6.
92. Sloan SS, Lebeis SL. Exercising influence: distinct biotic interactions shape root microbiomes. Curr Opin Plant Biol. 2015;26:32–6.
93. Sachdev DP, Cameotra SS. Biosurfactants in agriculture. Appl Microbiol Biotechnol. 2013;97(3):1005–16.
94. Rolli E, Marasco R, Vigani G, Ettoumi B, Mapelli F, Deangelis ML, Gandolfi C, Casati E, Previtali F, Gerbino R. Improved plant resistance to drought is promoted by the root-associated microbiome as a water stress-dependent trait. Environ Microbiol. 2015;17(2):316–31.
95. Martin M. Cutadapt removes adapter sequences from high-throughput sequencing reads. EMBnet J. 2011;17(1):10–2.
96. Zerbino DR, Birney E. Velvet: algorithms for de novo short read assembly using de Bruijn graphs. Genome Res. 2008;18(5):821–9.
97. Schulz MH, Zerbino DR, Vingron M, Birney E. Oases: robust de novo RNA-seq assembly across the dynamic range of expression levels. Bioinformatics. 2012;28(8):1086–92.
98. Luo R, Liu B, Xie Y, Li Z, Huang W, Yuan J, He G, Chen Y, Pan Q, Liu Y. SOAPdenovo2: an empirically improved memory-efficient short-read de novo assembler. Gigascience. 2012;1(1):18.

99. Grabherr MG, Haas BJ, Yassour M, Levin JZ, Thompson DA, Amit I, Adiconis X, Fan L, Raychowdhury R, Zeng Q. Trinity: reconstructing a full-length transcriptome without a genome from RNA-Seq data. Nat Biotechnol. 2011;29(7):644.
100. Nawrocki EP, Kolbe DL, Eddy SR. Infernal 1.0: inference of RNA alignments. Bioinformatics. 2009;25(10):1335–7.
101. Lomsadze A, Ter-Hovhannisyan V, Chernoff YO, Borodovsky M. Gene identification in novel eukaryotic genomes by self-training algorithm. Nucleic Acids Res. 2005;33(20):6494–506.
102. Gremme G, Brendel V, Sparks ME, Kurtz S. Engineering a software tool for gene structure prediction in higher organisms. Inf Softw Technol. 2005;47(15):965–78.
103. Lowe TM, Eddy SR. tRNAscan-SE: a program for improved detection of transfer RNA genes in genomic sequence. Nucleic Acids Res. 1997;25(5):0955–64.
104. Dobin A, Davis CA, Schlesinger F, Drenkow J, Zaleski C, Jha S, Batut P, Chaisson M, Gingeras TR. STAR: ultrafast universal RNA-seq aligner. Bioinformatics. 2013;29(1):15–21.
105. Trapnell C, Roberts A, Goff L, Pertea G, Kim D, Kelley DR, Pimentel H, Salzberg SL, Rinn JL, Pachter L. Differential gene and transcript expression analysis of RNA-seq experiments with TopHat and Cufflinks. Nat Protoc. 2012;7(3):562–78.
106. Schloss PD, Westcott SL, Ryabin T, Hall JR, Hartmann M, Hollister EB, Lesniewski RA, Oakley BB, Parks DH, Robinson CJ. Introducing mothur: open-source, platform-independent, community-supported software for describing and comparing microbial communities. Appl Environ Microbiol. 2009;75(23):7537–41.
107. Caporaso JG, Kuczynski J, Stombaugh J, Bittinger K, Bushman FD, Costello EK, Fierer N, Pena AG, Goodrich JK, Gordon JI. QIIME allows analysis of high-throughput community sequencing data. Nat Methods. 2010;7(5):335–6.
108. Cole JR, Wang Q, Cardenas E, Fish J, Chai B, Farris RJ, Kulam-Syed-Mohideen A, McGarrell DM, Marsh T, Garrity GM. The Ribosomal Database Project: improved alignments and new tools for rRNA analysis. Nucleic Acids Res. 2009;37(suppl 1):D141–5.
109. Goecks J, Nekrutenko A, Taylor J. Galaxy: a comprehensive approach for supporting accessible, reproducible, and transparent computational research in the life sciences. Genome Biol. 2010;11(8):R86.
110. Huson DH, Auch AF, Qi J, Schuster SC. MEGAN analysis of metagenomic data. Genome Res. 2007;17(3):377–86.
111. Treangen TJ, Koren S, Sommer DD, Liu B, Astrovskaya I, Ondov B, Darling AE, Phillippy AM, Pop M. MetAMOS: a modular and open source metagenomic assembly and analysis pipeline. Genome Biol. 2013;14(1):R2.
112. Meyer F, Paarmann D, D'Souza M, Olson R, Glass EM, Kubal M, Paczian T, Rodriguez A, Stevens R, Wilke A. The metagenomics RAST server–a public resource for the automatic phylogenetic and functional analysis of metagenomes. BMC Bioinform. 2008;9(1):386.
113. Aulchenko YS, Ripke S, Isaacs A, Van Duijn CM. GenABEL: an R library for genome-wide association analysis. Bioinformatics. 2007;23(10):1294–6.
114. Willer CJ, Li Y, Abecasis GR. METAL: fast and efficient meta-analysis of genomewide association scans. Bioinformatics. 2010;26(17):2190–1.
115. Purcell S, Neale B, Todd-Brown K, Thomas L, Ferreira MA, Bender D, Maller J, Sklar P, De Bakker PI, Daly MJ. PLINK: a tool set for whole-genome association and population-based linkage analyses. Am J Hum Genet. 2007;81(3):559–75.
116. Bradbury PJ, Zhang Z, Kroon DE, Casstevens TM, Ramdoss Y, Buckler ES. TASSEL: software for association mapping of complex traits in diverse samples. Bioinformatics. 2007;23(19):2633–5.
117. Danecek P, Auton A, Abecasis G, Albers CA, Banks E, DePristo MA, Handsaker RE, Lunter G, Marth GT, Sherry ST. The variant call format and VCFtools. Bioinformatics. 2011;27(15):2156–8.

118. Wang Q, Garrity GM, Tiedje JM, Cole JR. Naive Bayesian classifier for rapid assignment of rRNA sequences into the new bacterial taxonomy. Appl Environ Microbiol. 2007;73(16):5261–7.
119. Quast C, Pruesse E, Yilmaz P, Gerken J, Schweer T, Yarza P, Peplies J, Glöckner FO. The SILVA ribosomal RNA gene database project: improved data processing and web-based tools. Nucleic Acids Res. 2012:gks1219.
120. McDonald D, Price MN, Goodrich J, Nawrocki EP, DeSantis TZ, Probst A, Andersen GL, Knight R, Hugenholtz P. An improved Greengenes taxonomy with explicit ranks for ecological and evolutionary analyses of bacteria and archaea. ISME J. 2012;6(3):610–8.
121. Kanehisa M, Goto S. KEGG: kyoto encyclopedia of genes and genomes. Nucleic Acids Res. 2000;28(1):27–30.
122. Tatusov RL, Fedorova ND, Jackson JD, Jacobs AR, Kiryutin B, Koonin EV, Krylov DM, Mazumder R, Mekhedov SL, Nikolskaya AN. The COG database: an updated version includes eukaryotes. BMC Bioinform. 2003;4(1):41.
123. Overbeek R, Begley T, Butler RM, Choudhuri JV, Chuang HY, Cohoon M, de Crecy-Lagard V, Diaz N, Disz T, Edwards R, Fonstein M, Frank ED, Gerdes S, Glass EM, Goesmann A, Hanson A, Iwata-Reuyl D, Jensen R, Jamshidi N, Krause L, Kubal M, Larsen N, Linke B, McHardy AC, Meyer F, Neuweger H, Olsen G, Olson R, Osterman A, Portnoy V, et al. The subsystems approach to genome annotation and its use in the project to annotate 1000 genomes. Nucleic Acids Res. 2005;33(17):5691–702.
124. Burge SW, Daub J, Eberhardt R, Tate J, Barquist L, Nawrocki EP, Eddy SR, Gardner PP, Bateman A. Rfam 11.0: 10 years of RNA families. Nucleic Acids Res. 2012:gks1005.
125. Wheeler TJ, Clements J, Eddy SR, Hubley R, Jones TA, Jurka J, Smit AF, Finn RD. Dfam: a database of repetitive DNA based on profile hidden Markov models. Nucleic Acids. 2013;41:D70–82 (Database issue).
126. Chiusano ML. On the multifaceted aspects of bioinformatics in the next generation era: the run that must keep the quality. J Next Gener Seq Appl 2015.

CHAPTER 9

Genomics and bioinformatics resources for translational science in Rosaceae

Sook Jung[1], Dorrie Main[1]

Department of HorticultureWashington State UniversityPullmanUSA

ABSTRACT

Recent technological advances in biology promise unprecedented opportunities for rapid and sustainable advancement of crop quality. Following this trend, the Rosaceae research community continues to generate large amounts of genomic, genetic and breeding data. These include annotated whole genome sequences, transcriptome and expression data, proteomic and metabolomic data, genotypic and phenotypic data, and genetic and physical maps. Analysis, storage, integration and dissemination of these data using bioinformatics tools and databases are essential to provide utility of the data for basic, translational and applied research. This review discusses the currently available genomics and bioinformatics resources for the Rosaceae family.

Keywords: Rosaceae Bioinformatics Database Genomics Genetics Breeding

INTRODUCTION

Rosaceae, comprised of over 100 genera and 3,000 species, contains a variety of crop species that are both biologically and economically important. Fruit-producing crops include apple (*Malus*), pear (*Pyrus*), raspberries/blackberries (*Rubus*), strawberries (*Fragaria*), and stone fruits (*Prunus*), such as peach/nectarine, apricot, plum, cherry and almond. Rosaceae also contains a wide variety of ornamental plants including roses, flowering cherry, crabapple, quince and pear. Crop improvement has traditionally been performed by incorporating desired traits from wild relatives through conventional breeding. Recent advances in high-throughput technology have revolutionized biology and provide unprecedented opportunities for rapid advancement of crop improvement through marker or genomics-assisted breeding.

The first spike in data generation occurred in the early 1990s when large-scale sequencing became available through the use of Sanger technology for EST and BAC sequencing (Adams et al. 1991; Shizuya et al. 1992). In the last few years, the advent of next generation technologies, such as 454 and Illumina, have significantly enhanced the ability to generate large-scale transcriptome and genome sequence at a fraction of the cost of Sanger sequencing. Similar advances in DNA array, nuclear magnetic resonance (NMR), Fourier transform infrared spectroscopy, Fourier transform ion cyclotron resonance mass spectroscopy, high performance liquid chromatography and mass spectrometry have generated large-scale gene expression, proteome and metabolome data for many species. Other data types include molecular marker data along with genetic mapping data and/or large-scale genotyping data of various varieties. More recently, high-throughput phenotypic and genotypic data are also being generated to study the interaction between genotype, phenotype and environment as well as for breeding purposes. All of these large-scale data require proper analysis, storage, integration and dissemination to enhance our understanding of biology and to be utilized in further research. Bioinformatics tools and methodologies, therefore, have become an essential and integral part of the new era of "information-driven" biological research.

Large-scale genome and transcriptome data were initially accumulated for model species, but are now available for a wide range of species. Crop plants with sequenced genomes include rice (*Oryza sativa*) (International Rice Genome Sequencing Project 2005), grapevine (*Vitis vinifera*) (Jaillon et al. 2007), sorghum (*Sorghum bicolor*) (Paterson et al. 2009), cucumber (*Cucumis sativus*) (Huang et al. 2009), maize (*Zea mays*) (Schnable et al. 2009), soybean (*Glycine max*) (Schmutz et al. 2010), cotton (Wang et al. 2012), sweet orange (Xu et al. 2012) and five species in Rosaceae:

peach (*Prunus persica*), apple (*Malus domestica*) (Velasco et al. 2010), strawberry (*Fragaria vesca*) (Shulaev et al. 2011), pear (*Pyrus bretschneideri*) (Wu et al. 2013) and *Prunus mume* (Zhang et al. 2012), with red raspberry, black raspberry, apricot and plum, among others, currently being sequenced. In addition to the annotated whole genome sequences, a wealth of other genomic and genetic data is available for Rosaceae (Shulaev et al. 2008). These include BAC libraries, peach and apple physical maps, ESTs, numerous genetic maps in various species of Rosaceae and molecular markers that have been used for mapping and genotyping. Large-scale genotypic and phenotypic data are also being generated from various projects including HIDRAS (Gianfranceschi and Soglio 2004), ISAFRUIT (Audergon et al. 2009), GENBERRY (Diamanti et al. 2012) RosBREED (Iezzoni et al. 2010) and FruitBreedomics (http://www.fruitbreedomics.com). Currently, there are only limited proteomic and metabolomic data available for Rosaceae.

In this review, we discuss genomics and bioinformatics resources for the Rosaceae family with the corresponding database resources that can be utilized for Rosaceae researchers across disciplines (Figure 1; Table 1). Resources available in other model species are also discussed, since these present a valuable tool in conducting and future planning of research in Rosaceae.

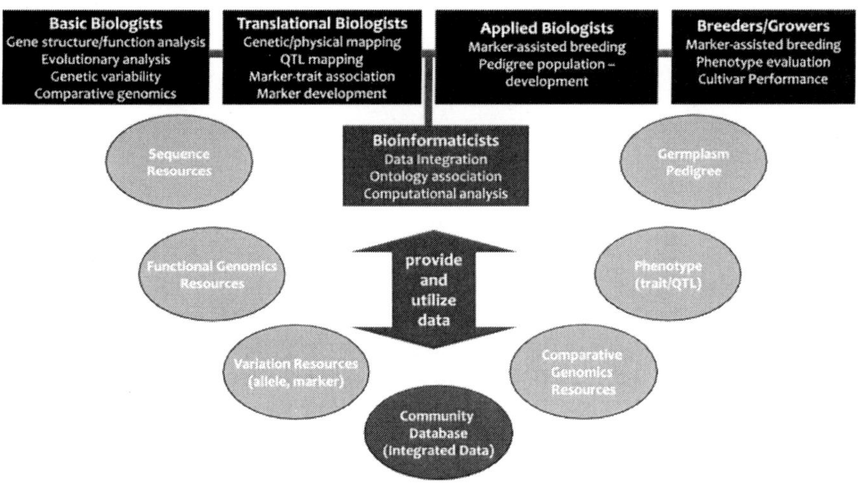

Figure 1. Genomics and bioinformatics resources that can be utilized by Rosaceae researchers across disciplines

Table 1. Database resources for the Rosaceae family

Name	URL	Reference
GenBank	http://www.ncbi.nlm.nih.gov/genbank/	Benson et al. (2013)
European Nucleotide Archive (ENA)	http://www.ebi.ac.uk/ena/	Cochrane et al. (2013)
DNA Data Bank of Japan (DDBJ)	http://www.ddbj.nig.ac.jp/	Kaminuma et al. (2011)
Phytozome	http://www.phytozome.org/	
CoGe (The Place to Compare Genomes)	http://synteny.cnr.berkeley.edu/CoGe/	Lyons and Freeling (2008)
The plant genome duplication database (PGDD)	http://chibba.agtec.uga.edu/duplication/	Lee et al. (2013)
Plaza	http://bioinformatics.psb.ugent.be/plaza/	Van Bel et al. (2012)
GreenPhylD	http://greenphyl.cirad.fr/v2/cgi-bin/index.cgi	Rouard et al. (2011)
SALAD database	http://salad.dna.affrc.go.jp/salad/en/	Mihara et al. (2010)
NCBI's Gene Expression Omnibus (GEO)	http://www.ncbi.nlm.nih.gov/geo/	Wheeler et al. (2007)
ESTree DB	http://www.itb.cnr.it/estree/	Lazzari et al. (2008)
UniProt	http://www.uniprot.org/	The UniProt Consortium (2012)
The Worldwide Protein Data Bank (wwPDB)	http://www.wwpdb.org/	Berman et al. (2007)
CATH database	http://www.cathdb.info/	Cuff et al. (2011)
SUPERFAMILY	http://supfam.cs.bris.ac.uk/SUPERFAMILY/	Wilson et al. (2009)
ARAMEMNON	http://aramemnon.uni-koeln.de/	Schwacke et al. (2003)
The Arabidopsis Interactions Viewer	http://bar.utoronto.ca/interactions/cgi-bin/arabidopsis_interactions_viewer.cgi	Geisler-Lee et al. (2007)
The Biomolecular Interaction Network Database (BIND)	http://download.baderlab.org/BINDTranslation/	Isserlin et al. (2011)
The Arabidopsis Subcellular Database (SUBA)	http://suba.plantenergy.uwa.edu.au/	Heazlewood et al. (2007)
The Plant Protein Phosphorylation DataBase (P3DB)	http://p3db.org/	Yao et al. (2012)
AtMetExpress	http://prime.psc.riken.jp/lcms/AtMetExpress/	Matsuda et al. (2010)
Golm Metabolome Database (GMD)	http://gmd.mpimp-golm.mpg.de/	Hummel et al. (2010)
Genome Database for Rosaceae (GDR)	http://www.rosaceae.org/	Jung et al. (2008)

SEQUENCE RESOURCES

The availability of extensive sequence data forms an essential genomics resource in designing various research platforms to understand the biology of crops and to apply the knowledge in their improvements. Three primary sequence databases, to one of which researchers seeking peer-reviewed publication of new genomic data are required to submit their genome or gene sequences, are GenBank (Benson et al. 2013), European Nucleotide Archive (Cochrane et al. 2013) and the DNA Data Bank of Japan (Kaminuma et al. 2011). These databases synchronize daily, and, consequently, are generally the most up-to-date source of genomic data for any species. A caveat to this rule is where genome sequences are released ahead of publication through other portals, as in the case of the peach, sweet orange, mandarin and cotton genomes, for example, which were released through Phytozome (http://www.phytozome) ahead of publication. GenBank is built and distributed by the National Center for Biotechnology (NCBI). They collect sequence data by submission from authors and bulk submission from high-throughput sequencing centers. The Entrez search site on the homepage of NCBI (http://www.ncbi.nih.gov) allows researchers to search all the different data that it houses. Searching for information on *Prunus* species, by typing Prunus[ORGN] in the search box, returns a results page categorized by the different databases it houses. As of February 1, 2013, this search returned 131 genes, 110,815 ESTs, 7,481 unigenes, 4,931 proteins, 48,261 genome survey sequences, 46,915 nucleotide sequences, 14,705 SNPs and 89 large-scale datasets in the short read archives (SRA). For *Malus*, 61 genes, 336,190 ESTs, 22,493 unigenes, 3,523 proteins, 76 genome survey sequences, 6,410 nucleotide sequences, 7,907 SNPs and 11 large-scale datasets in the SRA. For *Fragaria*, 609 genes, 58,795 ESTs, 3,055 proteins, 17 genome survey sequences, 3,212 nucleotide sequences and 18 large-scale datasets in the SRA. Users can also download the whole genome data from their FTP site and BLAST against the whole genome. NCBI, is therefore, a good source of sequence data for Rosaceae, although access to the whole genome data is somewhat limited, and some of the larger datasets such as the SRA are not generally very useful for biologists without access to bioinformatics expertise to help with analysis of data. Researchers can keep up to date on new data submitted to NCBI for their species of interest by creating a My NCBI account which saves search history and filters data, among other features. For more information on the data and tools available at NCBI (GenBank), DDJB or EMBL (ENA), readers should refer to the excellent online tutorials at these sites and the yearly update published in the Nucleic Acids Research Database Addition published every January.

Recently, several genomes of Rosaceae have been sequenced and some of these are available from the Genome Database for Rosaceae (GDR, http://www.rosaceae.org/peach/genome, Jung et al. 2008) and Phytozome (http://www.phytozome.org/peach.php). The peach genome v1.0 is a high quality plant genome. It currently consists of eight pseudomolecules (scaffolds) representing the eight chromosomes of peach, and are numbered according to their corresponding linkage groups. The genome sequencing of the double haploid cultivar 'Lovell' consisted of approximately 7.7-fold whole genome shotgun sequencing employing the highly accurate Sanger methodology, and was assembled using Arachne (Batzoglou et al. 2002). The assembled peach scaffolds cover nearly 99 % of the peach genome, with over 92 % having confirmed orientation. To further validate the quality of the assembly, 74,757 *Prunus* ESTs were queried against the genome at 90 % identity and 85 % coverage, and it was found that only ~2 % were missing. While gene prediction and annotation is an ongoing process, current estimates indicate that peach has 28,689 transcripts and 27,852 genes. The whole genome sequence of apple, *Malus* × *domestica*, was reported by sequencing and assembly of the 'Golden Delicious' apple genome followed the whole genome shotgun approach (Velasco et al. 2010). Of the 16.9-

fold genome coverage, 26 % was provided by Sanger dye primer sequencing of paired reads, and the remaining 74 % was from 454 sequencing by synthesis of paired and unpaired reads. The sequence data showed that a relatively recent (>50 million years ago) genome-wide duplication event occurred in Pyreae lineage. The domesticated apple genotypes are all highly heterozygous and the assembly produced overlapping contigs. Of 122,146 contigs, 103,076 were assembled into 1,629 metacontigs. Anchoring of metacontigs (598.3 Mb, or 71.2 % of genome) was based on the high-quality genetic map with 1,643 markers. In total, 17 linkage groups, or chromosomes, were reconstructed. The total number of genes predicted for the apple genome is 57,386, including some genes that may be present only in one of the two chromosomes of a pair. Efforts are on-going in the apple community to improve the assembly and annotation of the apple genome. A diploid strawberry, *Fragaria vesca*, has also been sequenced as a reference genome for the genus that contains the cultivated tetraploid strawberry, *Fragaria* × *ananassa* (Shulaev et al. 2011). The genome was sequenced to 39× coverage using second-generation technology, assembled de novo and then anchored to the genetic linkage map into seven pseudochromosomes. A total of 34,809 genes were predicted, with most being supported by transcript evidence. Newer version v1.1, which contains an updated pseudomolecule assembly of the original v1.0 is also available. Recently, whole genome sequencing of pear (*Pyrus bretschneideri*) and *Prunus mume* have also been reported. The pear genome was sequenced to 194× coverage using a combination of BAC-by-BAC and next generation sequencing (Wu et al. 2013). A 512.0 Mb sequence covered 97.1 % of the estimated genome size of this highly heterozygous species and 75.5 % of the sequence has been anchored to all 17 chromosomes by high density genetic maps comprising of 2,005 SNP markers. A total of 42,812 genes have been predicted with about 28.5 % encoding multiple isoforms. For *Prunus mume*, approximately 84.6 % (237 Mb) of its genome was assembled by combining 101× next-generation sequencing and optical mapping data (Zhang et al. 2012), while 83.9 % of scaffolds have been anchored to the eight chromosomes with genetic map constructed by restriction-site-associated DNA sequencing.

The whole genome sequence and annotation data of three Rosaceae species, peach, strawberry and apple, are available from GDR and Phytozome. Details on the resources and tools for the genomes in GDR are provided in the "Community database resources" section of this review. Phytozome is a joint project of the Department of Energy's Joint Genome Institute (JGI) and the Center for Integrative Genomics to facilitate comparative genomic studies amongst green plants. Families of orthologous and paralogous genes that represent the modern descendents of ancestral gene sets are constructed at key phylogenetic nodes. These families allow easy access to clade specific orthology/paralogy relationships as well as clade specific genes and gene expansions. As of release v9.0, Phytozome provides access to 31 sequenced and annotated green plant genomes which have been clustered into gene families at ten evolutionarily significant nodes. Where possible, each gene has been annotated with PFAM (Finn et al. 2010), KOG (Tatusov et al. 2003), KEGG (Aoki and Kanehisa 2005), and PANTHER (Mi et al. 2010) assignments, and publicly available annotations from RefSeq (Pruitt et al. 2007), UniProt (The UniProt Consortium 2012) and TAIR (Lamesch et al. 2012). Similar to the GDR genome sites, users can view the genome though the genome browser GBrowse, which provides access to transcripts, alternative transcripts, mapped ESTs, aligned plant peptides and genetic markers. Users can also search their sequences against the peach genome sequence, transcripts and proteome though BLAST and BLAT servers, and download all the genome annotation through a bulk data site. The Istituto di Genomica ApplicataIGA peach site (IGA, http://services.appliedgenomics.org/projects/drupomics/intro/) also contains a peach GBrowse and BLAST tools. The IGA peach GBrowse provides access to the same tracks as phytozome but in addition they have RNA-seq Illumina profiles from tissues sequenced from cotyledon and embryo, fruit, leaf and root; repetitive sequences and microsatellites; and a profile of exact genome 20-mers.

RESOURCES FOR VARIATION ANALYSIS

High-density genome scanning is now possible through use array platforms developed using comprising high-throughput and low-cost next generation sequencing (NGS) to identify single nucleotide polymorphisms (SNPs). The International RosBREED SNP Consortium (IRSC) *Malus* research community developed an Infinium® II WGG genotyping array (IRSC array, Chagne et al. 2012) for *Malus* and *Pyrus* using data from the re-sequencing of 27 *Malus* genotypes along with data from the 'Golden Delicious' genome sequence. The IRSC array contains a total of 7,867 *Malus* SNPs in addition to 921 *Pyrus* SNPs. A GoldenGate-based assay platform (Khan et al. 2012) is also available for *Malus* containing 12,299 SNPs identified from EST data from 14 genotypes. Of these, 1,411 SNPs were validated using four apple genotypes. The IRSC also developed Infinium® II WGG genotyping arrays for peach (Verde et al. 2012) and cherry (Peace et al. 2012). The cherry array used data from accessions of sweet cherry and of tart cherry, along with the peach genome sequence to develop a 6,000 SNP array, of which approximately one-third were informative for each crop (Peace et al. 2012).

The IRSC peach array comprised of 8,144 SNPs developed from re-sequencing of 53 peach genotypes along with data from the peach genome sequence (Verde et al. 2012).

FUNCTIONAL GENOMICS RESOURCES

EST resources

ESTs are valuable resources for marker development, genome annotation, co-expression studies and comparative analyses as well as for gene discovery. Currently, GenBank contains 527,240 ESTs for Rosaceae: 336,190 from *Malus*, 110,815 from *Prunus*, 58,795 from *Fragaria* and the rest from other genus. *Malus* has the largest collection of ESTs and the majority have been produced to aid discovery of genes involved in important agricultural traits with National Science Foundation funding (award no. 0321701; Schuyler Korban, principal investigator) and by HortResearch in New Zealand (Newcomb et al. 2006). Transcriptome analyses in *Prunus* focused on generating ESTs to identify candidate genes involved in different tissues, different stages of plant development and responding to abiotic and biotic stresses (Georgi et al. 2002; Horn et al. 2005; Vecchietti et al. 2009). Transcriptomics analyses have also been used to investigate fruit ripening and post-harvest physiology in *Prunus* (Trainotti et al. 2003; Vizoso et al. 2009). In *Fragaria*, 50,627 transcripts are from the diploid *Fragaria vesca* and 10,855 are from the cultivated tetraploid *Fragaria* × *ananassa*. Out of 143 SRA datasets for Rosaceae, 79 are from RNA sequences. The majority, 42 datasets, are from *Prunus* with other datasets from other genera: 12 from *Fragaria*, 10 from *Pyrus*, 9 from *Malus* and 6 from *Rosa*.

GDR contains all the publicly available Rosaceae ESTs and the unigene sets for the family Rosaceae and each genus. With the availability of the whole genome sequences, the ESTs from *Malus*, *Prunus* and *Fragaria* are anchored to the predicted gene transcripts from whole genome sequencing. The best matches between the ESTs and the predicted genes are available in a spreadsheet with links to the predicted genes in the GDR GBrowse. PlantGDB (http://www.planttgdb.org) also contains EST unigenes, assembled from GenBank ESTs, from various plant species. Unigene data is available from *Prunus persica*, *Prunus armeniaca*, *Malus domestica*, *Fragaria vesca* and *Fragaria ananassa*. ESTree DB (Lazzari et al. 2008) has a collection of ESTs from *Prunus persica* and *Prunus amygdalus*. The ESTs are from 12 peach libraries and 3 almond libraries produced in nine different laboratories. The peach unigene in the sixth release of ESTree DB contains 28,391 unigenes with 7,709 contigs and 20,682 singlets, assembled from 75,404 ESTs. The ESTree mapping project is underway to map about 200 additional ESTs on the peach transcriptome.

Microarrays

Malus has the largest collection of microarrays as well as ESTs which have been used to study environmental effects on tree-to-tree variability in the orchard, fruit aroma production (Schaffer et al. 2007), early development of apple fruit (Lee et al. 2007), apple fruit development from the floral bud to ripe fruit (Janssen et al. 2008), fruit development and ripening (Costa et al. 2010), fruitlet abscission (Botton et al. 2011), role of Ring finger gene family in fruit development (Li et al. 2011), fire blight susceptibility (Jensen et al. 2012) and apple fruit maturation and texture attributes (Zhu et al. 2012).

High-throughput transcriptome analysis using microarrays have been recently incorporated in various *Prunus* research projects. One of the platforms, microPEACH1.0, consists of 4,806 70-mer oligonucleotides designed from *Prunus persica* (peach) EST unigene clusters, mainly in the 3′ end terminal region. It was used to investigate transcriptome profiling of ripening nectarine (Ziliotto et al. 2008), stone formation in peach fruit (Dardick et al. 2010) and the role of Jasmonate in fruit ripening (Ziosi et al. 2008). GFNChile_Peach_0.9K_v1.0 contains 847 cDNAs from a cv. Loring ripe peach fruit (*Prunus persica*) cDNA library and used to study cold-responsive genes in peach fruit (Ogundiwin et al. 2008). Website (http://bioinfo.ibmcp.upv.es/genomics/ChillPeachDB) holding detailed information on the ChillPeach database was also created. Other investigations that used microarrays include the identification of key genes in almond adventitious shoot regeneration (Santos et al. 2009) and evaluation of expression of genes mostly engaged in fruit development between *Prunus mume* and *Prunus armeniaca* (Li et al. 2012). Recently, a 60-mer oligo-DNA microarray, constructed using *Prunus* Unigene V4 from GDR, was used to study fruit softening in peach (Tatsuki et al. 2013). Some of the results are available from NCBI's Gene Expression Omnibus (GEO) (Wheeler et al. 2007).

In *Fragaria*, cDNA microarray have been used to study maturation and non-climacteric ripening (Aharoni et al. 2002) and the evolution of fruit flavor compounds (Aharoni et al. 2004) in the cultivated octoploid *Fragaria* × *ananassa*. An oligo-based microarray with sequences from both diploid and octoploid has also been used to compare the transcriptome of the ripe receptacle in these species (Bombarely et al. 2010).

RNA-seq technology

The latest and most efficient tool for transcriptome analysis involves the use of high-throughput sequencing technologies to sequence cDNA. This technology is called Whole Transcriptome Shotgun Sequencing (WTSS) or RNA-seq. RNA-seq allows not only the assessment of the expression level of specific genes but also the detection of less-represented transcripts, allelic-specific expression of transcripts, post-transcriptional mutations and the expression of splice-variants. Even though it has been available for only a couple of years, this technology has been used extensively in humans, mammals and yeast and it is beginning to be used in plant species. *Arabidopsis* was the first plant species to be studied using this technology (Weber et al. 2007). Comparison of deep transcriptome sequencing with the EST database confirmed most of the annotated introns and identified thousands of novel alternatively spliced mRNA isoforms, suggesting at least 42 % of intron-containing genes of *Arabidopsis* are alternatively spliced (Filichkin et al. 2010). These RNA-seq data (Lister et al. 2008; Filichkin et al. 2010), along with proteomics data (Baerenfaller et al. 2008; Castellana et al. 2008), have been used to revise the *Arabidopsis* gene models and are incorporated in the TAIR10 release (Lamesch et al. 2012).

In *Malus*, RNA-seq experiments were performed to study genes associated with columnar phenotype, characterized by a compact growth habit, of apple trees (Krost et al. 2013). In *Prunus*, RNA-seq technology is currently being utilized in projects to study PPV (Plum Pox Virus) resistance, graft incompatibility fruit quality and flowering time in peach, apricot and plum, as summarized in Martínez-Gómez et al. (2011). The results are expected to be incorporated in GDR when the results are publicly available.

RESOURCES FOR PROTEOMICS AND METABOLOMICS

Proteomics and metabolomics refer to the large-scale study of proteins and metabolites. When transcriptomic, proteomic and metabolomic data are integrated, it can help to give us a more complete picture of a living organism in a specific condition. Due to the enormous complexity of the proteome, many different approaches are being taken to generate and catalogue proteomics data, such as structural and functional annotation of proteins, protein expression and dynamics, stress and developmental responses, post-translational protein modifications and protein interactions. UniProt (The UniProt Consortium 2012) provides a comprehensive, high-quality and freely accessible resource of protein sequence and functional information. Currently (searched in January 2013), 3,695, 2,495 and 1,982 protein entries are available for *Prunus, Malus* and *Fragaria*, respectively. Only 70, 45 and 34 entries of them are in Swiss-Prot and the rest (3,625, 2,450 and 1,948) are in TrEMBL.

The Worldwide Protein Data Bank (WwPDB) (Berman et al. 2007) is the main structural protein database. It includes RCSB (A Resource for Studying Biological Macromolecules) PDB in USA, PDBe in Europe (Velankar et al. 2011) and PDBj in Japan. Currently, PDB contains the coordinates and related information of more than 76,000 structures of proteins, nucleic acids, protein/nucleic acid complexes and other macromolecules that have been determined using X-ray crystallography, NMR and electron microscopy techniques. Only limited numbers of structures have been deposited in wwPDB. For *Prunus*, 11 structures are currently available from PDB including proteins from *Prunus persica*, *Prunus avium*, *Prunus dulcis* and *Prunus mume*. Other entries for the Rosaceae family include six from *Malus domestica* and one from *Fragaria* × *ananassa*. Other protein structure databases include CATH (Cuff et al. 2011) and SUPERFAMILIY (Wilson et al. 2009). The CATH (class, architecture, topology and homology) database provides a hierarchical classification of protein domain structures obtained from PDB. The classification class reflects the amino acid composition, architecture of the general shape of the protein domain and topology of the way in which the protein folds into this architecture. SUPERFAMILY provides the prediction of protein domains of known structure in amino acid sequences. The classification of domains is hierarchical, based on nature of the similarity (sequence, evolutionary and structural), class, fold, superfamily and family, following the structural classification of the protein (SCOP) database (Andreeva et al. 2008). SUPERFAMILY currently includes data for 2,476 distinct organisms, including 373 Eucaryotes. Data from *Prunus persica*, *Malus domestica* and *Fragaria vesca* are available from SUPERFAMILY.

Various web-based plant proteome-related databases are summarized in the database section of the proteomics subcommittee of the Multinational *Arabidopsis* Steering Committee Proteomics Subcommittee (MASCP) Web site (http://www.masc-proteomics.org/). Some of the databases are illustrated below. ARAMEMNON (Schwacke et al. 2003) is a database of plant membrane proteins with *Arabidopsis thaliana* as the reference model plant. Currently, the database holds all putative membrane proteins of five other plant species: grape (*Vitis vinifera*), poplar (*Populus trichocarpa*), rice (*Oryza sativa*), maize (*Zea mays*) and brachypodium (*Brachypodium distachyon*). The *Arabidopsis* Interactions Viewer (Geisler-Lee et al. 2007) is an interaction database for *Arabidopsis thaliana* predicted from interacting orthologs in yeast (*Saccharomyces cerevisiae*), nematode worm (*Caenorhabditis elegans*), fruit fly (*Drosophila melanogaster*) and human (*Homo sapiens*). The database includes 70,944 predicted and 22,156 confirmed *Arabidopsis* interacting proteins. The confirmed *Arabidopsis* interacting proteins come from BIND, the Biomolecular Interaction Network Database (Isserlin et al. 2011), high-density *Arabidopsis* protein microarrays (Popescu et al. 2007, 2009) and other literature sources. The interactions in BIND were identified using several different methods, such as yeast two-hybrid screens, but also via traditional biochemical methods. All subcellular localization data in the *Arabidopsis* Interactions Viewer are from SUBA, the *Arabidopsis* Subcellular Database (Heazlewood et al. 2007). Subcellular localization data in SUBA are brought together from various sources such as studies using chimeric fluorescent fusion proteins, proteomic surveys using mass spectrometry and literature. It also contains precompiled bioinformatic predictions for protein subcellular localizations from a set of ten different prediction tools. Complex relational queries can be performed between these experimental and predicted datasets to find and collate evidence for the subcellular location of *Arabidopsis* proteins. The Plant Protein Phosphorylation DataBase (P3DB) (Yao et al. 2012) hosts protein phosphorylation data for eight species: *Arabidopsis thaliana*, *Brassica napus*, *Glycine max*, *Medicago truncatula*, *Oryza sativa*, *Solanum tuberosum*, *Zea mays* and *Nicotiana tabacum*. AtMetExpress (Matsuda et al. 2010) and the Golm Metabolome Database (Hummel et al. 2007, 2010) are metabolome databases. AtMetExpress contains data from a study of phytochemical accumulation during development of the model plant *Arabidopsis thaliana* using liquid chromatography–mass spectrometry in samples covering many growth stages and organs. The Golm Database contains data of mass spectra from biologically active metabolites quantified using gas chromatography coupled to mass spectrometry. It covers data from mammals, yeast, corynebacterium, model plants, such as crop plants and related wild species, as well as required non-sample controls.

COMPARATIVE GENOMICS RESOURCES

Comparative genomics is an area of study where the structure and function of genomes from different species or varieties are compared. An important goal of comparative genomics is to obtain an insight into how genomes have evolved. Detection of duplicated regions within species and syntenic regions among species plays an important role in studies of genome evolution. The comparison of genomes and their contents among species allows us to identify genes and other sequence features that are conserved among species and/or genes that are specific to certain clades. These results help us to infer what genes and other sequence features are responsible for the similarities and differences among species. Comparative genomics also allows us to transfer knowledge from well-studied species to less-studied species.

With the increasing number of species with whole genome sequences, several web-based databases are available to compare whole genome data. A few contain the whole genome sequences of one or more species of the Rosaceae family: *Prunus persica*, *Malus domestica* and *Fragaria vesca*.

CoGe (The Place to Compare Genomes) (Lyons and Freeling 2008), contains whole genome data of 15,037 species across all domains of life, including *Prunus persica*, *Fragaria vesca* and *Malus domestica*. CoGe contains a series of web tools where users can select species of interest to view the genome data and/or perform some comparative analysis. The tools include OrganismView for searching organisms, CoGeBlast for blasting sequences against the genomes of interest, FeatView for searching genomic features by name, SynMap for whole genome syntenic dotplot analyses, SynFind for identifying syntenic regions across multiple genomes and GEvo for high-resolution sequence analysis of genomic regions. SynMap is the tool for comparative genomics which uses DAGChainer (Haas et al. 2004) as the underlying software for synteny detection.

The plant genome duplication database (PGDD) (Lee et al. 2013) contains 26 plant genomes including *Prunus persica*, *Malus domestica* and *Fragraria vesca*. PGDD contains similar tools to those in CoGE, where users can explore plant genes in terms of intra-genome or cross-genome syntenic relationships. Users can choose two species to view syntenic blocks in Dot Plot, can select a locus to view the regions that contain the locus along with the syntenic regions in multiple species, and can use BLAST to map the sequence of interest to selected genomes. PGDD uses the MCscan package (Tang et al. 2008), which uses DAGchainer to predict pairwise segments, as the underlying software for synteny detection.

Plaza (Van Bel et al. 2012) is another platform where users can access plant genome data to perform comparative analyses. Plaza contains 25 species including *Malus domestica* and *Fragraria vesca*. Data annotation includes primary gene annotation, gene family, orthologous genes, and functional annotation such as GO terms, InterPro Domains, and Reactome (pathway) data. In the Plaza Synteny Plot tool, users can start from a gene or a gene family to view the gene organization of all homologs of a gene family in selected species. In the Skyline plot, users can enter a gene locus to view the collinear regions that exist within a set of selected species. In WGDotplot, users can select two species to view syntenic blocks or select the same species twice to view all duplicated blocks within species. The collinear regions within species can also be viewed with Circle Plot. If available, the age of the collinear blocks, determined using Ks, is reported using a color code. From the Skyline plot and WGDotplot, users can access Multiplicon View to see the aligned gene strings of a set of homologous segments. The collinear regions are detected by i-ADHoRe (Simillion et al. 2008). In addition to those described above, Plaza provides other tools for genome evolution and collinearity research. The WGMapping tool allows users to choose all genes or selection of genes to display their location on the chromosomes along with the gene type. The functional clustering visualization tool provides an overview of the location and content of each functional cluster, detected using C-Hunter (Yi et al. 2007), on a chromosome-wide scale. Genomic sequences and genomic features can be viewed using Genome Browser. Tools such as Similarity heatmap, orthologous gene tool, Tree explorer and Gene family finder allows users to further explore gene family evolution.

The GreenPhylDB (Rouard et al. 2011) and SALAD databases (Mihara et al. 2010) are related databases. The GreenPhylDB contains data from *Malus domestica*, but the SALAD database does not yet contain any Rosaceae genome data. The difference between these and the comparative genome databases mentioned above is that GreenPhylDB and SALAD do not contain data or tools for synteny analysis, but focus on gene family, phylogeny and ortholog/paralog analyses. GreenPhylDB is a web resource for plant comparative and functional genomics. GreenPhylDB v3.0 contains 22 full genomes from the major phylum of plant evolution. The data include various lists of gene families, such as plant-, phylum- and species-specific lists, and tools to facilitate the comparisons. The SALAD database is a web-based resource for genome-wide comparative analysis of annotated protein sequences in plants. In SALAD, users can search for genes and view phylogenetic trees constructed from sequence alignment for a selected single motif or multiple motifs. Another functionality called sequence logo provides users with graphical representation of the sequence conservation of amino acids made from alignment of the conserved motif in each node. Users can also compare the sequence logos among members of distinct nodes to evaluate conservation and diversity of amino acid at any sites in the conserved motif.

GDR (Jung et al. 2008) uses GBrowse_Syn (McKay et al. 2010), a synteny browser, to show the orthologous regions among the three sequenced genomes of Rosaceae (Jung et al. 2012), as detected by the Mercator program (Dewey 2007). GBrowse_Syn is hyperlinked to GBrowse so that users can access various genome annotation data including markers from the conserved syntenic regions shown in GBrowse_Syn. Comparative genomics data made available in GDR thus allow users to explore other data, such as genomic features, anchored trait loci and genetic markers, in the orthologous regions.

COMMUNITY DATABASE RESOURCES

Plant community databases provide access to all or most of the datasets for individual or closely related species. Such databases include The *Arabidopsis* Information Resource (TAIR, Lamesch et al. 2012), the Genome Database for Rosaceae (GDR, Jung et al. 2008), the Solanaceae Genomics Network (SGN, Bombarely et al. 2011), Gramene (Youens-Clark et al. 2011), TreeGenes (Wegrzyn et al. 2008) and MaizeGDB (Schaeffer et al. 2011). Community databases generally store comprehensively integrated data such as annotated sequences of genomes and transcriptomes, genetic data, and molecular and phenotypic diversity data. As a result of data integration, the value of individual types of data increases exponentially, providing essential resources to accelerate the molecular understanding of phenotypic traits and the use of DNA information in crop improvement.

Initiated in 2002, GDR is the sole community database for the Rosaceae family, which includes economically important genera such as *Prunus*, *Fragaria*, *Malus*, *Pyrus*, *Rosa* and *Rubus*. The Rosaceae research community has generated data for the annotated genome sequences, physical maps, a large collection of ESTs, transcriptome map, numerous genetic maps, genetically mapped traits (MTL and QTL), genotypic diversity data, publicly available breeding data with both phenotypic and genotypic data, and cultivar evaluation data for growers. The integration and standardization of the data is crucial for the data to be utilized by different types of users including genomicists, geneticists, molecular biologists, evolutionary biologists, bioinformaticists, breeders and growers. The purpose of GDR is to collect, curate, analyze and integrate the data and provide efficient interfaces for user access to allow these numerous and complex data to be efficiently utilized. In the sections below, we describe the data and web interface, analysis tools and community tools available from GDR, together with work in progress and future plans. GDR tutorials are available at http://www.rosaceae.org/tutorials.

Data and web interface

Annotated whole genome sequence

The whole genome sequences of various versions of the three species, *Prunus persica* genome v1.0, *Malus domestica* genome v1.0 and v1.0p and *Fragaria vesca* genome v1.0 and v1.1, are available in GDR. *Malus domestica* genome v1.0 is represented as metacontigs, composed of assembled overlapping contigs that have been produced by the assembly of heterozygous genotypes, which have been anchored to the chromosomes. The v1.0p is the primary pseudo-haplogype assembly, which is composed of chromosome-anchored contigs that are non-overlapping. Users can access all the annotated genome data from each genome page. Through the graphic interface GBrowse (Donlin 2007; Stein et al. 2002), users can view various genomic features aligned to the genome, such as gene models, repeats, and SNPs, as well as alignments of ESTs, repeats, genetic markers and genes from other plant model species. Each feature has hyperlinks that lead to a page with sequences and other information, with further hyperlinks to external databases where applicable. The genetic marker feature in GBrowse is linked to the marker page in GDR where all the detailed marker information is available, such as primers, mapped positions, references, and link to CMap, the comparative map viewer in GDR. The mapped ESTs are also linked to the EST page in GDR, where all the detailed EST information is available. The genome pages in GDR contains various downloadable files, including the fasta files of predicted peach gene transcripts, CDS (coding sequences), and predicted gene peptides. Excel files of gene transcripts with homologs to *Arabidopsis* genes, Swiss-Prot entries and TrEMBL entries are also available with hyperlinks to external databases. Other Excel downladable files include various Rosaceae ESTs and genetic markers that map to the whole genome sequences and SNPs with hyperlinks to GBrowse in GDR. The orthologous regions among three species, detected by the Mercator program (Dewey 2007), are displayed using GBrowse_Syn (McKay et al. 2010). The whole genome data are also available to search by various categories, such as by name, interpro protein domain name or KEGG pathway terms, so that users can directly access the genes by querying. The predicted genes from the whole genome sequences have also been utilized in the construction of PlantCyc databases. Currently, three PlantCyc databases, peachCyc, appleCyc and FragariaCyc, are available in GDR for users to explore the pathway data.

Annotated EST unigene data

GDR contains all the publicly available Rosaceae ESTs, downloaded from the dbEST at NCBI (Gibney and Baxevanis 2011). Routine processing in GDR occurs in three stages: sequence filtering and trimming to obtain high-quality sequences, assembly into contigs to reduce the inherent redundancy and building unigene sets from the combined contigs and singlets, and sequence annotation. A unigene is available for *Prunus*, as well as *Malus*, *Fragaria*, *Rosa* and *Pyrus*. The assembled contigs and singlets for the four genera were assembled together to generate a putative unigene set for the entire Rosaceae ESTs. Other annotation includes putative function and Gene Ontology (The Gene Ontology Consortium 2013) association to contigs and ESTs by homology with SwissProt, TREMBL and InterPRO proteins (Hunter et al. 2012). Plant Structure Ontology (Ilic et al. 2007) is also utilized to annotate the ESTs with the tissue from which the ESTs are generated.

The unigene page is a good starting point for an overview of the various annotated data for the unigenes. It displays the overall results of the project with a sidebar containing links to the library information, putative homology, KEGG analysis, microsatellite analysis and downloads. A link to the gene ontology (GO) classification is also available for the genera assemblies as well as the Rosaceae family assembly. Downloadable data includes batch sequence in fasta format, homology results file in Excel format with links into GDR and external databases and SSR/ORF/primers results in Excel format.

The EST search site is for those users who are interested in a subset of ESTs. They can choose to search ESTs of the entire Rosaceae or the genus of interest by selecting the appropriate tabs. Users can also search either ESTs or contigs. In each search page, ESTs or contigs can be searched by their name(s), assembly results, sequence features such as SSR or SNP, taxonomy, tissue type and putative function including match description, match organism and GO term. Users can also perform a batch search by uploading a file with EST names. Previous unigene versions are also available for search to help those who have been using an older version in their research. The results can be downloaded in fasta format or as a tab-delimited file with SWISS-PROT homology results containing hyperlinks back into the data for each sequence retrieved. Instead of displaying all the details on one page, the EST details page initially displays the clone information and the sequence with a sidebar containing links to library details, unigene information, sequence homology, SSR/ORF information, map position and anchored BACs when applicable. A unigene information page provides the contig name and hyperlink for both the genus and family unigenes. The contig page gives similar annotation data for the contig with additional links to the SNP results and the comprising ESTs. For the ESTs anchored to peach BACs and/or to Rosaceae genetic maps, the EST detail page provides a link to view the ESTs' map positions using the GDR Map Viewer or CMap.

Genetic map

GDR currently contains data for 54 genetic maps for Rosaceae species. GDR uses CMap, the web-based comparative map tool, to allow users to compare maps from different cultivars and species. The comparative mapping facilitates the data transfer from well-studied species to less-studied ones. For example, the GDR map collection includes the TxE map (Dirlewanger et al. 2004; Howad et al. 2005), which is recognized as the reference map for *Prunus.* The TxE map, constructed from an almond × peach F2 population, contains 826 markers with a total distance of 524 cM. The TxE map contain many markers that are used in the construction of maps of other *Prunus* species such as peach, apricot, sour cherry, plum × almond–peach hybrid and almond × peach, but also other Rosaceae species such as apple and pear. The essential collinearity of the anchored markers in the *Prunus* maps and the presence of large collinear blocks among different genera in Rosaceae, such as *Prunus* and *Malus* (Dirlewanger et al. 2004), enable comparative mapping, an invaluable tool for cross-utilization of data in Rosaceae. In addition to the directly-mapped genetic markers, the TxE map in GDR-CMap displays peach transcriptome map data, major trait loci affecting agronomic characters found in various *Prunus* species (Dirlewanger et al. 2004), pathogen resistance loci (Lalli et al. 2005) and Rosaceae Conserved Orthologous Set (RosCOS) (Cabrera et al. 2009). CMap also contains the apple integrated map which was developed for anchoring metacontigs from whole genome sequencing. The integrated map were derived from six F1 populations totaling 720 individuals. The FV × FB diploid *Fragaria* reference map (Sargent et al. 2011), which played important role in the scaffold ordering of the *F. vesca* genome sequence and rose integrated consensus map, built based on the information of four diploid populations and more than 1,000 initial markers (Spiller et al. 2011) are other important resources for the community. GDR-CMap serves as an integrative tool in the utilization of the data anchored to the maps in Rosaceae. The anchored features, such as marker and ESTs, in the map are also linked to the corresponding GDR sites so that all the relevant information for the features can be viewed. Markers that are anchored to various genomes have hyperlinks to GBrowse. Another important resource in GDR is the peach physical map data. The peach physical map

(Zhebentyayeva et al. 2008) is constructed from two peach BAC libraries (Georgi et al. 2002) and the physical length of the map is estimated to be 303 Mb, which is 104.5 % of the peach genome. GDR uses two tools available from the WebAGCoL Package (Pampanwar et al. 2005) to display the current peach physical map.

Genetic markers and traits

To provide more details of the genetic markers and traits that have been used in genetic map development or genetic diversity studies, GDR contains an extensively annotated molecular marker database. Currently, over 44,439 extensively annotated markers, including SNPs from IRSC peach array v1 (Verde et al. 2012) and candidate SNPs, are available from GDR search engine. The marker annotation includes marker aliases, source cultivar, source description, primer sequences, PCR conditions, references, map position, associated ESTs and associated BACs. While annotation of trait data is at an initial stage, the traits are annotated in GDR with aliases, published symbol, curated trait category, taxon, trait description, screening method, map position and references.

The marker search site allows both a simple search by name and an advanced search with various search categories. The search category includes marker type, the species from which the marker is developed, the species to which the marker is mapped, map position, markers with associated BAC clones and markers with associated ESTs. Users can also upload a file of names to get the detailed data. In the trait search site, users can search trait by name, symbol, taxon or curated trait category. Other SNP markers included in arrays, such as IRSC apple 9K (Chagne et al. 2012), cherry 6K (Peace et al. 2012) and UC Davis peach 6K (Ahmad et al. 2011), as well as those in IRSC peach 9K, are available to download and view in GBrowse.

DNA polymorphism

GDR contains DNA polymorphism data from various projects on molecular diversity studies of Rosaceae species. Currently, data from nine different projects are available, three from *Prunus* diversity studies and the rest from *Malus* and *Pyrus* studies. All the current data are from projects with SSR markers. Users can query by the marker name or species to view the details of diversity projects and the genotype of the varieties used in the analyses. SNP markers from the RosBREED project (http://www.rosbreed.org) are being added.

Breeding data

One of the newest components of GDR is the search/browse site for breeding data. GDR contains password-protected private breeding data as well as publicly available breeding data. Private breeding data includes data from the Washington Apple Breeding Program and Pacific North West Sweet Cherry Breeding Program. Public data includes data from the federally funded RosBreed project, a program designed to establish a sustainable marker-assisted breeding infrastructure for US Rosaceae crops. The data includes the varieties and their pedigrees, phenotyping and genotyping data and experimental metadata. The current search interfaces allows users to search by datasets, variety name, trait values and pedigree. From the result page, users can view detailed results for a variety or download an Excel file with all the phenotyping/genotyping results that has been specified. Users can also select a pedigree by selecting a variety and number of ancestral and progeny generations to generate an input file for breeding software such as Pedimap and FlexQTL (Bink and van Eeuwijk 2009). A breeding decision tool called Cross Assist, which produces a list of parents to cross and the number of seedlings to screen to obtain certain number of seedlings above/within user-specified trait thresholds, is also available.

Analysis tools

GDR web-based tools include a BLAST server, FASTA server, CAP3 Assembly server and SSR server. The FASTA/BLAST servers allow users to conduct sequence homology analyses against various sequence databases including annotated sequences in GDR. The databases include whole genome nucleotide and protein sequences of peach, strawberry and apple; ESTs of the Rosaceae or each genus from NCBI, genera-specific unigene sets, as well as a family-wide unigene; Rosaceae genomic or protein sequences from NCBI, peach, apple and cherry SNP sequences; and *Arabidopsis* protein sequences from TAIR and the ESTs, unigene sets, SSR-containing ESTs from individual cDNA libraries of peach mesocarp, almond, octoploid strawberry and diploid strawberry. Peach mesocarp ESTs that are anchored to the peach BACs are also available for sequence analysis. Batch sequences can be uploaded for analysis and the results are returned as both raw aligned output and parsed out in Excel. The output in Excel has hyperlinks to the GDR and NCBI sites. FASTA formatted library files of both the sequences with or without matches are provided to allow the user to easily conduct further batch searches in GDR and other databases. An EST assembly server using the CAP3 program is also available so that users can assemble their own EST sets. The server returns the raw output, a summary report and a fasta file containing the combined contig sequences and singlet sequences, which are also available as individual files. The contig file lists the contig number and comprising clone names in the comment line for each assembled transcript. Also available is a SSR server that allows user-defined SSRs to be identified in uploaded sequences. Users can also choose to run Primer3 along with the SSR-detection program to generate primer sets for the SSRs. The results are returned in an Excel file containing all the SSRs, primers, ORFs in sequence and product size with summary information on motif number and type.

Community resources

GDR provides access to community-based news on various pages under the 'community' header bar, such as Rosaceae genomics, USRosEXEC, conferences, meetings, funding, employment, mailing lists and message boards. USRosEXEC stands for US Rosaceae Genomics, Genetics and Breeding Executive Committee, which serves as a communication and coordination focal point for the community. The USRosEXEC page provides the official documents, meeting minutes, membership and subcommittee information. Several mailing lists, in addition to the GDR mailing list, are available to serve the community with information for specific interests or purposes, and the archives can be viewed through the message board sites. All the publications in Rosaceae genomics and genetics are also available in GDR through the publication search site.

Future directions

With the availability of the whole genome sequences of apple, peach and strawberry, and other Rosaceae genomes and re-sequencing data to be added, the future direction of GDR will include further integrating the annotated whole genome data with other genomics, genetics and breeding data to accelerate the usage of DNA information in crop improvement as well as to improve our knowledge on various aspects of Rosaceae biology. The genetic markers will also be queried by the anchored genome position and the neighboring trait locus. The effort toward collecting and curating trait data, including MTL and QTL, will also be continued. When data for genetic markers, DNA polymorphism and trait data are integrated and easily searchable, users will be able to query for markers that are close to the trait of interest and polymorphic between the varieties of interest. The Genome Sequence Annotation Server (GenSAS) (Lee et al. 2011), an online annotation tool that provides a customizable automated pipeline for whole genome structural annotation, will be available for community for further curate the genome. In collaboration with RosBreed project, more breeding decision tools will be developed. Growers' gateway that allows growers to get information about cultivars and compare the cultivar performances will also be developed.

CONCLUSION

The recent addition of publicly available whole genome sequences of three Rosaceae species to the already existing wealth of genetic data has brought great opportunity for researchers to accelerate Rosaceae research. Multiple whole genome sequences in one family allow genome level comparative analysis to gain evolutionary insight as well as transfer knowledge among species. The second and third generations of high-throughput sequencing technology now allows resequencing of multiple varieties to catalogue sequence variations and quantitative gene expression analyses. The proteome and metabolome analyses are still in their infancy in Rosaceae, but the advances of tools and methodology with other model species will aid greatly in planning and adopting those analyses in Rosaceae research. Integration of different types data is critical in the interpretation and utilization of these data and hence the role of the community databases to bring together genomics, genetics, breeders and growers data will become increasingly important.

REFERENCES

1. Adams MD, Kelley JM, Gocayne JD, Dubnick M, Polymeropoulos MH, Xiao H, Merril CR, Wu A, Olde B, Moreno RF et al (1991) Complementary DNA sequencing: expressed sequence tags and human genome project. Science 252:1651–1656
2. Aharoni A, Keizer LC, Van Den Broeck HC, Blanco-Portales R, Muñoz-Blanco J, Bois G, Smit P, De Vos RC, O'Connell AP (2002) Novel insight into vascular, stress, and auxin-dependent and -independent gene expression programs in strawberry, a non-climacteric fruit. Plant Physiol 129:1019–1031
3. Aharoni A, Giri AP, Verstappen FW, Bertea CM, Sevenier R, Sun Z, Jongsma MA, Schwab W, Bouwmeester HJ (2004) Gain and loss of fruit flavor compounds produced by wild and cultivated strawberry species. Plant Cell 16:3110–3131
4. Ahmad R, Parfitt DE, Fass J, Ogundiwin E, Dhingra A, Gradziel TA, Lin D, Joshi NA, Martinez-Garcia PJ, Crisosto CH (2011) Whole genome sequencing of peach (Prunus persica L.) for SNP identification and selection. BMC Genomics 12:569
5. Andreeva A, Howorth D, Chandonia JM, Brenner SE, Hubbard TJP, Chothia C, Murzin AG (2008) Data growth and its impact on the SCOP database: new developments. Nucleic Acids Res 36:D419–D425

6. Aoki KF, Kanehisa M (2005) Using the KEGG database resource. Curr Protoc Bioinforma Chapter 1:Unit 1.12
7. Audergon JM, Ruiz D, Bachellez A, Blanc A, Corre MN, Croset C, Ferréol AM, Lambert P, Pascal T, Poëssel JL, Signoret V, Quilot B, Gouble B, Grotte M, Bogé M, Reiling P, Reich M, Bureau S, Boudehri K, Renaud C, Dirlewanger E, Deborde C, Maucourt M, Moing A, Monllor S, Dondini L, Arús P (2009) ISAFRUIT: study of the genetic basis of Prunus fruit quality in two peach and two apricot populations. Acta Hortic 814:523–528
8. Baerenfaller K, Grossmann J, Grobei MA, Hull R, Hirsch-Hoffmann M, Yalovsky S, Zimmermann P, Grossniklaus U, Gruissem W, Baginsky S (2008) Genome-scale proteomics reveals Arabidopsis thaliana gene models and proteome dynamics. Science 320:938–941
9. Batzoglou S, Jaffe DB, Stanley K, Butler J, Gnerre S, Mauceli E, Berger B, Mesirov JP, Lander ES (2002) ARACHNE: a whole genome shotgun assembler. Genome Res 13:91–96
10. Benson DA, Cavanaugh M, Clark K, Karsch-Mizrachi I, Lipman DJ, Ostell J, Sayers EW (2013) GenBank. Nucleic Acids Res 41:D36–D42
11. Berman H, Henrick K, Nakamura H, Markley JL (2007) The worldwide Protein Data Bank (wwPDB): ensuring a single, uniform archive of PDB data. Nucleic Acids Res 35:D301–D303
12. Bink MC, van Eeuwijk FA (2009) A Bayesian QTL linkage analysis of the common dataset from the 12th QTLMAS workshop. BMC Proc 3(Suppl 1):S4
13. Bombarely A, Merchante C, Csukasi F, Cruz-Rus E, Caballero JL, Medina-Escobar N, Blanco-Portales R, Botella MA, Muñoz-Blanco J, Sánchez-Sevilla JF, Valpuesta V (2010) Generation and analysis of ESTs from strawberry (Fragaria × ananassa) fruits and evaluation of their utility in genetic and molecular studies. BMC Genomics 11:503
14. Bombarely A, Menda N, Tecle IY, Buels RM, Strickler S, Fischer-York T, Pujar A, Leto J, Gosselin J, Mueller LA (2011) The sol genomics network (solgenomics.net): growing tomatoes using Perl. Nucleic Acids Res 39:D1149–D1155
15. Botton A, Eccher G, Forcato C, Ferrarini A, Begheldo M, Zermiani M, Moscatello S, Battistelli A, Velasco R, Ruperti B, Ramina A (2011) Signaling pathways mediating the induction of apple fruitlet abscission. Plant Physiol 155:185–208
16. Cabrera A, Kozik A, Howad W, Arus P, Iezzoni AF, van der Knaap E (2009) Development and bin mapping of a Rosaceae conserved ortholog set (COS) of markers. BMC Genomics 10:562
17. Castellana NE, Payne SH, Shen Z, Stanke M, Bafna V, Briggs SP (2008) Discovery and revision of Arabidopsis genes by proteogenomics. Proc Natl Acad Sci USA 105:21034–21038
18. Chagne D, Crowhurst R, Troggio M, Davey M, Gilmore B, Lawley C, Vanderzante S, Hellens R, Kumar S, Cestaro A, Velasco R, Main D, Rees J, Iezzoni A, Mockler T, Wilhelm L, van de Weg E, Gardiner S, Bassil N, Peace C (2012) Genome-wide SNP detection, validation, and development of an 8K SNP array for apple. PLoS ONE 7(2):e31745.
19. Cochrane G, Alako B, Amid C, Bower L, Cerdeño-Tárraga A, Cleland I, Gibson R, Goodgame N, Jang M, Kay S, Leinonen R, Lin X, Lopez R, McWilliam H, Oisel A, Pakseresht N, Pallreddy S, Park Y, Plaister S, Radhakrishnan R, Rivière S, Rossello M, Senf A, Silvester N, Smirnov D, Ten Hoopen P, Toribio A, Vaughan D, Zalunin V (2013) Facing growth in the European nucleotide archive. Nucleic Acids Res 41:D30–D35
20. Costa F, Alba R, Schouten H, Soglio V, Gianfranceschi L, Serra S, Musacchi S, Sansavini S, Costa G, Fei Z, Giovannoni J (2010) Use of homologous and heterologous gene expression profiling tools to characterize transcription dynamics during apple fruit maturation and ripening. BMC Plant Biol 10:229
21. Cuff AL, Sillitoe I, Lewis T, Clegg AB, Rentzsch R, Furnham N, Pellegrini-Calace M, Jones D, Thornton J, Orengo CA (2011) Extending CATH: increasing coverage of the

protein structure universe and linking structure with function. Nucleic Acids Res 39:D420–D426

22. Dardick CD, Callahan AM, Chiozzotto R, Schaffer RJ, Piagnani MC, Scorza R (2010) Stone formation in peach fruit exhibits spatial coordination of the lignin and flavonoid pathways and similarity to Arabidopsis dehiscence. BMC Biol 8:13
23. Dewey CN (2007) Aligning multiple whole genomes with Mercator and MAVID. Methods Mol Biol 395:221–236
24. Diamant J, Sguigna V, Mezzetti B, Faedi W, Maltoni ML, Denoyes B, Chartier P, Petit A (2012) European small berries genetic resources, GENBERRY: testing a protocol for detecting fruit nutritional quality in EU strawberry germplasm collections. Acta Hortic 926:33–37
25. Dirlewanger E, Graziano E, Joobeur T, Garriga-Caldere F, Cosson P, Howad W, Arus P (2004) Comparative mapping and marker-assisted selection in Rosaceae fruit crops. Proc Natl Acad Sci USA 101:9891–9896
26. Donlin MJ (2007) Using the Generic Genome Browser (GBrowse). Curr Protoc Bioinforma Chapter 9:Unit 9.9
27. Filichkin SA, Priest HD, Givan SA, Shen RK, Bryant DW, Fox SE, Wong WK, Mockler TC (2010) Genome-wide mapping of alternative splicing in Arabidopsis thaliana. Gen Res 20:45–58
28. Finn RD, Mistry J, Tate J, Coggill P, Heger A, Pollington JE, Gavin OL, Gunasekaran P, Ceric G, Forslund K, Holm L, Sonnhammer EL, Eddy SR, Bateman A (2010) The Pfam protein families database. Nucleic Acids Res 38:D211–D222
29. Geisler-Lee J, O'Toole N, Ammar R, Provart NJ, Millar AH, Geisler M (2007) A predicted interactome for Arabidopsis. Plant Physiol 145:317–329
30. Georgi LL, Wang Y, Yverggniaux D, Ormsbee T, Inigo M, Reighard GL, Abbott AG (2002) Construction of a BAC library and its application to the identification of simple sequence repeats in peach [Prunus persica (L.) Batsch]. Theor Appl Genet 105:1151–1158
31. Gianfranceschi L, Soglio V (2004) The European project HiDRAS: innovative multidisciplinary approaches to breeding high quality disease resistant apples. Acta Hortic 663:327–330
32. Gibney G, Baxevanis AD (2011) Searching NCBI databases using Entrez. Curr Protoc Bioinforma Chapter 1:Unit 1.3
33. Haas BJ, Delcher AL, Wortman JR, Salzberg SL (2004) DAGchainer: a tool for mining segmental genome duplications and synteny. Bioinformatics 20:3643–3646
34. Heazlewood JL, Verboom RE, Tonti-Filippini J, Small I, Millar AH (2007) SUBA: the Arabidopsis subcellular database. Nucleic Acids Res 35:D213–D218
35. Horn R, Lecouls AC, Callahan A, Dandekar A, Garay L, McCord P, Howad W, Chan H, Verde I, Main D, Jung S, Georgi L, Forrest S, Mook J, Zhebentyayeva T, Yu Y, Kim HR, Jesudurai C, Sosinski B, Arús P, Baird V, Parfitt D, Reighard G, Scorza R, Tomkins J, Wing R, Abbott AG (2005) Candidate gene database and transcript map for peach, a model species for fruit trees. Theor Appl Genet 110:1419–1428
36. Howad W, Yamamoto T, Dirlewanger E, Testolin R, Cosson P, Cipriani G, Monforte AJ, Georgi L, Abbott AG, Arús P (2005) Mapping with a few plants: using selective mapping for microsatellite saturation of the Prunus reference map. Genetics 171:1305–1309
37. Huang S, Li R, Zhang Z, Li L, Gu X, Fan W, Lucas WJ, Wang X, Xie B, Ni P, Ren Y, Zhu H, Li J, Lin K, Jin W, Fei Z, Li G, Staub J, Kilian A, van der Vossen EA, Wu Y, Guo J, He J, Jia Z, Ren Y, Tian G, Lu Y, Ruan J, Qian W, Wang M, Huang Q, Li B, Xuan Z, Cao J, Asan WuZ, Zhang J, Cai Q, Bai Y, Zhao B, Han Y, Li Y, Li X, Wang S, Shi Q, Liu S, Cho WK, Kim JY, Xu Y, Heller-Uszynska K, Miao H, Cheng Z, Zhang S, Wu J, Yang Y, Kang H, Li M, Liang H, Ren X, Shi Z, Wen M, Jian M, Yang H, Zhang G, Yang Z, Chen R, Liu S, Li J, Ma L, Liu H, Zhou Y, Zhao J, Fang X, Li G, Fang L, Li Y, Liu D, Zheng H, Zhang Y, Qin N, Li Z, Yang G, Yang S, Bolund L, Kristiansen K, Zheng H, Li S, Zhang X, Yang H, Wang J, Sun R, Zhang B, Jiang S, Wang J, Du Y, Li S (2009) The genome of the cucumber, Cucumis sativus L. Nat Genet 41:1275–1281

38. Hummel J, Selbig J, Walther D, Kopka J (2007) The Golm metabolome database: a database for GC–MS based metabolite profiling. In: Nielsen J, Jewett MC (eds) Metabolomics. Springer, Berlin, pp 75–96
39. Hummel J, Strehmel N, Selbig J, Walther D, Kopka J (2010) Decision tree supported substructure prediction of metabolites from GC–MS profiles. Metabolomics 6:322–333
40. Hunter S, Jones P, Mitchell A, Apweiler R, Attwood TK, Bateman A, Bernard T, Binns D, Bork P, Burge S, de Castro E, Coggill P, Corbett M, Das U, Daugherty L, Duquenne L, Finn RD, Fraser M, Gough J, Haft D, Hulo N, Kahn D, Kelly E, Letunic I, Lonsdale D, Lopez R, Madera M, Maslen J, McAnulla C, McDowall J, McMenamin C, Mi H, Mutowo-Muellenet P, Mulder N, Natale D, Orengo C, Pesseat S, Punta M, Quinn AF, Rivoire C, Sangrador-Vegas A, Selengut JD, Sigrist CJ, Scheremetjew M, Tate J, Thimmajanarthanan M, Thomas PD, Wu CH, Yeats C, Yong SY (2012) InterPro in 2011: new developments in the family and domain prediction database. Nucleic Acids Res 40:D306–D312
41. Iezzoni A, Weebadde C, Luby J, Yue C, Peace C, Bassil N, McFerson J (2010) RosBREED: enabling marker-assisted breeding in Rosaceae. Acta Hortic 859:389–394
42. Ilic K, Kellogg EA, Jaiswal P, Zapata F, Stevens PF, Vincent LP, Avraham S, Reiser L, Pujar A, Sachs MM, Whitman NT, McCouch SR, Schaeffer ML, Ware DH, Stein LD, Rhee SY (2007) The plant structure ontology, a unified vocabulary of anatomy and morphology of a flowering plant. Plant Physiol 143:587–599
43. International Rice Genome Sequencing Project (2005) The map-based sequence of the rice genome. Nature 436:793–800
44. Isserlin R, El-Badrawi RA, Bader GD (2011) The biomolecular interaction network database in PSI-MI 2.5. Database 2011:baq037
45. Jaillon O, Aury JM, Noel B, Policriti A, Clepet C, Casagrande A, Choisne N, Aubourg S, Vitulo N, Jubin C, Vezzi A, Legeai F, Hugueney P, Dasilva C, Horner D, Mica E, Jublot D, Poulain J, Bruyère C, Billault A, Segurens B, Gouyvenoux M, Ugarte E, Cattonaro F, Anthouard V, Vico V, Del Fabbro C, Alaux M, Di Gaspero G, Dumas V, Felice N, Paillard S, Juman I, Moroldo M, Scalabrin S, Canaguier A, Le Clainche I, Malacrida G, Durand E, Pesole G, Laucou V, Chatelet P, Merdinoglu D, Delledonne M, Pezzotti M, Lecharny A, Scarpelli C, Artiguenave F, Pè ME, Valle G, Morgante M, Caboche M, Adam-Blondon AF, Weissenbach J, Quétier F, Wincker P, French-Italian Public Consortium for Grapevine Genome Characterization (2007) The grapevine genome sequence suggests ancestral hexaploidization in major angiosperm phyla. Nature 449:463–467
46. Janssen BJ, Thodey K, Schaffer RJ, Alba R, Balakrishnan L, Bishop R, Bowen JH, Crowhurst RN, Gleave AP, Ledger S, McArtney S, Pichler FB, Snowden KC, Ward S (2008) Global gene expression analysis of apple fruit development from the floral bud to ripe fruit. BMC Plant Biol 8:16
47. Jensen PJ, Halbrendt N, Fazio G, Makalowska I, Altman N, Praul C, Maximova SN, Ngugi HK, Crassweller RM, Travis JW, McNellis TW (2012) Rootstock-regulated gene expression patterns associated with fire blight resistance in apple. BMC Genomics 13:9
48. Jung S, Staton M, Lee T, Blenda A, Svancara R, Abbott A, Main D (2008) GDR (Genome Database for Rosaceae): integrated web-database for Rosaceae genomics and genetics data. Nucleic Acids Res 36:D1034–D1040
49. Jung S, Cestaro A, Troggio M, Main D, Zheng P, Cho I, Folta KM, Sosinski B, Abbott A, Celton JM, Arús P, Shulaev V, Verde I, Morgante M, Rokhsar D, Velasco R, Sargent DJ (2012) Whole genome comparisons of Fragaria, Prunus and Malus reveal different modes of evolution between Rosaceous subfamilies. BMC Genomics 13:129
50. Kaminuma E, Kosuge T, Kodama Y, Aono H, Mashima J, Gojobori T, Sugawara H, Ogasawara O, Takagi T, Okubo K, Nakamura Y (2011) DDJB progress report. Nucleic Acids Res 39:D22–D27

51. Khan MA, Han Y, Zhao YF, Korban SS (2012) A high-throughput apple SNP genotyping platform using the GoldenGate™ assay. Gene 492:196–201
52. Krost C, Petersen R, Lokan S, Brauksiepe B, Braun P, Schmidt ER (2013) Evaluation of the hormonal state of columnar apple trees (Malus × domestica) based on high throughput gene expression studies. Plant Mol Biol 81:211–220
53. Lalli DA, Decroocq V, Blenda AV, Schurdi-Levraud V, Garay L, Le Gall O, Damsteegt V, Reighard GL, Abbott AG (2005) Identification and mapping of resistance gene analogs (RGAs) in Prunus: a resistance map for Prunus. Theor Appl Genet 111:1504–1513
54. Lamesch P, Berardini TZ, Li D, Swarbreck D, Wilks C, Sasidharan R, Muller R, Dreher K, Alexander DL, Garcia-Hernandez M, Karthikeyan AS, Lee CH, Nelson WD, Ploetz L, Singh S, Wensel A, Huala E (2012) The Arabidopsis information resource (TAIR): improved gene annotation and new tools. Nucleic Acids Res 40:D1202–D1210
55. Lazzari B, Caprera A, Vecchietti A, Merelli I, Barale F, Milanesi L, Stella A, Pozzi C (2008) Version VI of the ESTree db: an improved tool for peach transcriptome analysis. BMC Bioinforma 9(Suppl 2):S9
56. Lee YP, Yu GH, Seo YS, Han SE, Choi YO, Kim D, Mok IG, Kim WT, Sung SK (2007) Microarray analysis of apple gene expression engaged in early fruit development. Plant Cell Rep 26:917–926
57. Lee T, Cho I, Peace C, Jung S, Zheng P, Main D (2011) Development approach and architecture of GenSAS BIOCOMP'11, July 2011, Las Vegas
58. Lee TH, Tang H, Wang X, Paterson AH (2013) PGDD: a database of gene and genome duplication in plants. Nucleic Acids Res 41:D1152–D1158
59. Li Y, Wu B, Yu Y, Yang G, Wu C, Zheng C (2011) Genome-wide analysis of the RING finger gene family in apple. Mol Genet Genomics 286:81–94
60. Li X, Korir NK, Liu L, Shangguan L, Wang Y, Han J, Chen M, Fang J (2012) Microarray analysis of differentially expressed genes engaged in fruit development between Prunus mume and Prunus armeniaca. J Plant Physiol 169:1776–1788
61. Lister R, O'Malley RC, Tonti-Filippini J, Gregory BD, Berry CC, Millar AH, Ecker JR (2008) Highly integrated single-base resolution maps of the epigenome in Arabidopsis. Cell 133:523–536
62. Lyons E, Freeling M (2008) How to usefully compare homologous plant genes and chromosomes as DNA sequences. Plant J 53:661–673
63. Martínez-Gómez P, Crisosto CH, Bonghi C, Rubio M (2011) New approaches to Prunus transcriptome analysis. Genetica 139:755–769
64. Matsuda F, Hirai MY, Sasaki E, Akiyama K, Yonekura-Sakakibara K, Provart NJ, Sakurai T, Shimada Y, Saito K (2010) AtMetExpress development: a phytochemical atlas of Arabidopsis development. Plant Physiol 152:566–578
65. McKay SJ, Vergara IA, Stajich JE (2010) Using the Generic Synteny Browser (GBrowse_syn). Curr Protoc Bioinforma Chapter 9:Unit 9.12
66. Mi H, Dong Q, Muruganujan A, Gaudet P, Lewis S, Thomas PD (2010) PANTHER version 7: improved phylogenetic trees, orthologs and collaboration with the Gene Ontology Consortium. Nucleic Acids Res 38:D204–D210
67. Mihara M, Itoh T, Izawa T (2010) SALAD database: a motif-based database of protein annotations for plant comparative genomics. Nucleic Acids Res 38:D835–D842
68. Newcomb RD, Crowhurst RN, Gleave AP, Rikkerink EHA, Allan AC, Beuning LL, Bowen JH, Gera E, Jamieson KR, Janssen BJ et al (2006) Analyses of expressed sequence tags from apple. Plant Physiol 141:147–166
69. Ogundiwin EA, Martí C, Forment J, Pons C, Granell A, Gradziel TM, Peace CP, Crisosto CH (2008) Development of ChillPeach genomic tools and identification of cold-responsive genes in peach fruit. Plant Mol Biol 68:379–397
70. Pampanwar V, Engler F, Hatfield J, Blundy S, Gupta G, Soderlund C (2005) FPC web tools for rice, maize, and distribution. Plant Physiol 138:116–126
71. Paterson AH, Bowers JE, Bruggmann R, Dubchak I, Grimwood J, Gundlach H, Haberer G, Hellsten U, Mitros T, Poliakov A, Schmutz J, Spannagl M, Tang H, Wang X, Wicker

T, Bharti AK, Chapman J, Feltus FA, Gowik U, Grigoriev IV, Lyons E, Maher CA, Martis M, Narechania A, Otillar RP, Penning BW, Salamov AA, Wang Y, Zhang L, Carpita NC, Freeling M, Gingle AR, Hash CT, Keller B, Klein P, Kresovich S, McCann MC, Ming R, Peterson DG, Mehboob-ur-Rahman, Ware D, Westhoff P, Mayer KF, Messing J, Rokhsar DS (2009) The Sorghum bicolor genome and the diversification of grasses. Nature 457:551–556

72. Peace C, Bassil N, Main D, Ficklin S, Rosyara UR, Stegmeir T, Sebolt A, Gilmore B, Lawley C, Mockler TC, Bryant DW, Wilhelm L, Iezzoni A (2012) Development and evaluation of a genome-wide 6K SNP array for diploid sweet cherry and tetraploid sour cherry. PLoS ONE 7:e48305

73. Popescu SC, Popescu GV, Bachan S, Zhang Z, Seay M, Gerstein M, Snyder M, Dinesh-Kumar SP (2007) Differential binding of calmodulin-related proteins to their targets revealed through high-density Arabidopsis protein microarrays. Proc Natl Acad Sci USA 104:4730–4735

74. Popescu SC, Popescu GV, Bachan S, Zhang Z, Gerstein M, Snyder M, Dinesh-Kumar SP (2009) MAPK target networks in Arabidopsis thaliana revealed using functional protein microarrays. Genes Dev 23:80–92

75. Pruitt KD, Tatusova T, Maglott DR (2007) NCBI reference sequences (RefSeq): a curated non-redundant sequence database of genomes, transcripts and proteins. Nucleic Acids Res 35:D61–D65

76. Rouard M, Guignon V, Aluome C, Laporte MA, Droc G, Walde C, Zmasek CM, Périn C, Conte MG (2011) GreenPhylDB v2.0: comparative and functional genomics in plants. Nucleic Acids Res 39:D1095–D1102

77. Santos AM, Oliver MJ, Sánchez AM, Payton PR, Gomes JP, Miguel C, Oliveira MM (2009) An integrated strategy to identify key genes in almond adventitious shoot regeneration. J Exp Bot 60:4159–4173

78. Sargent DJ, Kuchta P, Lopez Girona E, Zhang H, Davis TM, Celton JM, Marchese A, Korbin M, Folta KM, Shulaev V, Simpson DW (2011) Simple sequence repeat marker development and mapping targeted to previously unmapped regions of the strawberry genome sequence. Plant Genome 4:165–177

79. Schaeffer ML, Harper LC, Gardiner JM, Andorf CM, Campbell DA, Cannon EK, Sen TZ, Lawrence CJ (2011) MaizeGDB: curation and outreach go hand-in-hand. Database 2011:bar022

80. Schaffer RJ, Friel EN, Souleyre EJ, Bolitho K, Thodey K, Ledger S, Bowen JH, Ma JH, Nain B, Cohen D, Gleave AP, Crowhurst RN, Janssen BJ, Yao JL, Newcomb RD (2007) A genomics approach reveals that aroma production in apple is controlled by ethylene predominantly at the final step in each biosynthetic pathway. Plant Physiol 144:1899–1912

81. Schmutz J, Cannon SB, Schlueter J, Ma J, Mitros T, Nelson W, Hyten DL, Song Q, Thelen JJ, Cheng J, Xu D, Hellsten U, May GD, Yu Y, Sakurai T, Umezawa T, Bhattacharyya MK, Sandhu D, Valliyodan B, Lindquist E, Peto M, Grant D, Shu S, Goodstein D, Barry K, Futrell-Griggs M, Abernathy B, Du J, Tian Z, Zhu L, Gill N, Joshi T, Libault M, Sethuraman A, Zhang XC, Shinozaki K, Nguyen HT, Wing RA, Cregan P, Specht J, Grimwood J, Rokhsar D, Stacey G, Shoemaker RC, Jackson SA (2010) Genome sequence of the palaeopolyploid soybean. Nature 463:178–83. Erratum in: Nature 465:120

82. Schnable PS, Ware D, Fulton RS, Stein JC, Wei F, Pasternak S, Liang C, Zhang J, Fulton L, Graves TA, Minx P, Reily AD, Courtney L, Kruchowski SS, Tomlinson C, Strong C, Delehaunty K, Fronick C, Courtney B, Rock SM, Belter E, Du F, Kim K, Abbott RM, Cotton M, Levy A, Marchetto P, Ochoa K, Jackson SM, Gillam B, Chen W, Yan L, Higginbotham J, Cardenas M, Waligorski J, Applebaum E, Phelps L, Falcone J, Kanchi K, Thane T, Scimone A, Thane N, Henke J, Wang T, Ruppert J, Shah N, Rotter K, Hodges J, Ingenthron E, Cordes M, Kohlberg S, Sgro J, Delgado B, Mead K, Chinwalla A, Leonard S, Crouse K, Collura K, Kudrna D, Currie J, He R, Angelova A, Rajasekar S, Mueller T, Lomeli R, Scara G, Ko A, Delaney K, Wissotski M, Lopez G, Campos D, Braidotti M, Ashley E, Golser W, Kim H, Lee S, Lin J, Dujmic Z, Kim W, Talag J, Zuccolo

A, Fan C, Sebastian A, Kramer M, Spiegel L, Nascimento L, Zutavern T, Miller B, Ambroise C, Muller S, Spooner W, Narechania A, Ren L, Wei S, Kumari S, Faga B, Levy MJ, McMahan L, Van Buren P, Vaughn MW, Ying K, Yeh CT, Emrich SJ, Jia Y, Kalyanaraman A, Hsia AP, Barbazuk WB, Baucom RS, Brutnell TP, Carpita NC, Chaparro C, Chia JM, Deragon JM, Estill JC, Fu Y, Jeddeloh JA, Han Y, Lee H, Li P, Lisch DR, Liu S, Liu Z, Nagel DH, McCann MC, SanMiguel P, Myers AM, Nettleton D, Nguyen J, Penning BW, Ponnala L, Schneider KL, Schwartz DC, Sharma A, Soderlund C, Springer NM, Sun Q, Wang H, Waterman M, Westerman R, Wolfgruber TK, Yang L, Yu Y, Zhang L, Zhou S, Zhu Q, Bennetzen JL, Dawe RK, Jiang J, Jiang N, Presting GG, Wessler SR, Aluru S, Martienssen RA, Clifton SW, McCombie WR, Wing RA, Wilson RK (2009) The B73 maize genome: complexity, diversity, and dynamics. Science 326:1112–1115

83. Schwacke R, Schneider A, van der Graaff E, Fischer K, Catoni E, Desimone M, Frommer WB, Flügge UI, Kunze R (2003) ARAMEMNON, a novel database for Arabidopsis integral membrane proteins. Plant Physiol 131:16–26

84. Shizuya H, Birren B, Kim UJ, Mancino V, Slepak T, Tachiiri Y, Simon M (1992) Cloning and stable maintenance of 300-kilobase-pair fragments of human DNA in Escherichia coli using an F-factor-based vector. Proc Natl Acad Sci USA 89:8794–8797

85. Shulaev V, Korban SS, Sosinski B, Abbott AG, Aldwinckle HS, Folta KM, Iezzoni A, Main D, Arús P, Dandekar AM, Lewers K, Brown SK, Davis TM, Gardiner SE, Potter D, Veilleux RE (2008) Multiple models for Rosaceae genomics. Plant Physiol 147:985–1003

86. Shulaev V, Sargent DJ, Crowhurst RN, Mockler TC, Folkerts O, Delcher AL, Jaiswal P, Mockaitis K, Liston A, Mane SP, Burns P, Davis TM, Slovin JP, Bassil N, Hellens RP, Evans C, Harkins T, Kodira C, Desany B, Crasta OR, Jensen RV, Allan AC, Michael TP, Setubal JC, Celton JM, Rees DJ, Williams KP, Holt SH, Ruiz Rojas JJ, Chatterjee M, Liu B, Silva H, Meisel L, Adato A, Filichkin SA, Troggio M, Viola R, Ashman TL, Wang H, Dharmawardhana P, Elser J, Raja R, Priest HD, Bryant DW Jr, Fox SE, Givan SA, Wilhelm LJ, Naithani S, Christoffels A, Salama DY, Carter J, Lopez Girona E, Zdepski A, Wang W, Kerstetter RA, Schwab W, Korban SS, Davik J, Monfort A, Denoyes-Rothan B, Arus P, Mittler R, Flinn B, Aharoni A, Bennetzen JL, Salzberg SL, Dickerman AW, Velasco R, Borodovsky M, Veilleux RE, Folta KM (2011) The genome of woodland strawberry (Fragaria vesca). Nat Genet 43:109–116

87. Simillion C, Janssens K, Sterck L, Van de Peer Y (2008) i-ADHoRe 2.0: an improved tool to detect degenerated genomic homology using genomic profiles. Bioinformatics 24:127–128

88. Spiller M, Linde M, Hibrand-Saint Oyant L, Tsai CJ, Byrne DH, Smulders MJ, Foucher F, Debener T (2011) Towards a unified genetic map for diploid roses. Theor Appl Genet 122:489–500

89. Stein LD, Mungall C, Shu S, Caudy M, Mangone M, Day A, Nickerson E, Stajich JE, Harris TW, Arva A, Lewis S (2002) The Generic Genome Browser: a building block for a model organism system database. Genome Res 12:1599–1610

90. Tang H, Bowers JE, Wang X, Ming R, Alam M, Paterson AH (2008) Synteny and collinearity in plant genomes. Science 320:486–488

91. Tatsuki M, Nakajima N, Fujii H, Shimada T, Nakano M, Hayashi KI, Hayama H, Yoshioka H, Nakamura Y (2013) Increased levels of IAA are required for system 2 ethylene synthesis causing fruit softening in peach (Prunus persica L. Batsch). J Exp Bot 64(4):1049–1059.

92. Tatusov RL, Fedorova ND, Jackson JD, Jacobs AR, Kiryutin B, Koonin EV, Krylov DM, Mazumder R, Mekhedov SL, Nikolskaya AN, Rao BS, Smirnov S, Sverdlov AV, Vasudevan S, Wolf YI, Yin JJ, Natale DA (2003) The COG database: an updated version includes eukaryotes. BMC Bioinforma 4:41

93. The Gene Ontology Consortium (2013) Gene ontology annotations and resources. Nucleic Acids Res 41:D530–D535PubMedCentral

94. The UniProt Consortium (2012) Reorganizing the protein space at the Universal Protein Resource (UniProt). Nucleic Acids Res 40:D71–D75PubMedCentral
95. Trainotti L, Zanin D, Casadoro G (2003) A cell wall-oriented genomic approach reveals a new and unexpected complexity of the softening in peaches. J Exp Bot 54:1821–1831
96. Van Bel M, Proost S, Wischnitzki E, Movahedi S, Scheerlinck C, Van de Peer Y, Vandepoele K (2012) Dissecting plant genomes with the PLAZA comparative genomics platform. Plant Physiol 158:590–600
97. Vecchietti A, Lazzari B, Ortugno C, Bianchi F, Malinverni R, Caprera A, Mignani I, Pozzi C (2009) Comparative analysis of expressed sequence tags from tissues in ripening stages of peach (Prunus persica L. Batsch). Tree Genet Genome 5:377–391
98. Velankar S, Alhroub Y, Alili A, Best C, Boutselakis HC, Caboche S, Conroy MJ, Dana JM, van Ginkel G, Golovin A, Gore SP, Gutmanas A, Haslam P, Hirshberg M, John M, Lagerstedt I, Mir S, Newman LE, Oldfield TJ, Penkett CJ, Pineda-Castillo J, Rinaldi L, Sahni G, Sawka G, Sen S, Slowley R, Sousa da Silva AW, Suarez-Uruena A, Swaminathan GJ, Symmons MF, Vranken WF, Wainwright M, Kleywegt GJ (2011) PDBe: Protein Data Bank in Europe. Nucleic Acids Res 39(Suppl 1):D402–D410
99. Velasco R, Zharkikh A, Affourtit J, Dhingra A, Cestaro A, Kalyanaraman A, Fontana P, Bhatnagar SK, Troggio M, Pruss D, Salvi S, Pindo M, Baldi P, Castelletti S, Cavaiuolo M, Coppola G, Costa F, Cova V, Dal Ri A, Goremykin V, Komjanc M, Longhi S, Magnago P, Malacarne G, Malnoy M, Micheletti D, Moretto M, Perazzolli M, Si-Ammour A, Vezzulli S, Zini E, Eldredge G, Fitzgerald LM, Gutin N, Lanchbury J, Macalma T, Mitchell JT, Reid J, Wardell B, Kodira C, Chen Z, Desany B, Niazi F, Palmer M, Koepke T, Jiwan D, Schaeffer S, Krishnan V, Wu C, Chu VT, King ST, Vick J, Tao Q, Mraz A, Stormo A, Stormo K, Bogden R, Ederle D, Stella A, Vecchietti A, Kater MM, Masiero S, Lasserre P, Lespinasse Y, Allan AC, Bus V, Chagné D, Crowhurst RN, Gleave AP, Lavezzo E, Fawcett JA, Proost S, Rouzé P, Sterck L, Toppo S, Lazzari B, Hellens RP, Durel CE, Gutin A, Bumgarner RE, Gardiner SE, Skolnick M, Egholm M, Van de Peer Y, Salamini F, Viola R (2010) The genome of the domesticated apple (Malus × domestica Borkh.). Nat Genet 42:833–839
100. Verde I, Bassil N, Scalabrin S, Gilmore B, Lawley CT, Gasic K, Micheletti D, Rosyara UR, Cattonaro F, Vendramin E, Main D, Aramini V, Blas AL, Mockler TC, Bryant DW, Wilhelm L, Troggio M, Sosinski B, Aranzana MJ, Arús P, Iezzoni A, Morgante M, Peace C (2012) Development and evaluation of a 9 K SNP array for peach by internationally coordinated SNP detection and validation in breeding germplasm. PLoS ONE 7(4):e35668.
101. Vizoso P, Meisel LA, Tittarelli A, Latorre M, Saba J, Caroca R, Maldonado J, Cambiazo V, Campos-Vargas R, Gonzalez M, Orellana A, Silva H (2009) Comparative EST transcript profiling of peach fruits under different post-harvest conditions reveals candidate genes associated with peach fruit quality. BMC Genomics 10:423
102. Wang K, Wang Z, Li F, Ye W, Wang J, Song G, Yue Z, Cong L, Shang H, Zhu S, Zou C, Li Q, Yuan Y, Lu C, Wei H, Gou C, Zheng Z, Yin Y, Zhang X, Liu K, Wang B, Song C, Shi N, Kohel RJ, Percy RG, Yu JZ, Zhu YX, Wang J, Yu S (2012) The draft genome of a diploid cotton Gossypium raimondii. Nat Genet 44:1098–1103
103. Weber APM, Weber KL, Carr K, Wilkerson C, Ohlrogge JB (2007) Sampling the Arabidopsis transcriptome with massively parallel pyrosequecing. Plant Physiol 144:32–42
104. Wegrzyn JL, Lee JM, Tearse BR, Neale DB (2008) TreeGenes: a forest tree genome database. Intl J Plant Genomics 2008:412875.
105. Wheeler DL, Barrett T, Benson DA, Bryant SH, Canese K, Chetvernin V, Church DM, DiCuccio M, Edgar R, Federhen S, Geer LY, Kapustin Y, Khovayko O, Landsman D, Lipman DJ, Madden TL, Maglott DR, Ostell J, Miller V, Pruitt KD, Schuler GD, Sequeira E, Sherry ST, Sirotkin K, Souvorov A, Starchenko G, Tatusov RL, Tatusova TA, Wagner L, Yaschenko E (2007) Database resources of the National Center for Biotechnology Information. Nucleic Acids Res 35:D5–D12

106. Wilson D, Pethica R, Zhou Y, Talbot C, Vogel C, Madera M, Chothia C, Gough J (2009) SUPERFAMILY: sophisticated comparative genomics, data mining, visualization and phylogeny. Nucleic Acids Res 37:D380–D386
107. Wu J, Wang Z, Shi Z, Zhang S, Ming R, Zhu S, Khan MA, Tao S, Korban SS, Wang H, Chen NJ, Nishio T, Xu X, Cong L, Qi K, Huang X, Wang Y, Zhao X, Wu J, Deng C, Gou C, Zhou W, Yin H, Qin G, Sha Y, Tao Y, Chen H, Yang Y, Song Y, Zhan D, Wang J, Li L, Dai M, Gu C, Wang Y, Shi D, Wang X, Zhang H, Zeng L, Zheng D, Wang C, Chen M, Wang G, Xie L, Sovero V, Sha S, Huang W, Zhang S, Zhang M, Sun J, Xu L, Li Y, Liu X, Li Q, Shen J, Wang J, Paull RE, Bennetzen JL, Wang J, Zhang S (2013) The genome of the pear (Pyrus bretschneideri Rehd.). Genome Res 23:396–408
108. Xu Q, Chen LL, Ruan X, Chen D, Zhu A, Chen C, Bertrand D, Jiao WB, Hao BH, Lyon MP, Chen J, Gao S, Xing F, Lan H, Chang JW, Ge X, Lei Y, Hu Q, Miao Y, Wang L, Xiao S, Biswas MK, Zeng W, Guo F, Cao H, Yang X, Xu XW, Cheng YJ, Xu J, Liu JH, Luo OJ, Tang Z, Guo WW, Kuang H, Zhang HY, Roose ML, Nagarajan N, Deng XX, Ruan Y (2012) The draft genome of sweet orange (Citrus sinensis). Nat Genet 45:59–66
109. Yao Q, Bollinger C, Gao J, Xu D, Thelen JJ (2012) P3DB: an integrated database for plant protein phosphorylation. Front Plant Sci 3:206
110. Yi G, Sze SH, Thon MR (2007) Identifying clusters of functionally related genes in genome. Bioinformatics 23:1053–1060
111. Youens-Clark K, Buckler E, Casstevens T, Chen C, Declerck G, Derwent P, Dharmawardhana P, Jaiswal P, Kersey P, Karthikeyan AS, Lu J, McCouch SR, Ren L, Spooner W, Stein JC, Thomason J, Wei S, Ware D (2011) Gramene database in 2010: updates and extensions. Nucleic Acids Res 39:D1085–D1094
112. Zhang Q, Chen W, Sun L, Zhao F, Huang B, Yang W, Tao Y, Wang J, Yuan Z, Fan G, Xing Z, Han C, Pan H, Zhong X, Shi W, Liang X, Du D, Sun F, Xu Z, Hao R, Lv T, Lv Y, Zheng Z, Sun M, Luo L, Cai M, Gao Y, Wang J, Yin Y, Xu X, Cheng T, Wang J (2012) The genome of Prunus mume. Nat Commun 3:1318
113. Zhebentyayeva T, Swire-Clark G, Georgi L, Garay L, Juns S, Forrest S, Blenda A, Blackmon B, Mook J, Horn R, Howad W, Arus P, Main D, Tomkins J, Sosinski B, Baird W, Reighard G, Abbott A (2008) A framework physical map for peach, a model Rosaceae species. Tree Genet Genome 4:745–756
114. Zhu Y, Varanasi V, Zheng P, Main D, Curry E, Mattheis J (2012) Multiple plant hormones and cell wall metabolism regulate apple fruit maturation patterns and texture attributes. Tree Genet Genome 8:1389–1406
115. Ziliotto F, Begheldo M, Rasori A, Bonghi C, Tonutti P (2008) Transcriptome profiling of ripening nectarine (Prunus persica L. Batsch) fruit treated with 1-MCP. J Exp Bot 59:2781–2791
116. Ziosi V, Bonghi C, Bregoli AM, Trainotti L, Biondi S, Sutthiwal S, Kondo S, Costa G, Torrigiani P (2008) Jasmonate-induced transcriptional changes suggest a negative interference with the ripening syndrome in peach fruit. J Exp Bot 59:563–573

CHAPTER 10

Bioinformatic Identification and Analysis of Extensins in the Plant Kingdom

Xiao Liu[1,2], Richard Wolfe[3] , Lonnie R. Welch[2,3], David S. Domozych[4] , Zoë A. Popper[5] , Allan M. Showalter[1,2*]

[1] *Department of Environmental and Plant Biology, Ohio University, Athens, Ohio, United States of America,*

[2] *Molecular and Cellular Biology Program, Ohio University, Athens, Ohio, United States of America,*

[3] *Russ College of Engineering and Technology, Center for Intelligent, Distributed and Dependable Systems, Ohio University, Athens, Ohio, United States of America,*

[4] *Department of Biology, Skidmore College, Saratoga Springs, New York, United States of America,*

[5] *Botany and Plant Science and Ryan Institute for Environmental, Marine, and Energy Research, School of Natural Sciences, National University of Ireland Galway, Galway, Ireland*

ABSTRACT

Extensins (EXTs) are a family of plant cell wall hydroxyproline-rich glycoproteins (HRGPs) that are implicated to play important roles in plant growth, development, and defense. Structurally, EXTs are characterized by the repeated occurrence of serine (Ser) followed by three to five prolines (Pro) residues, which are hydroxylated as hydroxyproline (Hyp) and glycosylated. Some EXTs have Tyrosine (Tyr)-X-Tyr (where X can be any amino acid) motifs that are responsible for intramolecular or intermolecular cross-linkings. EXTs can be divided into several classes: classical EXTs, short EXTs, leucine-rich repeat extensins (LRXs), proline-rich extensin-like receptor kinases (PERKs), formin-homolog EXTs (FH EXTs), chimeric EXTs, and long chimeric EXTs. To guide future research on the EXTs and understand evolutionary history of EXTs in the plant kingdom, a bioinformatics study was conducted to identify and classify EXTs from 16 fully sequenced plant genomes, including *Ostreococcus lucimarinus, Chlamydomonas reinhardtii, Volvox carteri, Klebsormidium flaccidum, Physcomitrella patens, Selaginella moellendorffii, Pinus taeda, Picea abies, Brachypodium distachyon, Zea mays, Oryza sativa, Glycine max, Medicago truncatula, Brassica rapa, Solanum lycopersicum*, and *Solanum tuberosum*, to supplement data previously obtained from *Arabidopsis thaliana* and *Populus trichocarpa*. A total of 758 EXTs were newly identified, including 87 classical EXTs, 97 short EXTs, 61 LRXs, 75 PERKs, 54 FH EXTs, 38 long chimeric EXTs, and 346 other chimeric EXTs. Several notable findings were made: (1) classical EXTs were likely derived after the terrestrialization of plants; (2) LRXs, PERKs, and FHs were derived earlier than classical EXTs; (3) monocots have few classical EXTs; (4) Eudicots have the greatest number of classical EXTs and Tyr-X-Tyr cross-linking motifs are predominantly in classical EXTs; (5) green algae have no classical EXTs but have a number of long chimeric EXTs that are absent in embryophytes. Furthermore, phylogenetic analysis was conducted of LRXs, PERKs and FH EXTs, which shed light on the evolution of three EXT classes.

INTRODUCTION

Extensins (EXTs) are a diverse family of hydroxyproline-rich glycoproteins (HRGPs) found only in the plant kingdom. They are cell wall proteins characterized by the repeated occurrence of serine (Ser) followed by several consecutive prolines (Pro) [1,2]. Some EXT molecules have Tyr-X-Tyr motifs (where X can be any amino acid) that are responsible for intramolecular or intermolecular cross-linking with other EXT molecules in forms of isodityrosine (Idt), di-Idt, and pulcherosine [3]. These cross-linking properties contribute to the extracellular matrix and play roles in plant development and defense mechanisms [4,5].

Besides cross-linking of Tyr motifs, post-translational modification of EXTs includes hydroxylation of Pro residues to hydroxyproline (Hyp), and subsequent glycosylation of Hyp and Ser residues. Peptidyl-Pro is hydroxylated by prolyl 4-hydroxylases (P4Hs). Plant P4Hs belong to a family of 2-oxoglutarate-dependent dioxygenases [6–8]. Characterization of P4Hs is reported for a number of plants, including *Arabidopsis thaliana* [6,9–11], *Nicotiana tabacum* [12], *Dianthus caryophyllus* [13], and *Chlamydomonas reinhardtii* [14].

O-glycosylation of EXTs predominantly occurs on Ser-Hyp_n motifs, with often four to five oligoarabinosides attached to Hyp residues and galactose (Gal) monosaccharides attached to Ser [1]. In Arabidopsis, the sequential addition of arabinose (Ara) residues is carried out by distinct arabinosyltransferases: hydroxyproline *O*-β-arabinosyltransferase (HPAT1-3) [15], reduced residual arabinose 1–3 (RRA1-3) [11,16], Xyloglucanase113 (XEG113) [17], and extensin arabinose deficient (ExAD) (Petersen et al., unpublished). The addition of Gal to Ser is carried out by Ser galactosyltransferase (SGT1) [18].

EXTs can be divided into several classes: classical EXTs, short EXTs, leucine-rich repeat extensins (LRXs), proline-rich extensin-like receptor kinases (PERKs), formin-homolog EXTs (FH EXTs), long chimeric EXTs and other chimeric EXTs [2]. Classical EXTs have signal peptide sequences which direct the proteins to the secretory system and ultimately the extracellular matrix. Most prominently, they have Ser-$Pro_{3\text{-}5}$ repeated motifs throughout their sequences. Moreover, some EXTs have Tyr-X-Tyr (YXY) motifs along with the Ser-$Pro_{3\text{-}5}$motifs. EXTs that are less than 200 amino acids in length are referred to as "Short EXTs". LRXs are a class of chimeric EXTs which usually have signal peptide sequences at the N terminus, followed by leucine-rich repeat (LRR) domains, and Ser-$Pro_{3\text{-}5}$ repeated modules near the C terminus. The LRR domain is known to be involved in protein-protein interactions [19], and the EXT domain is thought to contribute to the insolubility in the cell wall. These features make LRXs candidates for regulatory functions on the cell surface. In Arabidopsis, LRXs are implicated in root hair morphogenesis [20]. PERKs represent another class of chimeric EXTs. They lack a signal peptide sequence and their SP_n repeated motifs are predominately located at the N terminus; they have a protein kinase catalytic domain near their C terminus. In Arabidopsis, the PERK gene family contains 15 members, and PERK1 was localized at the plasma membrane [21]. Microarray data showed that there are two major groups of PERKs: those that are specifically expressed in the pollen and those that are generally expressed throughout all plant tissues [22]. Research has shown that PERKs may affect cell expansion and normal floral organ formation [23]. In Arabidopsis, they are associated with an abscisic acid response [24]. In *Brassica napus*, BnPERK1 is reported to be involved in signal perception and response to wound and/or pathogen stimuli [21]. A third class of chimeric EXTs is the FH EXTs. FH EXTs are characterized by significant homology to formins and the presence of repeated Ser-$Pro_{3\text{-}5}$ motifs. In eukaryotes, formins are associated with actin dynamics in that they control the assembly and elongation of unbranched actin filaments [25,26]. A fourth group of chimeric EXTs are termed "long chimeric EXT" because of their extraordinary sequence length that have more than 2,000 amino acids. Lastly, some EXTs were characterized as "other chimeric EXTs" as these EXTs have an EXT domain and one or more domain(s) not known to HRGPs or the above classes of chimeric EXTs.

Showalter et al. [2 and unpublished data] conducted the identification of the HRGP superfamily in *Arabidopsis thaliana* and *Populus trichocarpa* in which 59 and 60 EXTs were identified, respectively. In addition, Newman and Cooper [27] identified numerous proline-rich tandem repeat proteins (TRPs) including EXTs through a bioinformatics approach using EST and NCBI Non-Redundant protein sequence data of a number of plant species, but the search criteria for TRPs were not tailored for identifying EXTs. Nonetheless, knowledge about the number and distribution of EXTs in plant kingdom is still lacking.

BIO OHIO 2.0 is a newly revised and improved bioinformatics software program developed at Ohio University that was tailored to fulfill this task [2,28]. The program was designed and developed for protein identification based on amino acid signatures, such as biased amino acid composition and common HRGP amino acid motifs in the genome-encoded protein sequences (i.e., the predicted proteome). The program can also further analyze identified proteins by checking for the presence of potential signal peptide sequences and GPI anchor addition sequences and finding similar HRGPs via the Basic Local Alignment Search Tool (BLAST). Using this bioinformatics tool, Showalter et al. [2] identified and classified the HRGP superfamily in Arabidopsis (*Arabidopsis thaliana*) and poplar (*Populus trichocarpa*), two fully sequenced plant genomes ([2,28]; Showalter et al., unpublished).

Rapid advancement in the "next generation sequencing (NGS)" techniques is increasingly making genome sequences available. Thus, it is now feasible to conduct a more detailed analysis on the EXT family in the plant kingdom. Here, we analyzed 16 plant genomes: *Ostreococcus lucimarinus* [29], *Chlamydomonas reinhardtii* [30], *Volvox carteri* [31], *Klebsormidium flaccidum* [32], *Physcomitrella patens* [33], *Selaginella moellendorffii* [34], *Pinus taeda* [35], *Picea abies* [36], *Brachypodium distachyon* [37], *Zea mays* [38], *Oryza sativa* [39], *Glycine max* [40], *Medicago truncatula* [41], *Brassica rapa* [42], *Solanum lycopersicum* [43], *and Solanum tuberosum* [44]. We also integrated previously studied data on Arabidopsis and *P. trichocarpa* to determine the number and distribution of the EXT family members in the plant kingdom and examine the evolutionary history of this fundamental cell wall constituent [45].

MATERIALS AND METHODS

Identification of EXTs

The predicted protein data files from 16 plant species (*O. lucimarinus, C. reinhardtii, V. carteri, K. flaccidum, P. patens, S. moellendor ffii, P. taeda, P. abies, B. distachyon, Z. mays, O. sativa, G. max, M. truncatula, B . rapa, S. lycopersicum, and S. tuberosum*) were downloaded from the Phytozome website (www.phytozome.org). The protein database was searched for EXTs using BIO OHIO 2.0 software, which integrated more functional modules into the software compared to BIO OHIO 1.0 [2, 28]. Briefly, a regular expression of two or more SPPP repeats was used to search for candidate EXTs. Candidate EXT sequences were then analyzed for the positions of SP_n repeats and YXY cross-linking motifs, the presence of signal peptide sequences, the presence of GPI anchors, and for similar sequences using BLAST searches against known Arabidopsis EXTs. An EXT is determined by comparing all the above information with known features of each class of EXTs. If a sequence fails to fit in any class of EXTs, it is called a potential EXT. The BIO OHIO 2.0 program is freely available and can be downloaded from github: https://github.com/Showlaterlab/BIO-OHIO-2.0

BLAST Analysis

The functional module of BLAST was integrated into the BIO OHIO 2.0. All candidate EXTs identified were subjected to NCBI BLASTP analysis using the current Arabidopsis protein BLAST dataset (November 2010 TAIR10 Genome Release) downloaded from The Arabidopsis Information Resource (TAIR; www.arabidopsis.org).

Signal Peptides and GPI Anchors

The functional modules for signal peptides and GPI anchors were integrated into the BIO OHIO 2.0. All proteins were analyzed for signal peptides using SignalP (www.cbs.dtu.dk/services/SignalP/) [46] and for GPI anchor addition sequences using the big-PI plant predictor (mendel.imp.ac.at/gpi/plant_server.html) [47].

Sequence Alignment and Phylogenetic Analysis

Amino acid sequences were aligned by use of the Geneious software program (http://www.geneious.com/) to obtain conserved domains. Aligned sequences of LRXs, PERKs, and FH EXTs were input into Mega 6 for phylogenetic analysis using the maximum likelihood and the maximum parsimony methods [48]. For LRXs, the analysis involved 78 protein sequences. There were a total of 294 amino acid positions in the final dataset. The evolutionary history inferred by the Maximum Likelihood method was based on the JTT matrix-based model [49]. The evolutionary history inferred by the Maximum Parsimony method used the Tree-Bisection-Regrafting (TBR) algorithm with search level 1 in which the initial trees were obtained by the random addition of sequences (10 replicates) [50]. The bootstrap consensus tree inferred from 1000 replicates was shown and branches corresponding to partitions reproduced in less than 50% bootstrap replicates are collapsed. For the phylogenetic analysis of PERKs and FH EXTs, the same methods were used as for analysis of the LRXs. The analysis of PERKs involved 93 protein sequences, and a total of 283 amino acid positions were present in the final dataset. The analysis of FH EXTs involved 76 protein sequences, and a total of 377 amino acid positions were present in the final dataset.

RESULTS

In order to identify candidate EXTs, the BIO OHIO 2.0 program searched for protein sequences with two or more SPPP repeats from 16 plant proteomes: *O. lucimarinus, C. reinhardtii, V. carteri, K. flaccidum, P. patens, S. moellendorffii, P. taeda, P. abies, B. distachyon, Z. mays, O. sativa, G. max, M. tr uncatula, B. rapa, S. lycopersicum, and S. tuberosum.* This initial screening obtained 2563 candidate EXTs, among which 758 were determined as EXTs and 1804 as potential EXTs. The EXTs include 87 classical EXTs, 97 short EXTs, 61 LRXs, 75 PERKs, 54 FH EXTs, 38 long chimeric EXTs, and 346 other chimeric EXTs (Figure 1). In addition to having at least two SPPPs, these EXTs contain a HRGP domain that is rich in Pro, Alanine (Ala), Valine (Val), Ser, Glycine (Gly), and Threonine (Thr), and most proteins (76%) have predicted signal peptide sequences that direct them into the secretory pathway and ultimately to the cell wall. A representative EXT sequence from each class is shown in Figure 2. Detailed sequence feature analysis of identified EXTs for each species is shown in S1–S16 Tables. All the identified EXT sequences are shown in S1 Figure ; the sequences of potential EXTs are shown in S2 Figure . These potential EXTs have at least two SPPP repeat motifs but mostly lack a signal sequence and a HRGP domain.

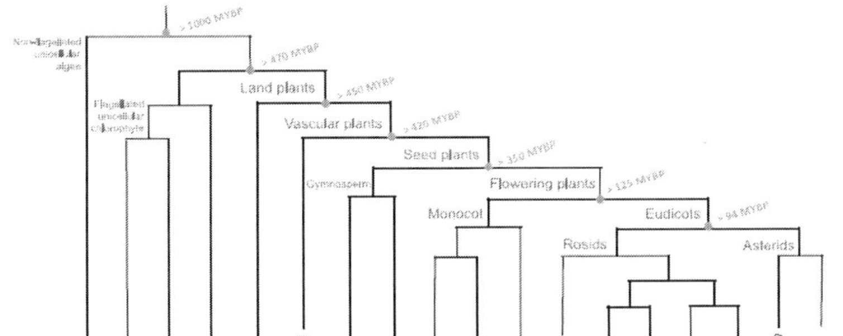

	O. lucimarinus	*C. reinhardtii*	*V. carteri*	*K. flaccidum*	*P. patens*	*S. moellendorffi*	*P. taeda*	*P. abies*	*B. distachyon*	*Z. mays*	*O. sativa*	*P. trichocarpa*	*G. max*	*M. truncatula*	*B. rapa*	*A. thaliana*	*S. lycopersicum*	*S. tuberosum*
Classical EXTs	0	0	0	0	0	13	0	1	1	0	0	8	6	4	16	20	15	31
Short EXTs	0	2	0	0	4	6	0	5	3	8	7	22	20	9	12	12	14	7
LRXs	0	0	0	0	5	2	0	2	5	5	5	10	6	4	15	11	5	7
PERKs	0	0	0	0	6	0	2	0	5	10	7	12	14	4	14	13	7	6
FH EXTs	0	0	1	0	2	2	3	2	3	5	3	5	9	2	11	6	7	4
Long Chimeric EXTs	4	27	7	0	0	0	0	0	0	0	0	0	0	0	0	0	0	0
Other Chimeric EXTs	4	153	84	4	1	21	0	0	6	15	5	3	6	8	20	3	10	9
Total	8	182	92	4	18	44	5	10	23	43	27	60	61	31	88	65	58	64

Figure 1. Phylogenetic distribution of EXTs in selected plant genomes.

(A) Dendrogram showing the evolutionary relationships of species selected representing major plant divisions. (B) The distribution of EXTs identified in this study and in the previous literature (Showalter et al. 2010, and unpublished data). EXTs are divided into seven subclasses including classical EXTs, short EXTs, LRXs, PERKs, FH EXTs, chimeric EXTs, and long chimeric EXTs.

Green colored sequences at the N terminus indicate predicted signal peptides. SP_3(blue), SP_4 (red), SP_5 (purple), and YXY (dark red) repeats are also indicated. Sequences typical of AGPs, AP, PA, SP, TP, VP, and GP repeats, are also indicated (yellow).

Classical and short EXTs

Classical EXTs are categorized as having EXT domains throughout the protein sequence, except at the N terminus where there is usually a signal peptide that directs the protein into the secretory pathway and ultimately to the cell wall. The EXT domains contain repeated motifs of SP_n, where n ≥3. Moreover, most classical EXTs have cross-linking YXY motifs in addition to the SP_n motifs.

Figure 2. Protein sequences encoded by representative EXT gene classes in *Brassica rapa* and *Chlamydomonas reinhardtii.*

In this study, no classical EXTs were identified in the five earliest diverging species (*O. lucimarinus*, *C. reinhardtii*, *V. carteri*, *K. flaccidum*, and *P. patens*), indicating that classical EXTs are absent in these non-vascular species that are either aquatic green algae (*O. lucimarinus*, *C. reinhardtii*, *V. carteri*, *K. flaccidium*) or land plants that are dependent on water for reproduction, lacking roots, and predominantly living in humid habitats (*P. patens*). However, classical EXTs were found in tracheophytes (vascular plants), including early diverging members as 13 were identified in the lycophyte *S. moellendorffii*, 11 of which contained YXY motifs. *S. moellendorffii* EXT11 was found to share high similarity with all 12 of the other EXTs, indicating the likely occurrence of gene duplication events (**DATA NOT SHOWN**).

Despite the presence of classical EXTs in tracheophytes dating back to more than 420 million years before present (MYBP), classical EXTs were nearly absent from the genomes of the two gymnosperm species and the three monocot species examined here. No classical EXTs were identified in loblolly pine (*P. taeda*), and only one classical EXT (MA_74039g0010) was identified in Norway spruce (*P. abies*). Furthermore, while this protein contains 35 $SPPP_{3\text{-}5}$, it lacks a signal peptide. Similarly, no classical EXTs were identified in corn (*Z. mays*), or rice (*O. sativa*), while *B. distachyon*, a non-crop species, only contained one apparent classical EXT, Bradi3g10280. This protein contains 11 SP_3 and three SP_4 repeats, along with 19 QAAA repeats, which is not known to be associated with any other EXTs or HRGPs. In addition, a BLAST search with Bradi3g10280 as query found no hits of significant similarity to other protein sequences (**DATA NOT SHOWN**). These findings are consistent with two previous studies in monocots that found a lack of the SP_n repeat motif in *Z. mays* [51,52].

Classical EXTs, however, were ubiquitous in eudicots. In this project, five species were chosen for analysis: *G. max*, *M. truncatula*, *B. rapa*, *S. lycopersicum*, and *S. tuberosum*. Classical EXTs were found in all these species. In addition, previous research reported on the identification of classical EXTs in Arabidopsis and poplar, respectively [2 and unpublished data]. An overview of the number of classical EXTs identified in these plants is shown in Figure 1.

The frequency of SP_3, SP_4 and SP_5 (or more) in classical EXTs among the above species was calculated to determine which of these repeat motifs is dominant in classical EXTs. The results showed that SP_4 repeats universally predominated in EXT sequences, with the lowest being in the lycophyte *S. moellendorffii* (56%) and the highest being in *S. tuberosum* (80%) (Figure 3). However, the dominance of the SP_4 repeated motif is not seen in other categories of EXTs (**DATA NOT SHOWN**).

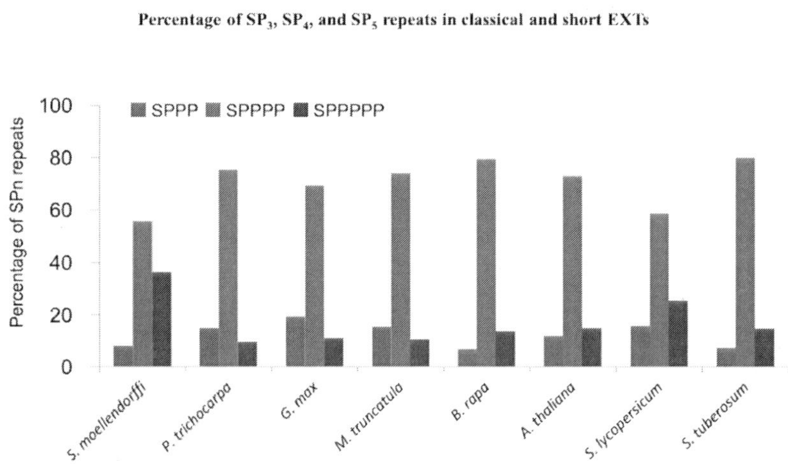

Figure 3. The frequency of SP_3, SP_4, and SP_5 repeats in classical EXTs of selected genomes.

The frequency was calculated by the total number of each type of repeat divided by the total number of SP_3, SP_4, and SP_5 adding together in each species.

The average number of YXY motifs in classical EXT and non-classical EXT (i.e. all other classes of EXTs) was calculated to confirm the observation that YXY motifs are abundant exclusively in classical EXTs. As is shown in Figure 4, the average number of YXY motifs in classical EXTs ranges from 5.7 (in *S. lycopersicum*) to 27.5 (in *B. rapa*), whereas less than two occurrences of the YXY motif were found in non-classical EXTs in all species studied.

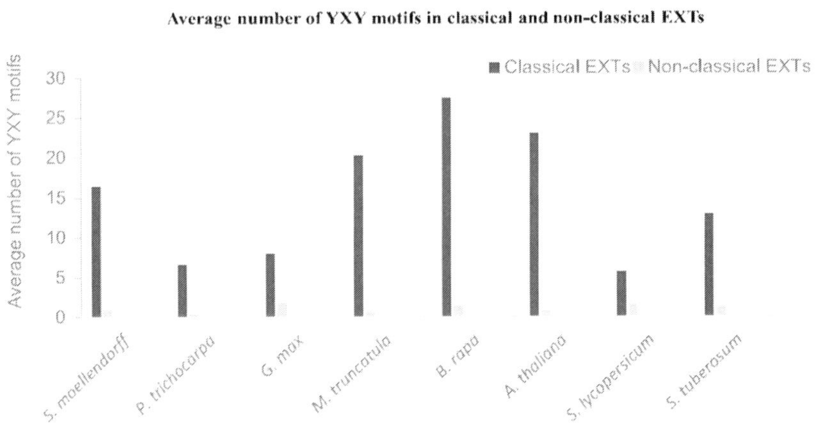

Figure 4. Average number of YXY motif in classical and non-classical EXTs.

The frequency was calculated as the total number of YXY repeats divided by the total number of classical and non-classical EXTs in chosen species.

Short EXTs were found in most species in this study but no short EXTs were identified from the *O. lucimarinus, V. carteri, K. flaccidum, and P. taeda* genomes. Interestingly, two short EXTs were identified in the aquatic species, *C. reinhardtii*. Unlike classical EXTs, short EXTs were also found in *P. patens*. Overall, there is a slight increase in the number of short EXTs in the embryophytes, which indicates the importance of this group of proteins in plant growth, development, and defense (Figure 1).

LRXs, PERKs, and FHs

Leucine-rich repeat extensins (LRXs) are a group of chimeric EXTs. A typical LRX has an N terminal signal peptide, followed by a leucine-rich repeat (LRR) domain, and a C terminal EXT domain. A representative structure of an LRX is shown in Figure 5. In this study, LRXs were found in all but the four algal species (*O. lucimarinus, C. reinhardtii, V. carteri, K. flaccidum*) and one gymnosperm (*P. taeda*). *B. rapa*, found to contain sequences for 15, had the highest number of LRXs of any species. However, most species contain two to seven LRXs (Figure 1 and S1 Figure). Five LRXs were identified in *P. patens*, suggesting the possibility that LRXs were derived during plant terrestrialisation (and subsequently lost from some species e.g. *P. tadea*). A BLAST search against all the LRXs identified in this study revealed that these five LRXs share more homology among themselves than any LRXs identified in other species, indicating they are likely paralogs derived from gene duplication events. Interestingly, for the two gymnosperm species included in this study, two LRXs were identified in *P. abies* while none were found in *P. taeda*. According to evidence from chloroplast, mitochondrial and nuclear genes although they are closely related members of the Pinaceae the genera *Picea* and *Pinus* diverged ~140 million years ago and it is reasonable that differences in LRXs may exist between them [53]. LRXs were identified in all flowering plants in this study, with eudicots having more LRXs in general.

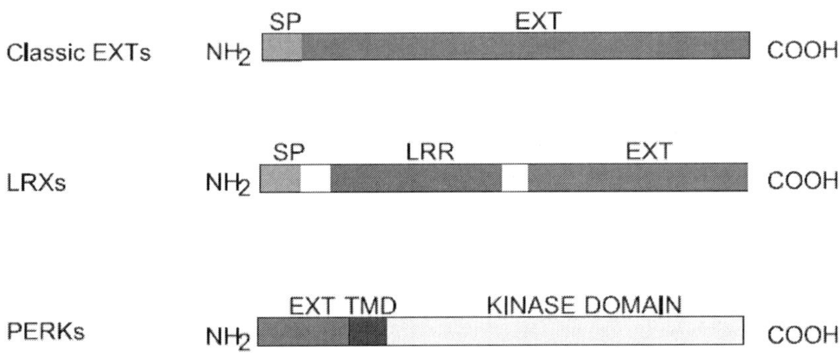

Figure 5. Structural schemes of classical EXTs, LRX, and PERKs.

Classical EXTs have an N-terminal signal peptide (green) followed only by EXT domain (red). LRXs have an N-terminal signal peptide (green) followed by a leucine-rich region (LRR, blue). The EXT domain (red) of LRXs is located at the C terminus. The EXT domain of PERKs is located at N terminus followed by a transmembrane domain (TMD, black) and a kinase domain (yellow).

To explore the evolutionary relationship of LRXs in different species, phylogenetic analysis was conducted using the maximum likelihood method based on the JTT matrix-based model. The phylogenetic analysis showed that all LRXs in the moss *P. patens* were clustered together as the outgroup. The rest of the LRXs fell into five major clades. Among them, all eudicot LRXs fell into clades A and B (*G. max, M. truncatula, P. trichocarpa, A. thaliana, B. rapa, S. lycopersicum,* and *S. tuberosum*), while clades C and D contained all the monocot LRXs (*O. sativa*, *Z. mays*, and *B. distachyon*). This topology indicates that either LRXs in monocots and eudicots went through quite different changes, or one or more clades of LRXs were derived after the divergence of monocots and eudicots. Notably, previously reported PEXs were found only in clade E, which contained all PEXs from Arabidopsis (AtPEX1-4), two *O. sativa* PEXs (OsPEX1 and OsPEX3), one *Z. mays* PEX (ZmPEX1), and one *S. lycopersicum* PEX (LePEX1). Therefore, it is likely that ancestral PEX gene(s) existed before the division of monocots and eudicots, and that the rest of the LRXs in clade E may also be PEXs (Figure 6). Phylogenetic analysis was also conducted using the maximum parsimony method (S3 Figure), which showed almost identical tree topology as inferred by the maximum likelihood method. The aligned sequences of LRXs are shown in S4 Figure .

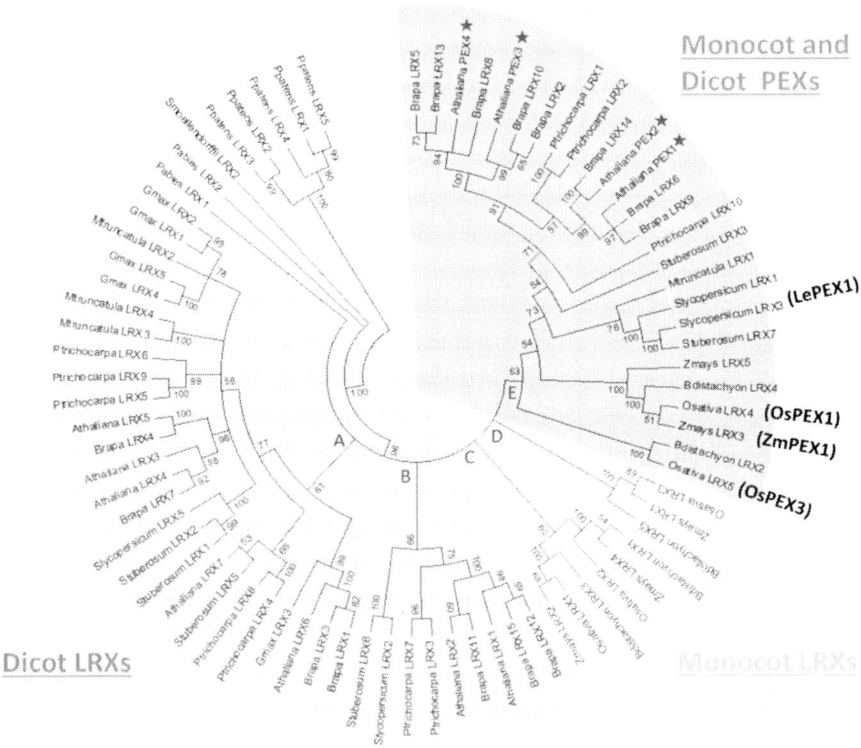

Figure 6. Maximum Likelihood Analysis of LRXs.

The evolutionary history was inferred by using the Maximum Likelihood method based on the JTT matrix-based model. The bootstrap consensus tree inferred from 1000 replicates is taken to represent the evolutionary history of the taxa analyzed. Branches corresponding to partitions reproduced in less than 50% bootstrap replicates are collapsed. The analysis involved 78 amino acid sequences. There were a total of 294 positions in the final dataset. The green-colored fan area (clades A and B) indicates LRXs from eudicot species. The yellow-colored fan area (clades C and D) indicates LRXs from monocot species. The blue-colored fan area indicates possible PEX clade. Previously reported PEXs were either marked with red "☆" or in parentheses.

Proline-rich extensin-like receptor kinases (PERKs) represent another group of chimeric EXTs. They have an extracellular EXT domain at the N terminus followed by a transmembrane domain and an intracellular receptor kinase domain (Figure 5). In this study, PERKs were identified in most of the species in this study but not in *O. lucimarinus, C. reinhardtii, V. carteri, K. flaccidum, S. moellendorffii and P. abies.* Notably, BLAST analysis for two *K. flaccidum* proteins (kfl00031_0230 and kfl00671_0010p) revealed that they were similar to Arabidopsis PERKs, but a closer look found that they differed from the general PERK structure and were thus classified as chimeric EXTs (S4 Table and Figure 5). A number of PERKs were identified in *P. patens*, suggesting PERKs were derived after the terrestrialization of plants. With the exception in *P. taeda*, at least five PERKs were identified in each of other species with *G. max* and *B. rapa* having as many as 14 PERKs. To explore the evolutionary history of PERKs in these species, phylogenetic analysis was conducted using the maximum likelihood method based on the JTT matrix-based model (Figure 7). The phylogenetic tree shows that PERKs from these species form two dominant clades (clade A and B). The expression pattern analysis of PERKs in Arabidopsis showed that some of the Arabidopsis PERK members were pollen-specific genes while others were more broadly expressed [22]. In this tree, five AtPERKs that are pollen-specific were clustered in one sub-clade under clade B, namely, PERKs 3–7. However, other pollen-specific PERKs were also seen elsewhere in the tree (AtPERKs 11 and 12). The phylogenetic tree showed that most groups at the tips were either the same or closely related species, and the tree had high confidence at the tips in general. PERKs from monocot plants did not form a single branch, but they tended to cluster together. Similar to LRXs, tree topology inferred by the maximum parsimony method was nearly identical to that inferred by the maximum likelihood method (S5 Figure). The aligned sequences of PERKs are shown in S6 Figure .

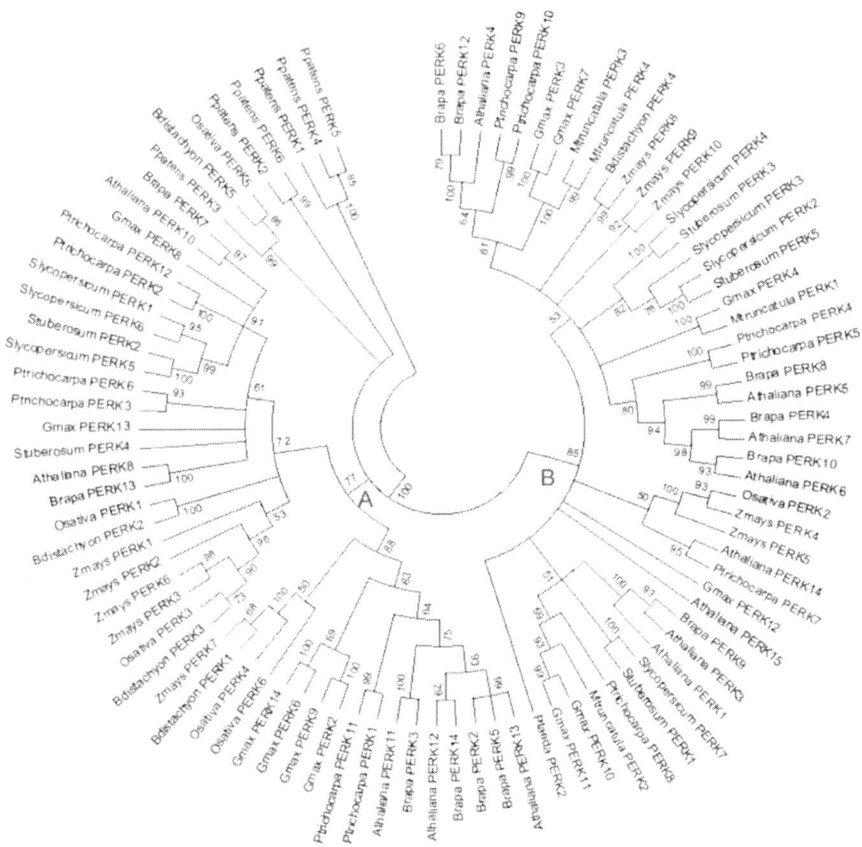

Figure 7. Maximum Likelihood Analysis of PERKs.

The evolutionary history was inferred by using the Maximum Likelihood method based on the JTT matrix-based model. The bootstrap consensus tree inferred from 1000 replicates is taken to represent the evolutionary history of the taxa analyzed. Branches corresponding to partitions reproduced in less than 50% bootstrap replicates are collapsed. The analysis involved 93 amino acid sequences. There were a total of 283 positions in the final dataset. A and B represent two major clades.

Formins were first found in animal cells as cytoplasmic proteins that are associated with the organization of the actin cytoskeleton. Plant formin homologs are quite different in structure and function [54]. Plant formins may play important roles in cell cortex organization, including cortical actin, microtubule cytoskeletons, and the attachment to the plasma membrane [55]. In this study, formin homolog EXTs (FH EXTs) were categorized as a third group of chimeric EXTs, as some of the formin homologs were found to have an N terminal signal peptide and contain a number of SP_n repeats. FH EXTs were found in all but three of the algal species investigated: *O. lucimarinus, C. reinhardtii, and K. flaccidum*. Interestingly, one FH EXT (Vocar20008550m) was found in *V. carteri*, the remaining algal species examined, suggesting that plant formin homologs were derived before divergence of the embryophytes. In general, an increase in number of FH EXTs was found in higher plants, with *B. rapa* having as many as 11 FH EXTs. To explore the evolutionary history of FH EXTs in these species, phylogenetic analysis was conducted using the maximum likelihood method based on the JTT matrix-based model (Figure 8). Notably, the FH EXT identified in *V. carteri* was placed as the outgroup given the considerable difference between this sequence and other FH EXTs upon alignment. Similar to the analysis done by Deeks et al. [54], phylogenetic analysis revealed that FH EXTs were clustered into two major clades with 100% confidence. Clade A contained Arabidopsis FH1-11, while clade B contained AtFH 12–21. In general, FH EXTs from the same or closely related species tended to cluster together, and higher confidence was shown in deeper branches (with over 80% bootstrap support). Phylogenetic analysis was also conducted using the maximum parsimony method (S7 Figure), which showed almost identical tree topology as inferred by the maximum likelihood method. The aligned sequences of FH EXTs are shown in S8 Figure .

The evolutionary history was inferred by using the Maximum Likelihood method based on the JTT matrix-based model. The bootstrap consensus tree inferred from 1000 replicates is taken to represent the evolutionary history of the taxa analyzed. Branches corresponding to partitions reproduced in less than 50% bootstrap replicates are collapsed. The analysis involved 73 amino acid sequences. There were a total of 377 positions in the final dataset.

Long chimeric EXTs and other chimeric EXTs

A group of chimeric EXTs were characterized to contain an enormous number of amino acids, usually over 2,000 amino acids per protein. This group of EXTs was referred to as “long chimeric EXTs”. A few such EXTs were previously reported in *C. reinhardtii* [56,57]. In this project, 27 long chimeric EXTs were found in the *C. reinhardtii* genome. In addition, four and 84 long chimeric EXTs were identified in *O. lucimarinus* and *V. carteri*, respectively (Figure s 1, 2 and S1 Figure ; S1–S3 Tables). Long chimeric EXTs, however, were not found in any other species analyzed in this study. Other chimeric EXTs were categorized as being chimeric EXTs but did not belong to LRXs, PERKs, FH EXTs, or long chimeric EXTs. In this study, chimeric EXTs were identified in all but the two gymnosperm species; however, the number of chimeric EXTs in each species varied greatly (Figure 1). While most other species have no more than 21 chimeric EXTs, the two flagellated unicellular chlorophyte species, *C. reinhardtii* and *V. carteri*, contained 153 and 84 chimeric EXTs, respectively.

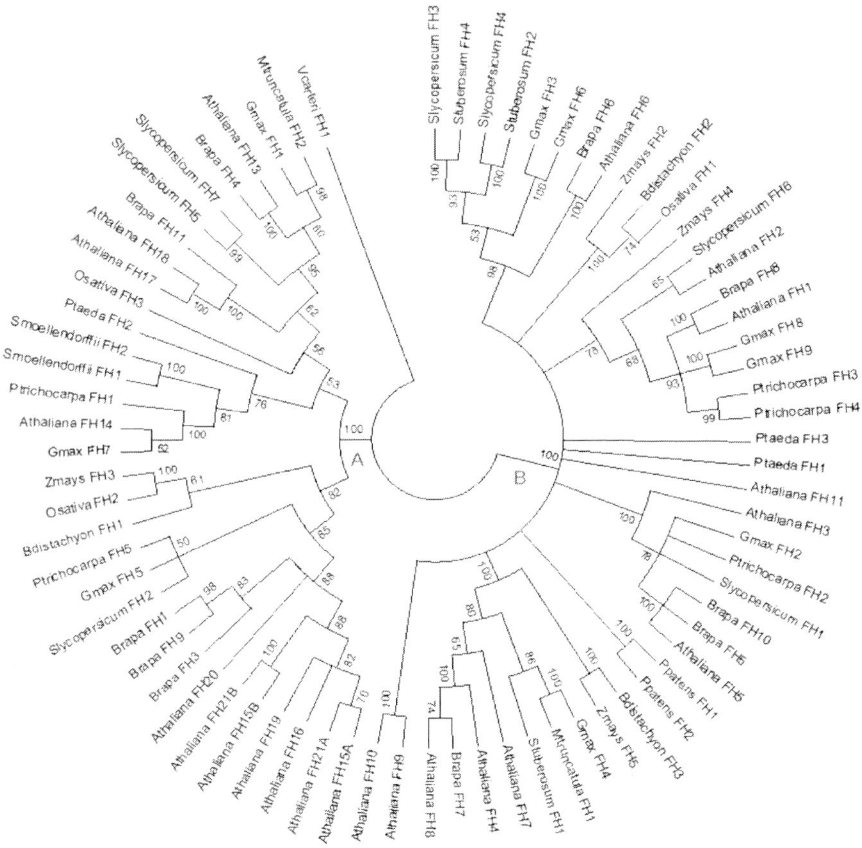

Figure 8. Maximum Likelihood Analysis of FHs.

Comparison with previously identified EXTs

The literature was searched to compare previously characterized EXTs with EXTs identified in this study. A total of 54 EXTs were obtained, of which 32 were identified in species included in this study. Among the 32 EXTs, 24 EXTs were found to be identical or similar to previously identified EXTs. The list of previously identified EXTs and their counterparts in this study is listed in S17 Table [1,4,20,21,51,52,56,58–89].

DISCUSSION

Bioinformatic identification of plant EXTs using BIO OHIO 2.0

The first comprehensive bioinformatics identification of EXTs was done in Arabidopsis in which 59 EXTs were identified [2], followed by poplar where 60 EXTs were found (Showalter et al., unpublished data). In another study, Newman and Cooper [27] identified numerous proline-rich TRPs including EXTs from proteomic databases, but they adopted more stringent search criteria which resulted in fewer EXTs identified in their study. Up to now, there is still a lack of understanding of the number and distribution of EXTs in the plant kingdom. With a rapidly increasing number of plant genomes fully sequenced, vast amounts of data are being produced. A need to process, mine and analyze genome data to provide biological meanings awaits.

The Bio OHIO 2.0 software program provides an efficient and reliable tool to identify proteins with biased amino acid compositions and known repetitive motifs [2,28]. The newly revised and improved 2.0 version integrated more functional modules that include searching for the presence of a signal peptide, GPI anchor, as well as automated BLAST searches against the Arabidopsis proteome. These improvements make the program an ideal bioinformatic tool to study cell wall components, and gain insight about evolution of protein families.

In this project, a total of 758 EXTs including 87 classical EXTs, 97 short EXTs, 61 LRXs, 75 PERKs, 54 FH EXTs, 38 long chimeric EXTs, and 346 other chimeric EXTs were identified among over half a million predicted protein sequences in 16 plant genomes ranging from primitive aquatic plants to eudicots. Moreover, the functions of searching for signal peptide, GPI anchor, and BLAST searches against the model plant Arabidopsis were incorporated in the program, making it more robust and efficient for identification.

The origin and evolution of EXTs

The EXTs identified in this study showed that the EXT prototypes might occur early in evolution of green photosynthetic organisms, most likely in the form of chimeric EXTs, as they are found both in the early diverging species and later diverging species. This suggests that ancestral EXTs might originate as a stretch of amino acids or a small EXT domain that functioned in favor of evolution, and thus had more widely spread over time in evolution. On the other hand, long chimeric EXTs were identified in only three of the most primitive species (*O. lucimarinus*, *C. reinhardtii*, and *V. carteri*), but are absent in all land plants, indicating the lack of evolutionary advantage for these long molecules. LRXs, PERKs, and FH EXTs predated the evolution of classical EXTs, as they are found in the land plant *P. patens*, where no classical EXTs were found (Figure 1 and S5 Figure). The origin of classical EXTs was possibly associated with plant vascularization, as although they are absent from the (non-vascular) bryophyte *P. patens* they are present in the (vascular) lycophyte *S. moellendorffii*. Further analysis is needed to help address whether they are indeed associated with vascularization or alternatively with terrestrialization, since *P. patens* is the only bryophyte examined in this study. Interestingly, classical EXTs are absent or nearly absent in gymnosperms *P. taeda* (0) and *P. abies* (1), as well as monocot plants such as *B. distachyon* (1), *Z. mays* (0), and *O. sativa* (0). However, further analysis is needed to provide greater support for this conclusion, as it is possible that classical EXTs may be more abundant in other gymnosperm and monocot species. Eudicots have the greatest numbers of EXTs and the largest number of Tyr-X-Tyr cross-linking motifs, which largely occurs in the classical EXTs.

Combining previously identified EXTs in Arabidopsis and poplar, phylogenetic analysis was conducted for LRXs, PERKs and FH EXTs, but not for classical EXTs as the latter failed to align in the EXT domain due to the highly varied number of SP_n repeat motifs and the Tyr-X-Tyr motifs, which is necessary for a meaningful outcome. Similarly, phylogenetic analysis was not done for short EXTs, chimeric and long chimeric EXTs due the heterogeneity of their non-EXT domains.

The phylogenetic analysis of LRXs included the LRR regions of 78 LRXs from 13 plant genomes. Five *P. patens* LRXs were clustered in the phylogenetic trees generated by both maximum likelihood and maximum parsimony methods, indicating that one ancestral gene duplicated multiple times. LRXs were found in almost all of the more advanced species in this study, demonstrating the widespread nature of this protein family. Comparable to Baumberger et al. [90] who reported LRXs form two clades: reproductive pollen-expressed LRXs referred to as PEXs and a vegetative LRXs, analysis here showed that LRXs form three major groups, including one group of eudicot-specific LRXs, one group of monocot-specific LRXs, and one group of likely PEXs. The existence of eudicot and monocot specific clades indicates that LRXs in monocot and eudicot evolved quite distinctly from ancient LRXs. The PEX specific clade contains LRXs from both monocots and eudicots, indicating that ancestral LRXs duplicated and diversified to become pollen-specific LRXs before the division of monocot and eudicot [90].

The PERK phylogenetic analysis included the receptor kinase domain of 93 protein sequences from 12 plant genomes. Similar to LRXs, PERKs identified in *P. patens* were clustered at the root of the phylogenetic tree by both maximum likelihood and maximum parsimony methods. However, PERKs were not found in the lycophyte *S. moellendorffii* or the gymnosperm *P. abies*. The phylogenetic tree shows that PERKs form two major clades, both included monocots and eudicots. This indicates that ancestral PERK genes existed before the division of monocots and eudicots. Expression pattern analysis of PERKs in Arabidopsis showed that some of the AtPERK members were pollen-specific genes while others were more broadly expressed [22]. Here, pollen-specific AtPERK3-7 were clustered in clade B, but other pollen-specific PERKs were also seen in other branches. The phylogenetic tree shows most groups at the tips are either the same or closely related species, and the tree has high confidence at the tips in general. This may be due to gene duplication events that lead to gene redundancy as is seen in Arabidopsis [91,92].

Phylogenetic analysis of FH EXTs included the FH domain of 76 proteins from eleven plant species. All 23 formin homologs in Arabidopsis were included in this analysis, as it was interesting to see their distributions in the phylogenetic tree. Notably, only AtFH1, AtFH5, AtFH8, AtFH13, AtFH16, and AtFH20 contain two or more SPPP. The phylogenetic tree showed two major clades (clades A and B) with high confidence, which was consistent with the study of Deeks et al. [54] who showed that Arabidopsis FHs form two major types.

A major gap in our understanding of the evolution of EXTs in green plants is a result of the lack of significant genomic information for charophytes, i.e., the group of green algae ancestral and most closely related to modern day land plants. However, immunological based screening has revealed the presence of EXTs as well other HRGPs including arabinogalactan-proteins in many of the charophyte taxa [93–95]. A previous study on cell wall biosynthetic pathways in charophytes [96] and this report have also shown that the presence of EXT-like macromolecules in the charophyte *K. flaccidum* and charophyte ancestors including the prasinophyte (*Ostreococcus*) and the charophyte sister clade, the chlorophytes (*Chlamydomonas* and *Volvox*), thereby supporting the presence of EXTs in charophytes. Similarly, the lack of genomic information for bryophytes and ferns, with only one bryophyte sequence and no fern sequences available, makes pinpointing some of the EXT distribution patterns difficult to interpret. For instance, classical EXTs were found in *S. moellendorffii* but not in *P. patens*, suggesting that the origin of classical EXTs might be associated with vascularization. However, since only one bryophyte and one lycophyte genome was investigated it may be possible that *P. patens* is an exception and that classical EXTs are present in other bryophytes, in which case classical EXTs may be associated instead with terrestrialization. A significantly more resolved interpretation will soon be possible as more genomic data for charophytes, bryophytes, and ferns become available.

CONCLUSIONS

A revised and newly improved bioinformatics software program BIO OHIO 2.0 was utilized to identify and classify EXTs from predicted proteomes of 16 plant species. A total of 758 EXTs were identified, including 87 classical EXTs, 97 short EXTs, 61 LRXs, 75 PERKs, 54 FH EXTs, 38 long chimeric EXTs, and 346 other chimeric EXTs. Analysis of these data revealed that: (1) classical EXTs were likely derived after the terrestrialization of plants; (2) LRXs, PERKs, and FHs were likely derived earlier than classical EXTs; (3) gymnosperms and monocots have few classical EXTs; (4) Eudicots have the greatest number of classical EXTs and Tyr-X-Tyr cross-linking motifs are predominantly in classical extensins; (5) green algae lack classical EXTs but have a number of long chimeric EXTs that are absent in embryophytes. Furthermore, phylogenetic analysis was conducted for LRXs, PERKs and FH EXTs, which shed light on the evolution of the EXTs.

ACKNOWLEDGMENTS

The authors thank Dr. Melanie Schori for providing valuable comments and suggestions.

AUTHOR CONTRIBUTIONS

Conceived and designed the experiments: AMS XL. Performed the experiments: XL. Analyzed the data: XL AMS DSD ZAP. Contributed reagents/materials/analysis tools: RW LW. Wrote the paper: XL AMS DSD ZAP.

REFERENCES

1. 1.Kieliszewski M, Lamport D. Extensin—repetitive motifs, functional sites, posttranslational codes, and phylogeny. Plant J. 1994;5(2):157–72. pmid:8148875

2. 2.Showalter AM, Keppler B, Lichtenberg J, Gu D, Welch LR. A bioinformatics approach to the identification, classification, and analysis of hydroxyproline-rich glycoproteins. Plant Physiol. 2010;153(2):485–513. pmid:20395450
3. 3.Held M, Tan L, Kamyab A, Hare M, Shpak E, Kieliszewski M. Di-isodityrosine is the intermolecular cross-link of isodityrosine-rich extensin analogs cross-linked in vitro. J Biol Chem. 2004;279(53):55474–82. pmid:15465824
4. 4.Merkouropoulos G, Barnett D, Shirsat A. The Arabidopsis extensin gene is developmentally regulated, is induced by wounding, methyl jasmonate, abscisic and salicylic acid, and codes for a protein with unusual motifs. Planta. 1999;208(2):212–9. pmid:10333585
5. 5.Cannon MC, Terneus K, Hall Q, Tan L, Wang Y, Wegenhart BL, et al. Self-assembly of the plant cell wall requires an extensin scaffold. Proc Natl Acad Sci U S A. 2008;105(6):2226–31. pmid:18256186
6. 6.Hieta R, Myllyharju J. Cloning and characterization of a low molecular weight prolyl 4-hydroxylase from Arabidopsis thaliana. Effective hydroxylation of proline-rich collagen-like, and hypoxia-inducible transcription factor alpha-like peptides. J Biol Chem. 2002;277:23965–71. pmid:11976332
7. 7.Koski MK, Hieta R, Bollner C, Kivirikko KI, Myllyharju J, Wierenga RK. The active site of an algal prolyl 4-hydroxylase has a large structural plasticity. J Biol Chem. 2007;282:37112–23. pmid:17940281
8. 8.Koski MK, Hieta R, Hirsila M, Ronka A, Myllyharju J, Wierenga RK. The crystal structure of an algal prolyl 4-hydroxylase complexed with a proline-rich peptide reveals a novel buried tripeptide binding motif. J Biol Chem.2009;284:25290–301. pmid:19553701
9. 9.Tiainen P, Myllyharju J, Koivunen P. Characterization of a second Arabidopsis thaliana prolyl 4-hydroxylase with distinct substrate specificity. J Biol Chem. 2005;280:1142–8. pmid:15528200
10. 10.Vlad F, Spano T, Vlad D, Daher FB, Ouelhadj A, Fragkostefanakis S, et al. Involvement of Arabidopsis prolyl 4 hydroxylases in hypoxia, anaoxia and mechanical wounding. Plant Signal Behav. 2007;2:368–9. pmid:19704601
11. 11.Velasquez SM, Ricardi MM, Dorosz JG, Fernandez PV, Nadra AD, Pol-Fachin L, et al. O-glycosylated cell wall proteins are essential in root hair growth. Science. 2011;332:1401–3. pmid:21680836
12. 12.Yuasa K, Toyooka K, Fukuda H, Matsuoka K. Membrane-anchored prolyl hydroxylase with an export signal from the endoplasmic reticulum. Plant J. 2005;41:81–94. pmid:15610351
13. 13.Vlad F, Tiainen P, Owen C, Spano T, Daher FB, Oualid F, et al. Characterization of two carnation petal prolyl 4 hydroxylases. Physiol Plant. 2010;140:199–207. pmid:20553416
14. 14.Keskiaho K, Hieta R, Sormunen R, Myllyharju J. Chlamydomonas reinhardtii has multiple prolyl 4-hydroxylases, one of which is essential for proper cell wall assembly. Plant Cell. 2007;19:256–269. pmid:17220203
15. 15.Ogawa-ohnishi M, Matsushita W, Matsubayashi Y. Identification of three hydroxyproline O-arabinosyltransferases in Arabidopsis thaliana. Nat Chem Biol. 2013;9:726–30. pmid:24036508
16. 16.Egelund J, Obel N, Ulvskov P, Geshi N, Pauly M, Bacic A, et al. Molecular characterization of two Arabidopsis thaliana glycosyltransferase mutants, rra1 and rra2, which have a reduced residual arabinose content in a polymer tightly associated with the cellulosic wall residue. Plant Mol Biol. 2007;64:439–49. pmid:17401635
17. 17.Gille S, Hansel U, Ziemann M, Pauly M. Identification of plant cell wall mutants by means of a forward chemical genetic approach using hydrolases. Proc Natl Acad Sci U S A. 2009;106:14699–704. pmid:19667208
18. 18.Saito F, Suyama A, Oka T, Yoko-O T, Matsuoka K, Jigami Y, et al. Identification of novel peptidyl serine α-galactosyltransferase gene family in plants. J Biol Chem. 2014;289(30):20405–20. pmid:24914209

19. 19.Kobe B, Deisenhofer J. The leucine-rich repeat: A versatile binding motif. Trends Biochem Sci. 1994;19:415–21. pmid:7817399

20. 20.Baumberger N, Ringli C, Keller B. The chimeric leucine-rich repeat/extensin cell wall protein LRX1 is required for root hair morphogenesis in Arabidopsis thaliana. Genes Dev. 2001;15(9):1128–39. pmid:11331608

21. 21.Silva N, Goring D. The proline-rich, extensin-like receptor kinase-1 (PERK1) gene is rapidly induced by wounding. Plant Mol Biol. 2002;50(4–5):667–85. pmid:12374299

22. 22.Nakhamchik A, Zhao Z, Provart NJ, Shiu S, Keatley SK, Cameron RK, et al. A comprehensive expression analysis of the Arabidopsis proline-rich extensin-like receptor kinase gene family using bioinformatic and experimental approaches. Plant Cell Physiol. 2004;45:1875–81. pmid:15653807

23. 23.Haffani YZ, Silva-Gagliardi NF, Sewter SK, Aldea MG, Zhao Z, Nakhamchik A, et al. Altered expression of perk receptor kinases in Arabidopsis leads to changes in growth and floral organ formation. Plant Signal Behav. 2006;1:251–60. pmid:19516986

24. 24.Bai L, Zhang G, Zhou Y, Zhang Z, Wang W, Du Y, et al. Plasma membrane-associated proline-rich extensin-like receptor kinase 4, a novel regulator of Ca2+ signaling, is required for abscisic acid responses in Arabidopsis thaliana. Plant J, 2009;60(2):314–27. pmid:19566594

25. 25.Mao Y. Formin: a link between kinetochores and microtubule ends. Trends Cell Biol. 2011;21(11):625–9. pmid:21920754

26. 26.Chalkia D, Nikolaidis N, Makalowski W, Klein J, Nei M. Origins and evolution of the formin multigene family that is involved in the formation of actin filaments. Mol Biol Evol. 2008;25(12):2717–33. pmid:18840602

27. 27.Newman AM, Cooper JB. Global analysis of proline-rich tandem repeat proteins reveals broad phylogenetic diversity in plant secretomes. PLOS One. 2011;6(8):e23167. pmid:21829715

28. 28.Lichtenberg J, Keppler B, Conley T, Gu D, Burns P, Welch L, et al. Prot-Class: A bioinformatics tool for protein classification based on amino acid signatures. Nat Science. 2012;4:1161–4.

29. 29.Palenik B, Grimwood J, Aerts A, Rouzé P, Salamov A, Putnam N, et al. The tiny eukaryote Ostreococcus provides genomic insights into the paradox of plankton speciation. Proc Natl Acad Sci U S A. 2007;104(18):7705–10. pmid:17460045

30. 30.Merchant SS, Prochnik SE, Vallon O, Harris EH, Karpowicz SJ, Witman G, et al. The Chlamydomonas genome reveals the evolution of key animal and plant functions. Science. 2007;318(5848):245–50. pmid:17932292

31. 31.Prochnik SE, Umen J, Nedelcu AM, Hallmann A, Miller SM, Nishii I, et al. Genomic analysis of organismal complexity in the multicellular green alga Volvox carteri. Science. 2010;329(5988):223–6. pmid:20616280

32. 32.Hori K, Maruyama F, Fujisawa T, Togashi T, Yamamoto N, Seo M, et al. Klebsormidium flaccidum genome reveals primary factors for plant terrestrial adaptation. Nat Commun. 2014;5:3978. pmid:24865297

33. 33.Physcomitrella patens v3.0, DOE-JGI, Available: http://www. phytozome. net/Physcomitrella

34. 34.Banks JA, Nishiyama T, Hasebe M, Bowman JL, Gribskov M, dePamphilis C, et al. The Selaginella genome identifies genetic changes associated with the evolution of vascular plants. Science. 2011;332(6032):960–3. pmid:21551031

35. 35.Zimin A, Stevens KA, Crepeau MW, Holtz-Morris A, Koriabine M, Marçais G, et al. Sequencing and assembly of the 22-gb loblolly pine genome. Genetics. 2014;196(3):875–90. pmid:24653210

36. 36.Nystedt B, Street NR, Wetterbom A, Zuccolo A, Lin YC, Scofield DG, et al. The Norway spruce genome sequence and conifer genome evolution. Nature. 2013;497:579–84. pmid:23698360

37. 37.International Brachypodium Initiative. Genome sequencing and analysis of the model grass Brachypodium distachyon. Nature. 2010;463(7282):763–8. pmid:20148030
38. 38.Schnable PS, Ware D, Fulton RS, Stein JC, Wei F, Pasternak S, et al. The B73 maize genome: complexity, diversity, and dynamics. Science. 2009;326(5956):1112–5. pmid:19965430
39. 39.Ouyang S, Zhu W, Hamilton J, Lin H, Campbell M, Childs K, et al. The TIGR rice genome annotation resource: improvements and new features. Nucleic Acids Res. 2007;35:D883–7. pmid:17145706
40. 40.Schmutz J, Cannon SB, Schlueter J, Ma J, Mitros T, Nelson W, et al. Genome sequence of the palaeopolyploid soybean. Nature. 2010;463(7278):178–83. pmid:20075913
41. 41.Young ND, Debellé F, Oldroyd GE, Geurts R, Cannon SB, Udvardi MK, et al. The Medicago genome provides insight into the evolution of rhizobial symbioses. Nature. 2011;480 (7378):520–4. pmid:22089132
42. 42.Brassica rapa FPsc v1.3, DOE-JGI. Available: http://www. phytozome. net/BrapaFPsc
43. 43.Tomato Genome Consortium. The tomato genome sequence provides insights into fleshy fruit evolution. Nature. 2012;485(7400):635–41. pmid:22660326
44. 44.Xu X, Pan S, Cheng S, Zhang B, Mu D, Ni P, et al. Genome sequence and analysis of the tuber crop potato. Nature. 2011;475(7355):189–95. pmid:21743474
45. 45.Tan L, Showalter AM, Egelund J, Hernandez-Sanchez A, Doblin MS, Bacic A. Arabinogalactan-proteins and the research challenges for these enigmatic plant cell surface proteoglycans. Front Plant Sci. 2012;3:140. pmid:22754559
46. 46.Petersen TN, Brunak S, von Heijne G, Nielsen H. SignalP 4.0: discriminating signal peptides from transmembrane regions. Nat Methods. 2011;8(10):785–6. pmid:21959131
47. 47.Eisenhaber B, Wildpaner M, Schultz CJ, Borner GHH, Dupree P, Eisenhaber F. Glycosylphosphatidylinositol lipid anchoring of plant proteins. Sensitive prediction from sequence- and genome-wide studies for Arabidopsis and rice. Plant Physiol. 2003;133(4):1691–701. pmid:14681532
48. 48.Tamura K, Stecher G, Peterson D, Filipski A, Kumar S. MEGA6: molecular evolutionary genetics analysis version 6.0. Mol Biol Evol. 2013;30:2725–9. pmid:24132122
49. 49.Jones DT, Taylor WR, Thornton JM. The rapid generation of mutation data matrices from protein sequences. Comput Appl Biosci. 1992;8:275–82. pmid:1633570
50. 50.Nei M, Kumar S. Molecular evolution and phylogenetics. New York: Oxford University Press; 2000.
51. 51.Stiefel V, Pérez-Grau L, Albericio F, Giralt E, Ruiz-Avila L, Ludevid MD, et al. Molecular cloning of cDNAs encoding a putative cell wall protein from Zea mays and immunological identification of related polypeptides. Plant Mol Biol. 1988;11(4):483–93. pmid:24272405
52. 52.Kieliszewski MJ, Leykam JF, Lamport DTA. Structure of the threonine-rich extensin from Zea mays. Plant Physiol. 1990;92:316–26. pmid:16667276
53. 53.Wang XQ, Tank DC, Sang T. Phylogeny and divergence times in the Pinaceae: evidence from three genomes. Mol Biol Evol. 2000;17:773–81. pmid:10779538
54. 54.Deeks MJ, Hussey PJ, Davies B. Formins: Intermediates in signal-transduction cascades that affect cytoskeletal reorganization. Trends Plant Sci. 2002;7:492–8. pmid:12417149
55. 55.Fatima Cvrčková. Formins: emerging players in the dynamic plant cell cortex. Scientifica. 2012;2012:712605. pmid:24278734
56. 56.Woessner JP, Goodenough UW. Molecular characterization of a zygote wall protein: an extensin-like molecule in Chlamydomonas reínhardtíí. Plant Cell. 1989;1:901–11. pmid:2535530

57. 57.Lee J-H, Waffenschmidt S, Small L, Goodenough U. Between-species analysis of short-repeat modules in cell wall and sex-related hydroxyproline-rich glycoproteins. Plant Physiol. 2007;144:1813–26. pmid:17573538

58. 58.Chen J, Varner JE. An extracellular matrix protein in plants: characterization of a genomic clone for carrot extensin. EMBO J. 1985;4:2145–51. pmid:15938047

59. 59.Chen J, Varner JE. Isolation and characterization of cDNA clones for carrot extensin and a proline-rich 33-kDa protein. Proc Natl Acad Sci U S A. 1985;82(13):4399–403. pmid:16593579

60. 60.Smith JJ, Muldoon EP, Willard JJ, Lamport DTA. Tomato extensin precursors P1 and P2 are highly periodic structures. Phytochemistry. 1986;5(5):1021–1030.

61. 61.Epstein L, Lamport DTA. An intramolecular linkage involving isodityrosine in extensin. Phytochemistry. 1984;23(6):1241–6.

62. 62.Li X, Kieliszewski M, Lamport DTA. A chenopod extensin lacks repetitive tetrahydroxyproline blocks. Plant Physiol. 1990;92(2):327–33. pmid:16667277

63. 63.Ertl H, Hallmann A, Wenzl S, Sumper M. A novel extensin that may organize extracellular matrix biogenesis in Volvox carteri. EMBO J. 1992;11(6):2055–62. pmid:1600938

64. 64.Showalter AM, Zhou J, Rumeau D, Worst SG, Varner JE. Tomato extensin and extensin-like cDNAs: structure and expression in response to wounding. Plant Mol Biol. 1991;16:547–65. pmid:1714316

65. 65.Garcia-Mas J, Messeguer R, Arús P, Puigdomènech P. The extensin from Prunus amygdalus. Plant Physiol. 1992;100(3):1603–4. pmid:16653168

66. 66.Fong C, Kieliszewski MJ, Zacks R, Leykam JF, Lamport DTA. A gymnosperm extensin contains the serine-tetrahydroxyproline motif. Plant Physiol. 1992;99(2):548–52. pmid:16668921

67. 67.Zhou J, Rumeau D, Showalter AM. Isolation and characterization of two wound-regulated tomato extensin genes. Plant Mol Biol. 1992;20:5–17. pmid:1381233

68. 68.Chen C, Cornish EC, Clarke AE. Specific expression of an extensin-like gene in the style of Nicotiana alata. Plant Cell. 1992;4:1053–62. pmid:1392608

69. 69.Goldman MH, Pezzotti M, Seurinck J, Mariani C. Developmental expression of tobacco pistil-specific genes encoding novel extensin-like proteins. Plant Cell. 1992;4:1041–51. pmid:1392607

70. 70.Bown DP, Bolwell P, Gatehouse JA. Characterisation of potato (Solanum tuberosum L.) extensins: a novel extensin-like cDNA from dormant tubers. Gene. 1993;134:229–33. pmid:8262381

71. 71.Memelink J, Swords KM, de Kam RJ, Schilperoort RA, Hoge JH, Staehelin LA. Structure and regulation of tobacco extensin. Plant J. 1993;4(6):1011–22. pmid:7506607

72. 72.Brownleader MD, Dey PM. Purification of extensin from cell walls of tomato (hybrid of Lycopersicon esculentum and L. peruvianum) cells in suspension culture. Planta.1993;191:457–69. pmid:7764229

73. 73.Parmentier Y, Durr A, Marbach J, Hirsinger C, Criqui MC, Fleck J, et al. A novel wound-inducible extensin gene is expressed early in newly isolated protoplasts of Nieotiana sylvestris. Plant Mol Biol. 1995;29:279–92. pmid:7579179

74. 74.Rubinstein AL, Broadwater AH, Lowrey KB, Bedinger PA. Pex1, a pollen-specific gene with an extensin-like domain. Proc Natl Acad Sci U S A. 1995;92(8):3086–90. pmid:7724520

75. 75.Hong JC, Cheong YH, Nagao RT, Bahk JD, Choand MJ, Key JL. Isolation and characterization of three soybean extensin cDNAs. Plant Physiol. 1994;104(2):793–6. pmid:8159793

76. 76.Arsenijević-Maksimović L, Broughton WJ, Krause A. Rhizobia modulate root-hair-specific expression of extensin genes. Mol Plant Microbe Interact. 1997;10(1):95–101. pmid:9002273

77. 77.Ahn JH, Choi Y, Kwon YM, Kim SG, Choi YD, Lee JS. A novel extensin gene encoding a hydroxyproline-rich glycoprotein requires sucrose for its wound-

inducible expression in transgenic plants. Plant Cell. 1996;8:1477–90. pmid:8837503

78. 78.Bucher M, Schroeer B, Willmitzer L, Riesmeier JW. Two genes encoding extension-like proteins are predominantly expressed in tomato root hair cells. Plant Mol Biol. 1997;35(4):497–508. pmid:9349272
79. 79.Hirsinger C, Parmentier Y, Durr A, Fleck J, Jamet E. Characterization of a tobacco extensin gene and regulation of its gene family in healthy plants and under various stress conditions. Plant Mol Biol. 1997;33(2):279–89. pmid:9037146
80. 80.Sturaro M, Linnestad C, Kleinhofs A, Olsen OA, Doan DNP. Characterization of a cDNA encoding a putative extensin from developing barley grains (Hordeum vulgare L.). J Exp Bot. 1998;49(329):1935–44.
81. 81.Goodrum LJ, Patel A, Leykam JF, Kieliszewski MJ. Gum arabic glycoprotein contains glycomodules of both extensin and arabinogalactan-glycoproteins. Phytochemistry. 2000;54(1):99–106. pmid:10846754
82. 82.Rathbun EA, Naldrett MJ, Brewin NJ. Identification of a family of extensin-like glycoproteins in the lumen of rhizobium-induced infection threads in pea root nodules. Mol Plant Microbe Interact. 2002;15(4):350–9. pmid:12026173
83. 83.Shimizu T, Inoue T, Shiraishi H. Cloning and characterization of novel extensin-like cDNAs that are expressed during late somatic cell phase in the green alga Volvox carteri. Gene. 2002;284(1–2):179–87. pmid:11891059
84. 84.Yoshiba Y, Aoki C, Iuchi S, Nanjo T, Seki M, Sekiguchi F, et al. Characterization of four extensin genes in Arabidopsis thaliana by differential gene expression under stress and non-stress conditions. DNA Res. 2001;8(3):115–22. pmid:11475326
85. 85.Stratford S, Barne W, Hohorst DL, Sagert JG, Cotter R, Golubiewski A, et al. A leucine-rich repeat region is conserved in pollen extensin-like (Pex) proteins in monocots and dicots. Plant Mol Biol. 2001;46(1):43–56. pmid:11437249
86. 86.Chida H, Yazawa K, Hasezawa S, Iwai H, Satoh S. Involvement of a tobacco leucine-rich repeat-extensin in cell morphogenesis. Plant Biotec. 2007;24:171–7.
87. 87.Nuñez A, Fishman ML, Fortis LL, Cooke PH, Hotchkiss AT Jr. Identification of extensin protein associated with sugar beet pectin. J Agric Food Chem. 2009;57(22):10951–8. pmid:19860469
88. 88.Guzzardi P, Genot G, Jamet E. The Nicotiana sylvestris extensin gene, Ext 1.2A, is expressed in the root transition zone and upon wounding. Biochim Biophys Acta. 2004;1680(2):83–92. pmid:15488988
89. 89.Corbin DR, Sauer N, Lamb CJ. Differential regulation of a hydroxyproline-rich glycoprotein gene family in wounded and infected plants. Mol Cell Biol. 1987;7(12):4337–44. pmid:3437892
90. 90.Baumberger N, Doesseger B, Guyot R, Diet A, Parsons RL, Clark MA, et al. Whole-genome comparison of leucine-rich repeat extensins in Arabidopsis and rice: a conserved family of cell wall proteins form a vegetative and a reproductive clade. Plant Physiol. 2003;131:1313–26. pmid:12644681
91. 91.Shiu SH, Bleecker AB. Expansion of the receptor-like kinase/pelle gene family and receptor-like proteins in Arabidopsis. Plant Physiol. 2003;132:530–43. pmid:12805585
92. 92.Champion A, Kreis M, Mockaitis K, Picaud A, Henry Y. Arabidopsis kinome: after the casting. Funct Integr Genomics. 2004;4:163–87. pmid:14740254
93. 93.Sørensen I, Pettolino F, Bacic A, Ralph J, Lu F, O'Neill M, et al. The Charophycean green algae provide insights into the early origins of plant cell walls. Plant J. 2011;68:201–11. pmid:21707800
94. 94.Eder M, Tenhaken R, Driouich A, Lutz-Meindl U. Occurrence and characterization of arabinogalactan-like proteins and hemicelluloses in Micrasterias (Streptophyta). J Phycol. 2008;44:1221–34.
95. 95.Domozych DS, Sørensen I, Willats WGT. The distribution of cell wall polymers during antheridium development and spermatogenesis in the charophycean green alga, Chara corallina. Ann Bot. 2009;104:1045–56. pmid:19696037

96. 96.Mikkelsen MD, Harholt J, Ulvskov P, Johansen IE, Fangel JU, Doblin MS, et al. Evidence for land plant cell wall biosynthetic mechanisms in charophyte green algae. Ann Bot. 2014;114:1217–36. pmid:25204387

CHAPTER 11

Feature Selection Has a Large Impact on One-Class Classification Accuracy for MicroRNAs in Plants

Malik Yousef,[1,2] Müşerref Duygu Saçar Demirci,[3] Waleed Khalifa,[1,2] and Jens Allmer[3,4]

[1]*Computer Science, The College of Sakhnin, 30810 Sakhnin, Israel*

[2]*The Institute of Applied Research, The Galilee Society, P.O. Box 437, 20200 Shefa Amr, Israel*

[3]*Molecular Biology and Genetics, Izmir Institute of Technology, Urla, 35430 Izmir, Turkey*

[4]*Bionia Incorporated, IZTEKGEB A8, Urla, 35430 Izmir, Turkey*

ABSTRACT

MicroRNAs (miRNAs) are short RNA sequences involved in posttranscriptional gene regulation. Their experimental analysis is complicated and, therefore, needs to be supplemented with computational miRNA detection. Currently computational miRNA detection is mainly performed using machine learning and in particular two-class classification. For machine learning, the miRNAs need to be parametrized and more than 700 features have been described. Positive training examples for machine learning are readily available, but negative data is hard to come by. Therefore, it seems prerogative to use one-class classification instead of two-class classification. Previously, we were able to almost reach two-class classification accuracy using one-class classifiers. In this work, we employ feature selection procedures in conjunction with one-class classification and show that there is up to 36% difference in accuracy among these feature selection methods. The best feature set allowed the training of a one-class classifier which achieved an average accuracy of ~95.6% thereby outperforming previous two-class-based plant miRNA detection approaches by about 0.5%. We believe that this can be improved upon in the future by rigorous filtering of the positive training examples and by improving current feature clustering algorithms to better target pre-miRNA feature selection.

1. INTRODUCTION

MicroRNAs (miRNAs) are short regulatory nucleotide sequences which were discovered about 2 decades ago [1]. Since then, they were shown to exist in organisms ranging from sponges [2] to human [3] and also in plants where they may have evolved independently [4]. Due to their involvement in posttranscriptional regulation, miRNAs have been implicated in human diseases [5, 6] and, for example, in plant stress response [7]. Regulation of gene expression is of great interest and, therefore, miRNAs have received increasing interest.

Some miRNAs have been experimentally detected and they are stored in databases such as miRBase [8] and miRTarBase [9]. Unfortunately, experimental detection of miRNAs is quite involved and further suffers from the fact that some miRNA-target combinations may be expressed only in response to specific external or internal stresses so that it seems impossible to experimentally detect all miRNAs of any higher eukaryotic organism [10–13]. Additionally, current approaches may not use the full potential of available experimental techniques [14]. Even among the experimentally validated miRNAs in miRBase, there may be entries which are dubious and we have shown that some may not represent true miRNAs [15]. Thus, computational methods for miRNA detection are required to complement experimental approaches.

Computational approaches for miRNA detection and miRNA's target detection [16] are generally based on machine learning [17]. In order to use miRNAs in machine learning, they must be parameterized and many different features have been described [18]. For successful machine learning, known examples need to be provided to the algorithms and these are generally collected from miRBase. Except for a few examples [10], two-class classification is used for machine learning. In this case in addition to the positive learning examples, negative ones have to be provided.

Unfortunately, the negative class cannot be defined for any larger eukaryotic organism since any part of a genome must be coexpressed with any other part of the genome to prove that they cannot interact as miRNAs, a futile endeavor. Therefore, examples for the negative class have been arbitrarily created in studies performing machine learning for miRNA detection, but the most abundantly used negative dataset is from [19] although new datasets are being proposed [20]. While in our experience the dataset is a significant improvement over using randomly selected or generated examples, it is likely to also contain positive examples.

To overcome the problem of needing to define artificial negative examples, one-class classification can be used instead [21, 22]. One-class classification (OCC) only needs the target class for learning and then classifies unknown examples into target or unknown (outlier) class. We have recently analyzed the use of one-class classification for miRNA detection in plants and found that it is competitive in comparison to two-class classification [23].

We also found that some features among the hundreds of features described for miRNA parameterization [24] are more discriminative than others. Unfortunately, feature selection is NP-hard [25] and selecting the best subset from more than 700 features on a per-dataset basis is not possible in an acceptable amount of time. Apart from the computational complexity of feature selection, methods have only been well established for two-class classification [26–28], while only few approaches deal with one-class classification [29–31].

In this study, we used different feature selection approaches and evaluated the effectiveness of each one against the others for the classification of miRNAs from different plant species using one-class classification. We found that clustering of features is important for classification performance and helps to improve classification accuracy by up to 36% (Oryza sativa) and on average by about 30% compared to selecting features by the lowest information gain. Clustering in its present form, however, is ill suited for selection from pre-miRNA features, since logically correlated features may be uncorrelated when examining the feature vectors in isolation and vice versa. Therefore, the clustering methods performed best, but not the expected clustering method which selected the feature with the highest information gain from each cluster. With a peak performance of 98.80% accuracy for Zea mays, feature selection using clustering methods was able to increase the accuracy achieved in our previous studies by 1–4% and the two-class classification benchmark PlantMiRNAPred by ~0.5%. This study showed that feature selection is effective despite varying data quality. Concluding, it seems important to devise effective data selection methodologies to ensure that only true miRNAs are used as positive examples. Novel feature clustering approaches taking into account logical feature correlation need to be coupling this data filtering strategy. When both are achieved, one-class classification may significantly outperform two-class classification for pre-miRNA detection in all cases (for plant) since the gap is already closed with this study.

2. MATERIALS AND METHODS

2.1. Data

We downloaded all available microRNAs from selected plant species from miRBase [8] (Releases 20 and 21). The selected species were Glycine max (gma), Zea mays (zma), Sorghum bicolor (sbi), Physcomitrella patens (ppt), Arabidopsis thaliana (ath), Populus trichocarpa (ptc), and Oryza sativa (osa). Our negative data pool consisted of the 980 pseudo pre-miRNAs constituting the PlantMiRNAPred dataset [32]. For these data, all pre-miRNA features were calculated as described previously [24, 33, 34].

2.2. One-Class Classification

One-class classification was performed using randomly sampled 90% of the positive data for training and 10% for testing (Supplementary File 1 in Supplementary Material available online at, Figure). In addition to the 10% positive data during testing, the pseudo negative sequences were used as unknown class. To determine the stability of the feature set, the process was repeated 100 times. We used the DDtools [35] implementation of a OCC.

2.3. Feature Selection Strategies

Feature selection is not a straightforward process since two features which individually may not be very discriminative may have high selective power in combination. Additionally, we have more than 700 features so trying all combination of features from 1 to 700 for all datasets considered in this study is not possible. In order to test the impact of feature selection on the classification accuracy, four negative and four positive feature selection methods were designed. According to our previous research [24], 50 features seem sufficient for miRNA detection [24] and this limit was used in this study. Among the feature selection methods used in this study, only PCF is using only the positive class whereas all others are using both classes.

2.3.1. Selecting Features with Low Information Gain (LIG)

Information gain was calculated on a per-dataset level using KNIME [36] and for each dataset the process had to be iteratively repeated until the 50 features with lowest information gain. First information gain was calculated and since most features received a score of zero they could not be differentiated. After removing all features with information gain above zero and recalculating information gain for the remaining features, the pool of features that receive zero became successively smaller until only 50 features remained. These remaining features were selected and used for classification (Supplementary File 1, LIG).

2.3.2. Random Feature Selection (RFS)

50 features, from the overall list of features, were randomly selected for each dataset (Supplementary File 1, RFS).

2.3.3. Selecting Random Feature from Feature Clusters (RFC)

Since correlated features may not much contribute to the discriminative power of a feature set, features were clustered into 100 clusters and from each cluster a random feature was selected from which 50 random features were selected for classification (Supplementary File 1, RFC).

2.3.4. Selecting Features from Clusters (SFC)

Instead of selecting a representative feature per cluster (as in RFC and HIC), this approach selects up to three clusters and uses all features therein to see the impact of correlation among features on the classification (Supplementary File 1, SFC).

2.3.5. Selecting Features with High Information Gain (HIG)

Contrary to the selection of features with low information gain, we selected the features with the highest information gain on a per-dataset basis (Supplementary File 1, HIG).

2.3.6. Selecting Feature with the Highest Information Gain from Feature Clusters (HIC)

Among the 700 features describing miRNAs are likely some which describe very similar information and, to avoid overrepresentation of such information, it may be beneficial to cluster such features and use single features to represent each cluster. 100 clusters were produced per dataset and from each cluster the feature with the highest information gain was selected. The selected features were again ranked via information gain and the top 50 were selected (Supplementary File 1, HIC).

2.3.7. Zero-Norm Feature Selection (ZNF)

For each feature vector, we define the zero-norm to be the nonzero values for all positive examples. First features whose vector consists of zero values are removed. Additionally, we defined () as the number of values with nonzero value, for example, () = 5. Moreover, we define different thresholds for () to determine the relevance of a feature and remove the ones below a given threshold (Supplementary File 1, ZNF).

2.3.8. Pearson Correlation-Based Feature Selection (PCF)

The Pearson correlation-based feature selection method was introduced by Lorena et al. [29]. The Pearson correlation measure allows detection of the linearity relation among features. The pairwise distances among features were calculated using Pearson correlation. Features with lower correlation were preferred during the feature selection process (Supplementary File 1, PCF).

3. RESULTS AND DISCUSSION

3.1. Feature Selection

Feature selection is an important process in machine learning and not all of the more than 700 features which have been proposed to describe a pre-miRNA might be useful. We used eight feature selection methods and performed one-class classification using each of the selected feature sets. 50 features were used as a target and we did not try different sized feature sets. Additionally, for each feature selection method a combined feature set was created containing the most commonly selected features (denoted by suffix _comb). To judge feature selection success, one-class classification was performed using the selected features. The average accuracy of 100 times cross validation served as performance measure to evaluate the feature selection methods. The selected features by the different methodologies are presented in Supplementary Table .

For the combined feature selection method, SFC and HIC outperform all other feature selection methods consistently for all datasets (Figure 1). However, the performance varies strongly with the organism and the variation is similar for all feature selection methods. This synchronized variation indicates that the available training data is either not sufficient or not accurate enough or both. This is in line with a previous study where we showed that not all data on miRBase may represent true miRNAs [15]. Nonetheless, since the feature selection methods SFC and HIC consistently achieved the highest accuracy compared to all other feature selection methodologies, the assessment of feature selection methods is not hampered by the shortcomings of the datasets. The average accuracy over all feature selection methods can further help define dataset quality. ppt has the best result (~88%) followed by zma (~87%), while gma (~80%) and osa (~78%) are tailing the list (Supplementary Table).

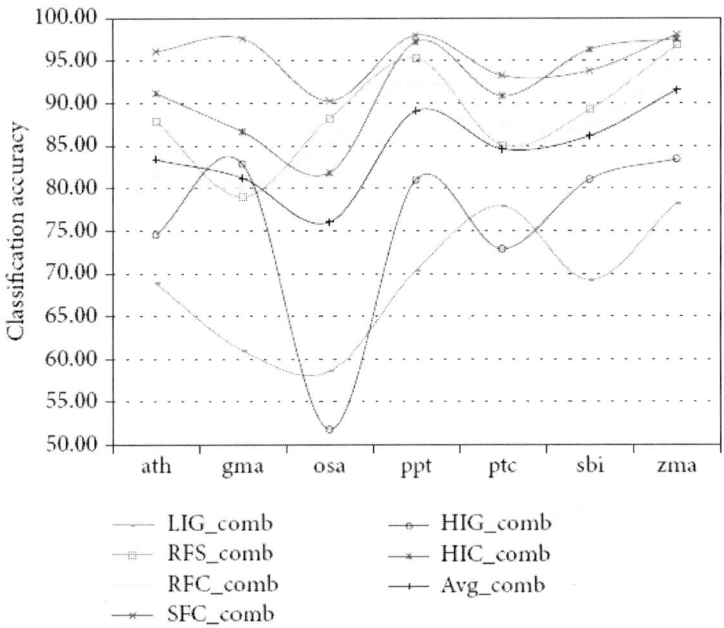

Figure 1: Classification accuracy of the combined feature sets on a per-organism basis. Note that there is no mathematical relationship that supports the connection among points (measurements) and that the lines were only added for visual guidance to enhance the synchronized variation on a per-organism basis. Supplementary Table contains the underlying data (including sensitivity and specificity) and a plot for the individual feature selection on a per-species basis.

Data in Figure 1 is in respect to combined feature sets and it may be informative to investigate accuracy in respect to feature selection on a per-organism basis (Figure 2) since the average accuracy of average accuracies for all feature selected methods is slightly higher for individual feature selection (84.69%) as compared to combined feature selection (84.58); see Supplementary Table for more information.

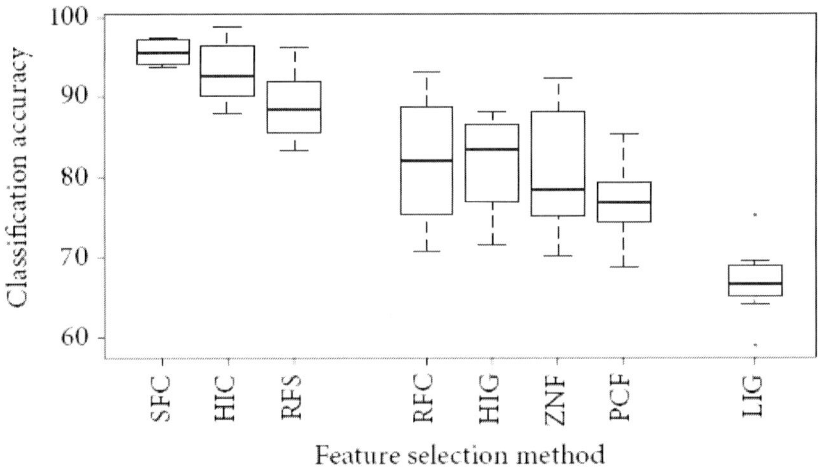

Figure 2: Accuracy distribution for the seven selected organisms assessed using one-class classification.

Figure 2 shows that the feature selection methods fall into three categories, improved accuracy (SFC, HIC, and RFS), no or little effect (RFC, HIG, ZNF, and PCF), and finally LIG, which was chosen as a negative control along with RFS and SFC. LIG which selects the features with the lowest information gain performs as expected and leads to the lowest effectiveness of the resulting classifiers. Interestingly, RFS and SFC which we expected to provide low performance lead to effective classifiers. Many feature clusters exist in the large feature space and RFS (random feature selection) likely picked from different clusters. This is supported by the observation that the spread of RFS is much larger than for SFC and HIC. SFC merely selects features from three clusters and was expected to have low performance. However, it turned out to be the best performing algorithm which we attribute to a critical problem in clustering for feature selection. We used -means clustering on the feature vector containing calculations for positive examples. Unfortunately, features like % A and % G are strongly correlated although their vectors are anticorrelated. This is true for many of the features among the more than 700 described features and the opposite is also true where two feature vectors appear highly correlated although they describe logically uncorrelated information. For these reasons, HIC which selects the feature with the highest information gain from each cluster performed not as well as would be possible if correlation would be perfectly following logical constraints. RFC (random feature selection from clusters) performed better than HIG, ZNF, and PCF, which supports the notion of why HIC did not consistently show the best performance among all algorithms. Nonetheless, HIC showed the overall best performance for zma with 98.80% accuracy. Among the feature selection methods, there is a large performance gap (~38%). On a per-organism basis, the gap reaches ~36% (O. sativa) but on average it is close to 30% (Supplementary Table), which highlights the importance of proper feature selection.

3.2. Organism-Based Feature Selection

Feature selection as described above was performed on a per-organism basis and from the selected features the most common ones were used to define a combined feature selection (Supplementary Table). It was our belief that using feature selection on a per-organism basis would have an effect on the accuracy that can be achieved. On average this is true and a slight advantage of about 0.2% accuracy when comparing the averages of the averages of all feature selection methods can be observed (Supplementary Table). Unfortunately, it turns out that the number of positive results and their correctness are more important than the feature selection method and whether it is performed on a per-organism basis or not. For A. thaliana, G. max, and O. sativa the assumption holds but, for the other species used here, the results were not so clear. Perhaps the strategy of selecting the 50 most common features among the first 100 features into the combined feature set is not actually selecting a feature set that would describe the selected clade.

In order to investigate whether feature selection has to be made on a per-species basis, we used the interactive tree of life [37] to establish the phylogenetic relationship among the selected plant species (Figure 3). Three larger groups are visible one of which consists of a single species (ppt) while the other two contain three species each.

We checked the deviation of the accuracy achieved using feature selection on a per-organism basis to the average for the organisms and found that for SFC and HIC the distance of P. patens to the average was different from the distances of the other organisms which were alike in their groups (Supplementary Table) as expected from their placement in the phylogenetic tree (Figure 3).

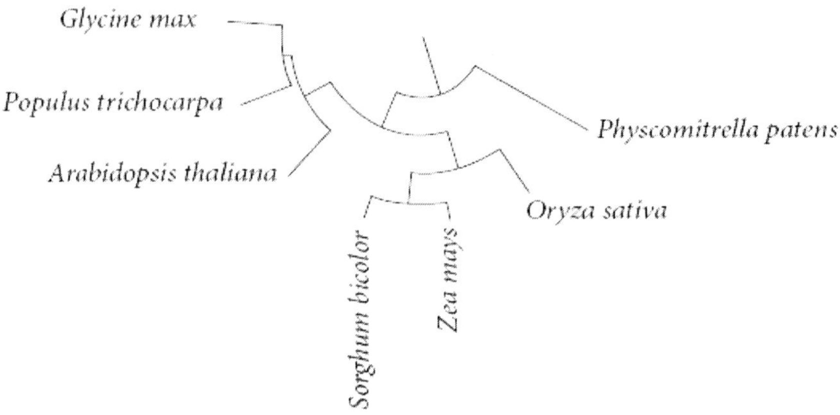

Figure 3: Phylogenetic relationship among the selected plant species.

3.3. Performance Comparison

In a previous study we introduced sequence motifs as features for pre-miRNA detection and compared one-class classification to two-class classification [23]. Compared to the results we achieved with MotifmiRNAPred, this study shows that feature selection is essential for classification performance (Table 1).

Table 1: Comparison of classification accuracies among this and other published methods (PlantMiRNAPred [32], Triplet-SVM [38], microPred [39], and MotifmiRNAPred [33]). The best performance per organism is highlighted in bold. ACC: accuracy.

Organism	PlantMiRNAPred	Triplet-SVM	microPred	MotifmiRNAPred	This study
	ACC	ACC	ACC	ACC	ACC
gma	**98.50**	74.10	86.70	89.80	97.38
zma	**98.30**	66.90	93.80	94.80	93.59
sbi	**98.40**	69.50	94.60	93.50	94.25
ath	92.20	76.00	89.40	93.30	**97.39**
ppt	92.40	71.40	89.50	90.20	**97.20**
ptc	91.80	75.20	84.90	92.20	**94.00**
osa	94.20	75.50	90.40	90.30	**95.32**
Avg	95.11	72.66	89.90	92.01	**95.59**

For some of the organisms (gma, zma, and sbi), PlantMiRNAPred, which employs two-class classification, performs better than the one-class classification performed in this study. On the other hand, on average one-class classification is slightly better and performs better than PlantMiRNAPred on some organisms (ath, ppt, ptc, and osa). The improvement to MotifmiRNAPred which also uses one-class classification is about 3.5% on average and this again highlights the importance of feature selection. The improvement of classification accuracy in this study (using one-class classification) over PlantMiRNAPred (using two-class classification) is about 0.5% accuracy, but it is an important step since for the first time OCC based pre-miRNA detection outperforms two-class classification.

4. CONCLUSIONS

One-class classifiers were trained using different feature sets selected by various feature selection methods (see Supplementary Table for selected features). Feature clustering seems essential since the best performing methods use clustering (Figure 2). On average 30% accuracy can be gained using feature selection (Supplementary Table). Compared to previously published methods, the OCC using SFC filtered features performs better (~3.5%) than our previously established OCC and slightly better (~0.5%, ~96%) than the best two-class classifier for plant pre-miRNA detection (Table 1).

A peak accuracy of close to 99% may seem sufficient for miRNA detection, but this Figure needs to be put into context with millions of hairpins in a genome which could constitute a pre-miRNA. Thus accuracies above 99.99% (about 1000 false positives in a genome of 3Gb) are needed if all prediction ought to be tested experimentally.

To achieve such high accuracies, the following points need to be tackled:(1)In the future, it will be important to devise a method to filter the positive training data since not all datasets seem to represent a pure class and this strongly influences classification accuracy (Figure 1).(2)Clustering is essential for feature selection, but current clustering methods do not take into account logical correlation (Figure 2).(3)A switch to OCC needs to be performed since negative data for pre-miRNAs does not exist [10, 40].(4)Feature selection should be performed on a per-organism basis and how classifiers perform for groups of organisms when features are selected on the group basis should be investigated.Finally, all feature selection methods in this study, save PCF, are based on two classes and in the future additional feature selection methods that are independent of the negative class should be investigated. Since there is a large interest in the field of feature selection spawned by big data problems, many new methods have been devised and we plan to use some of them [41, 42] in future research to select a stable and performant set of features.

COMPETING INTERESTS

The authors declare that there are no competing interests regarding the publication of this paper.

ACKNOWLEDGMENTS

The work was supported by the Scientific and Technological Research Council of Turkey (Grant no. 113E326) to Jens Allmer.

REFERENCES

1. R. C. Lee, R. L. Feinbaum, and V. Ambros, "The C. elegans heterochronic gene lin-4 encodes small RNAs with antisense complementarity to lin-14," Cell, vol. 75, no. 5, pp. 843–854, 1993.
2. V. N. Kim, J. Han, and M. C. Siomi, "Biogenesis of small RNAs in animals," Nature Reviews Molecular Cell Biology, vol. 10, no. 2, pp. 126–139, 2009.
3. K. U. Tüfekci, M. G. Öner, R. L. J. Meuwissen, and Ş. Genç, "The role of microRNAs in human diseases," Methods in Molecular Biology, vol. 1107, pp. 33–50, 2014.
4. E. J. Chapman and J. C. Carrington, "Specialization and evolution of endogenous small RNA pathways," Nature Reviews Genetics, vol. 8, no. 11, pp. 884–896, 2007.
5. B. Alural, G. A. Duran, K. U. Tufekci et al., "EPO mediates neurotrophic, neuroprotective, anti-oxidant, and anti-apoptotic effects via downregulation of miR-451 and miR-885-5p in SH-SY5Y neuron-like cells," Frontiers in Immunology, vol. 5, article 475, 2014.
6. B. Alural, A. Ozerdem, J. Allmer, K. Genc, and S. Genc, "Lithium protects against paraquat neurotoxicity by NRF2 activation and miR-34a inhibition in SH-SY5Y cells," Frontiers in Cellular Neuroscience, vol. 9, article 209, 2015.
7. Z. Zhang, J. Yu, D. Li et al., "PMRD: plant microRNA database," Nucleic Acids Research, vol. 38, no. 1, Article ID gkp818, pp. D806–D813, 2009.
8. S. Griffiths-Jones, H. K. Saini, S. van Dongen, and A. J. Enright, "miRBase: tools for microRNA genomics," Nucleic Acids Research, vol. 36, no. 1, pp. D154–D158, 2008.
9. S.-D. Hsu, F.-M. Lin, W.-Y. Wu et al., "MiRTarBase: a database curates experimentally validated microRNA-target interactions," Nucleic Acids Research, vol. 39, no. 1, pp. D163–D169, 2011.
10. M. Yousef, S. Jung, L. C. Showe, and M. K. Showe, "Learning from positive examples when the negative class is undetermined—microRNA gene identification," Algorithms for Molecular Biology, vol. 3, article 2, 2008.
11. J. Ding, S. Zhou, and J. Guan, "MiRenSVM: towards better prediction of microRNA precursors using an ensemble SVM classifier with multi-loop features," BMC Bioinformatics, vol. 11, supplement 11, p. S11, 2010.

12. Y. Wu, B. Wei, H. Liu, T. Li, and S. Rayner, "MiRPara: a SVM-based software tool for prediction of most probable microRNA coding regions in genome scale sequences," BMC Bioinformatics, vol. 12, no. 1, article 107, 2011.
13. W. Ritchie, D. Gao, and J. E. J. Rasko, "Defining and providing robust controls for microRNA prediction," Bioinformatics, vol. 28, no. 8, pp. 1058–1061, 2012.
14. M. D. Saçar and J. Allmer, "Current limitations for computational analysis of miRNAs in cancer," Pakistan Journal of Clinical and Biomedical Research, vol. 1, no. 2, pp. 3–5, 2013.
15. M. D. Saçar, H. Hamzeiy, and J. Allmer, "Can MiRBase provide positive data for machine learning for the detection of MiRNA hairpins?" Journal of Integrative Bioinformatics, vol. 10, no. 2, article 215, 2013.
16. M. Yousef, N. Najami, and W. Khalifav, "A comparison study between one-class and two-class machine learning for MicroRNA target detection," Journal of Biomedical Science and Engineering, vol. 3, no. 3, pp. 247–252, 2010.
17. M. D. Saçar and J. Allmer, "Machine learning methods for microRNA gene prediction," Methods in Molecular Biology, vol. 1107, pp. 177–187, 2014.
18. M. D. Saçar and J. Allmer, "Comparison of four Ab initio MicroRNA prediction tools," in Proceedings of the International Conference on Bioinformatics Models, Methods and Algorithms (BIOINFORMATICS '13), pp. 190–195, Barcelona, Spain, February 2013. View at Scopus
19. K. L. S. Ng and S. K. Mishra, "De novo SVM classification of precursor microRNAs from genomic pseudo hairpins using global and intrinsic folding measures," Bioinformatics, vol. 23, no. 11, pp. 1321–1330, 2007.
20. L. Wei, M. Liao, Y. Gao, R. Ji, Z. He, and Q. Zou, "Improved and promising identificationof human microRNAs by incorporatinga high-quality negative set," IEEE/ACM Transactions on Computational Biology and Bioinformatics, vol. 11, no. 1, pp. 192–201, 2014.
21. L. M. Manevitz and M. Yousef, "One-class SVMs for document classification," Journal of Machine Learning Research, vol. 2, pp. 139–154, 2002.
22. L. Manevitz and M. Yousef, "One-class document classification via Neural Networks," Neurocomputing, vol. 70, no. 7–9, pp. 1466–1481, 2007.
23. M. Yousef, J. Allmer, and W. Khalifa, "Accurate plant MicroRNA prediction can be achieved using sequence motif features," Journal of Intelligent Learning Systems and Applications, vol. 8, no. 1, pp. 9–22, 2016.
24. M. D. Saçar and J. Allmer, "Data mining for microrna gene prediction: on the impact of class imbalance and feature number for microrna gene prediction," in Proceedings of the 8th International Symposium on Health Informatics and Bioinformatics (HIBIT '13), pp. 1–6, IEEE, Ankara,Turkey, September 2013.
25. E. Amaldi and V. Kann, "On the approximability of minimizing nonzero variables or unsatisfied relations in linear systems," Theoretical Computer Science, vol. 209, no. 1-2, pp. 237–260, 1998.
26. S. Paul, M. Magdon-Ismail, and P. Drineas, "Feature selection for linear SVM with provable guarantees," Journal of Machine Learning Research, vol. 38, pp. 735–743, 2015.
27. I. Guyon, J. Weston, S. Barnhill, and V. Vapnik, "Gene selection for cancer classification using support vector machines," Machine Learning, vol. 46, no. 1–3, pp. 389–422, 2002.
28. M. E. Ahsen, N. K. Singh, T. Boren, M. Vidyasagar, and M. A. White, "A new feature selection algorithm for two-class classification problems and application to endometrial cancer," in Proceedings of the 51st IEEE Conference on Decision and Control (CDC '12), pp. 2976–2982, Maui, Hawaii, USA, December 2012.
29. L. H. N. Lorena, A. C. P. L. F. Carvalho, and A. C. Lorena, "Filter feature selection for one-class classification," Journal of Intelligent & Robotic Systems, vol. 80, supplement 1, pp. 227–243, 2015.

30. P. Xuan, M. Guo, Y. Huang, W. Li, and Y. Huang, "Maturepred: efficient identification of microRNAs within novel plant Pre-miRNAs," PLoS ONE, vol. 6, no. 11, Article ID e27422, 2011.
31. M. Hall, E. Frank, G. Holmes, B. Pfahringer, P. Reutemann, and I. H. Witten, "The WEKA data mining software," ACM SIGKDD Explorations Newsletter, vol. 11, no. 1, pp. 10–18, 2009.
32. P. Xuan, M. Guo, X. Liu, Y. Huang, W. Li, and Y. Huang, "PlantMiRNAPred: efficient classification of real and pseudo plant pre-miRNAs," Bioinformatics, vol. 27, no. 10, pp. 1368–1376, 2011.
33. M. Yousef, J. Allmer, and W. Khalifa, "Sequence motif-based one-class classifiers can achieve comparable accuracy to two-class learners for plant microRNA detection," Journal of Biomedical Science and Engineering, vol. 8, no. 10, pp. 684–694, 2015.
34. M. D. Saçar, C. Bağcı, and J. Allmer, "Computational prediction of MicroRNAs from toxoplasma gondii potentially regulating the hosts' gene expression," Genomics, Proteomics and Bioinformatics, vol. 12, no. 5, pp. 228–238, 2014.
35. D. M. J. Tax, DDTools, the Data Description Toolbox for Matlab, 2015.
36. M. R. Berthold, N. Cebron, F. Dill et al., "KNIME: the Konstanz Information Miner," in Data Analysis, Machine Learning and Applications, C. Preisach, H. Burkhardt, L. Schmidt-Thieme, and R. Decker, Eds., Studies in Classification, Data Analysis, and Knowledge Organization, pp. 319–326, 2008.
37. I. Letunic and P. Bork, "Interactive Tree of Life v2: online annotation and display of phylogenetic trees made easy," Nucleic Acids Research, vol. 39, supplement 2, pp. W475–W478, 2011.
38. P. Jiang, H. Wu, W. Wang, W. Ma, X. Sun, and Z. Lu, "MiPred: classification of real and pseudo microRNA precursors using random forest prediction model with combined features," Nucleic Acids Research, vol. 35, no. 2, pp. W339–W344, 2007.
39. R. Batuwita and V. Palade, "microPred: effective classification of pre-miRNAs for human miRNA gene prediction," Bioinformatics, vol. 25, no. 8, pp. 989–995, 2009.
40. J. Allmer, "Computational and bioinformatics methods for microRNA gene prediction," Methods in Molecular Biology, vol. 1107, pp. 157–175, 2014.
41. Q. Zou, J. Zeng, L. Cao, and R. Ji, "A novel features ranking metric with application to scalable visual and bioinformatics data classification," Neurocomputing, vol. 173, part 2, pp. 346–354, 2016.
42. X. Zeng, Y. Liao, Y. Liu, and Q. Zou, "Prediction and validation of disease genes using HeteSim Scores," IEEE/ACM Transactions on Computational Biology and Bioinformatics, 2016.

CHAPTER 12

An Efficient Protocol for *Agrobacterium*-Mediated Transformation of the Biofuel Plant *Jatropha Curcas* by Optimizing Kanamycin Concentration and Duration of Delayed Selection

Qiantang Fu[1] • Chaoqiong Li1,[2] • Mingyong Tang[1,3] • Yan-Bin Tao[1] • Bang-Zhen Pan[1] • Lu Zhang[1] • Longjian Niu[1] • Huiying He[1] • Xiulan Wang[1] • Zeng-Fu Xu

[1] *Key Laboratory of Tropical Plant Resources and Sustainable Use, Xishuangbanna Tropical Botanical Garden, Chinese Academy of Sciences, Menglun 666303, Yunnan, China*

[2] *College of Life Science and Agriculture, Zhoukou Normal University, Zhoukou 466001, Henan, China*

[3] *University of Chinese Academy of Sciences, Beijing 100049, China*

ABSTRACT

Jatropha curcas is considered a potential biodiesel feedstock crop. Currently, the value of *J. curcas* is limited because its seed yield is generally low. Transgenic modification is a promising approach to improve the seed yield of *J. curcas*. Although *Agrobacterium*-mediated genetic transformation of *J. curcas* has been pursued for several years, the transformation efficiency remains unsatisfying. Therefore, a highly efficient and simple *Agrobacterium*-mediated genetic transformation method for *J. curcas* should be developed. We examined and optimized several key factors that affect genetic transformation of *J. curcas* in this study. The results showed that the EHA105 strain was superior to the other three *Agrobacterium tumefaciens* strains for infecting *J. curcas* cotyledons, and the supplementation of 100 mM acetosyringone slightly increased the transient transformation frequency. Use of the appropriate inoculation method, optimal kanamycin concentration and appropriate duration of delayed selection also improved the efficiency of stable genetic transformation of *J. curcas*. The percentage of β-glucuronidase positive *J. curcas* shoots reached as high as 56.0 %, and 1.70 transformants per explant were obtained with this protocol. Furthermore, we optimized the root-inducing medium to achieve a rooting rate of 84.9 %. Stable integration of the T-DNA into the genomes of putative transgenic lines was confirmed by PCR and Southern blot analysis. Using this improved protocol, a large number of transgenic *J. curcas* plantlets can be routinely obtained within approximately 4 months. The detailed information provided here for each step of *J. curcas* transformation should enable successful implementation of this transgenic technology in other laboratories.

Keywords: Biofuel Cotyledon Delayed selection Physic nut Transformation Kanamycin concentration

INTRODUCTION

With the decreasing supplies of fossil fuels and the worsening of environmental pollution, biodiesel has received increasing attention as an alternative fuel. *Jatropha curcas* is considered a potential oilseed crop for biofuel production because its seeds contain up to 40 % oil, which can be easily converted to biodiesel or bio-jet fuel and used to partially or fully replace fossil fuel (Fairless 2007; Juan et al. 2011; Makkar and Becker 2009). In addition, *J. curcas* can grow under climate and soil conditions that are unsuitable for food production (Maghuly and Laimer 2013), and it is a renewable feedstock for producing soap, green fertilizers, pesticides and medicines (Kumar and Sharma 2008). However, at present, *J. curcas* does not contribute to the biofuel industry because the yield of its seeds is generally low in many areas, and the investment in large-scale cultivation of *J. curcas* had moved ahead of the scientific studies aiming to fully explore this plant and understand its limitations (Sanderson 2009).

Breeding of varieties that produce high and stable yields is one of the most efficient approaches to making *J. curcas* into a successful biofuel crop. Because of the low genetic diversity of *J. curcas* (Rosado et al. 2010; Sun et al. 2008), conventional breeding technology has limited potential for use in the genetic improvement of this plant (Sujatha et al. 2008). Transgenic breeding techniques can complement conventional breeding technology and have many advantages, such as directional cultivation of new breeds, reduced costs, a shorter breeding period, and the ability to introduce genes for traits that may not be available within the species or may be difficult to introduce via conventional breeding methods (Herr and Carlson 2013; Li et al. 1996; Visarada et al. 2009). Moreover, with the completion of genomic sequencing analysis (Hirakawa et al. 2012; Sato et al. 2011), expressed sequence tag analysis (Chen et al. 2011; Eswaran et al. 2012; Natarajan et al. 2010), and transcriptomic studies (Costa et al. 2010; Natarajan and Parani 2011; Zhang et al. 2014), *J. curcas* is poised to become a new model woody plant. Genetic transformation may become an important method for molecular breeding and gene function analysis of *J. curcas*.

The efficiency of the in vitro plant regeneration system, the efficiency of *Agrobacterium tumefaciens* transformation, and the specific antibiotic selection procedure are key factors in plant genetic transformation (Kajikawa et al. 2012; Li et al. 2008; Mao et al. 2009). To date, shoot regeneration systems for *J. curcas* have been successfully established using various explants such as cotyledons, epicotyls, hypocotyls, leaves, petioles, nodes, and stems (Khurana-Kaul et al. 2010; Kumar and Reddy 2010; Sharma et al. 2011; Singh et al. 2010; Sujatha and Mukta 1996; Toppo et al. 2012). Some regeneration systems have been utilized in genetic transformation protocols employing *A. tumefaciens* (Khemkladngoen et al. 2011; Kumar et al. 2010; Li et al. 2008; Mao et al. 2009; Misra et al. 2012; Pan et al. 2010) or particle bombardment (Joshi et al. 2011; Purkayastha et al. 2010). Due to several advantages of *Agrobacterium*-mediated genetic transformation, including the defined integration of transgenes, potentially low copy number, and preferential integration into transcriptionally active regions of the chromosome (Newell 2000), this method is the most widely used to generate transgenic *J. curcas*. Several *A. tumefaciens* strains have been used in the genetic transformation of *J. curcas*, including LBA4404 (Kumar et al. 2010; Li et al. 2008; Misra et al. 2012), EHA105 (He et al. 2009; Jaganath et al. 2014; Mazumdar et al. 2010), GV3101 (Qin et al. 2014), and AGL1 (Mao et al. 2009). The types of selectable markers and the selection pressure also play important roles in the genetic transformation of different explants. Using the herbicide phosphinothricin as a selective agent, Li et al. (2008) first reported the generation of transgenic *J. curcas*. Thereafter, several groups found that transgenic *J. curcas* can be regenerated by selection using hygromycin (Joshi et al. 2011; Kumar et al. 2010; Mao et al. 2009), kanamycin (Khemkladngoen et al. 2011; Misra et al. 2012; Pan et al. 2010), or bispyribac-sodium salt (Kajikawa et al. 2012). Many protocols required multiple handling steps and frequent changes of different medium, which not only require additional time but also may introduce contamination. The regeneration of transformed shoots takes 4–7 months, and the transformation efficiency is 4–53 % (Kajikawa et al. 2012; Khemkladngoen et al. 2011; Kumar et al. 2010; Li et al. 2008; Misra et al. 2012). Therefore, a simple, highly efficient, and rapid *Agrobacterium*-mediated genetic transformation system for *J. curcas* should be developed. Notably, root induction of regenerated shoots is also an important step for obtaining transgenic plants. Several root-inducing mediums (RIM) have been successfully used in the root induction of regenerated *J. curcas* shoots, but the rooting efficiency varied greatly, from 16 to 86 % in 1/2 MS medium supplemented with 0.2–0.5 mg L^{-1} IBA (Kajikawa et al. 2012; Khemkladngoen et al. 2011; Li et al. 2008; Toppo et al. 2012). Mazumdar et al. (2010) reported that the rooting induction efficiency of the *J. curcas* shoots was 75 % in 1/2 MS containing 5.0 μM (0.9 mg L^{-1}) NAA within 17–20 days of cultivation. So RIM should also be optimized to ensure high and stable rooting efficiency of transformed shoots.

We have previously developed a simple and efficient *Agrobacterium*-mediated genetic transformation system for *J. curcas* using the cotyledons from mature seeds as explants (Pan et al. 2010). In this simple protocol, all stages of cultivation, including co-cultivation, callus initiation and proliferation, selection of transgenic calli, and regeneration of shoots, were performed on a single medium (MS-Jc1: MS medium supplemented with 3 mg L^{-1} BA and 0.01 mg L^{-1} IBA) with different antibiotics (Pan et al. 2010). However, the transformation efficiency of this method is low and variable (Kajikawa et al. 2012); thus, it requires substantial improvement.

In the present study, we further improved the protocol described previously by Pan et al. (2010) by optimizing several key factors including the supplementation of acetosyringone (AS), selection of *A. tumefaciens* strains, scheme of transformant selection, determination of antibiotic selection pressure, and formulation of RIM. This improved protocol substantially increases the regeneration efficiency of transgenic *J. curcas*; the percentage of β-glucuronidase (GUS)-positive shoots reached an average of 56.0, and 84.9 % of the transgenic shoots successfully rooted. Moreover, this protocol includes additional details that were not previously described by Pan et al. (2010).

MATERIALS AND METHODS

Preparation of explants

Mature *J. curcas* L. seeds were collected from trees grown in Xishuangbanna Tropical Botanical Garden (21°54′N, 101°46′E, 580 m asl) at the Chinese Academy of Sciences, located in Mengla County, Yunnan Province, China. The husked seeds were soaked in sterile water for 10–12 h at room temperature. The seeds were then surface sterilized with 75 % (v/v) ethanol for 1 min, followed by soaking for 15 min in 1.1 % (w/v) sodium hypochlorite and 0.1 % (v/v) Tween-20 with occasional agitation. Subsequently, the seeds were rinsed three times with sterile distilled water. The sterilized seeds were carefully dissected with a scalpel, and the embryos were removed from the seeds. Approximately 3/4 of the papery cotyledons were excised by cutting off the base of the cotyledons and the embryo axes, as shown in Pan et al. (2010). These cotyledons were used as explants for *Agrobacterium*-mediated genetic transformation.

Agrobacterium tumefaciens strains and vector used for transformation

The *A. tumefaciens* strains LBA4404, EHA105, GV3101, and AGL1, all of which harbor the binary vector pCAMBIA2301 (CAMBIA, Canberra, Australia), were used for transformation experiments. EHA105 carrying *35S:AtFT* was used to evaluate the transformation method in subsequent transformation experiments. A single colony of *A. tumefaciens* was inoculated into 5 mL of liquid yeast extract and beef (YEB) medium containing 50 mg L^{-1} kanamycin and 25 mg L^{-1} streptomycin or 50 mg L^{-1} rifampicin and grown overnight at 28 °C on a rotary shaker at 200 rpm. An aliquot (0.5 mL) of the overnight culture was inoculated into 50 mL of liquid YEB medium containing the same antibiotics and allowed to grow at 28 °C with vigorous shaking until the OD_{600} reached approximately 0.6–1.0. *A. tumefaciens* cells were then collected by centrifugation at 5000 rpm for 5 min at 4 °C and re-suspended at $OD_{600} = 0.4$ with liquid Murashige and Skoog (MS) medium (Murashige and Skoog 1962) supplemented with 100 μM AS. The bacterial suspension was allowed to stand for approximately an hour before use.

Transformation, selection, and regeneration of transformants

The excised cotyledon explants were inoculated with the *A. tumefaciens* suspension by co-incubation with shaking on a rotary shaker for 20 min at 150 rpm. The temperature was maintained at 28 °C. The explants were then transferred onto sterile filter paper to remove excess *A. tumefaciens* before being cultured on co-cultivation medium (CCM: MS-Jc1 supplemented with 100 μM AS) for 3 days at 26 ± 2 °C in the dark. Following co-cultivation, the explants were washed three times with sterile water containing 500 mg L^{-1} cefotaxime to eliminate *A. tumefaciens* and blotted dry on sterile filter paper. Then, they were transferred initially to callus-inducing medium [CIM: MS-Jc1 supplemented with 300 mg L^{-1} cefotaxime and 100 mg L^{-1} ticarcillin and clavulanate potassium (timentin)] for callus initiation and growth and subsequently cultured for 1, 2, or 3 weeks (Table 1). The explants were cultured under a 14 h light/10 h dark cycle with fluorescent light (100 μ Em^{-2} s^{-1}) at 26 ± 2 °C. Then, the swollen cotyledon explants with small calli were placed on shoot-inducing medium (SIM: MS-Jc1 supplemented with 300 mg L^{-1} cefotaxime, 100 mg L^{-1} timentin, and 20, 30, or 40 mg L^{-1}kanamycin) with 1 cm of the cut end of each cotyledon inserted into the medium. After the 3-week culture period, the explants were individually subcultured to fresh SIM. Resistant calli and shoots were subcultured at 3-week intervals. The number of shoots per explant, frequency of positive GUS staining and transgene incorporation were scored after 12 weeks of cultivation on SIM.

Table 1. Effects of kanamycin concentration and duration of delayed selection on the regeneration and transformation efficiency of *J. curcas*

Kanamycin concentration (mg L^{-1})	Duration of delayed selection (weeks)	Number of explants*	Number of resistant shoots/explant**	Frequency of GUS-positive shoots (%)***	Number of GUS-positive shoots/explant
20	1	120	5.56 ± 0.33^{Aa}	12.7 ± 3.1^{Ded}	0.70 ± 0.15^{Cc}
20	2	116	4.06 ± 0.35^{Bb}	5.3 ± 1.2^{Ee}	0.21 ± 0.04^{Dde}
20	3	100	3.46 ± 0.26^{BCbc}	1.3 ± 1.2^{Ee}	0.05 ± 0.04^{De}
30	1	111	3.35 ± 0.33^{BCbc}	36.0 ± 2.0^{Bb}	1.19 ± 0.18^{Bb}
30	2	112	2.67 ± 0.29^{CDcd}	27.3 ± 2.3^{Cc}	0.70 ± 0.08^{CDcd}
30	3	109	2.09 ± 0.18^{CDd}	14.0 ± 2.0^{Dd}	0.29 ± 0.03^{Dde}
40	1	98	3.03 ± 0.22^{Cc}	56.0 ± 3.5^{Aa}	1.70 ± 0.15^{Aa}
40	2	110	2.39 ± 0.21^{CDcd}	33.3 ± 2.3^{BCbc}	0.80 ± 0.03^{BCc}
40	3	110	1.88 ± 0.17^{Dd}	20.7 ± 3.1^{CDcd}	0.39 ± 0.05^{CDd}

* Each value is the mean ± SE of three independent experiments, each with 30–40 cotyledons. The cotyledons that were not green or that were contaminated by *A. tumefaciens* were not counted. ** Shoots of at least 1.5 cm in length were counted. ***Percentage of GUS-positive shoots calculated as the number of GUS-positive shoots/total number of resistant shoots × 100. Values with different lowercase letters are significantly different ($P < 0.05$, Tukey's test). Values with different uppercase letters are significantly different ($P < 0.01$, Tukey's test)

Rooting and acclimation

Green and healthy regenerated shoots with three to four leaves were excised and transferred to root-inducing medium (RIM: half-strength (1/2) MS medium supplemented with 100 mg L^{-1}timentin, 1.0 % sucrose, 0.25 % carrageenan, and different combinations of IBA and NAA) (Table 2). The percentage of root induction was recorded and evaluated after 4 weeks and was calculated by comparing the number of rooting shoots to the total number of shoots. Rooted plantlets were carefully removed from the medium, washed thoroughly in running water to remove RIM attached to the roots, soaked in 0.1 % w/v carbendazim for 2 h, and then planted in polythene cups filled with sterilized soil consisting of humus:peat:vermiculite (3:1:1), covered with transparent plastic film to maintain humidity, and grown in the greenhouse at 22 ± 2 °C under a 16/8 h (light/dark) photoperiod with fluorescent light (100 μ Em^{-2} s^{-1}). After 2 weeks, the transparent plastic film was removed, and the plants were sprayed with water to avoid wilting of the leaves in the first 2 days. The number of surviving plants was recorded after an additional 1–2 weeks.

Table 2. Effects of different concentrations of IBA and NAA in root-inducing medium (RIM) on the induction of roots from *J. curcas* shoots

IBA (mg L^{-1})	NAA (mg L^{-1})	Number of tested shoots*	Efficiency of root induction (%)**
0	0	86	5.8 ± 1.3^{De}
0	0.1	85	32.8 ± 3.9^{Cd}
0	0.2	88	65.0 ± 3.5^{Bb}
0.1	0	86	9.2 ± 2.7^{De}
0.1	0.1	87	58.5 ± 3.4^{BCbc}
0.1	0.2	87	48.2 ± 2.6^{Cc}
0.2	0	90	13.3 ± 2.3^{De}
0.2	0.1	85	84.9 ± 3.7^{Aa}
0.2	0.2	87	41.3 ± 3.2^{Ccd}

* Each value is the mean ± SE of two independent experiments, each with 40–50 shoots. IBA, indole-3-butyric acid; NAA, α-naphthaleneacetic acid. ** Data were acquired after 4 weeks. Percentage root induction was calculated as the number of rooting shoots/total number of shoots × 100. Values with different lowercase letters are significantly different ($P < 0.05$, Tukey's test). Values with different uppercase letters are significantly different ($P < 0.01$, Tukey's test)

Molecular analysis of transgenic plantlets

Genomic DNA was isolated from the leaves of putative transgenic plants and controls using the cetyl trimethylammonium bromide method (Allen et al. 2006). PCR analysis of the isolated genomic DNA was performed to identify transgenes in putative transformants using primers (Table S1) for the *GUS* and *NPTII* genes. A *35S CaMV* promoter primer and a *GUS*-specific primer were used to detect the *GUS* gene, which produced a 778-bp fragment. The *NPTII* gene was amplified with the *NPTII*-specific primer pair, producing a 679-bp fragment. PCR amplification reactions were performed in a 20-μL volume, and the PCR parameters were as follows: 94 °C for 3 min, followed by 30 cycles of 94 °C for 30 s, 61 °C (for *GUS*)/58 °C (for *NPTII*) for 30 s and 72 °C for 1 min with a final extension of 10 min at 72 °C. The plasmid pCAMBIA2301 was used as a positive control, and non-transgenic plant DNA was used as a negative control. PCR amplification products were separated on 1.0 % (w/v) agarose gels and visualized by ethidium bromide staining under UV light.

For Southern blot analysis, 5 μg of genomic DNA from the PCR-positive leaves was digested with *Eco*RI and *Xho*I (*Eco*RI and *Xho*I cut the pCAMBIA2301 plasmid once outside of the *GUS* gene) overnight at 37 °C to cleave a unique site in the T-DNA. Following digestion, the DNA fragments were separated on 0.8 % (w/v) agarose gels, which were then processed and transferred to Hybond-N^+ membranes (Roche Diagnostics, Mannheim, Germany) following the standard procedure (Sambrook et al. 2001). The 483-bp fragment of the *GUS* probe was labeled with digoxigenin (DIG) using a Roche PCR DIG Probe Synthesis Kit. Hybridization and chemiluminescent detection of the blots were performed following the manufacturer's instructions (Roche). Total RNA of *35S:AtFT* transgenic lines and controls was extracted from frozen tissue as described by Ding et al. (Ding et al. 2008). First-strand cDNA was synthesized using the PrimeScript® RT Reagent Kit with gDNA Eraser (TAKARA, Dalian, China) according to the manufacturer's instructions. The primers used for RT-PCR are listed in Table S1.

Histochemical staining of β-glucuronidase (GUS) activity

Histochemical staining of GUS activity was performed according to the method of Jefferson et al. (1987) with some modifications. Transient *GUS* expression in cotyledon explants was monitored at 3 days after *A. tumefaciens* transformation, and stable *GUS* expression was examined in kanamycin-resistant calli, leaf lamina of positive shoots, stems, roots, flowers, young fruits of putative T_0 transgenic adult plants, T_1 cotyledons from GUS-positive plants, and the same tissues of the control plants. These samples were immersed in GUS staining solution (50 mM sodium phosphate (pH 7.0), 0.5 mM $K_3Fe(CN)_6$, 0.5 mM $K_4Fe(CN)_6 \cdot 3H_2O$, 0.5 % Triton X-100 and 1 mM X-gluc) and subjected to vacuum for 20 min. After incubation in the GUS staining solution at 37 °C overnight, the samples were destained in 70 % ethanol for several hours to visualize the GUS-stained areas. The efficiency of transient transformation was calculated by comparing the number of GUS-stained cotyledons to the total number of cotyledons infected by *A. tumefaciens*. The percentage of GUS-positive shoots was calculated by comparing the number of GUS-stained leaves to the total number of leaves from kanamycin-resistant shoots.

Statistical analysis

In this study, 30–50 explants were used for each experimental condition, and each experiment was repeated three times. The efficiency of transient transformation, the percentage of GUS-positive shoots and root induction were analyzed using the Statistical Product and Service Solution software (version 16.0; SPSS Inc., Chicago, IL, USA), and the data are presented as the mean ± standard error of the repeated experiments. Analysis of variance among the means was determined using one-way ANOVA with Tukey's post hoc test.

RESULTS

Influence of AS and *A. tumefaciens* strains on the efficiency of transient transformation

To determine the effects of AS on the transformation efficiency of cotyledon explants of *J. curcas* infected with *A. tumefaciens*, the efficiency of transient transformation was analyzed by examining the GUS-stained cotyledons after 3 days of co-cultivation with *A. tumefaciens*. The addition of 100 mM AS during infection and co-cultivation slightly increased the transient transformation efficiency from 84 to 92 % using EHA105, and from 48 to 54 % using LBA4404, respectively (Figure 1a). Therefore, 100 mM AS was used in the subsequent infection and co-cultivation experiments. To select the best *A. tumefaciens* strain for *J. curcas* transgenic studies, the transformation efficiencies of four *A. tumefaciens* strains harboring pCAMBIA2301—EHA105, LBA4404, GV3101, and AGL1—were examined. As shown in Figure 1b, different *A. tumefaciens* strains exhibited significant differences in their ability to transform cotyledon explants of *J. curcas*. Compared with the other *A. tumefaciens* strains, EHA105 showed the highest transient transformation efficiency (92 %). The transient transformation efficiencies of LBA4404, GV3101, and AGL1 were approximately 54, 19 and 14 %, respectively (Figure 1b). An intense blue color was observed in the incisions in cotyledons (Figure 1c). The number of stained areas per EHA105-infected explant was significantly higher than that observed with other *A. tumefaciens* strains, whereas no GUS staining was detected in control explants transformed with EHA105 without the binary vector (Figure 1c). Therefore, EHA105 was superior to the other three *A. tumefaciens* strains for infecting *J. curcas* cotyledons, and was selected for subsequent transformation experiments in this study.

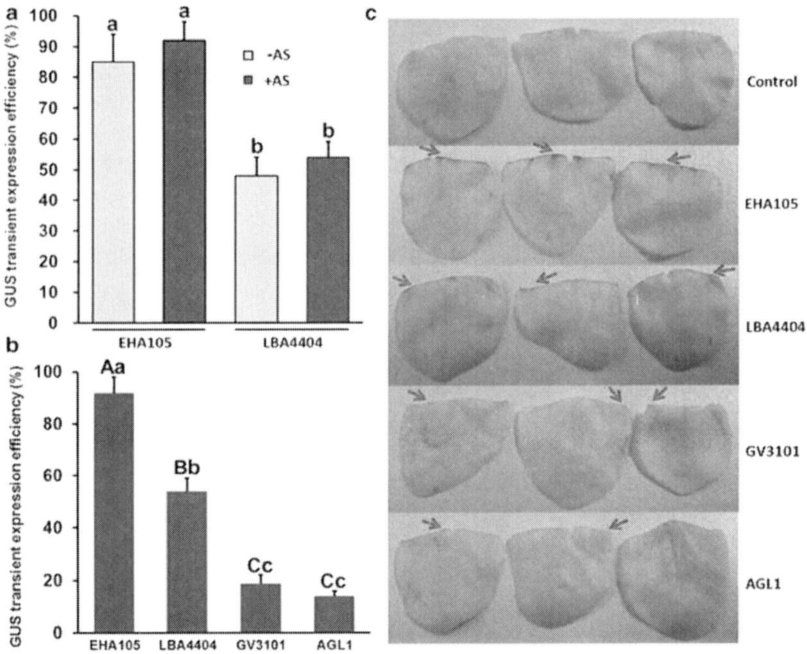

Figure 1. Efficiency of transient transformation of *J. curcas* cotyledon explants after 3 days of co-cultivation with *Agrobacterium tumefaciens* . **A** Influence of AS on the efficiency of transient transformation. **B** Influence of *Agrobacterium* strains on the efficiency of transient transformation. **C** Transient GUS expression differed in the incisions of cotyledons after co-cultivation with different *Agrobacterium* strains. Each value is the mean ± SE of three independent experiments, each including 30 cotyledons. Values with different *lowercase letters* are significantly different ($P < 0.05$, Tukey's test). Values with different *uppercase letters* are significantly different ($P < 0.01$, Tukey's test). *Red arrows* indicate GUS staining on cotyledon explants

Optimization of kanamycin concentration and duration of delayed selection

As shown in our previous report of *J. curcas* transformation, cotyledon explants excised from mature seeds of *J. curcas* were very sensitive to the selection agent kanamycin, and 5 mg L^{-1}kanamycin in the medium completely inhibited callus formation from cotyledon explants (Pan et al. 2010). The regeneration frequency and positive transformation rate were relatively low when cotyledon explants were cultivated on CIM without kanamycin for 4 weeks after co-cultivation with *Agrobacterium* (Pan et al. 2010). To improve the efficiency of kanamycin selection in *J. curcas*, nine independent selection experiments were carried out with different kanamycin concentrations and duration of delayed selection (Table 1). After co-cultivation with *A. tumefaciens*, the cotyledon explants were then transferred onto CIM and cultured for 1, 2, or 3 weeks without selection. The cotyledon explants began to turn green and grow larger after 1 week (Figure 2a), and some small calli were produced on the cut ends of the swollen cotyledon explants. The cotyledon explants were inserted into the SIM with the cut ends embedded in the medium (Figure 2b). A clear difference was observed between the transformed and untransformed cotyledon explants during the kanamycin selection process: there were many green calli formed at the incisions of the transformed cotyledon explants (Additional file 1: Figure S1 a and c), whereas no such resistant calli were formed from the untransformed cotyledon explants (Additional file 1: Figure S1 b and d). The base of cotyledon explants embedded in the medium turned yellow, and they gradually formed many resistant calli at the incisions after 3 weeks under the condition of 1-week delayed selection with 40 mg L^{-1} kanamycin, whereas the upper section of cotyledon explants (above the medium) stayed green and continued to grow (Figure 2c, d). Then, the resistant calli were excised and transferred onto new SIM for the second cycle of selection; during this period, some adventitious buds were found in the regions of the resistant calli. Resistant calli and shoots were subcultured at 3-week intervals. After three passages of selection, many resistant shoots were regenerated, as shown in Figure 2e. Additional adventitious shoots were obtained during the fourth selection cycle. Resistant shoots that were 1.5–2 cm in length with 3–4 leaves were excised and cultured on RIM for inducing roots (Figure 2f).

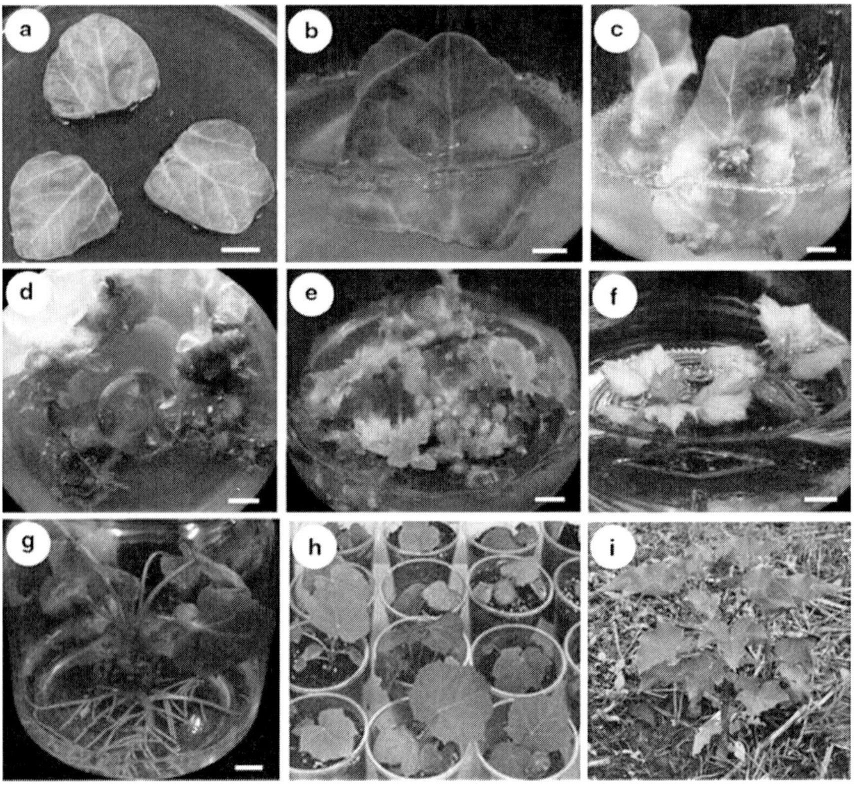

Figure 2. Different stages of *Agrobacterium*-mediated genetic transformation of *J. curcas*. **A** Swollen cotyledon explants incubated on CIM for 1 week. **B** Inserting cotyledon explants into SIM after cultivation on CIM for 7 days to induce resistant calli. **C** Three-week-old resistant calli in SIM under the condition of 1-week delayed selection with 40 mg L^{-1}kanamycin. **D** Resistant calli (indicated by *red arrows*) in SIM. **E** Resistant shoot buds on SIM under the condition of 1-week delayed selection with 40 mg L^{-1} kanamycin in the third cycle of selection. **F** Resistant plantlets were excised and transferred to RIM. **G** Putative transgenic plantlets rooted after incubation on RIM for 4 weeks. **H** Transgenic plants in pots acclimatized in the greenhouse for 4 weeks. **I** Growth of transgenic plantlets in the field. *Scale bars* represent 5 mm

We utilized GUS staining as an indicator of successful transformation, and the *GUS* expression rate was calculated as follows: GUS-positive shoots/total kanamycin-resistant shoots (Table 1). The number of shoots per explant was measured to calculate the regeneration frequency. The regenerated shoots from the third and fourth cycles of kanamycin selection were statistically analyzed, and the leaves of these regenerated shoots were examined by GUS staining. As shown in Table 1, the regeneration frequency of shoots decreased significantly as the kanamycin concentration and the duration of delayed selection after co-cultivation increased. The highest regeneration frequency (5.56 shoots per explant) was observed under the condition of 1-week delayed selection with 20 mg L^{-1} kanamycin, whereas the lowest regeneration frequency (1.88 shoots per explant) was observed under the condition of 3-week delayed selection with 40 mg L^{-1} kanamycin (Table 1). More importantly, significant differences in the number of GUS-positive shoots per explant and the frequency of GUS-positive shoots were observed among different kanamycin concentrations and duration of delayed selection. The number of GUS-positive shoots per explant and the frequency of GUS-positive shoots were only 0.05 and 1.3 %, respectively, under the condition of a 3-week delayed selection with 20 mg L^{-1}kanamycin. When the kanamycin concentration was increased and the duration of delayed selection was reduced, the number of GUS-positive shoots per explant and the frequency of GUS-positive shoots increased gradually. Under the condition of a 1-week delayed selection with 40 mg L^{-1} kanamycin, the number of GUS-positive shoots per explant and the frequency of GUS-positive shoots reached maxima of 1.70 and 56.0 %, respectively (Table 1).

Optimization of the root-inducing medium

To obtain high rooting efficiency, we optimized RIM used for root induction from regenerated shoots by adding different concentrations of IBA and NAA to 1/2 MS. As shown in Table 2, the rooting frequency increased with increases in the concentrations of IBA or NAA. The lowest rooting frequency was observed on 1/2 MS basal medium, only 5.8 %. A rooting frequency of approximately 65.0 % was observed on RIM with 0.2 mg L^{-1} NAA. The combination of 0.2 mg L^{-1} IBA and 0.1 mg L^{-1} NAA in RIM further increased the rooting percentage to its highest value (84.9 %) (Table 2). Adventitious roots were initiated on regenerated shoots incubated on RIM within 2 weeks, and the roots were much more numerous and stronger after incubation on RIM for 4 weeks (Figure 2g). Rooted plantlets were planted in sterilized soil consisting of humus:peat:vermiculite (3:1:1) and acclimatized in the greenhouse at 22 ± 2 °C. After 4–5 weeks, more than 85 % of the acclimatized plantlets survived, and there was no obvious phenotypic difference among GUS transgenic *J. curcas* plants (Figure 2h). Then, the plantlets were transferred to the fields for further growth (Figure 2i).

Confirmation of transgenic plants

To confirm the presence of the transgenes in the putative transgenic plants, genomic DNA was extracted from the leaves of non-transformed and putative transgenic plants in pots, which were selected with 40 mg L^{-1} kanamycin after cultivation on CIM for 1 week. Then, the reporter gene *GUS* and the selective marker gene *NPTII* were identified by PCR amplification. A 778-bp fragment corresponding to the *GUS* gene was detected in most of the putative transgenic plants (Figure 3a, upper panel, lanes 1 – 10) and the positive control (Figure 3a, upper panel, lane P), but no amplification product was found in the non-transformed control plant (Figure 3a, upper panel, lane C). The expected 679-bp fragment of the *NPTII* gene was also detected in most of the putative transgenic plants (Figure 3a, lower panel, lanes 1–10), whereas the control plant showed no amplification products (Figure 3a, lower panel, lane C).

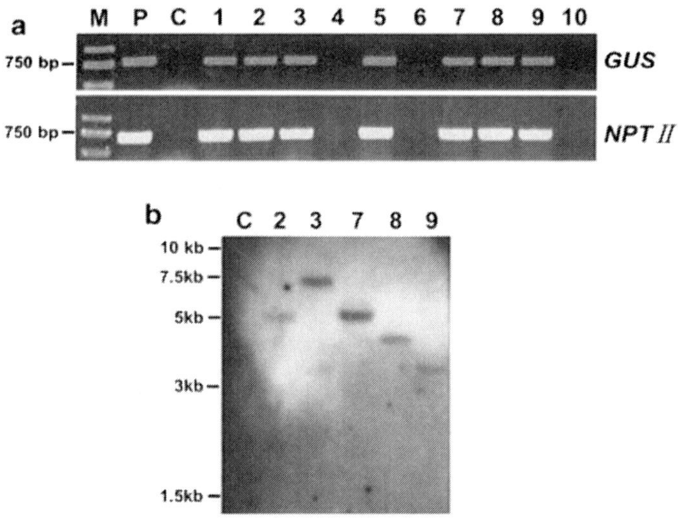

Figure 3. PCR and Southern blot analysis of transgenic *J. curcas* plants. **A** Amplification of the 778-bp *GUS* gene (*upper panel*) and the 679-bp *NPTII* gene (*lower panel*) in transgenic plants. Lanes: M, molecular size marker (Trans2 K DNA ladder); P, positive control (pCAMBIA2301 plasmid); C, negative control (non-transformed plant); 1–10, putative transgenic *J. curcas* lines. **B** Southern hybridization of genomic DNA isolated from negative control (a non-transformed plant) and PCR-positive shoots. Genomic DNA was digested with *Eco*RI and hybridized with a DIG-labeled *GUS* probe. Lanes: C, negative control (a non-transformed plant); 2, 3, 7, 8, and 9, five independent transgenic plants

The integration of the *GUS* gene into the genome of transgenic *J. curcas* plants was further verified by Southern blot analysis of genomic DNA from PCR-positive transgenic plants. As shown in Figure 3b, all of the tested plants showed one band after hybridization (Figure 3b, lanes 2, 3, 7, 8 and 9), and no hybridization band was detected when control plant DNA was used (Figure 3b, lane C). This result indicates that a single copy of the *GUS* gene had integrated into the genomes of the tested transgenic plants.

GUS histochemical analysis

The stable transformation events were further monitored using histochemical analysis of GUS activity. As shown in Figure 4a and b, intense blue staining was visibly detected in a resistant callus and young leaf from a regenerated shoot under the condition of 1-week delayed selection with 40 mg L^{-1} kanamycin, whereas no such blue staining was detected in a non-transformed callus and leaf (control) (Figure 4h, i). To visualize stable GUS expression in adult transgenic plants, we next analyzed the stems, roots, flowers, and fruits. Strong GUS staining was observed in the stems, roots, flowers and fruits of T_0 adult transgenic plants (Figure 4c–f), whereas no GUS expression was observed in the negative control plants (Figure 4j–m). Positive GUS expression was also observed in cotyledons of the progeny (T_1) of transgenic plants (Figure 4g), whereas GUS activity was not detected in the cotyledons of the control plants (Figure 4n), indicating that the transgene was stably inherited in T_1 progeny.

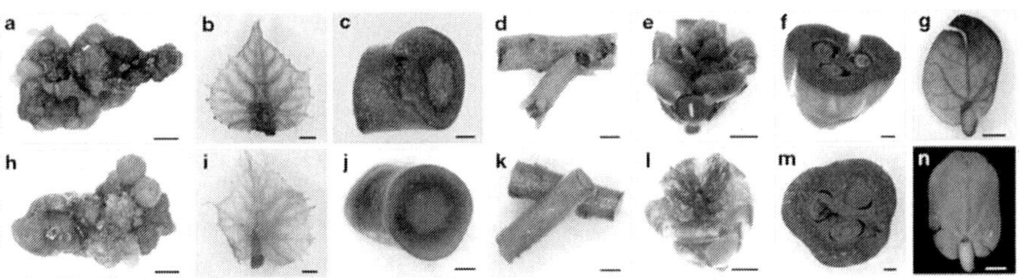

Figure 4. Histochemical GUS staining of various tissues of transgenic *J. curcas* Representative photos of GUS-stained calli (**A** and **H**), leaves (**B** and **I**), stems (**C** and **J**), roots (**D** and **K**), flowers (**E** and **L**), and fruits (**F** and **M**) from T_0 transgenic plants (**A-F**) and control (**H-M**). Cotyledons from the progeny (T_1) of transgenic (**G**) and control (**N**) plants are also shown. *Scale bars* represent 5 mm

Overexpression of *Arabidopsis FLOWERING LOCUS T* in *J. curcas* causes early flowering in vitro

FLOWERING LOCUS T (*FT*) is a key gene mediating the vegetative-to-reproductive transition and is conserved in many flowering plants (Kardailsky et al. 1999; Li et al. 2014; Ye et al. 2014). Using the transformation protocol described above, we transferred *Arabidopsis FT* (*AtFT*) cDNA under the control of the CaMV 35S promoter into 30 cotyledon explants of *J. curcas*, and we obtained a total of 17 independent transgenic shoots in 2.5 months. Flower buds formed directly on most of the *35S:AtFT* transgenic shoots of *J. curcas* in vitro, whereas the control explants did not produce flower buds under the same conditions (Figure 5a, b). Semi-quantitative reverse transcription PCR (RT-PCR) analysis of gene expression related to flower bud development in *J. curcas* was performed on the shoot apexes of the wild-type (WT) and *35S:AtFT* transgenic lines cultured in vitro. Consistent with the phenotype of the transgenic shoots, which showed early flowering in vitro, the transcript levels of *AtFT, JcSOC1, JcLFY, JcAP1* and *JcAP3* were significantly upregulated in the transgenic lines (Figure 5c). These results demonstrate that the improved transformation protocol provides a reliable method for the introduction of functional genes into *J. curcas*.

Figure 5. Phenotypes of WT and *35S:AtFT* transgenic *J. curcas* shoots cultured in vitro and semi-quantitative RT-PCR analysis of flowering genes in these lines. **A** WT *J. curcas* shoot cultured in vitro. **B** Flower buds of *35S:AtFT* transgenic *J. curcas* cultured in vitro. **C** Semi-quantitative RT-PCR analysis of flowering genes in the shoot apexes of WT and *35S:AtFT* transgenic *J. curcas* cultured in vitro. *Red arrows* indicate flower buds. *Scale bars* represent 1 cm

DISCUSSION

As a potent inducer of *A. tumefaciens* for infecting plant cells, AS has been successfully employed to improve transformation efficiency in many plants (Bull et al. 2009; Ramamoorthy and Kumar 2012; Wang et al. 2012). Similar to the results obtained by Li et al. (2008) and Kumar et al. (2010), the addition of 100 mM AS in this study slightly enhanced GUS transient expression efficiency in *J. curcas* compared with the control (Figure 1a). The *A. tumefaciens* strain is one of the most important factors potentially affecting the efficiency of genetic transformation. Different *Agrobacterium* strains have been successfully used to transform *J. curcas*; however, the transformation efficiency varied greatly among the different transformation methods (Jaganath et al. 2014; Kajikawa et al. 2012; Li et al. 2008). We found that EHA105 was superior to other *Agrobacterium* strains for genetic transformation of *J. curcas* cotyledons (Figure 1b, c). Our Southern blot analysis showed that the 5 transgenic *J. curcas* lines that we tested each had a single copy of the transgene (Figure 3b), which is consistent with the report that *Agrobacterium* strain EHA105 appeared to be capable of generating single-copy transgenic tomato plants (Chetty et al. 2013).

The in vitro regeneration ability of the explants is crucial for establishing a successful *Agrobacterium*-mediated transformation system. Different explants of *J. curcas* have been reported to be used in transformation (Kajikawa et al. 2012; Khemkladngoen et al. 2011; Kumar et al. 2010; Li et al. 2008; Mao et al. 2009; Misra et al. 2012; Pan et al. 2010). In many plants, cotyledon explants are better suited to regeneration and transformation than other explants (Li et al. 2008, 2006; Murthy et al. 2003). Cotyledons excised from mature seeds of *J. curcas*, which are available in large quantities for transformation year-round, were found susceptible to *A. tumefaciens* infection (Li et al. 2014; Pan et al. 2010; Tao et al. 2015). In the current study, multiple resistant calli could be rapidly induced from the cut edges of cotyledon explants after 2–3 weeks of incubation on SIM (Figure 2c), and transgenic shoots were regenerated from these resistant calli (Figure 2e).

The selection strategy is also a key factor in improving transformation efficiency. Many plant species, including *J. curcas*, have been found to be hypersensitive to kanamycin, and this hypersensitivity is most likely the source of the reduced transformation efficiencies observed in previous studies (Kajikawa et al. 2012; Pan et al. 2010). Alternative selection strategies, including reduced kanamycin concentrations and delayed selection, have been successful for obtaining transgenic plants in kanamycin–hypersensitive species, including apple (Yao et al. 1995) and almond (Miguel and Oliveira 1999; Ramesh et al. 2006). In this study, with increasing kanamycin concentrations and decreasing durations of delayed selection, the number of GUS-positive shoots increased (Table 1). A stable transformation rate of approximately 56.0 % and 1.70 GUS-positive shoots per explant were obtained under the condition of 1-week delayed selection with 40 mg L^{-1} kanamycin (Table 1). The success of selection depends not only on kanamycin concentrations and delayed selection but also on improving the contact between explants and medium during inoculation (Bhatia et al. 2005). The cotyledon explants transferred from CIM were inserted into SIM with approximately 1 cm of their cut ends embedded in the medium (Figure 2b). After 2–3 weeks, green calli and adventitious buds developed at the cut ends of cotyledon explants in the medium (Figure 2c, d). This improved inoculation method might also result in a reduced number of escapees from kanamycin selection. To our knowledge, this is the first report describing the insertion of cut ends of explants into selection medium during *Agrobacterium*-mediated transformation of *J. curcas*.

We found that the rooting efficiency of regenerated *J. curcas* shoots varied greatly among different RIMs, which agrees with previous studies (Kajikawa et al. 2012; Khemkladngoen et al. 2011; Li et al. 2008; Mazumdar et al. 2010). The differences in rooting efficiency might depend not only on the concentrations of IBA and NAA in RIMs but also on the condition of shoots and/or the hormone levels in SIMs in these reports. Using different combinations of IBA and NAA, we optimized the conditions for root induction of the regenerated shoots. The highest rooting efficiency (84.9 %) was observed in 1/2 MS supplemented with 0.2 mg L^{-1} IBA and 0.1 mg L^{-1} NAA (Table 2). For the regenerated shoots that are difficult to root, in vivo grafting could be employed to obtain transgenic *J. curcas* plants (Jaganath et al. 2014; Li et al. 2014). In addition, transgenic *J. curcas* shoots overexpressing *AtFT* driven by the strong constitutive 35S promoter, which were obtained in this study and produced flower buds in vitro (Figure 5b), were not able to develop into normal plants. As we showed previously with *J. curcas FT* homolog (Li et al. 2014), the use of weaker constitutive promoters, tissue specific promoters, or inducible promoters will be required to produce normal transgenic *J. curcas* plants overexpressing *AtFT*.

In conclusion, several important factors that affect the efficiency of *Agrobacterium*-mediated genetic transformation of *J. curcas* were optimized in this study, including the supplementation of AS, *A. tumefaciens* strain, kanamycin concentration, duration of delayed selection, inoculation method, and root-inducing medium, which have not been tested in our previous study (Pan et al. 2010). Compared to our previous protocol, genetic transformation and rooting efficiency of the improved protocol were greatly enhanced and reached an average of 56.0

and 84.9 %, respectively. The overall scheme is presented in Figure 6. The protocol requires only two basal media and takes approximately 4 months from the start of co-cultivation to the rooting of plantlets. This improved transformation protocol is simple and reproducible; thus, it may be useful for studies of functional genes and genetic improvement of *J. curcas*. And the genetic transformation strategy reported here may be applied to other plants with various selective agents and explant types.

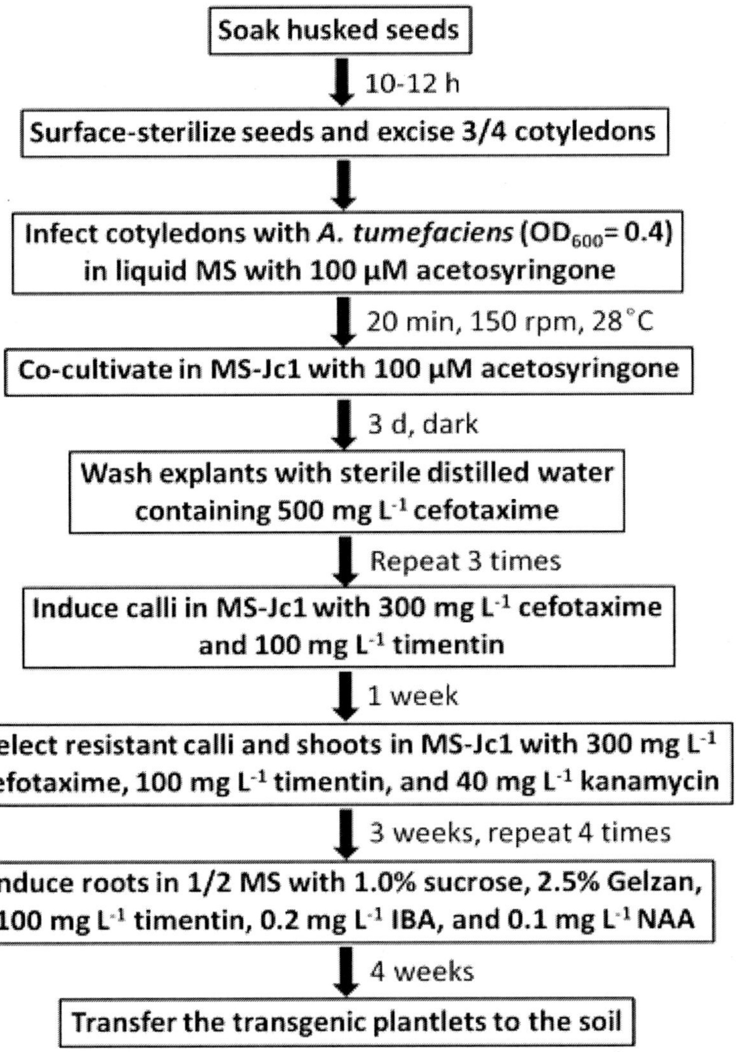

Figure 6. Schematic representation of the improved protocol for *Agrobacterium*-mediated genetic transformation and regeneration of *J. curcas* using cotyledons as explants

ACKNOWLEDGMENTS

This work was supported by the Natural Science Foundation of China (31370595 and 31300568), the Natural Science Foundation of Yunnan Province (2011FA034), the West Light Foundation of CAS, and the CAS 135 program (XTBG-T02). The authors gratefully acknowledge the Central Laboratory of the Xishuangbanna Tropical Botanical Garden for providing research facilities.

REFERENCES

1. Allen GC, Flores-Vergara MA, Krasnyanski S, Kumar S, Thompson WF (2006) A modified protocol for rapid DNA isolation from plant tissues using cetyltrimethylammonium bromide. Nat Protoc 1:2320–2325
2. Bhatia P, Ashwath N, Midmore DJ (2005) Effects of genotype, explant orientation, and wounding on shoot regeneration in tomato. In Vitro Cell Dev Biol Plant 41:457–464
3. Bull SE, Owiti JA, Niklaus M, Beeching JR, Gruissem W, Vanderschuren H (2009) Agrobacterium-mediated transformation of friable embryogenic calli and regeneration of transgenic cassava. Nat Protoc 4:1845–1854
4. Chen M-S, Wang G-J, Wang R-L, Wang J, Song S-Q, Xu Z-F (2011) Analysis of expressed sequence tags from biodiesel plant Jatropha curcas embryos at different developmental stages. Plant Sci 181:696–700
5. Chetty VJ, Ceballos N, Garcia D, Narvaez-Vasquez J, Lopez W, Orozco-Cardenas ML (2013) Evaluation of four Agrobacterium tumefaciens strains for the genetic transformation of tomato (Solanum lycopersicum L.) cultivar Micro-Tom. Plant Cell Rep 32:239–247
6. Costa GGL, Cardoso KC, Del Bem LEV, Lima AC, Cunha MAS, de Campos-Leite L, Vicentini R, Papes F, Moreira RC, Yunes JA, Campos FAP, Da Silva MJ (2010) Transcriptome analysis of the oil-rich seed of the bioenergy crop Jatropha curcas L. BMC Genom 11:462
7. Ding L-W, Sun Q-Y, Wang Z-Y, Sun Y-B, Xu Z-F (2008) Using silica particles to isolate total RNA from plant tissues recalcitrant to extraction in guanidine thiocyanate. Anal Biochem 374:426–428
8. Eswaran N, Parameswaran S, Anantharaman B, Kumar GRK, Sathram B, Johnson TS (2012) Generation of an expressed sequence tag (EST) library from salt-stressed roots of Jatropha curcas for identification of abiotic stress-responsive genes. Plant Biol 14:428–437
9. Fairless D (2007) Biofuel: the little shrub that could—maybe. Nature 449:652–655
10. He Y, Pasapula V, Li X, Lu R, Niu B, Hou P, Wang Y, Xu Y, Chen F (2009) Agrobacterium tumefaciens-mediated transformation of Jatropha curcas: factors affecting transient transformation efficiency and morphology analysis of transgenic calli. Silvae Genet 58:123
11. Herr J, Carlson J (2013) Traditional breeding, genomics-assisted breeding, and biotechnological modification of forest trees and short rotation woody crops. In: Jacobson M, Ciolkosz D (eds) Wood-based energy in the northern forests. Springer, New York, pp 79–99
12. Hirakawa H, Tsuchimoto S, Sakai H, Nakayama S, Fujishiro T, Kishida Y, Kohara M, Watanabe A, Yamada M, Aizu T, Toyoda A, Fujiyama A, Tabata S, Fukui K, Sato S

(2012) Upgraded genomic information of Jatropha curcas L. Plant Biotechnol 29:123–130
13. Jaganath B, Subramanyam K, Mayavan S, Karthik S, Elayaraja D, Udayakumar R, Manickavasagam M, Ganapathi A (2014) An efficient in planta transformation of Jatropha curcas (L.) and multiplication of transformed plants through in vivo grafting. Protoplasma 251:591–601
14. Jefferson RA, Kavanagh TA, Bevan MW (1987) GUS fusions: beta-glucuronidase as a sensitive and versatile gene fusion marker in higher plants. EMBO J 6:3901–3907PubMedCentralPubMed
15. Joshi M, Mishra A, Jha B (2011) Efficient genetic transformation of Jatropha curcas L. by microprojectile bombardment using embryo axes. Ind Crop Prod 33:67–77
16. Juan JC, Kartika DA, Wu TY, Hin T-YY (2011) Biodiesel production from jatropha oil by catalytic and non-catalytic approaches: an overview. Bioresour Technol 102:452–460
17. Kajikawa M, Morikawa K, Inoue M, Widyastuti U, Suharsono S, Yokota A, Akashi K (2012) Establishment of bispyribac selection protocols for Agrobacterium tumefaciens-and Agrobacterium rhizogenes-mediated transformation of the oil seed plant Jatropha curcas L. Plant Biotechnol 29:145–153
18. Kardailsky I, Shukla V, Ahn J, Dagenais N, Christensen S, Nguyen J, Chory J, Harrison M, Weigel D (1999) Activation tagging of the floral inducer FT. Science 286:1962
19. Khemkladngoen N, Cartagena JA, Fukui K (2011) Physical wounding-assisted Agrobacterium-mediated transformation of juvenile cotyledons of a biodiesel-producing plant, Jatropha curcas L. Plant Biotechnol Rep 5:235–243
20. Khurana-Kaul V, Kachhwaha S, Kothari S (2010) Direct shoot regeneration from leaf explants of Jatropha curcas in response to thidiazuron and high copper contents in the medium. Biol Plant 54:369–372
21. Kumar N, Reddy M (2010) Plant regeneration through the direct induction of shoot buds from petiole explants of Jatropha curcas: a biofuel plant. Ann Appl Biol 156:367–375
22. Kumar A, Sharma S (2008) An evaluation of multipurpose oil seed crop for industrial uses (Jatropha curcas L.): a review. Ind Crop Prod 28:1–10
23. Kumar N, Vijay Anand K, Pamidimarri D, Sarkar T, Reddy MP, Radhakrishnan T, Kaul T, Reddy M, Sopori SK (2010) Stable genetic transformation of Jatropha curcas via Agrobacterium tumefaciens-mediated gene transfer using leaf explants. Ind Crop Prod 32:41–47
24. Li HQ, Sautter C, Potrykus I, PuontiKaerlas J (1996) Genetic transformation of cassava (Manihot esculenta Crantz). Nat Biotechnol 14:736–740
25. Li MR, Li HQ, Wu GJ (2006) Study on factors influencing Agrobacterium-mediated transformation of Jatropha curcas. J Mol Cell Biol 39:83–89
26. Li MR, Li HQ, Jiang HW, Pan XP, Wu GJ (2008) Establishment of an Agrobacterium-mediated cotyledon disc transformation method for Jatropha curcas. Plant Cell Tissue Organ Cult 92:173–181
27. Li C, Luo L, Fu Q, Niu L, Xu Z-F (2014) Isolation and functional characterization of JcFT, a FLOWERING LOCUS T (FT) homologous gene from the biofuel plant Jatropha curcas. BMC Plant Biol 14:125
28. Maghuly F, Laimer M (2013) Jatropha curcas, a biofuel crop: functional genomics for understanding metabolic pathways and genetic improvement. Biotechnol J 8:1172–1182
29. Makkar HP, Becker K (2009) Jatropha curcas, a promising crop for the generation of biodiesel and value-added coproducts. Eur J Lipid Sci Technol 111:773–787
30. Mao HZ, Ye J, Hai CN (2009) Genetic transformation of Jatropha curcas. International Publication No: WO2010071608A9
31. Mazumdar P, Basu A, Paul A, Mahanta C, Sahoo L (2010) Age and orientation of the cotyledonary leaf explants determine the efficiency of de novo plant regeneration and Agrobacterium tumefaciens-mediated transformation in Jatropha curcas L. S Afr J Bot 76:337–344

32. Miguel CM, Oliveira MM (1999) Transgenic almond (Prunus dulcis Mill.) plants obtained by Agrobacterium-mediated transformation of leaf explants. Plant Cell Rep 18:387–393
33. Misra P, Toppo DD, Mishra MK, Saema S, Singh G (2012) Agrobacterium tumefaciens-mediated transformation protocol of Jatropha curcas L. using leaf and hypocotyl segments. J Plant Biochem Biotechnol 21:128–133
34. Murashige T, Skoog F (1962) A revised medium for rapid growth and bioassays with tobacco tissue cultures. Physiol Plant 15:473–497
35. Murthy HN, Jeong JH, Choi YE, Paek KY (2003) Agrobacterium-mediated transformation of niger [Guizotia abyssinica (L. f.) Cass.] using seedling explants. Plant Cell Rep 21:1183–1187
36. Natarajan P, Parani M (2011) De novo assembly and transcriptome analysis of five major tissues of Jatropha curcas L. using GS FLX titanium platform of 454 pyrosequencing. BMC Genom 12:191
37. Natarajan P, Kanagasabapathy D, Gunadayalan G, Panchalingam J, Shree N, Sugantham PA, Singh KK, Madasamy P (2010) Gene discovery from Jatropha curcas by sequencing of ESTs from normalized and full-length enriched cDNA library from developing seeds. BMC Genom 11:606
38. Newell CA (2000) Plant transformation technology. Mol Biotechnol 16:53–65
39. Pan J, Fu Q, Xu Z-F (2010) Agrobacterium tumefaciens-mediated transformation of biofuel plant Jatropha curcas using kanamycin selection. Afr J Biotech 9:6477–6481
40. Purkayastha J, Sugla T, Paul A, Solleti S, Mazumdar P, Basu A, Mohommad A, Ahmed Z, Sahoo L (2010) Efficient in vitro plant regeneration from shoot apices and gene transfer by particle bombardment in Jatropha curcas. Biol Plant 54:13–20
41. Qin X, Zheng X, Huang X, Lii Y, Shao C, Xu Y, Chen F (2014) A novel transcription factor JcNAC1 response to stress in new model woody plant Jatropha curcas. Planta 239:511–520
42. Ramamoorthy R, Kumar PP (2012) A simplified protocol for genetic transformation of switchgrass (Panicum virgatum L.). Plant Cell Rep 31:1923–1931
43. Ramesh SA, Kaiser BN, Franks T, Collins G, Sedgley M (2006) Improved methods in Agrobacterium-mediated transformation of almond using positive (mannose/pmi) or negative (kanamycin resistance) selection-based protocols. Plant Cell Rep 25:821–828
44. Rosado TB, Laviola BG, Faria DA, Pappas MR, Bhering LL, Quirino B, Grattapaglia D (2010) Molecular markers reveal limited genetic diversity in a large germplasm collection of the biofuel crop Jatropha curcas L. in Brazil. Crop Sci 50:2372–2382
45. Sambrook J, Russell DW, Russell DW (2001) Molecular cloning: a laboratory manual. Cold Spring Harbor Laboratory Press, Cold Spring Harbor
46. Sanderson K (2009) Wonder weed plans fail to flourish. Nature 461:328–329
47. Sato S, Hirakawa H, Isobe S, Fukai E, Watanabe A, Kato M, Kawashima K, Minami C, Muraki A, Nakazaki N (2011) Sequence analysis of the genome of an oil-bearing tree, Jatropha curcas L. DNA Res 18:65–76
48. Sharma S, Kumar N, Reddy MP (2011) Regeneration in Jatropha curcas: factors affecting the efficiency of in vitro regeneration. Ind Crop Prod 34:943–951
49. Singh A, Reddy MP, Chikara J, Singh S (2010) A simple regeneration protocol from stem explants of Jatropha curcas—A biodiesel plant. Ind Crop Prod 31:209–213
50. Sujatha M, Mukta N (1996) Morphogenesis and plant regeneration from tissue cultures of Jatropha curcas. Plant Cell Tissue Organ Cult 44:135–141
51. Sujatha M, Reddy TP, Mahasi M (2008) Role of biotechnological interventions in the improvement of castor (Ricinus communis L.) and Jatropha curcas L. Biotechnol Adv 26:424–435
52. Sun QB, Li LF, Li Y, Wu GJ, Ge XJ (2008) SSR and AFLP markers reveal low genetic diversity in the biofuel plant Jatropha curcas in China. Crop Sci 48:1865–1871
53. Tao Y-B, He L-L, Niu L-J, Xu Z-F (2015) Isolation and characterization of an ubiquitin extension protein gene (JcUEP) promoter from Jatropha curcas. Planta 241:823–836

54. Toppo DD, Singh G, Purshottam D, Misra P (2012) Improved in vitro rooting and acclimatization of Jatropha curcas plantlets. Biomass Bioenerg 44:42–46
55. Visarada K, Meena K, Aruna C, Srujana S, Saikishore N, Seetharama N (2009) Transgenic breeding: perspectives and prospects. Crop Sci 49:1555–1563
56. Wang Q, Xing S, Pan Q, Yuan F, Zhao J, Tian Y, Chen Y, Wang G, Tang K (2012) Development of efficient Catharanthus roseus regeneration and transformation system using Agrobacterium tumefaciens and hypocotyls as explants. BMC Biotechnol 12:34
57. Yao JL, Cohen D, Atkinson R, Richardson K, Morris B (1995) Regeneration of transgenic plants from the commercial apple cultivar Royal Gala. Plant Cell Rep 14:407–412
58. Ye J, Geng Y, Zhang B, Mao H, Qu J, Chua N-H (2014) The Jatropha FT ortholog is a systemic signal regulating growth and flowering time. Biotechnol Biofuels 7:91
59. Zhang L, Zhang C, Wu P, Chen Y, Li M, Jiang H, Wu G (2014) Global analysis of gene expression profiles in physic nut (Jatropha curcas L.) seedlings exposed to salt stress. PLoS ONE 9:e97878

CHAPTER 13

Bioinformatic Characterization of Desaturase Partial Gene in Sesamum Indicum

Spandana B[1*], DSRS Prakash[2] and Nagalakshmi S[1]

[1]*Institute of Biotechnology (IBT), Acharya NG Ranga Agricultural University (ANGRAU), Guntur, Andhra Pradesh, India*

[2]*Department of Human Genetics, Andhra University, Visakhapatnam, Andhra Pradesh, India*

ABSTRACT

By using Bioinformatic tools three sets of codehop, webprimer and manually, gene specific degenerate primers were designed for Sesamum desaturase. Through Gradient PCR approach maximum homology among desaturase sequences was observed to be with Arachis hypogaea than the other oil seed crops compared. Primers designed through Codehop and webprimershowed amplification of the gene sequence. The aim of this present study was to utilize bioinformatic tools for sesame improvement programs towards diversifying the fatty acid composition of the seed oil.

Keywords: Degenerate primers; Desaturase; Sesamum; Fatty acids; Conserved regions

INTRODUCTION

Fatty acid desaturases are the enzymes that are responsible for the insertion of cis-double bonds into pre-formed fatty acid chains in reactions that require oxygen and reducing equivalents. In the last few decades considerable work has been carried out in several plant species to alter the fatty acid composition of the oil by classical plant breeding methods and were able to generate low erucic acid. Brassica (canola oil) and sunflower oil containing high oleic acid. Currently Canola oil isthe lowest saturated fat vegetable oil. Fatty acids are synthesized from acetyl-Co A by a series of reactions that are localized in the plastids. Desaturases play a key role in polyunsaturated fatty acid homeostasis i.e., mostly the unsaturation which is the initial step in the essential fatty acid metabolism. Depending upon the intermediate product, Desaturases are of different types i.e., Stearoyl-ACP desaturases, oleate desaturases, linoleate desaturases etc. By understanding the enzymes involved in the synthesis and accumulation of seed- specific fatty acids, it may be possible to manipulate the composition of seed oil by generating transgenic plants by molecular biology techniques. It is necessary first to identify plant species which synthesize economically valuable oils and characterize the proteins/enzymes involved in the triacylglycerol biosynthesis.

Material and Methods

Genomic DNA Isolation: Total Genomic DNA of sesamum was extracted using a modified Cetyl trimethyl ammonium bromide (CTAB) method and quantified. All the reagents required for CTAB buffer preparation were obtained from Sigma-Aldrich (USA)

Designing of degenerate primers

The oil plant species containing the desaturase genes which have already been reported were collected from NCBI and EBI data banks. Desaturase nucleotide sequences of B. juncea, G.hirsutum, H.anus, S.indicum, A. hypogeae, R. communis, G. max. were taken from NCBI database and they are converted into protein sequences by web tool http://www. expasy.ch/tools/dna.html (Figure ures S1 and S2) The consensus amino acid and nucleotide sequences of desaturases (Fad D) were selected by using the Protein Domain database, Pro Dom (http:// prodes. toulouse. inra. Fr /prodom /doc/ prodom.html), NCBI (http:// ncbi.nlm.nih.gov) and EBI (www.ebi.ac.uk) tools and desaturase as the query. Clustal W software is used for multiple sequence alignment to identify the most conserved regions (Figure ures S3 and S4). The similarity matrix and the phylogenetic trees were produced using DNA Star (version 2004). The use of software in biological applications has given a new dimension to the field of bioinformatics. Many different programs for the design of primers are now available [1].

Degenerate primers for conserved regions of desaturases were designed manually and by using few software programmes like Web Primer and Codehop [2,3]. The consensus of both these protein and DNA sequence alignments gave rise to 3 sets of primer pairs which are listed in **TABLE 1**. Primer sequence degeneracy was also taken into consideration. In the present study Apart from the above mentioned softwares, the universal site for most of the bioinformatics tools were also used. (www.expasy.org). All the sequences were given for multiple sequence analysis by Clustal W. The resulted aligned sequences were given to BLOCKMAKER (http:// bioinformatics.weizmann.Ac.il/ blocks/blockmkr/www/make_blocks.html) with each block at a time to BLOCKMAKER. The blocks with maximum similarity and minimum degeneracy were selected from the results. These blocks were given to web programme Codehop (http://bioinformatics.weizmann.ac.il/ blocks/ codehop. html) for designing of primers. Primers were selected based on minimum degeneracy minimum degeneracy, length and percentage of GC. With one DNA sample three sets of gene specific degenerate primers (Manual, Web primer, Codehop) synthesized by Biotech Desk, India Pvt. Ltd.) were tested for PCR amplification of the gene PCR reactions were carried out in a DNA thermocycler (Eppendorf, USA) with a heated lid. Each 20 μl reaction volume contained about 100 ng of template DNA, 1X PCR buffer (10 mM Tris-HCl pH 5.3, 50 mM KCl), 2 mM MgCl2 (Invitrogen, USA), 2 mM dNTP, (Sigma -Aldrich., USA), 5 pmol each of Forward and Reverse primer and 1U of Taq DNA polymerase (Stratagene, Germany) and programmed for an initial denaturation step of 5 min at 94°C, followed by 35 cycles of 1 min at 94°C, 1 min at 50°C, extension was carried out at 72°C for 1 min and a final extension at 72°C for 10 min and a hold temperature of 14°C at the end. Negative controls were also run without template DNA to avoid non-specific amplification.

On 1.4% agarose gels PCR were electrophoresed, in 1X TAE Buffer at 50 V for 3 hrs and then stained with ethidium bromide (0.5 ug/ ul). To ascertain reproducibility of the raeaction PCR reactions were repeated three times reaction.

Table 1: PCR primers designed and targeted fragments of the Desaturase gene in Sesamum indicum.

Primer pair	Primer sequence (5'-3')	bp	Product size(bp)
Des M	F -ACYGGNRTNTGGGTNMTNGCNCAYG	25	400
	R- NGCYTGYTAYTGNGGCATNGTNGAC	25	
Des W	F -GGGCGTGTCACTAAGATTGAA	21	400
	R- TGTACCAGAGCACACCTTTGT	21	
Des C1	F- TGGACATCATGCTTTCTCTGAT-3'	24	400
	R- GGTACGATACCTCCGATGATTCC-5'	24	
Des 2	F -AGTGCTTTCACGCGATTTCT	20	400
	R -TCTCTCGAAGCAGTGTGGTG5'	20	

Degenerate bases information

N=A/T/G/C R=A/G M=A/C W=A/T S=G/C Y=C/T K=G/T V=A/G/C H=A/C/T D=A/G/T B=C/G/T

GEL ELUTION

Fragments ranging from 400-700 bp were eluted by using Column method (Qiagen Kit)

LIGATION

A 10 µl Ligation reaction volume was set up containing 1 µl of pGEM easy vector (Promega, USA); 5 µl of PCR enriched (gel eluted) product; 1 µl 10 X T4 DNA ligation buffer (New England Biolabs, USA), 2.4 U/µl of T4 DNA ligase (New England Biolabs, USA) and incubated at 14°C overnight. The samples were stored at - 20°C until used.

TRANSFORMATION

Ligated product was transformed. The plasmid isolation was done by Alkaline lysis method. After PCR amplified plasmidsamplification range (400 bp desaturase) was selected and was given for sequencing. Using Vecscreen (NCBI) the vector sequences were removed form the sequenced clones using and homology search was carried.

SEQUENCE ALIGNMENT AND PHYLOGENETIC ANALYSIS

The nucleotide sequences and deduced aminoacid sequenceswere analysed using comparitive and bioinformatic analysis that are available online at the websites (http://www.ncbi.nlm.nih.gov/) and (http://cn.expasy.org/). The nucleotide sequence, deduced aminoacid sequence, and open reading frame were analyzed and the sequence comparision was conducted through a database search using BLAST programs (Figure S5). The phylogenetic analysis of desaturase gene from other oilcrop species was aligned with Clustal W program using default parameters [4].

From the edible oil yielding species Glycine max, Brassica rapa, Helianthus annuus, Arachis hypogaea the sequences were selected for characterization and converted to amino acid sequences to identify the conserved regions which are given Figure S1.

DESATURASE

The Degenerate primers were designed by manually, webprimer and codehop methods which are given in **TABLES 1 AND 2**.

Table 2: Homology results of Desaturase.

Accession	Organism	Max score
AB 262185.1	*Phanerochaete chrysosporium* Pc-ole1 gene for delta9-fatty acid desaturase, complete cds	87.8
AY500378.1	*Lilium longiflorum* phytoene desaturase mRNA, partial cds	87.8
AY 234124.1	*Primula farinosa* sphingolipid delta-8 desaturase mRNA, complete cds	41.0
EU 424174.1	*Sorghum bicolor* fatty acid desaturase mRNA, complete cds	31.9

RESULTS

Three sets of gene specific degenerate primers (manual, Web primer, Codehop) synthesized by Biotech Desk, India Pvt. Ltd.) were tested for PCR amplification by gradient PCR with one sesame sample **FIGURE 1** The amplified products of each primer is eluted and transformed and isolated plasmids from selected colonies are sequenced. A total of 10 positive clones of which 6 clones were sequenced. The sequences obtained was subjected to BLAST search.

Lanes 1-11: *S. indicum* DNA amplified with Codehop designed primer using a gradient from 45 -55°C with a 0.9°C raise at each lane

Lanes 12-21: *S. indicum* DNA amplified with Web designed primer using a gradient from 45 -55°C with a 0.9°C raise at each lane

Lanes 22-31: *S. indicum* DNA amplified with manually designed primer using a gradient from 45 -55°C with a 0.9°C raise at each lane

LaneM: 100 bp DNA ladder

Figure 1: Gradient PCR amplification with degenerate primers for desaturase.

Figure 1: Gradient PCR amplification with degenerate primers for desaturase.

SEARCH RESULTS OF DESATURASE IN THE NCBI FOR HOMOLOGS

The Six sequenced desaturase sequences were submitted to the NCBI (National Centre for Biotechnology Information) website for BLAST-N (http://www. ncbi.nlm. nih.gov/ BLAST/) which showed high level of sequence homology both at the nucleotide and amino acid level with those from *P. crysosporium, L. longiflorum, P. farinose* and *S. bicolor* as shown in **FIGURE 2**.

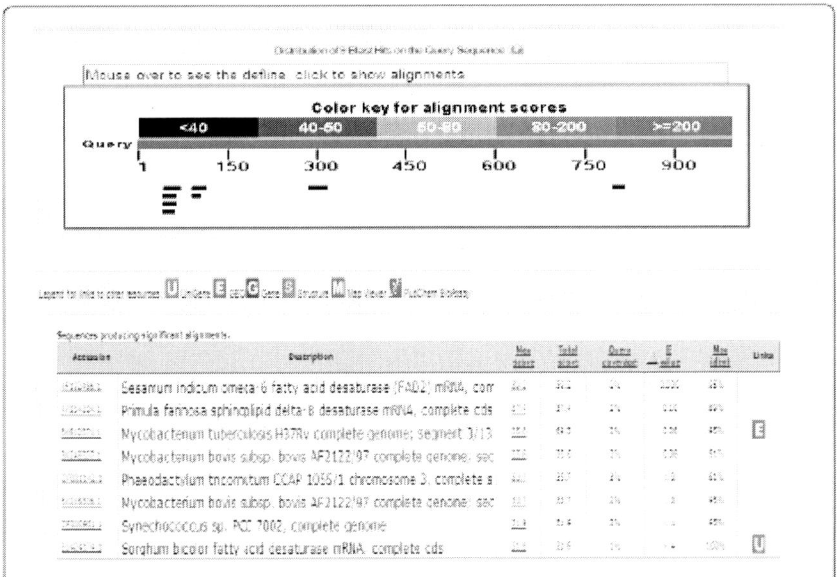

Figure 2: Blast results with P-11_M13-20F sequence.

Figure 2: Blast results with P-11_M13-20F sequence.

Most of the sequences showed high homology with Sesamum indicum omega-6 fatty acid desaturase (FAD2) mRNA, complete cds and also with *Phanerochaete chrysosporium* Pc-ole1 gene for delta9-fatty acid desaturase, complete cds, *Lilium longiflorum*phytoene desaturase mRNA, partial cds, *Primula farinosa* sphingolipid delta-8 desaturase mRNA, complete cds *Mortierella sp.* strain M10 mRNA for delta-9 fatty acid desaturase, partial *Sorghum bicolor* fatty acid desaturase mRNA, complete cds, *Chlamydomonas reinhardtii* fatty acid desaturase (CHLREDRAFT_32523) mRNA, partial cds.

A comparison of Fatty acid desaturase sequences showing phylogenetic relationship at the deduced amino acid level with those of other oilseed crops is shown in the dendrogram (**FIGURE 3**). The homology was based on small portion of entire coding sequence, as the sequence length of the obtained sequence was 181 to 538 bp, where as the length of the selected sequences was 1149 to 1095 bp.The expected DNA fragment range was 400 bp and the primers were designed excluding N terminal region with C terminal region only as N terminal region was having more INDELS (Insertions and deletions).

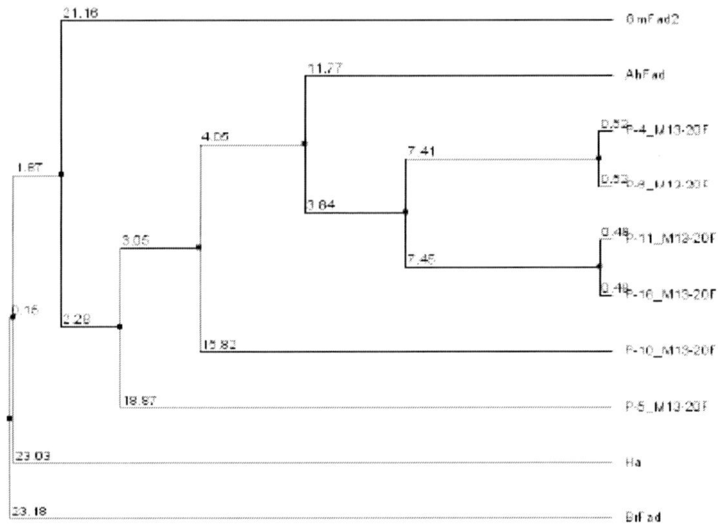

Figure 3: Dendrogram showing relationship of the derived amino acid sequence of desaturase of sesame with various oil crops.

Figure 3: Dendrogram showing relationship of the derived amino acid sequence of desaturase of sesame with various oil crops.

A sequence homology search revealed that the putative protein had high homology with the desaturase sequences of *Arachis hypogaea* indicating that the desaturase isolated from *S. indicum* was closer to A. hypogaea than the other oil crops compared.

DISCUSSION

Degenerate primers based on the amino acid sequence of conserved regions were also used to search for members of a gene family [2]. Computer programs have also been developed specifically for degenerate primer design [3].

Bioinformatics approaches to lipid research have recently begun using large amounts of mass spectrometry and microarray data. Phylogenetic analysis of protein families related to fatty acids has also been performed. However, to our knowledge, there is no report that describes the investigation of fatty acid structures based on the comprehensive analysis of the gene contents in the genomes. The fatty acid structures is hard to predict from genomic information due to the the functional diversity of the key enzymes. Many studies in plant systems were also conducted, which eventually led to the purification of a soluble stearoyl-acyl carrier protein (ACP) 9-desaturase from safflower [5]. Later studies allowed for the full nucleotide sequence of stearoyl- ACP 9-desaturases from castor and cucumber to be reported [6]. Various genes encoding enzymes involved in fatty acid synthesis have been isolated from the species and characterized. Carotene desaturase gene was cloned and characterized from chlorella protothecoides cs- 41 to elucidate lutein biosynthesis pathway [7]. Cloning of Psy1, Pds and Zds cDNAs encoding the enzymes responsible for lycopene biosynthesis, namely phytoene synthase 1 (PSY1), phytoene desaturase (PDS) and ζ -carotene desaturase (ZDS), respectively, from high lycopene tomato cultivar, *Solanum lycopersicum* KKU-T34003 [8]. Although the pathway for sesame is not documented, the fatty acid profile suggests synthesis via the known route common to most major oil crops Desaturase, This enzyme in fatty acid biosynthesis in sesame appear to be showing homology to different oilseed helianthus and arachis. The reasons are unknown but a more detailed study of these enzymes across the different oilcrops will through some light on the fatty acid biosynthesis pathway.

In the present study after analysis of available sequences of desaturase related proteins, retrieved desaturase subgroup clusters and selected six representative sequences of most members of desaturase families which specifically belonged to dicots as *Sesamum* is a eudicot species. In the present study codehop and Webprimer designed primers showed amplification for desaturase whereas manual designed primer gave no amplification. Desaturase derived sequences showed high homology with desaturase genes of *Phanerochaete, lillium, primula sp.*

REFERENCES

1. Kamel A, Abd-Elsalam A (2003) Bioinformatic tools and guideline for PCR primer design. African Journal of Biotechnology 2: 91-95.
2. Wilks AF, Kurban RR, Hovens CM, Ralph SJ (1989) The application of the polymerase chain reaction to cloning members of the protein tyrosine kinase family. Gene 85: 67-74.
3. Chen H, Zhu G (1997) Computer program for calculating the melting temperature of degenerate oligonucleotides used in PCR or hybridization. Biotechniques 22: 1158-1160.
4. Thompson JD, GibsonTJ, Plewniak F, Jeanmougin F, Higgins DG (1997) The CLUSTAL X windows interface: flexible strategies for multiple sequence alignment aided by quality analysis tools. Nucleic Acids Research 25: 4876-4882.

5. McKeon TA, Stumpf PK (1982) Purification and characterization of the stearoyl-acyl carrier protein desaturase and the acyl-acyl carrier protein thioesterase from maturing seeds of safflower. Journal of Biological Chemistry 257: 1214-12147.
6. Shanklin J, Sommerville CH (1991) Stearoyl- acyl-carrier-protein desaturase from higher plants is structurally unrelated to the animal and fungal homologs. Proc Natl Acad Sci88: 2510-2514.
7. Meiya Li, Zhibing Gan, Yan Cui, Chunlei Shi, Xianming Shi (2011) Cloning and Characterization of the ζ-Carotene Desaturase Gene from Chlorella protothecoides CS-41. Journal of Biomedicine and Biotechnology 2011: 731542.
8. Krittaya S, Preekamol K (2013) cDNA cloning and expression analyses of phytoene synthase 1, phytoene desaturase and ζ -carotene desaturase genes from Solanum lycopersicumKKU-T34003 Songklanakarin. Journal of Science and Technology 35: 517-527.

CHAPTER 14

Enhanced Resistance to Bacterial and Fungal Pathogens by Overexpression of a Human Cathelicidin Antimicrobial Peptide (Hcap18/LL-37) in Chinese Cabbage

Yu-Jin Jung[1],Soon-Youl Lee[2],Yong-Sun Moon[3],Kwon-Kyoo Kang[1]

[1].*Department of HorticultureHankyong National UniversityAnsungSouth Korea*

[2].*Korea School of Applied Life Sciences Genetic Informatics CenterGRRC Hankyong National UniversityAnsungSouth Korea*

[3].*Department of HorticultureYeungnam UniversityGyeongsanSouth Korea*

ABSTRACT

The human cathelicidin antimicrobial protein hCAP18, which includes the C-terminal peptide LL-37, is a multifunctional protein. As a possible approach to enhancing the resistance to plant disease, a DNA fragment coding for hCAP18/LL-37 was fused at the C-terminal end of the leader sequence of endopolygalacturonase-inhibiting protein under the control of the cauliflower mosaic virus 35S promoter region. The construct was then introduced into *Brassica rapa*. LL-37 expression was confirmed in transgenic plants by reverse transcription-polymerase chain reaction and western blot analysis. Transgenic plants exhibited varying levels of resistance to bacterial and fungal pathogens. The average size of disease lesions in the transgenic plants was reduced to less than half of that in wild-type plants. Our results suggest that the antimicrobial LL-37 peptide is involved in wide-spectrum resistance to bacterial and fungal pathogen infection.

Keywords: Antimicrobial peptide Cathelicidin Disease resistance Transgenic Chinese cabbage

INTRODUCTION

Bacterial and fungal plant pathogens severely affect crop productivity. For example, *Xanthomonas campestris* pv. *campestris* and *Pectobacterium carotovorum* subsp. *carotovorum*, which cause black rot and soft rot, respectively, are present worldwide and severely damage plants and reduce their yields, especially in cruciferous plants (Boman 2003). Therefore, the development of cruciferous plants that are resistant to black and soft rot diseases has been a major goal of researchers for several decades. Strategies based on transgenic approaches to enhance plant disease resistance involve the use of genes associated with plant defense pathways (Makandar et al. 2006; Zhang et al. 2007) and genes encoding plant or fungal hydrolytic enzymes (Bieri et al. 2003), defense-related transcription factors (Chen and Chen 2002; Sohn et al. 2006) and antimicrobial peptides (Alan et al. 2004).

A large number of antimicrobial peptides from different organisms have been characterized (Simmaco et al. 1998). The human cathelicidin antimicrobial protein hCAP18 is the only member of the mammalian cathelicidin family of proteins that is present in humans (Gudmundsson et al. 1996). The holoprotein consists of a conserved prodomain, a cathelin domain, and the non-conserved C-terminal peptide LL-37, which is enzymatically cleaved after secretion (Sorensen et al. 2001; Yamasaki et al. 2006). Its precursor molecule, an 18 kDa human cationic antimicrobial protein (hCAP-18), is secreted by activated neutrophil granulocytes. After release, the helical C-terminal end of this precursor comprising 37 amino acids is cleaved off, thereby forming the functional antimicrobial peptide LL-37 (Sorensen et al. 2001). Since LL-37 is the only human antimicrobial peptide that is active at physiological or elevated salt concentration conditions, there is a significant interest in using this peptide for pharmaceutical applications (De Smet and Contreras 2005; Reddy et al. 2004; Travis et al. 2000).

In the present study, we report the transgenic expression of human cathelicidin antimicrobial peptide carrying the substitution Met37Leu in Chinese cabbage. The expression of this peptide in cabbage plants significantly inhibited the growth of *Pectobacterium carotovorum* subsp. *carotovorum* on the plant leaves, and it conferred resistance to several fungal pathogens. These results further support the assignment of a defense role to LL-37 and highlight its plant biotechnological potential.

MATERIALS AND METHODS

Expression vector construction

The leader sequence of the gene encoding *Phaseolus vulgaris* endopolygalacturonase-inhibiting protein (PGIP) (GenBank Accession No. X64769) was fused upstream of an LL-37-coding DNA fragment to cause the extracellular localization of the mature protein. The PGIP signal peptide (87 bp) was amplified by PCR with primers A linked with *Bam*HI site (5′-CCGGATCCATGACTCAATTCAATATCCCA-3′) and B (5′-AGAGAGT GCAGTTCTCAA-3′). The coding region of *LL-37* (111 bp) was substituted with Met-LL37-Leu and amplified by PCR from *pFALL37* DNA using primers C (5′-ATGCTGCTGGGTGATTTCTTC-3′) and D with *Sac*I site (5′-CGAGAGCTCCTAGGACTCTGTCC TGGG-3′). The two products were ligated into pBlueScript-SK (Stratagene, La Jolla, CA, USA) at the *Bam*HI and *Sac*I restriction sites. The generated *LL-37* fragment was further amplified using primers A and D. For *Agrobacterium* transformation, the PCR product was subcloned into pBI121 binary vector driven by cauliflower mosaic virus 35S (CaMV35S) promoter (Gelvin 1998). The Ti plasmid vector construct pBI–LL37 was confirmed by DNA sequencing (ABI 377 DNA sequencer; Perkin-Elmer, Cypress, CA, USA).

Plant transformation and regeneration

The prepared construct was transformed into Chinese cabbage using the protocol described in Min et al. (2007). A total of 168 hypocotyls from in vitro grown seedlings of Chinese cabbage (*Brassica rapa* cv. Osome) were inoculated with *Agrobacterium tumefaciens* strain LBA4404 carrying either pBI-LL37 or pBI121. Green shoots that developed in the selective medium were transferred to a rooting medium containing

100 mg L^{-1} kanamycin and 500 mg L^{-1}carbenicillin. Rooted shoots were screened by PCR for the presence of the transgene before transfer to plastic pots.

Estimation of transformants and generation of homozygous lines

Self-pollinated seeds obtained from T_0 plants were sown into plastic pots in the greenhouse. Two weeks after germination, seedlings were sprayed with 400 mg L^{-1} kanamycin in water; they were sprayed again 2 days later. Three days after the second spray, the ratios of green seedlings to bleached seedlings were determined, and the results were analyzed by a Chi-square test for goodness of fit to the ratios 3:1, 15:1, or 63:1. In order to obtain transformants homozygous for the *LL-37* gene, kanamycin-resistant T_1 progenies were grown to produce selfed T_2 seeds. T_2 lines that had no bleached segregants after kanamycin sprays were assumed to be homozygous for the *LL-37* and *nptII* genes.

Molecular analysis of transformants

PCR analyses were conducted to detect the presence of *LL-37*- or *nptII*-specific fragments. Primers 35SF (5′-TCCACTGACGTAAGGGATGA-3′) and LL-37R (5′-CGAGAGCTCCTAGGACTCTGTCCTGGG-3′), which amplified a fragment of size approximately 750 bp, including sequences from the 3′ end of the 35S promoter, signal peptide (SP), and part of the *LL-37* gene, were used to screen for putative *LL-37* transformants. Putative transformants were screened using primers *nptII*F (5′-TCGGCTATGA CTGGGCACAACAGC-3′) and *nptII*R (5′-AAGA AGGCGATAGAAGGCGATGCG-3′), which amplified a 722-bp *nptII*-specific fragment. Genomic DNA was isolated from young leaves of Chinese cabbage plants using a DNeasy Plant kit (Qiagen, Germantown, MD, USA.) as per the manufacturer's instructions. PCR was performed and the reaction conditions were followed (1 cycle of 94°C for 1 min; 30 cycles of 94°C for 30 s, 55°C for 30 s, and 72°C for 1 min; and 1 cycle of 72°C for 10 min). The reaction products were electrophoresed on a 2% (w/v) agarose gel with 1× TAE buffer and visualized by staining with ethidium bromide.

In order to analyze gene expression in transgenic plants by RT-PCR, total RNA from wild-type and transgenic plants was reverse transcribed using AMV reverse transcriptase (Roche, USA) with oligo (dT) primers for 1 h at 42°C. The expression level of actin mRNA was used as a quantitative control.

Western blot analysis was performed by following standard molecular techniques (Sambrook et al. 1998). Briefly, for western blotting, 200 ng of purified protein or synthetic peptide was electroblotted onto a polyvinylidene difluoride (PVDF) membrane. Blots were probed with 1:1,000 dilution of polyclonal Hbt Ab antiserum raised in rabbit and then with 1:10,000 dilution of goat anti-rabbit immunoglobulin-peroxidase conjugate (Vector, Burlingame, CA, USA). The blots were then developed using the enhanced chemiluminescence (ECL) developing system (GE Health Care, USA). The Low Range BioRad 161-0304 markers (BioRad, Hercules, CA, USA) were used as the molecular size markers.

Pathogen inoculation

P. carotovorum subsp. *carotovorum* KACC 10057 obtained from the Korean Agricultural Culture Collection (http://kacc.rda.go.kr) was grown in Luria–Bertani (LB) medium until the absorbance at 600 nm (A_{600}) was 0.2, which corresponds to a concentration of approximately 2×10^8 CFU/mL. Three different concentrations of the culture (10^4, 10^6, and 10^8 CFU/mL) with 10 mM $MgCl_2$ were inoculated by syringe infiltration. The inoculated plants were transferred to a growth chamber and incubated at 28°C under continuous light. They were examined for 12–96 h after inoculation. Lesion length (cm) and disease index (DI) were recorded for each individual plant; the disease index ranged from 0 to 6 on the basis of the development of the disease lesions: 0, no lesion; 1, lesion size 0.1–0.5 cm; 2, 0.5–1.5 cm; 3, 1.5–3.5 cm; 4, 3.5–5.5 cm; 5, 5.5–8.5 cm; and 6, over 8.5 cm or plant dead.

The fungi *Fusarium oxysporum* f. sp. *Lycopersici* (KACC 40032), *Colletotrichum higginsianum* (KACC 40807), and *Rhizoctonia solani* (KACC 40107) were inoculated on plant leaves by placing 10 μl of an aqueous suspension containing 10^6 spores/mL on the leaves. Plants were maintained in highly humidified conditions (100% RH) at 25°C with 16 h of light in a growth chamber.

Evaluation of in vitro inhibition assays

Total and extracellular fluids were extracted by the methods described in Alan et al. (2004). Protein concentrations in the leaflet fluids were determined by the Bradford assay (Bradford 1976). The in vitro inhibition assays were evaluated by the protocol from Alan et al. (2004). Briefly, the experiments tested whether total fluid (TF) and extracellular fluid (EF) from three homozygous lines possessed antimicrobial activity. Assays were also performed using TF and EF from wild-type plant and LB medium as controls. A volume of 248 μL of TF, EF, and LB was mixed with

2.5 μL of 10^8 CFU/mL *P.carotovorum* subsp. *carotovorum* in Eppendorf tubes and incubated on a shaker for 4 h. The samples remaining in the tubes were mixed in the ratio 1:9 with LB, returned to the shaker, and incubated overnight at 37°C. The bacterial growth in these tubes was determined by a spectrophotometer at 600 nm.

RESULTS

Generation and characterization of *LL-37* transgenic plants

The sequence of mature peptide of Met-LL37-Leu (GenBank accession No. NM-004345) is shown in Figure 1a and the construct of binary pBI121 in Figure 1b. Morphological characteristics of non-transformed Chinese cabbage (wt) and transgenic plants are shown in Figure 1c. Transformation experiments with *Agrobacterium* carrying pBI-LL37 yielded 17 independent kanamycin-resistant putative transformants. Three out of 17 transformants showed abnormal phenotypes and eliminated for further experiment. PCR analysis with 35SF and LL-37R primers revealed that 11 of 14 transformants contained the expected amplified product (Figure 2a). The PCR-positive plants (T0) were transferred to the greenhouse where they were further observed for their phenotypic characters and were grown for recovery of self-pollinated seeds. Most transformants were phenotypically similar to non-transformed plants. Segregation analysis was indicated that T1 transformants with a single copy showed 3:1 kanamycin resistance. Four of them were randomly selected for selfing, and named lines B11, B12, B13, and B14 (T2). Ten plants per line after selfing were screened on kanamycin to get homozygotics (T3). Additionally, we confirmed the integration of the genes in four homozygous lines (T3; B21, 22, 23, and 24 from T2; B11, 12, 13, and 14) by PCR using the 35S-specific, *nptII*-specific, and *LL-37*-specific primer sets (Figure 2b). All of these contained the expected amplified product.

Expression of *LL-37* in homozygous lines

In order to confirm *LL-37* expression in the transformants, we performed quantitative RT-PCR using total RNA from 4 homozygous lines. The results of RT-PCR indicated that LL-37 RNA was expressed in the homozygous lines, and the expression levels were similar across all the transformants (Figure 3). Western blot analysis also confirmed that the four homozygous lines expressed the 4-kDa peptide at varying levels, whereas the control plants did not show the LL-37-specific band (Figure

4). Therefore, we concluded that the antimicrobial *LL-37* gene was stably integrated into the genome of the transformants and was transcribed into mRNA that yielded the LL-37 peptide.

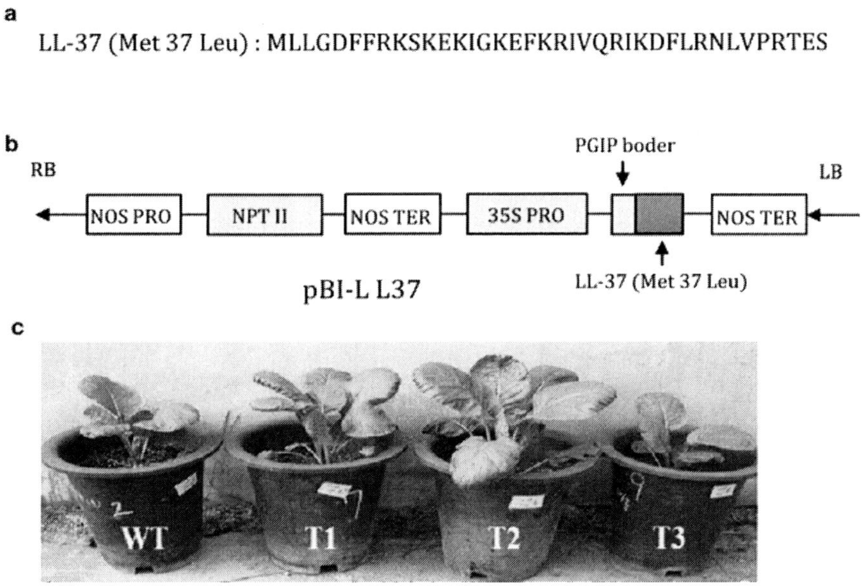

Figure 1. Construction of a binary plant expression vector pBI121 for the transformation of Chinese cabbage plants. **A** Sequence of the peptide (GenBank Accession No. NM-004345). **B** Schematic diagram of the expression construct pBI-LL37. **C** Morphological characteristics of non-transformed Chinese cabbage (wt) and transgenic plants

Increased resistance of the transgenic plants to soft rot

The resistance to several rot pathogens was evaluated for the four homozygous lines harboring the human *LL-37* gene by comparing the severity of the disease to that seen with the nontransgenic plants. Transgenic and control plants were inoculated with conidia, and the sizes of the disease lesions were determined. In order to test the resistance of the transgenic plants, we first cultivated the cabbage soft rot pathogen *P.carotovorum* subsp. *carotovorum* in LB medium for 1–2 days. We tested the resistance of the plants to three different concentrations of the bacterial pathogen: 10^4, 10^6, and 10^8 CFU/mL. We challenged the leaf body and leaf vein with 10^4 and 10^6 CFU/mL of the bacterial culture, respectively; subsequently, the pathogenesis at the leaf was examined after 24, 48, and 72 h of inoculation. The control cabbage plants were susceptible to the bacterial pathogen and their leaves became softer with visible lesions 12 h after inoculation. After 48 h, the symptom spread all over the leaves, and the plants died owing to softening 72 h after inoculation of the pathogen. On the other hand, the softening symptom was not observed in three of the four transgenic lines, B21, B22, and B24, by 72 h after bacterial inoculation (Figure 5). When 10^8 CFU/mL of the bacterial culture was inoculated, the leaves of the control plants showed the softening symptom 12 h after inoculation; the condition worsened and the leaves became amorphous after 24 h of inoculation. The transgenic lines only showed mild softening of the leaves around the area of inoculation (Figure 6). Therefore, it could be concluded that transgenic plants showed significantly increased resistance to the bacterial pathogen *P. carotovorum subsp. carotovorum*, which causes soft rot.

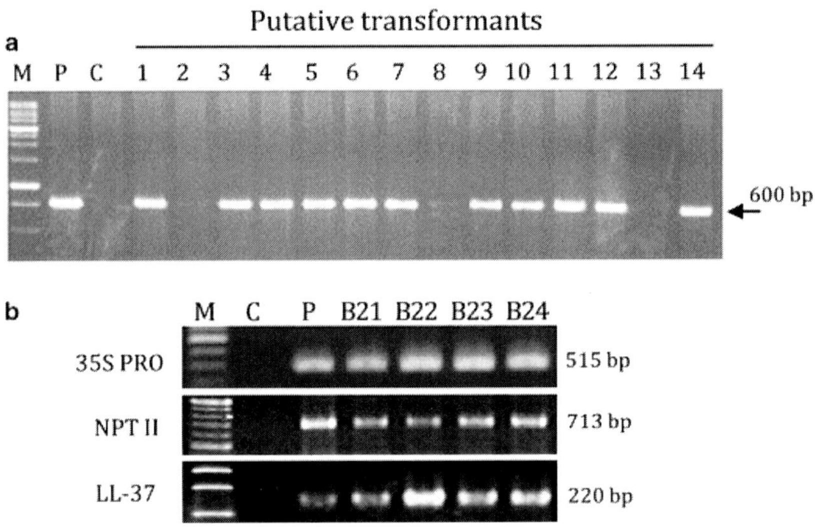

Figure 2.. PCR amplification of the foreign genes in the transgenic Chinese cabbage lines. **A** PCR analysis of 14 putative LL-37 Chinese cabbage lines (T0) using the 35SF (5′-ATGGAGTCAAAGATTCAAATAGAG-3′) and LL-37R (5′-CGAGAGCTCCTAGGACTCTGTCCTGGG-3′) primers. *Lanes*: *M* 1 kb ladder, *P* positive control (pBI LL-37), *C* non-transformed plant, *1–14* putative transformants. **B** Four T3 (B21, 22, 23, 24) of single transgenic homozygous lines from T2 (B11, 12, 13, 14) were reconfirmed by PCR after genetic segregation events by kanamycin resistance. Amplification products of the 35S promoter, *NptII*, and *LL-37* were separated using a 1.5% agarose gel. *Lanes*: *M* DNA ladder, *C* non-transformed plant, *P* pBI*LL-37*, *B21–B24* independent T3 transgenic homozygous lines

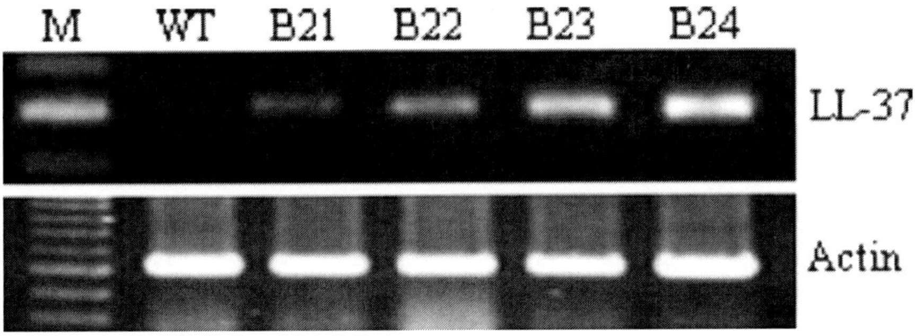

Figure 3. Analysis of the *LL-37* gene expression in the transgenic homozygous lines by using reverse transcription polymerase chain reaction. Total RNA was isolated from each plant, and 0.5 μg of this RNA was amplified with LL-37-specific primers. The Chinese cabbage actin gene was amplified as a loading control. The amplification products of the *LL-37* gene were separated using a 1.5% agarose gel. *Lanes*: *M* DNA ladder, *WT* wild-type plant, *B21–B24* independent transgenic homozygous lines

Figure 4. Western blot analysis of transgenic homozygous lines. *Lanes*: *WT* wild-type plant, *B21–B24*, transgenic homozygous lines containing the *LL-37* gene, *arrow* indicates 4 kDa protein

Increased resistance of the transgenic plants to fungal pathogens

The leaves of the transgenic plants were challenged with several fungal pathogens such as *F. oxysporum, C. higginsianum,* and *R. solani* in order to test the resistance of the transgenic plants to fungal pathogens. Three of the four transgenic homozygous lines, B21, B22, and B23, showed higher resistance to *F. oxysporum* than the control plants (Figure 7). In particular, B21 showed remarkably high resistance to *F. oxysporum.* After being challenged with *C. higginsianum* and *R. solani*, the control plants showed typical symptoms and died after 6 days of inoculation, while the transgenic plants were resistant, as shown in Figure 7. Therefore, we can conclude that the LL-37 peptide expressed in the transgenic plants acts as an effective antimicrobial peptide.

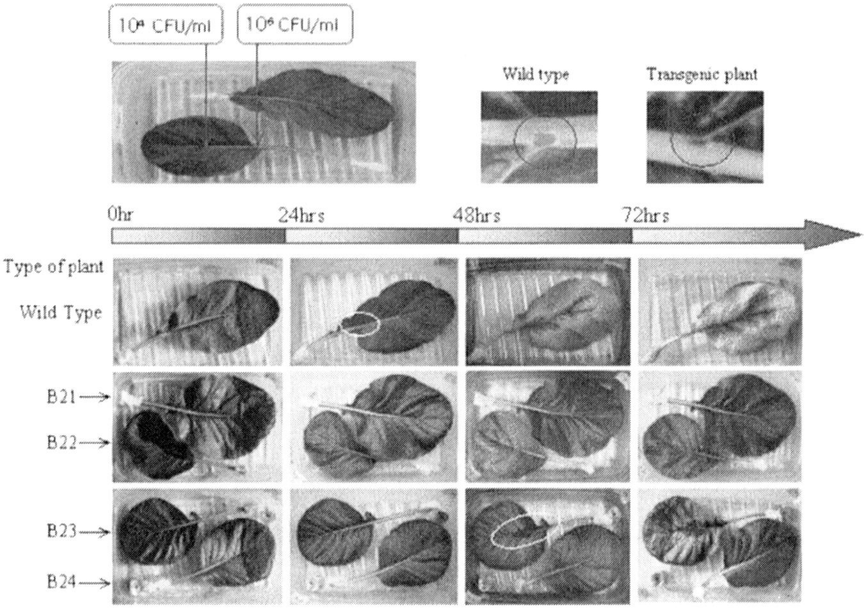

Figure 5. Temporal development of soft rot in Chinese cabbage ("Osome") plants when the leaves were inoculated with *Pectobacteriumcarotovorum* subsp. *carotovorum*. *B21–B24* Transgenic homozygous lines

Bacterial inhibition assays in transgenic leaflets

We collected TF and EF from the leaflets of three transgenic homozygous lines and a non-transformed line in order to study their effects on the activity of LL-37. The inhibition of the bacterial (*P.carotovorum* subsp. *carotovorum*) growth by TF and EF was evaluated using the spectrophotometric method. As expected, bacteria grew normally when they were incubated in LB medium only and in TF or EF obtained from the non-transformed line, while the growth was completely inhibited when EF obtained from the three transgenic lines was added to the LB medium (Figure 8). Bacteria also grew well in the TF obtained from the three transgenic lines, although their growth was lower than that in LB medium alone.

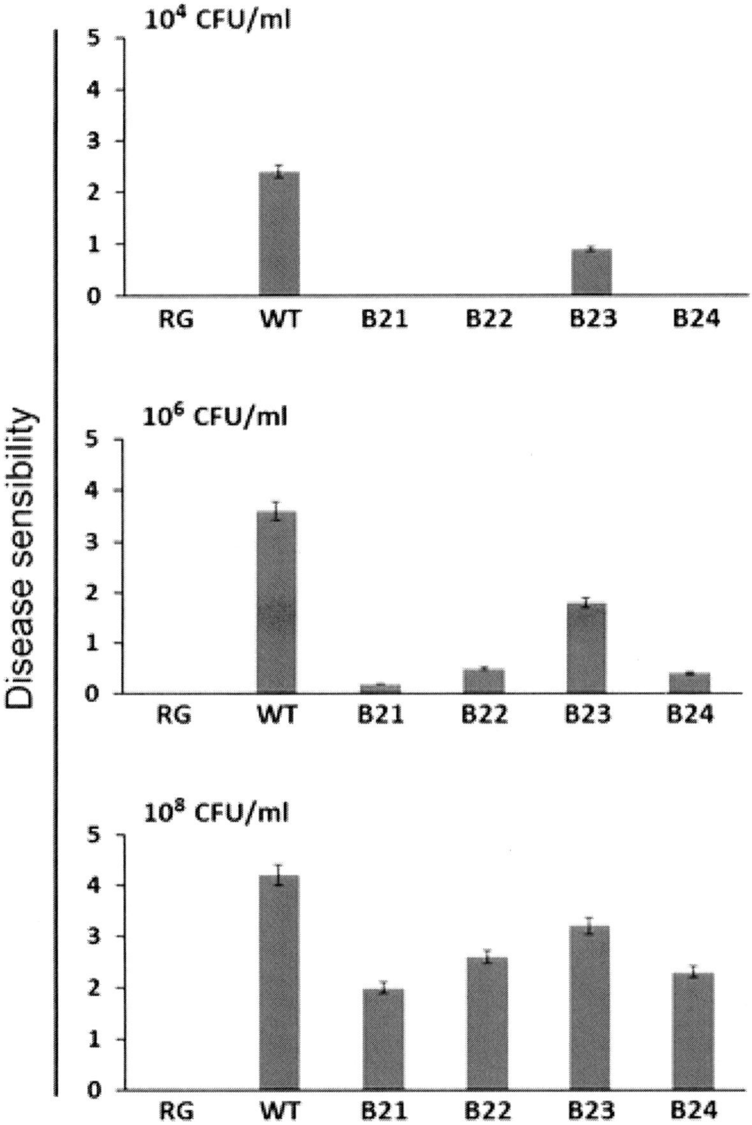

Figure 6. Disease development in transgenic homozygous lines containing the *LL-37* gene, which were inoculated with 10^4, 10^6, or 10^8 CFU/mL of *P. carotovorum* subsp. *carotovorum*. Disease manifestations were scored 3 days after inoculation. *RG* water, *WT* wild-type plant, *B21–B24* LL-37 transgenic homozygous lines. Disease lesions: *0*, no lesion; *1*, lesion size 0.1–0.5 cm; *2*, 0.5–1.5 cm; *3*, 1.5–3.5 cm; *4*, 3.5–5.5 cm; *5*, 5.5–8.5 cm; and *6*, over 8.5 cm or plant dead

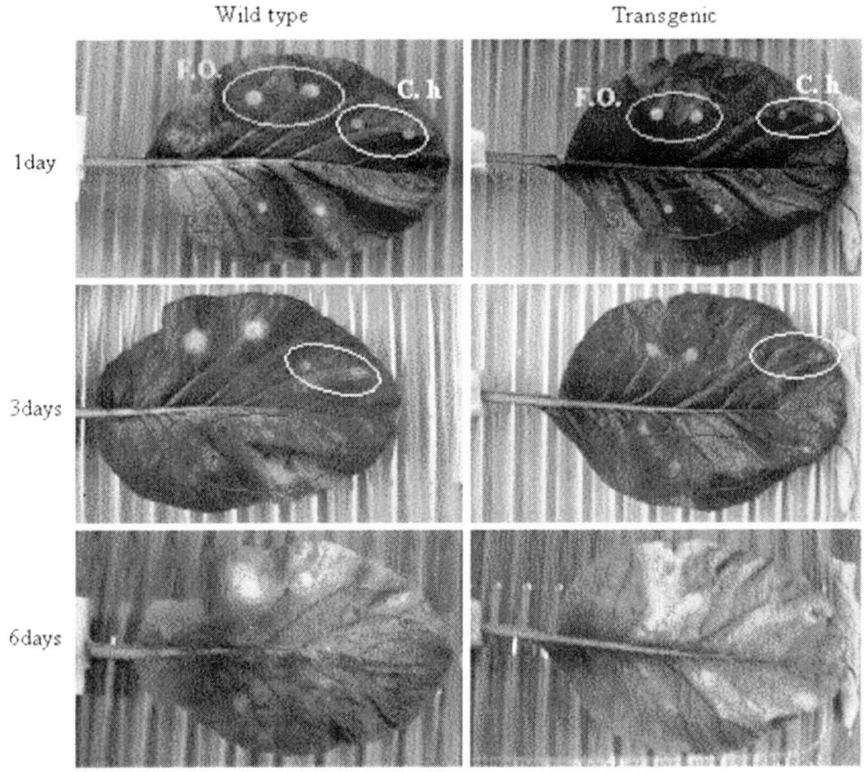

Figure 7. Enhanced disease resistance of transgenic homozygous lines to several fungal pathogens. F.O., *Fusarium oxysporum*; C.h., *Colletotrichum higginsianum*; R.S., *Rhizoctonia solani*. Disease manifestations assessed until 6 days after inoculation

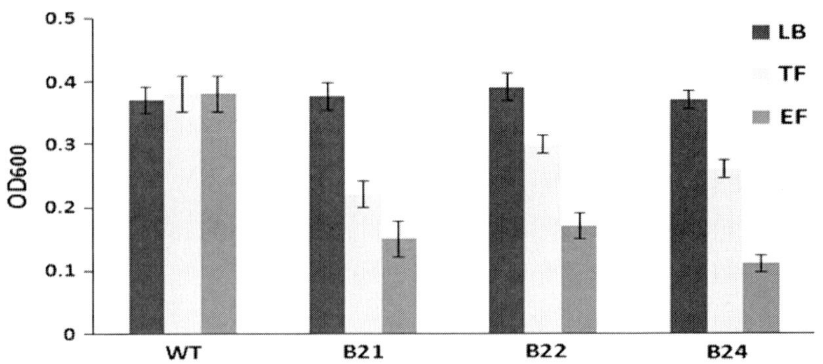

Figure 8. Bacterial inhibition assays of total fluid (*TF*) and extracellular fluid (*EF*) obtained from the leaflets of transgenic homozygous lines; *LB* Luria–Bertani medium

DISCUSSION

In response to a microbial attack, plants activate a complex series of responses that lead to the local and systemic induction of a broad spectrum of antimicrobial defenses (Kim and Martin 2004; Kunkel 2002). When induced defense responses are rapidly and coordinately triggered during a given plant–pathogen interaction, plants become broadly resistant to diseases. These defense responses include the strengthening of mechanical barriers, oxidative burst, and production of antimicrobial compounds (Hammond-Kosack and Parker 2003; Park 2005). Some research has been performed to bolster plant defenses against bacteria and fungi by genetically engineering plants to express antimicrobial peptides (Lee et al. 2008; Prasad et al. 2008).

In this study, we have chosen a variant of the antimicrobial peptide LL-37 as an interesting candidate for transgenic plant expression; this variant is designed to target the peptide into extracellular spaces. Extracellular targeting was originally intended to prevent possible deleterious effects of the peptide in plant cells. Moreover, secretion into extracellular spaces allows the plant-produced peptide to come into direct contact with pathogens growing and multiplying extracellularly before attacking the cells (Ponti et al. 2003). We developed homozygous Chinese cabbage lines stably expressing LL-37, which did not cause adverse effects on the plant phenotypes. These transgenic homozygous lines were tested with four important pathogens of this crop and found to inhibit the growth of the bacterial pathogen causing Chinese cabbage rot and also that of three fungal pathogens.

The results of our pathogenicity assays suggest that the expression of LL-37 provides a moderate level of resistance against a bacterial pathogen (*P. carotovorum* subsp. *carotovorum*) at the inoculum concentration of 10^4 CFU/mL. However, the extent of disease suppression provided by LL-37 expression was reduced as the inoculum concentrations were increased to 10^6 and 10^8 CFU/mL. Moreover, we observed reduction in the survival of *P. carotovorum* subsp. *carotovorum* cells incubated with EF from LL-37-expressing Chinese cabbage cell lines. However, bacteria grew normally when they were incubated in LB medium only and in TF or EF obtained from a non-transformed cell line. Therefore, leaves appeared to express LL-37 at sufficient levels in the extracellular spaces to retard bacterial multiplication and hence decrease disease severity. The functionality of this transgene and the presence of antimicrobial activity in the EF indicates that the LL-37 peptide was properly targeted to the extracellular space even with a foreign plant signal peptide. Furthermore, the mammalian peptide was not subjected to a processing step in the foreign plant cell environment that rendered it inactive.

Until now, studies involving the enhancement of resistance to various bacterial, fungal, and oomycete pathogens by the expression of antimicrobial peptides have been reported for rice, tobacco, poinsettia, banana, and more host species (Chakrabarti et al. 2003; Liang et al. 2002; Smith et al. 1998). However, progress on identifying the defense mechanisms in Chinese cabbage (*B. rapa*), an important vegetable crop in Asia, has been very slow. Previously, we reported an enhancement in the resistance to bacterial soft rot by the expression of the bromelain gene in Chinese cabbage (Jung et al. 2008). Here, we demonstrate that expression of the human LL-37 peptide has antimicrobial activity toward both bacterial and fungal pathogens of Chinese cabbage. Since pathogens have the ability to overcome gene-for-gene host defense mechanisms in the field by undergoing mutations in the cognate avirulence genes, any transgenic model based on such resistance-conferring genes may easily be evaded by pathogens. On the other hand, this problem is less likely to occur in transgenic plants overexpressing genes with more general antimicrobial activity. Therefore, the expression of the human LL-37 peptide is expected to confer durable resistance (i.e., field resistance) to a wide variety of pathogens infecting Chinese cabbage plants.

In order to obtain a high level of resistance as observed in the case of R gene-mediated resistance, the overexpression of multiple antifungal proteins with different functions may be necessary. In a *R. solani* infection assay, tobacco plants coexpressing the barley transgenes (a class II chitinase, a class II β-1,3-glucanase, and a type I ribosome-inactivating protein) were reported to elicit significantly enhanced protection against fungal attack as compared to that of the corresponding isogenic lines expressing a single barley transgene at a similar level (Brogue et al. 1991). Thus, other antimicrobial genes such as the bromelain gene (Jung et al. 2008) could be stacked with the human *LL-37* gene by crossing different transgenic lines and this may be able to strongly and durably inhibit the growth of pathogens.

ACKNOWLEDGMENTS

We thank to Dr. Philip Becraft for critical reviews and comments. This work was supported by a grant from the Next-Generation BioGreen 21 Program (No. PJ008085), Rural Development Administration, Republic of Korea.

Open Access

REFERENCES

1. Alan AR, Blowers A, Earle ED (2004) Expression of a magainin-type antimicrobial peptide gene (MSI-99) in tomato enhances resistance to bacterial speck disease. Plant Cell Rep 22:388–396
2. Bieri S, Potrykus I, Fütterer J (2003) Effects of combined expression of antifungal barley seed proteins in transgenic wheat on powdery mildew infection. Mol Breed 11:37–48
3. Boman HG (2003) Antibacterial peptides: basic facts and emerging concepts. J Intern Med 254:197–215
4. Bradford MM (1976) A rapid and sensitive method for the quantification of microgram quantities of protein utilizing the principle of protein-dye binding. Anal Biochem 72:248–254
5. Brogue K, Chet I, Holliday M, Cressman R, Biddle P, Knowlton S, Mauvais CJ, Broglie R (1991) Transgenic plants with enhanced resistance to the fungal pathogen Rhizoctonia solani. Science 254:1194–1197
6. Chakrabarti A, Ganapathi TR, Mukherjee PK, Bapat VA (2003) MSI-99, a magainin analogue, imparts enhanced disease resistance in transgenic tobacco and banana. Planta 216:587–596
7. Chen C, Chen Z (2002) Potentiation of developmentally regulated plant defense response by AtWRKY18, a pathogen-induced Arabidopsis transcription factor. Plant Physiol 129:706–716
8. De Smet K, Contreras R (2005) Human antimicrobial peptides: defensins, cathelicidins and histatins. Biotechnol Lett 27:1337–1347
9. Gelvin SB (1998) The introduction and expression of transgenes in plants. Curr Opin Biotech 9:227–232
10. Gudmundsson GH, Agerberth B, Odeberg J, Bergman T, Olsson B, Salcedo R (1996) The human gene FALL39 and processing of the cathelin precursor to the antibacterial peptide LL-37 in granulocytes. Eur J Biochem 238:325–332
11. Hammond-Kosack KE, Parker JE (2003) Deciphering plant-pathogen communication: fresh perspectives for molecular resistance breeding. Curr Opin Biotech 14:177–193
12. Jung YJ, Choi CS, Park JH, Kang HW, Choi JE, Nou IS, Lee SY, Kang KK (2008) Overexpression of the pineapple fruit bromelain gene (BAA) in transgenic Chinese cabbage (Brassica rapa) results in enhanced resistance to bacterial soft rot. Electron J Biotechnol 11:1–8
13. Kim YJ, Martin GB (2004) Molecular mechanisms involved in bacterial speck disease resistance of tomato. Plant Pathol J 20:7–12
14. Kunkel BN (2002) Brooks cross talk between signaling pathways in pathogen defense. Curr Opin Plant Biol 5:325–331
15. Lee SC, Hwang IS, Choi HW, Hwang BK (2008) Involvement of the pepper antimicrobial protein CaAMP1 gene in broad spectrum disease resistance. Plant Physiol 143:1004–1020

16. Liang H, Catranis CM, Maynard CE, Powell WA (2002) Enhanced resistance to the poplar fungal pathogen, Septoria musiva, in hybrid poplar clones transformed with genes encoding antimicrobial peptides. Biotechnol Lett 24:383–389
17. Makandar R, Essig JS, Schapaugh MA, Trick HN, Shah J (2006) Genetically engineered resistance to Fusarium head blight in wheat by expression of Arabidopsis NPR1. Mol Plant Microbe Interact 19:123–129
18. Min BW, Cho YN, Song MJ, Noh TK, Kim BK, Chae WK, Park YS, Choi YD, Harn CH (2007) Successful genetic transformation of Chinese cabbage using phosphomannose isomerase as a selection marker. Plant Cell Rep 26:337–344
19. Park JM (2005) The hypersensitive response: a cell death during disease resistance. Plant Pathol J 21:99–101
20. Ponti D, Mangoni ML, Mignogna G, Simmaco M, Barra D (2003) An amphibian antimicrobial peptide variant expressed in Nicotiana tabacum confers resistance to phytopathogens. Biochem J 370:121–127
21. Prasad BD, Jha S, Chattoo BB (2008) Transgenic indica rice expressing Mirabilis jalapa antimicrobial protein (Mj-AMP2) shows enhanced resistance to the rice blast fungus Magnaporthe oryzae. Plant Sci 175:364–371
22. Reddy KV, Yedery RD, Aranha C (2004) Antimicrobial peptides: premises and promises. Int J Antimicrob Ag 24:536–547
23. Sambrook J, Fritsch EF, Maniaatis T (1998) Molecular cloning: a laboratory manual, 2nd edn. Cold Spring Harbor Laboratory Press, Cold Spring Harbor, New York
24. Simmaco M, Mignogna G, Barra D (1998) Antimicrobial peptides from amphibian skin: what do they tell us? Biopolymers 47:435–450
25. Smith FD, Gadoury DM, Vaneck JM, Blowers A, Sanford JC, Van der Meij J, Eisenreich R (1998) Enhanced resistance to powdery mildew in transgenic poinsettia conferred by antimicrobial peptides. Phytopathology 88:S83
26. Sohn KH, Lee SC, Jung HW, Hong JK, Hwang BK (2006) Expression and functional roles of the pepper pathogen-induced transcription factor RAV1 in bacterial disease resistance, and drought and salt stress tolerance. Plant Mol Biol 61:897–915
27. Sorensen OE, Follin P, Johnsen AH, Calafat J, Tjabringa GS, Hiemstra PS, Borregaard N (2001) Human cathelicidin, hCAP-18, is processed to the antimicrobial peptide LL-37 by extracellular cleavage with proteinase 3. Blood 97:3951–3959
28. Travis SM, Anderson NN, Forsyth WR, Espiritu C, Conway BD, Greenberg EP, McCray PB, Lehrer RI, Welsh MJ, Tack BF (2000) Bactericidal activity of mammalian cathelicidin-derived peptides. Infect Immun 68:2748–2755
29. Yamasaki K, Schauber J, Coda A, Lin H, Dorschner RA, Schechter NM, Bonnart C, Descargues P, Hovnanian A, Gallo RL (2006) Kallikrein-mediated proteolysis regulates the antimicrobial effects of cathelicidins in skin. Fed Am Soc Exp Biol J 20:2068–2080
30. Zhang X, Dai Y, Xiong Y, DeFraia C, Li J, Dong X, Mou Z (2007) Overexpression of ArabidopsisMAP kinase kinase 7 leads to activation of plant basal and systemic acquired resistance. Plant J 52:1066–1079

CHAPTER 15

Cloning and Bioinformatics Analysis of Rosa Rugosa β-1,3-Glucanase Gene (RrGlu)

Yanan Fu[1*], Juanjuan Sun[1*], Yan Ma[1], Shutang Xing[2], Lanyong Zhao[1], Zongda Xu[1], Xiaoyan Yu[1#]

[1]*Shandong Provincial Research Center of Landscape Engineering Technology for Urban and Rural, Taian, China*

[2]*College of Horticulture Science and Engineering, Shandong Agricultural University, Taian, China*

ABSTRACT

In order to reveal which role the callose played in R. rugosa pollination incompatibility, the full- length cDNA sequence of β-1,3-glucanase gene was cloned for the first time from the stylus of Rosa rugosa "Tanghong" with RT-PCR and RACE methods and named as RrGlu. The full-length cDNA is 1380 bp with an open reading frame of 1041 bp, encoding 346 amino acids. The derived protein has a molecular weight of 37.85 kD, a calculated pI of 9.12, a pfam00332 conserved domain at position 36 - 345, and belongs to glycosyl hydrolase family 17. The derived protein is a hydrophilic protein secreted into the vacuole. There is a signal peptide cleavage site at position 34 - 35, a transmembrane domain at position 13 - 32, six Ser phosphorylation sites, three Thr phosphorylation sites, three Tyr phosphorylation sites, one N-glycosylation site, and five O-glycosylation sites. There are 31.50% α-helixes, 30.92% random coil, 25.14% extended peptide chain, and 12.43% β-corner struc- ture. This protein and the Glu protein from eight other species, including Prunus persica, share a sequence homology of greater than 72%; all of the proteins contain a pfam00332 conserved domain and a β-1,3-glucanase active center sequence (LIVM)-X-(LIVMFYW)3-(STAG)-E-(ST)-G-W-P-(ST)-X-G. Furthermore, their phylogenetic relationships are consistent with their traditional classifications. These results were meaningful to reveal the molecular mechanism of R. rugosa pollination incompatibility and improve the theory and techniques of breeding ornamental R. rugosa.

Keywords: Rosa rugosa, β-1,3-Glucanase Gene, Clone, Bioinformatics

1. INTRODUCTION

The Rosa rugosa is a famous traditional Chinese flower. It is fragrant as well as resistant to cold, drought, pest, disease, salt, and alkali [1] -[3] . However, due to its short bloom phase and uniform color and pattern, it has not been widely used in landscaping. In contrast, the Rosa hybrid belongs to the same species but comes in many varieties, has a year-round bloom, rich color, and ornate patterns. The hybridization of the traditional rose with the R. hybrid resulted in new varieties of roses that were aromatic, highly resistant, and highly ornamental. Nevertheless, the incompatibility between the two plants seriously hindered the breeding process [4] - [6] . In addition, the rose fruit is highly ornamental as well. However, its complete gametophytic self-incompatibility causes it to be very inconvenient for landscaping [7] . Thus far, the self-incompatibility mechanism has not been reported. Xiaoyan Yu et al. found that both the interspecific hybridization incompatibility between the R. rugosa and the R. hybrid as well as the self-incompatibility of the R. rugosa caused the pollen tube in the style to stop growing. In addition, the large amount of callose deposition at the top of the pollen tube and the intercellular space of the style channel may be the critical factors hindering the growth of the pollen tube [8] [9] . Furthermore, it is important to note that a similar phenomenon appeared in the outcrossing pollination of the Rhododendron, Solanum, Actinidia, and other Gramineae plants as well as the self-pollination of the Rosaceae, Solanaceae, and, Theaceae plants [10] -[16] . Therefore, callose may play a very important role in plant pollination incompatibility.

The hydrolysis and synthesis of callose are catalyzed by β-1,3-glucanase (Glu) and β-1,3-glucan synthase (Cals) [17] [18] , respectively. Glu is an important class of pathogenesis-related proteins that can be induced in pathological or related conditions, and Glu is involved in the development of a variety of plant growth processes, such as cell division, microsporogenesis, fruit ripening, and seed germination [19] -[23] .

According to previous studies, we speculate that the Glu gene is likely involved in regulating the affinity of plant pollination. However, at the present, Glu gene research primarily focuses on the disease resistance genetic engineering of plants [24] -[28] . Its regulation of pollination affinity has not been reported nor has any study of the R. rugosa Glu gene. Therefore, the aim of this study is to clone the R. rugosa Glu gene from R. rugosa styles and perform a bioinformatics analysis to establish a foundation for further exploring the role of callose in R. rugosa pollination incompatibility from the molecular point of view.

2. MATERIAL AND METHODS

2.1. Plant Material

The plant material, Chinese representative R. rugosa "Tanghong", was from the rose germplasm resources garden at Shandong Agricultural College. R. rugosa "Tanghong" is the most representative traditional rose in China.

2.2. Methods

2.2.1. Pollination Drawn

Between May 2015 and June 2015, the robust "Tanghong" buds were parchment isolated at 5:00-6:00 pm the day before blooming. At 6:00-7:00 am the next morning, self-pollination was performed, and the buds were parchment isolated when the anthers started to release pollen. Twelve hours after pollination, the styles were collected and flash frozen with liquid nitrogen and then stored in a −80°C freezer.

2.2.2. Total RNA Extraction and cDNA Synthesis

An EASYspin plant RNA Rapid Extraction Kit from Adlai Biotechnology Co., Ltd. was used to extract the total RNA from the R. rugosa style tissue. Agarose gel electrophoresis and spectrophotometer were used to determine the quality and concentration of the RNA. An EasyScript First-Strand cDNA Synthesis SuperMix Kit from Beijing TransGen Biotech Co., Ltd. was used to synthesize the first-strand cDNA.

2.2.3. Cloning of the Middle Fragment

According to the reported Glu sequences of Prunus persica, Prunus mume, Malus domestica, Malus hupehensis, and Pyrus bretschneideri, the degenerate primers F1 (5'-TACATYGCBGTWGGAAAYGAA-3') and R1 (5'-GGCCAACCRSTYTCSGATA-3') were designed with Primer Premier 5.0. PCR amplification was conducted using the synthesized cDNA in Section 2.2.2 as a template and F1 and R1 as the primers. The reaction system included 1 μL cDNA, 1 μL F1 primer (10 μmol/L), 1 μL R1 primer (10 μmol/L), and 12.5 μL PCR MIX, with ddH_2O added to a total volume of 25 μL. The reaction conditions were: 94°C for 3 min; 94°C for 30 s, 55°C for 30 s, and 72°C for 30 s for a total of 36 cycles; and then extension at 72°C for 10 min. Next, 1% agarose gel electrophoresis was used to detect the PCR products. The target PCR fragment was recovered with the MiniBEST Agarose Gel DNA Extraction Kit Ver. 3.0 (TaKaRa). The recovered fragment was ligated to the pMD18-T vector and then transformed into E. coli DH5a. The positive clones were selected and sent to BGI for sequencing.

2.2.4. 3' RACE and 5' RACE

The 3' RACE specific primers MG1 (5'-GCGCTGCTCGATCCCATTATACGCT-3') and MG2 (5'- CGATGCCATGTTGGACGCTGTGTAT-3') and the 5' RACE specific primers GSP1 (5'- CAGACTTGAAGGAACC-3'), GSP2 (5'-GTGTCGATGGCTGTGGAAAC-3'), and GSP3 (5'- CAGCATTGGAAATTGCGGTT-3') were all designed with Primer Premier 5.0. Nested PCR was conducted using MG-1, MG-2, and the SMARTer™ RACE cDNA Amplification Kit (Clontech) in order to obtain the 3'- terminal sequence of the target gene. Nested PCR was also conducted using GSP1, GSP2, GSP3, and the 5' RACE System for Rapid Amplification of cDNA Ends (Version 2.0, Invitrogen) in order to obtain the 5'-ter- minal sequence of the target gene.

2.2.5. Full-Length Gene Sequence Splicing and Verification

DNAstar software was used to splice the middle fragment, the 5'-terminal sequence, and the 3'-terminal sequence in order to obtain the full-length cDNA sequence of the gene. The 5'- and 3'-primers for the spliced sequence were designed with Primer Premier 5 as follows: F2 (5'-GCTCTAGAATGTCTAAATGCAATTCTTCAG-3') and R2 (5'-CGGGATCCATTGAAATTGATAGGGTATTTTGG-3'). The spliced sequence was amplified using the reverse transcription product of cDNA as a template, and then, it was further validated and verified.

2.2.6. Bioinformatics Analysis of Gene

BLASTX (NCBI) was used to study the homology of the nucleotide sequence and the deduced amino acid sequence. The ORF finder (NCBI) was used to search for an open reading frame, and the Conserved Domains database (NCBI) was used to analyze the conserved domains. The ProtParam Tool was used to analyze protein physical and chemical properties. Post Prediction, WOLF PSORT, and SubLocv were used to predict protein sub-cellular localization. Furthermore, ProtScale was used to predict hydrophilic or hydrophobic protein properties. The SignalP 4.0 Server was used to predict the protein signal peptide. The TMHMM Server v2.0 was used to predict the protein transmembrane domain. The NetPhos 2.0 Server was used to predict potential protein phosphorylation sites, and the NetNGlyc 1.0 Server and NetOGlyc 4.0 Server were used to predict potential protein glycosylation sites. ExPaSy-SOPMA was used to predict protein secondary structure. DNAMAN5. 2.2 was used to conduct multiple sequence alignment. The Neighbor-Joining method from Mega5 was used to create the phylogenetic tree.

3. RESULTS AND ANALYSIS

3.1. Cloning of the Rosa rugosa Glu Gene

The cloned middle fragment is 560 bp (Figure 1(a)), the cloned 3'-terminal fragment is 515 bp (Figure 1(b)), and the cloned 5'-terminal fragment is 591 bp (Figure 1(c)). These three fragments were spliced together with DNAstar in order to obtain a 1380 bp cDNA sequence. The spliced sequence was then validated by PCR amplification (Figure 1(d)). In addition, the Blast analysis confirmed that all its homologous genes are the Glu gene. Thus, the obtained gene is the R. rugosa β-1,3-glucanase gene, which was named RrGlu (GenBank accession number: KU144821).

3.2. Bioinformatics Analysis of the RrGlu Gene

The RrGlu gene has a full length of 1380 bp, an open reading frame of 1041 bp, a 5' UTR of 131 bp, and a 3' UTR of 208 bp, encoding 346 amino acids. The derived protein (the RrGlu protein) has a molecular weight of 37.85 kD, an isoelectric point of 9.12, a pfam00332 conserved domain at position 36 - 345, and a conserved β-1,3-glucanase motif LEIVVSDSGWPTAG in the activity center at position 264 - 277. Thus RrGlu protein

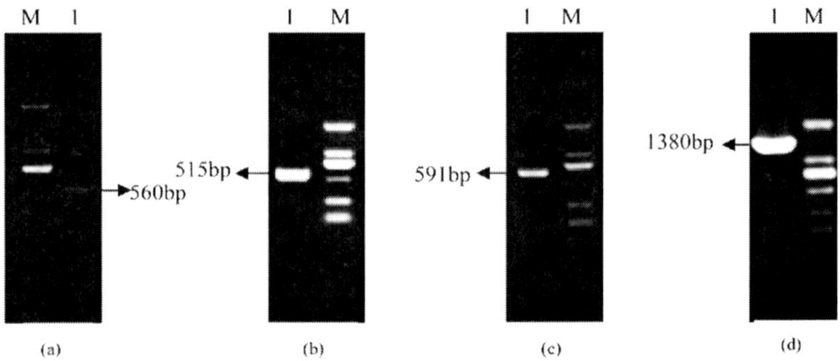

Figure 1. PCR amplification of β-1,3-glucanase cDNA. (a) Intermediate fragment; (b) 3'-RACE; (c) 5'-RACE; (d) Full-length fragment.

belongs to the glycosyl hydrolase family 17. Furthermore, the subcellular localization prediction result indicated that the protein is a secreted protein and is probably located at the vacuole. The hydrophilicity analysis further showed that the overall average hydrophobic index is −0.191, thus indicating a hydrophilic protein. The signal peptide prediction result demonstrated that a signal peptide cleavage site (ADA-QI) exists at position 34 - 35. The transmembrane domain analysis showed that a transmembrane domain exists at position 13 - 32. The phosphorylation site prediction results demonstrated that there are six Ser phosphorylation sites, three Thr phosphorylation sites, and three Tyr phosphorylation sites, thereby providing a reference for the future study of the regulation of gene expression and protein modification. The glycosylation site prediction results showed that there is one N-glycosylation site and five O-glycosylation sites. The secondary structure prediction result demonstrated that there is 31.50% α-helix, 30.92% random coil, 25.14% extended peptide chain, and 12.43% β-corner. The BLAST results showed that the protein shares 72% - 82% homology with the Glu amino acid sequences of Prunus persica (AAL30426.1), Prunus mume (XP 008240769.1), Malus hupehensis (ADR71671.1), Malus domestica (XP 008351633.1), Pyrus bretschneideri (XP 009363234.1), Morus notabilis (XP 010090235.1), Eucalyptus grandis (XP 010056683.1), and Citrus sinensis (CAA03908.1). The multiple sequence alignment result demonstrated that the RrGlu protein and the above plant Glu amino acid sequences all have a pfam00332 conserved domain and a conserved β-1,3-glucanase enzyme active site sequence (LIVM)-X-(LIVMFYW)3-(STAG)- E-(ST)-GWP-(ST)-XG (Figure 2). The above results further confirm that RrGlu indeed belongs to β-1,3-gluca- nase. Furthermore, the constructed phylogenetic tree revealed that RrGlu is closely related to Glu from the same family member Malus domestica, Malus hupehensis, and Pyrus bretschneideri, whereas it is relatively distant from

Eucalyptus grandis and Citrus sinensis, which are from different families, consistent with the traditional classification results (Figure 3).

4. DISCUSSION

4.1. Classification of β-1,3-Glucanase Gene

Nearly all higher plants contain β-1,3-glucanase genes, and they belong to a large gene family. At present, the Glu genes of dozens of plants have been cloned, and different species have different Glu genes [29] -[32] . Thus far, Glu genes can be divided into four types according to their isoelectric point, positioning, mRNA expression patterns, and sequence homology: type I Glu is a basic protein that is primarily located in the vacuole; type II Glu is an acidic protein that is primarily located in the intercellular space; type III Glu is also an acidic protein that is primarily located in the extracellular space and is an inducer releasing β-1,3-glucanase; and type IV Glu is an acidic protein as well that is primarily located in the extracellular space but is a non-inducible β-1,3-gluca- nase [33] -[36] . Our study revealed that the RrGlu gene encodes a derived protein (RrGlu) with an isoelectric point of 9.12 and is located in the vacuole. Therefore, the RrGlu may belong to type I Glu. However, it is necessary to further investigate the specific function of the RrGlu gene by transgenic experiments. This study has laid a foundation for the further exploration of the function of the RrGlu gene.

4.2. The Function of β-1,3-Glucanase Gene

β-1,3-glucanase gene is an important class of pathogenesis-related proteins that can be induced in pathological or related conditions. Previous studies showed that the plants of β-1,3-glucanase gene expression level is very low in normal circumstances. The gene expression increased, activity was significantly enhanced when it is induced by abiotic factors such as ethylene, cytokinins, mechanical damage and metal ions as well as subjected to biological factors, for example, pathogens, insect feeding and so on [37] . At the same time, β-1,3-glucanase gene is involved in the development of a variety of plant growth processes, such as cell division, microsporogenesis, fruit ripening, and seed germination.

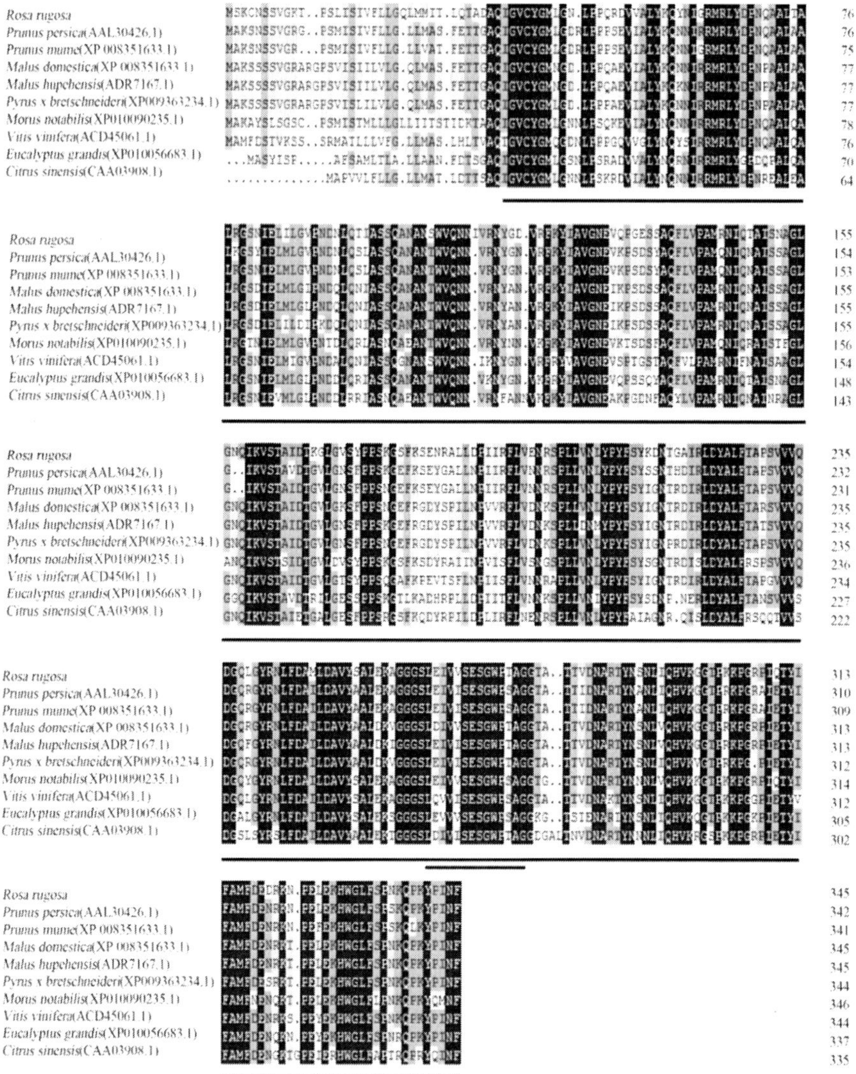

Figure 2. Multiple alignment of the RrGlu with other Glus. Notes: The underlined line is a conserved domain of RrGlu, and the double line is the common sequence of the enzyme active center of Glus.

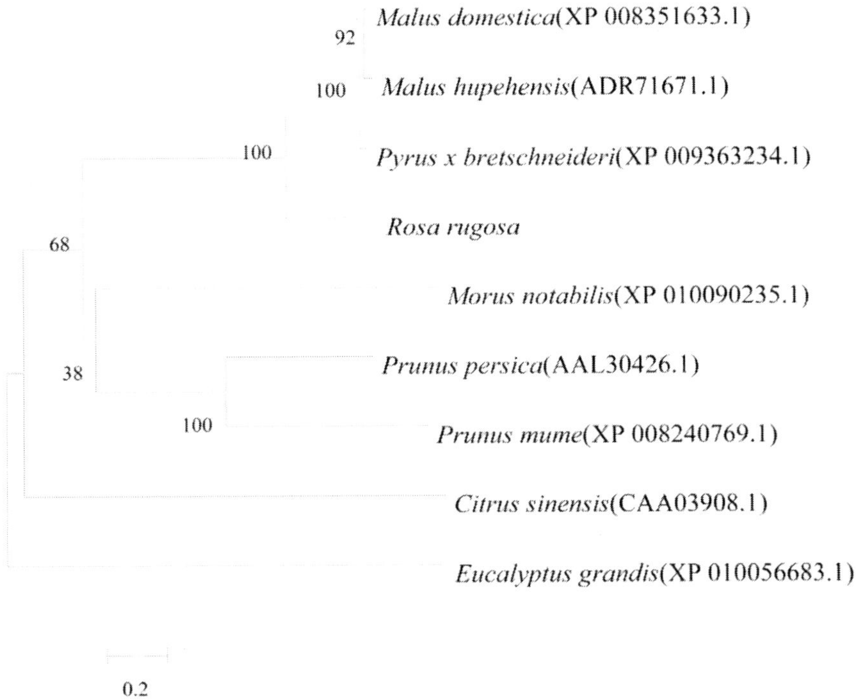

Figure 3. The phylogenetic tree derived from the alignment of amino acid secquences of RrGlu and other Glus.

For example, Arabidopsis anther specifically expressed β-1,3-glucanase gene expressed in advance can cause pre-dissolution of the microsporocyte callose wall at anther development process, which affected the microspores further to develop into mature pollen grains, result in male sterility [33] . Previous studies in our laboratory have found that the gene is likely to be involved in the regulation of the pollination affinity of R. rugosa. Therefore, we cloned the gene from the stylus of R. rugosa, and the gene was analyzed by bioinformatics. Next, we will further through the transgenic to verify whether the gene has pollination compatibility regulation function.

4.3. Evolution of β-1,3-Glucanase Gene

In addition, previous studies on the Glu phylogenetic relationships of bananas, cotton, tobacco, grapes, and other plants have demonstrated highly conserved Glu sequences in evolution and consistency between the Glu evolution and the kinship of the plant sources. This study also found that RrGlu shares 72% sequence homology with nine other species, including Prunus persica. Furthermore, RrGlu is closely related to Glu from the same family Malus domestica, Malus hupehensis, and Pyrus bretschneideri, whereas it is relatively distant from Eucalyptus grandis and Citrus sinensis, which are from different families. This indicates that the results of this study are consistent with previous studies.

In this study, the full-length cDNA sequence of β-1,3-glucanase gene was cloned successfully from R. rugosa and named as RrGlu. The bioinformatics characteristics of the RrGlu gene were analyzed, which provided the basis for further research on the function of RrGlu gene.

ACKNOWLEDGEMENTS

This work was funded by the National Science Foundation of China (NSFC) (31200524).

CITE THIS PAPER

Yanan Fu, Juanjuan Sun*, Yan Ma, Shutang Xing, Lanyong Zhao, Zongda Xu, Xiaoyan Yu# (2016) Cloning and Bioinformatics Analysis of *Rosa rugosa β*-1,3-Glucanase Gene (*RrGlu*). *American Journal of Plant Sciences*,**07**,461-468.

REFERENCES

1. Feng, L.G., Sheng, L.X., Zhao, L.Y., et al. (2008) Changes of Aromatic Components and Contents in the Process of Rosa rugosa Development. Chinese Agricultural Science, 41, 4341-4351.
2. Wen, J.L., Xue, Z.D. and Hu, J.J. (2000) Cold Resistance and Plasma Membrane Permeability of Rosa rugosa. Journal of Northwest Forestry College, 15, 16-20.
3. Liu, J. (2013) Study on the Cross Breeding and Hybrid Compatibility between Rosa rugosa and Modern Rosa hybrid. Beijing Forestry University, Beijing.
4. Wang, Q. (2010) Preliminary Study on Cross of Rosa cvs. and Rosa rugosa and Resistance to Black Spot of Rose. Beijing Foresty University, Beijing.
5. Zhang, Y., Liu, T., Zhang, H.X. and Bai, S.L. (2010) Advances in Breeding and Tissue Culture of Rosa hybrid. Chinese Agricultural Bulletin, 26, 250-254.
6. Yu, X.Y. (2010) Study on Embryology and Pollination Compatibility of Rosa rugosa. Shandong Agricultural University, Taian.
7. Yu, X.Y., Zhao, L.Y., Feng, Z., et al. (2009) 22 Parts of Self-Compatibility Study of Domestic Rosa rugosa Resources. Chinese Agricultural Science, 42, 3236-3242.

8. Yu, X.Y., Xing, S.T. and Zhao, L.Y. (2014) Analysis on the Causes of the Cross between Rosa rugosa and Rosa hybrid. Chinese Agricultural Science, 47, 3112-3120.
9. Ding, Y.Y., Zhao, L.Y., Zhang,L., et al. (2012) Effect of Rose Styles of Extract Solution on the Growth of Pollen Tube in Vitro of Rosa rugosa and Rosa hybrid. Chinese Agricultural Science, 45, 1791-1797.
10. Kuboyama, T. and Takeda, G. (2000) Genomic Factors Responsible for Abnormal Morphology of Pollen Tubes in the Interspecific Cross Nicotiana tabacum X N. Rustica. Sexual Plant Reproduction, 12, 333-337.
11. Williams, E., Knox, R.B. and Rouse, J.L. (1982) Pollination Sub-Systems Distinguished by Pollen Tube Arrest after Interspecific Crosses in Rhododendron (Ericaceae). Journal of Cell Science, 53, 255-277.
12. Natalie, K., Fritz, R.E. and Hanneman, J.R. (1989) Interspecific Incompatibility Due to Stylar Barriers in Tuber-Bearing and Closely Related Non-Tuber-Bearing Solanums. Sexual Plant Reproduction, 2, 184-192.
13. Ling, T.B. and Mu, X.J. (1995) Actinidia arguta and Actinidia deliciosa Interspecific Hybridization of Pollen Tube Behavior and Early Embryogenesis of Observation. Journal of Plant Science, 37, 607-612.
14. Qiu, J., Yang, B.J., Jiang, L.R., et al. (2005) Observation of Distant Hybridization between Rice and Sorghum Pollen Behavior. Molecular Plant Breeding, 3, 835-840.
15. Li, X., Zhang, S.L., Tao, S.T., et al. (2007) Differences in Pollen Germi-nation in Situ and Pollen Tube Growth of Chinese Cherry and Sweet Cherry. Journal of Northwest Plants, 27, 429-434.
16. Wang, Y., Jiang, C.J. and Zhang, H.Y. (2008) Observation of Incompatibility of Self Pollen Tube of Tea Tree in the Style of Living Body. Tea Science, 28, 429-435.
17. Tang, S. (2014) Analysis of Callose Degradation Related Genes MSR66 in Anthers of Rape. Huazhong Agricultural University, Wuhan.
18. Doblin, M.S., De Melis, L., Newbigin, E., et al. (2001) Pollen Tubes of Nicotiana alata Express Two Genes from Different Beta-Glucan Synthase Families. Plant Physiology, 25, 2040-2052.
19. Van Loon, L.C. and Strien, E.A. (1999) The Families of Pathogenesis-Related Proteins, Their Activities, and Comparative Analysis of PR-1 Type Proteins. Physiological and Molecular Plant Pathology, 55, 85-97.
20. Hinton, D.M., et al. (1980) Glucanase in Fruits and Vegetables. Journal of the American Society for Horticultural Science, 105, 499-502.
21. Morohashi, Y. and Matsushima, H. (2000) Development of β-1,3-Glucanase Activity in Germinated Tomato Seeds. Journal of Experimental Botany, 51, 1381-1387.
22. Buchner, P., Rochat, C., Wuilleme, S., et al. (2002) Characterization of Atissue-Specific and Developmentally Regulated β-1,3-Glucanase Gene in Pea (Pisum sativum). Plant Molecular Biology, 49, 171-186.
23. Akiyama, T., Pillai, M.A. and Sentoku, N. (2004) Cloning, Characterization and Expression of OsGLN2, a Rice Endo-1,3-β-Glucanase Gene Regulated Developmentally in Flowers and Hormonally in Germinating Seeds. Planta, 220, 129-139.
24. Su, X.Y., Gao, L., Zhang, Y.J., et al. (2014) Expression Analysis of Wheat TcLr35 in β-1,3-Glucanase Genes and Signaling Molecules to Induce Leaf Rust. Chinese Agricultural Science, 47, 2774-2783.
25. Zhang, Y.N., Qu, M., Xu, X.L., et al. (2003) Agrobacterium rhizogenes Mediated Chitinase Gene and Beta-1, 3-Transformation of Tobacco Glucanase Gene. Plant Research, 23, 198-203.
26. Li, Y.F., Zhu, R., Xie, L.X., et al. (2005) Induced Expression of Barley Beta-1,3-(GIII) Promoter in Transgenic Rice. Journal of Plant Physiology and Molecular Biology, 31, 143-148.
27. Zhao, J., Wang, B., Yu, T., et al. (2007) Study on Agrobacterium Mediated Transformation of -1, 3-, and Gene into Peanut. Journal of Peanut, 36, 20-23.

28. Chen, X.F., Sui, H.L., Ma, Q.H., et al. (2015) The Expression of Downy Mildew Induced Chitinase and Glucanase Gene of Chinese Cabbage. Shandong Agricultural Sciences, 47, 96-99.
29. Paul, B., Katharina, K., Alexander, G., Marilyn, P., Grant, C., Kim, F., Samuel, C. and Alison, M. (2011) Callose Synthase GSL7 Is Necessary for Normal Phloem Transport and Inflorescence Growth in Arabidopsis. Plant Physiology, 155, 328-341.
30. Zhang, Y., Gao, J. and Xu, Y.M. (2010) Cloning and Sequencing Analysis of β-1,3-Glucan Synthasease Gene from Moso Bamboo. Molecular Plant Breeding, 8, 533-541.
31. Zhai, G.H., Ruan, X.L., Wu, L.T., et al. (2011) Cloning and Sequence Analysis of a Full-Length cDNA of β-1,3-Glucanase Gene from Musa pisangawake. Chinese Agricultural Science, 44, 3134-3141.
32. Dong, J., Cai, C.P., Wang, L.K., Zhao, L., Zhang, T.Z. and Guo, W.Z. (2011) Structure and Expression Analysis of Two Novel Genes Encoding β-Glucanase in Cotton. Acta Agronomica Sinica, 37, 40-47.
33. Liao, X.F. (2012) Functional Analysis of Arabidopsis anther Specific Expression of β-1,3-Gene. Huazhong Agricultural University, Wuhan.
34. Leubner-Metzger, G. and Meins, F. (1999) Function and Regulation of Plant β-1,3-Glucanases(PR-2). In: Datta, S.K. and Muthukrishnan, S., Eds., Pathogenesis-Related Proteins in Plants, Boca Raton, 49-76.
35. ucciaglia, P.A. and Smith, A.G. (1994) Cloning and Characterization of Tag1,a Tobacco Anther Beta-1,3-Glucanase Expressed during Tetrad Dissolution. Plant Molecular Biology, 24, 903-914.
36. Domingo, C., Conejero, V. and Vera, P. (1994) Genes Encoding Acidic and Basic Class III Beta-1,3-Glucanases Are Expressed in Tomato Plants upon Viroid Infection. Plant Molecular Biology, 24, 725-732.
37. Wan, P.P. (2004) Purification, Properties and Antifungal Activity of β-1,3-Glucanase. Shandong Agricultural University, Taian.

CHAPTER 16

Ngsmirplant: Comprehensive Characterization of the Small RNA Transcriptomes of Plants

Jie Bai[1,2], Chen Dan[1], Yi Zhang[1], Guoping Zhao[1,2,3]&, Xiaoming Ding[1]

[1] *Department of Microbiology and Microbial Engineering, School of Life Sciences, Fudan University, Shanghai 200433, China*

[2] *Key Laboratory of Synthetic Biology, Institute of Plant Physiology and Ecology, Shanghai Institutes for Biological Sciences, Chinese Academy of Sciences, Shanghai 200032, China*

[3] *Shanghai-MOST Key Laboratory of Health and Disease Genomics, Chinese National Human Genome Center at Shanghai, Shanghai 201203, China*

MicroRNAs (miRNAs) are short, endogenously expressed RNAs (~21–22 nt) processed from stem-loop regions of longer RNA precursors by a Dicer-like enzyme (Jones-Rhoades et al., 2006). In plants, microRNAs play an essential role in regulating many fundamental molecular interactions, including plant growth, development and response to environmental stress (Rogers and Chen, 2013). In comparison with the roles of microRNAs in animals, biogenesis and function of microRNAs in plants appears to be drastically different (Llave et al., 2002; Zhang et al., 2007). For example, prevalent in animal, microRNAs negatively regulate gene expression by attenuating translation. In contrast, direct microRNA-guided mRNA cleavage occurs in plant kingdom. In addition, for microRNA induced regulation in animal, translation is inhibited by microRNA with imperfect base-pairing to target mRNA. In plant, however, microRNAs inhibit the translation of target mRNA through extensively complementarity to its targets (Wu, 2013).

MicroRNAs can be identified by the approaches of hybridization-based experimental and bioinformatics prediction followed by experimental validation (Jones-Rhoades et al., 2006). With the limitations of experimental methods, microRNAs of low expression level are very difficult to be identified, causing certain microRNA to be missed. Bioinformatics approaches are often associated with a relatively high false negative and false positive rate. Thus, so far, due to limit of these methods, only a small number of plant microRNAs have been identified. Nowadays, with the advent of next-generation sequencing technologies, such as Roche 454, Illumina Solexa and ABI SOLiD, small RNA transcriptome research has been revolutionized. Compared to previously reported methods, it has distinct advantages of in-depth analysis of small RNA populations and the ability to identify novel microRNAs (Creighton et al., 2009; Fahlgren et al., 2009; Tam et al., 2014). However, deep sequencing technologies have generated hundreds of thousands of reads with large amounts of data, which give us substantial bioinformatics challenges to mine the hidden biological significance.

Currently, several useful tools have been developed to allow users to investigate the microRNAs from deep sequencing data, such as miRanalyzer (http:// web.bioinformatics.cicbiogune.es/microRNA/), miRExpress (http:// miRExpress.mbc.nctu.edu.tw), mirTools (http://centre. bioinformatics.zj. cn/mirtools/), miRNAkey (http://ibis.tau.ac.il/ miRNAkey), DSAP (http://dsap.cgu.edu.tw), mirDeep* (http://www.australianprostatecentre.org/research/ software /mirdeep-star), miREval 2.0 (http://mimirna.centenary.org.au/mireval/). However, except the UEA small RNA tool (http://srna-tools.cmp. uea.ac.uk), few tools are available for identifying plant microRNAs due to important differences of existed reference resources, mode of action and search criteria. Herein, a central motivation for this study is to specifically develop a comprehensive tool in an effort to easily investigate the large amount of NGS data for plant microRNAs. We describe a new web-based resource NGSmirPlant for the identification, characterization and differential expression analysis of microRNAs, as well as detection of novel microRNAs. Compared with the UEA small RNA tool, NGSmirPlant provides detailed annotation information for the comparison between mature microRNAs and users-input microRNAs, such as absolute counts, relative counts and sequences. In addition, NGSmirPlant implements an evolutional conservation analysis for all known microRNA families included in the result among a set of plant species. The interface of NGSmirPlant is easy to use and will be extremely helpful to plant molecular biologists with amateur bioinformatics skills to investigate the plant small RNA transcriptome.

Different small RNA has different length, which is due to distinct mechanisms of its biogenesis process. Through a read filter process, NGSmirPlant will obtain the unique length distribution of small RNA transcriptome obtained from deep sequencing, which can be used to distinguish the different classes of small RNAs and estimate the performance of sequencing library constructed. In this study, NGSmirPlant provides two kinds of length distribution (Figure 1): one is based on unique tag and the other is based on tag abundance. The unique tag distribution is only used for estimation of the diversity of unique tag obtained from the library. The tag abundance distribution can be used to estimate the performance of sequencing library. For example,microRNAs are the most abundant class of small RNAs in plant, which will has a peak at the 21 or 22 nucleotides lengths from Dicer digestion products (Jones-Rhoades et al., 2006).

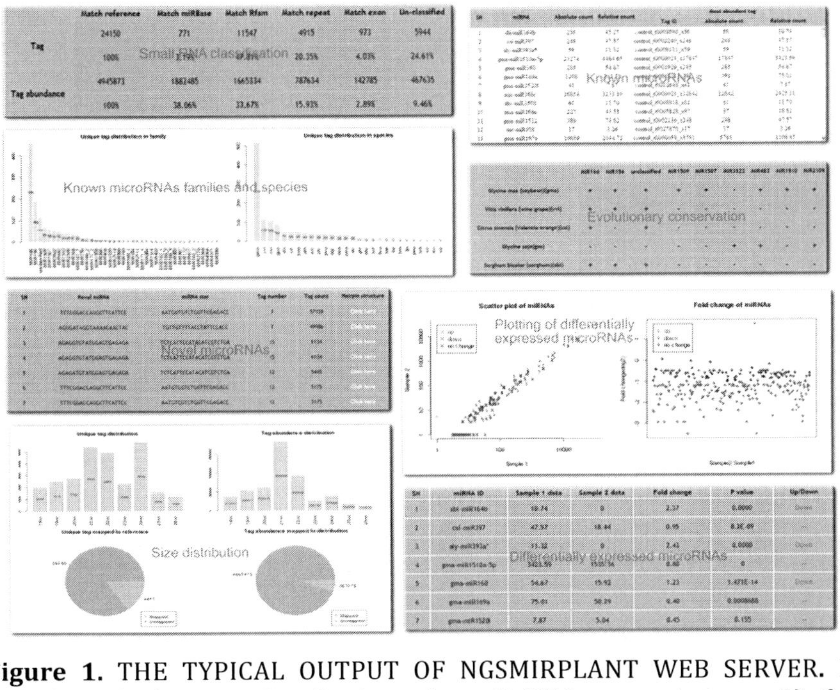

Figure 1. THE TYPICAL OUTPUT OF NGSMIRPLANT WEB SERVER. It includes: 1) the size distribution of small RNA transcriptome; 2) the classification of small RNAs known microRNAs, repeat-associated RNA, the ncRNAs reads annotated by Rfam (rRNA, tRNA, snRNAs and snoRNA et al.) and mRNA degradation products; 3) the known and novel microRNAs identification; 4) evolutionary conservation of individual microRNA in various plant species; and 5) differentially expressed microRNAs between samples

Recent studies have revealed that small RNA transcriptome is far more complex than previously thought, which might play important roles in the physiology of the cell (Creighton et al., 2009; Fahlgren et al., 2009; Tam et al., 2014). Deep sequencing has offered an exciting opportunity to determine the complexity of small RNA transcriptome in the mean of single-base resolution, genome-wide scale and cost-effectiveness. Initially, NGSmirPlant performed the mapping of sequence tags mapping to the reference genome or EST using SOAP 2.0 (http://miRExpress.mbc.nctu.edu.tw). It is found that, besides high abundance of microRNA, non-coding small RNAs (snRNAs, snoRNAs and tRNAs et al. obtained from the Rfam database) will also occupy many sequence tags even for the specific designed microRNA sequencing library (http://122.228.158.106/NGSmirPlant/download1.php?id=1414376897). In addition, a small proportion of sequencing tags could be mapped to exon sequences, which are likely to be RNA degradation products. Therefore, the aligned reads are classified into known microRNAs, repeat-associated RNA, the ncRNAs reads annotated by Rfam (rRNA, tRNA, snRNAs and snoRNA et al.) and mRNA degradation products (Figure 1). If a sequence tag has multiple hits, we only assign it a single category based on the following priority: ncRNA from Rfam, known plant microRNA, repeat-associated RNA and mRNA degradation product. Through such a way, NGSmirPlant will identify the expression level of each kind of small RNA by calculating the number of total tags it belongs to.

One of the most important advantages for deep sequencing of small RNA transcriptome is that it can be used for the discovery of novel microRNAs (Creighton et al., 2009). In NGSmirPlant, to identify candidate novel microRNAs, we first filter out sequence tags that have been classified into annotated categories, such as non-coding RNA and known microRNAs. With the implement of the miRDeep program (Friedlander et al., 2008), NGSmirPlant provides candidate novel microRNAs with the RNA secondary structure, minimal free energy, tag counts of mature/mature* sequence et al. (Figure 1). Users can use other method, such as hybridization-based experimental method, to validate the reliability of novel microRNA identified.

Since many microRNA families are evolutionarily conserved across many lineages in the plant kingdom and the fact that lots of microRNAs in different plant species shared the same mature sequence in miRBase 19.0 (http://microrna.sanger.ac.uk/), microRNA genes in one species may have orthologs or homologs existed in other species. Given that, we enlarged the annotation scale to an expansion of multiple species of known and novel microRNAs. For the identified microRNAs, the measurement of its evolutionary conservation was performed by similarity search against the currently known microRNAs in miRBase 19.0

in other plant species (Figure 1). If a read matched one or more known microRNAs with >95% identity, it was regarded as homologous in evolution. Through the evolutionary conservation profile, the lineage-specific and species-specific microRNAs can be identified and visualized easily.

NGSmirPlant employs two different strategies to investigate the microRNAs expression profile and detect the differentially expressed microRNAs between different libraries (Figure 1). One is based on the total tag count (#specific microRNA tags/#total microRNA sequence tags) and the other is based on the most abundant tag count (#the most abundant tag of specific microRNA/#total microRNA sequence tags). Each identified microRNA sequence numbers were normalized to the total number of microRNA read counts followed by multiplying one million to obtain reads per million (RPM) value. The statistical significance (*P*-value) is inferred based on the Bayesian method (Audic and Claverie, 1997). Flexibly, the threshold for fold change as differentially expressed microRNA is user-definable. NGSmirPlant provides differentially expressed microRNAs with microRNA ID, relative sequenced count, fold change, increase/decrease and *P* value.

The web server of NGSmirPlant is well organized in a user-friendly access manner. It runs with the input of a single sample or multiple samples. For "single" mode, NGSmirPlant will analyze and annotate small RNAs embed in the sample and give a list of novel microRNA details. For "multiple" mode, the differentially expressed microRNAs between the pair samples will be in-depth investigated additionally.

The web server input of NGSmirPlant is pre-processed FASTA format where all the unique sequence tags aggregated with a number behind the name to represent the expression level and cleaned into a non-redundant FASTA file to reduce the input size. The users can download the Perl script in the filter column of web server to a local client for generating the proper input format of NGSmirPlant. The size of input file should be limited to 10 MB, which can be in FASTA format or compressed in zip or gz format of FASTA file (Figure 2). A demo for the web server is provided. Users can download the example input file for "single" or "multiple" mode and run the tools with default parameters. In general, a typical "single" run takes about 20 min and a typical "multiple" run takes about 40 min.

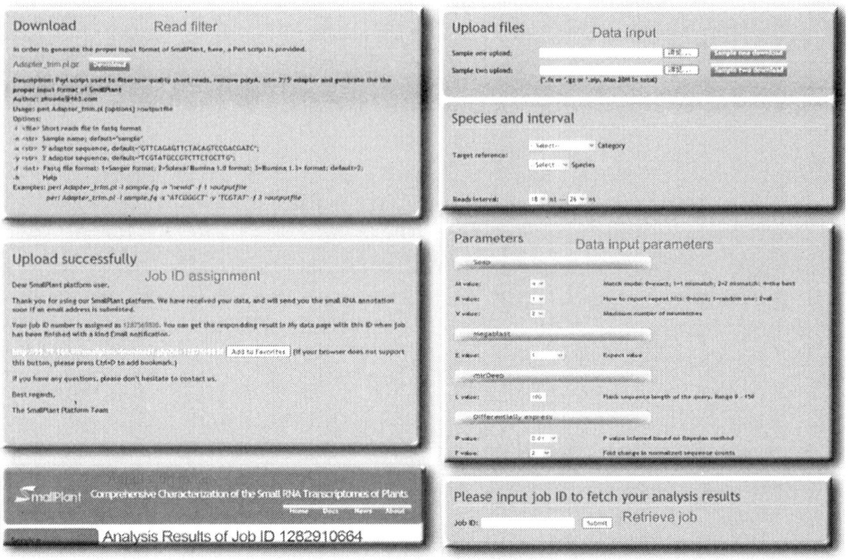

Figure 2. The typical input of ngsmirplant showing the reads filter requirement, file input, job id assignment and job retirement

NGSmirPlant provide a number of important parameters in both of “single” mode and “multiple” mode, so users can set these parameters by themselves (Figure 2). Users can set the reads length interval (default is 18~30 nt) thus the tag sequences longer or shorter than the threshold will be discarded. During the alignment mapping process, a set of parameters are available for users to define. With the “R value” parameter, users can freely define the strategy for reporting repeat hits. It should be noticed that the “M value” parameter denotes the maximum number of allowed mismatches for the seed part of a read (at most two mismatches) while the “V value” parameter means the total allowed mismatches in one read. When predicting novel microRNAs, the expected value and the flanking sequence length of the query can be defined by users. For “multiple” mode, two parameters for differentially expression are provided. By setting the threshold of *P* value inferred based on Bayesian method and fold change in normalized sequence counts, the specific microRNA satisfied both the *P*-value and the fold change will be deemed to be significantly differentially expressed in the two samples (Figure 3).

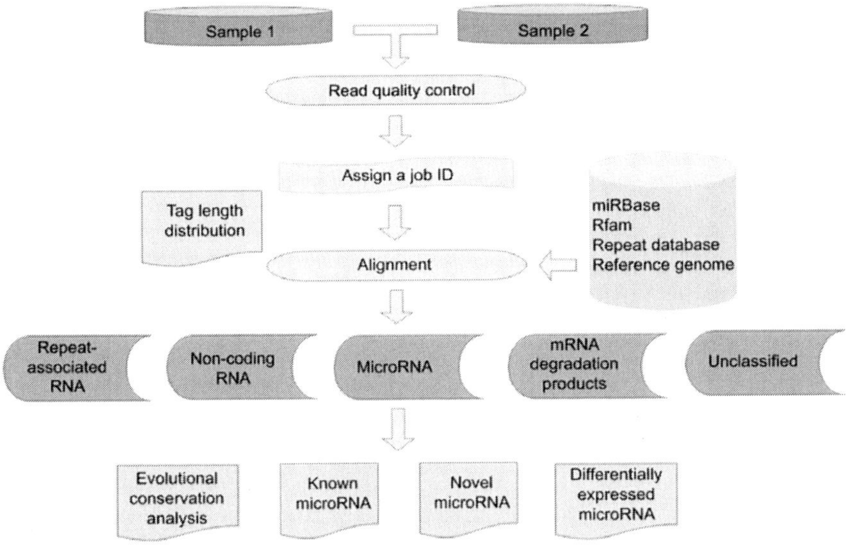

Figure 3. The workflow of NGSmirPlant. In brief, NGSmirPlant firstly performs the read quality control and aligns the clean reads to a number of reference databases, such as genomic sequences or ESTs, miRBase, Rfam and repeat database produced by RepeatMasker. The small RNA annotation categories will be then identified, include microRNAs, repeat-associated RNA, non-coding RNA and mRNA degradation products. Meanwhile, NGSmirPlant will provide detailed annotation of known and novel microRNAs, differentially expressed microRNA between samples and its evolutional conservation

Once a job is submitted successfully, NGSmirPlant will be assigned with a unique job ID and users need to wait for the result by e-mail provided. Then, users will receive a notification e-mail with a URL link automatically. Users can use this unique ID to visit the annotation results or retrieve previous job, which will be kept one month in NGSmirPlant. The job ID submitted older than one month will be deleted. Subsequently, NGSmirPlant will show: 1) basic summary information of library, such the length distribution of unique sequence size, which can be used to indicate the most abundant of small RNA category in the library; 2) alignment information indicating the tag abundance matched to reference genome, miRBase, Rfam, repeat, exon and unannotation respectively; 3) detailed annotation information of the known microRNA identified, such species and families distribution, the absolute (total number of sequence tags mapped to a specific microRNA) and relative (normalization of matched read counts to the total number of microRNA reads and then multiplied by a million) expression level and evolutionary conservation of individual microRNA in various plant species; 4) detailed annotation information of other non-coding RNA identified, such as rRNA, tRNA, snRNAs and snoRNA; 5) detailed annotation information of novel microRNA identified, such as novel microRNA sequence, microRNA star and hairpin structure; 6) detailed information of microRNA different expression. All these results can be downloaded from NGSmirPlant to allow users to perform easy visualization or further analysis in the local computer.

The small RNAs, especially the microRNAs, play important roles in different biological processes in plant, such as plant growth, development and response to environmental stress. However, currently, only a small number of plant microRNAs have been detected due to the limitation of hybridization-based experimental and bioinformatics prediction methods. However, the recently developed deep sequencing technologies have been proved to have distinct advantages of unbiased and in-depth analysis of the complexity of small RNA transcriptome and its powerful ability to identify known and novel microRNAs. However, next-generating sequencing technologies have produced hundreds of thousands of short reads with large amounts of data, which give us substantial bioinformatics challenges to discover the hidden biological significance. Therefore, in this study, NGSmirPlant (http://122.228.158.106/NGSmirPlant/) is specifically designed towards a web-based server to fulfill the urgent need for bioinformatics analysis of plant small RNA transcriptome by deep sequencing. In general, NGSmirPlant can be used to facilitate a genome-wide analysis of microRNAs profiling and discovery, exploring evolutionary conservation of individual microRNA in various plant species.

In comparison with other microRNA profiling tools, such as mirTools (http://centre.bioinformatics.zj.cn/mirtools/) and CPSS (http://mcg.ustc.edu.cn/db/cpss/index.html) which are either specialized for animal small RNAs analysis or incomplete analysis of plant small RNAs, NGSmirPlant is specialized for comprehensive characterization of the small RNA transcriptome of plants. Currently, it covers a broad range of 85 plant organisms from the KEGG database (http://www.genomejp/kegg/catalog/org_list.html, includes completely finished genomic sequences and the available ESTs) to allow users to effectively identify known and novel microRNAs. It is believed that the comprehensive integration of plant species could meet the need of most researchers who involve in plant small RNA analysis from deep sequencing. Moreover, NGSmirPlant performs evolutionary conservation analysis for each microRNA family while mirTools and CPSS do not. In this part, those relatively conserved microRNA families among different species are listed with a symbol "+". In addition, the detailed annotation of known microRNAs is performed for the comparison between mature microRNAs and users-input microRNAs, such as absolute counts, relative counts and sequences. Compared with CPSS, NGSmirPlant exhibits more readable graphs, such as the detailed distribution of unique tags in families and species, which provides a more concise result for users to visualize the results. As with other microRNA profiling tools, such as the UEA small RNA tool, NGSmirPlant also perform an analysis of differentially expressed microRNAs between samples. Meanwhile, it is very useful that NGSmirPlant draws scatter plots for total tag count and the most abundant tag, which makes the expression differentiation extremely clear.

In the future version, we will continually integrate more number of finished plant genomes or ESTs data as they become increasingly available. In addition, the pre-processing script provided in the current version only supports with Illumina sequencing data and as such would require users to develop their own pre-processing steps in order to use 454 or SOLiD data. In the future, a more general and flexible pre-processing script will be made available that would accept the raw data obtained from the three common used high-throughput sequencing platforms and generate the required format for NGSmirPlant input. To further improve on the power of NGSmirPlant, future development will implement the microRNA target prediction tools to predict the target genes of identified microRNAs in plant. Moreover, in addition to the microRNA target prediction tools, NGSmirPlant will plan to perform detailed functional annotation of microRNA targeted genes, including Gene Ontology, KEGG pathway and protein-protein interaction. Through which, NGSmirPlant will provide the opportunities to elucidate the complex microRNAs mediated regulatory networks. NGSmirPlant welcomes any questions, comments and suggestions, which will be a useful feedback for future updating. In conclusion, we believe that NGSmirPlant will be a straight-forward valuable resource that are opening fascinating opportunities for in-depth investigation of small RNA transcriptome generated from next-generation sequencing technologies.

NOTES

This work was supported by National Basic Research Program (973 Program) (No. 2012CB721102), National Natural Science Foundation of China (Grant No. 30830002) and the China Postdoctoral Science Foundation funded project (2012T50444 and 2012M520947).

Jie Bai, Chen Dan, Yi Zhang, Guoping Zhao and Xiaoming Ding declare that they have no conflict of interest.

This article does not contain any studies with human or animal subjects performed by the any of the authors.

REFERENCES

1. Audic S, Claverie JM (1997) The significance of digital gene expression profiles. Genome Res 7:986–995
2. Creighton CJ, Reid JG, Gunaratne PH (2009) Expression profiling of microRNAs by deep sequencing. Brief Bioinform 10:490–497
3. Fahlgren N, Sullivan CM, Kasschau KD, Chapman EJ, Cumbie JS, Montgomery TA, Gilbert SD, Dasenko M, Backman TW, Givan SA, Carrington JC (2009) Computational and analytical framework for small RNA profiling by high-throughput sequencing. Rna 15:992–1002
4. Friedlander MR, Chen W, Adamidi C, Maaskola J, Einspanier R, Knespel S, Rajewsky N (2008) Discovering microRNAs from deep sequencing data using miRDeep. Nat Biotechnol 26:407–415
5. Jones-Rhoades MW, Bartel DP, Bartel B (2006) MicroRNAS and their regulatory roles in plants. Annu Rev Plant Biol 57:19–53
6. Llave C, Kasschau KD, Rector MA, Carrington JC (2002) Endogenous and silencing-associated small RNAs in plants. Plant Cell 14:1605–1619
7. Rogers K, Chen X (2013) Biogenesis, turnover, and mode of action of plant microRNAs. Plant Cell 25:2383–2399
8. Tam S, de Borja R, Tsao MS, McPherson JD (2014) Robust global microRNA expression profiling using next-generation sequencing technologies. Lab Investig 94:350–358
9. Wu G (2013) Plant microRNAs and development. J Genet Genomics 40:217–230
10. Zhang B, Wang Q, Pan X (2007) MicroRNAs and their regulatory roles in animals and plants. J Cell Physiol 210:279–289

CHAPTER 17

Cadmium Resistance in Tobacco Plants Expressing the *Musi* Gene

Young-Nam Kim[1], Ji-Seoung Kim[1], Sang-Gyu Seo[1], Youngwoo Lee[1], Seung-Woo Baek[1], Il-Sup Kim[2], Ho-Sung Yoon[2], Kwon-Rae Kim[3],Sun-Hyung Kim[1], Kye-Hoon Kim[1]

[1].*Department of Environmental HorticultureThe University of Seoul Seoul Republic of Korea*

[2].*Department of Biology Kyungpook National University Daegu Republic of Korea*

[3].*Han-River Environment Research Center National Institute of Environmental Research Kyounggi-do Republic of Korea*

ABSTRACT

MuSI, a gene that corresponds to a domain that contains the rubber elongation factor (REF), is highly homologous to many stress-related proteins in plants. Since *MuSI* is up-regulated in the roots of plants treated with cadmium or copper, the involvement of *MuSI* in cadmium tolerance was investigated in this study. *Escherichia coli* cells overexpressing *MuSI* were more resistant to Cd than wild-type cells transfected with vector alone. *MuSI* transgenic plants were also more resistant to Cd. *MuSI* transgenic tobacco plants absorbed less Cd than wild-type plants. Cd translocation from roots to shoots was reduced in the transgenic plants, thereby avoiding Cd toxicity. The number of short trichomes in the leaves of wild-type tobacco plants was increased by Cd treatment, while this was unchanged in *MuSI* transgenic tobacco. These results suggest that *MuSI* transgenic tobacco plants have enhanced tolerance to Cd via reduced Cd uptake and/or increased Cd immobilization in the roots, resulting in less Cd translocation to the shoots.

Keywords: *MuSI* Cadmium tolerance Rubber elongation factor (REF) Tobacco Cadmium accumulation Trichomes

INTRODUCTION

High biomass plants such as willows and poplars have been introduced into ecosystems for phytoextraction (Robinson et al. 2000). However, this approach is limited by the need to overcome the low heavy metal accumulation in the plants due to their low heavy metal tolerance. Transgenic plants created by genetic engineering have stronger tolerance to heavy metals or accumulate more heavy metals than non-transgenic plants (Cherian and Oliveira 2005). Some genes of interest involved in Cd tolerance have been studied. *MT1* transgenic tobacco seedlings can tolerate up to 200 mM Cd (Pan et al. 1994). The roots of *gshll* transgenic Indian mustard grow longer and accumulate three times more total Cd per plant than wild-type plants in the presence of 0.15 mM Cd (Zhu et al. 1999). *OASTL* transgenic *Arabidopsis* can tolerate up to 300 mM Cd and produced more biomass in agar (Kawashima et al. 2004). *ZntA* transgenic *Arabidopsis* grew better than wild-type plants in the presence of 70 mM Cd (Lee et al. 2003). The genes of these transgenic plants express proteins including ones involving metallothionein and phytochelatin, which are cysteine-rich peptides that chelate heavy metals and reduce Cd toxicity by forming peptide-metal complexes (Zenk 1996; Cobbett 2000; Cherian and Oliveira 2005; Espen and D'Souza 2005).

Transcription of *MuSI*, a previously identified gene, is induced in response to dehydration in sweet potato roots (*Ipomoea batatas* L. cv. Yulmi) (Kim et al. 2009). MuSI protein is significantly homologous to putative stress-related proteins in soybean and *Arabidopsis* (Seo et al. 2010). Hence, the designation of *MuSI* is an abbreviation for multiple stress response gene I. The predicted amino acid sequence of the full-length MuSI protein is similar to that of small rubber particle proteins (SRPPs) found in rubber trees (Seo et al. 2010). *MuSI* expression is also induced by various stress signals including dehydration, high salt, heavy metals, oxidation, and plant hormones (Seo et al. 2010). In addition, 35S::*MuSI* transgenic tobacco plants exhibit markedly enhanced tolerance to heavy metal stress compared to control plants (Seo et al. 2010).

Altogether, these studies imply that *MuSI* may play a role in heavy metal stress tolerance. The present study sought to deduce the underlying mechanism of *MuSI* transgenic tobacco tolerance to Cd stress. Our results may be used to promote the use of *MuSI* transgenic plants in phytoremediation of Cd-contaminated soils.

MATERIALS AND METHODS

Construction and expression of the *MuSI* expression plasmid

The *MuSI* gene was amplified by polymerase chain reaction (PCR) from sweet potato cDNA as previously described (Seo et al. 2010) and cloned into a T&A cloning vector (Real Biotech, Taiwan). After confirming that the sequence was correct, the *MuSI* gene was digested with *Bam*HI and *Eco*RI, and ligated into the corresponding sites of the *Escherichia coli* expression vector pET28a(+). *E. coli* cells were treated with 0.2 mM isopropyl β-D-thiogalactoside (IPTG) for 3 h at 37°C. Proteins were then extracted from the bacteria cells by sonication in cold phosphate-buffered saline (PBS) buffer containing 1 mM dithiothreitol, 1 mM phenylmethylsulfonyl fluoride (PMSF), 2 μM pepstatin A, and protease inhibitor cocktail. Crude extract was collected by centrifugation (15,000*g* at 4°C for 30 min). Total protein concentration was determined using protein dye reagent (Bio-Rad, USA). The proteins (20 μg) were resolved by 12% sodium dodecyl sulfate-polyacrylamide gel electrophoresis (SDS-PAGE). The bands in the gels were electrophoretically transferred onto polyvinylidene fluoride membranes (Bio-Rad) for 3 h at 4°C at 180 mA. The membranes were blocked with 5% skim milk in TBST buffer containing 25 mM Tris–HCl, pH 7.4; 150 mM NaCl, and 0.05% Tween-20 for 1 h at room temperature, and incubated overnight at 4°C with an anti-His-tag antibody (1:1,000 dilution) (Millipore, Billerica, MA, USA). Blots were washed with TBST for 40 min, and incubated with an anti-mouse IgG antibody coupled to horseradish peroxidase (1:1,000 dilution) (Millipore) for 90 min at room temperature. After washing with TBST, antibody binding was visualized using ECL western blotting detection reagent (GE Healthcare, Little Chalfont, UK). Anti-DnaJ antibody (Stressgen Biotechnologies, Victoria, BC, Canada) was used as an *E. coli* standard control.

Recombinant protein purification

E. coli BL21 (DE3) harboring the *MuSI* plasmid was grown in LB medium supplemented with 100 μg/mL ampicillin with vigorous shaking (160*g*) at 37°C. When the culture reached the mid-log phase (*A*600 = 0.6), IPTG was added at a final concentration of 0.2 mM, and the culture was grown for another 4 h at 37°C. The cells were harvested by centrifugation at 5,000*g* for 10 min, and then washed twice with cold-PBS buffer. The resulting cell pellet was resuspended in lysis buffer (50 mM phosphate buffer, pH 8.0; 0.3 M NaCl, 10 mM imidazole, and 1 mM PMSF) and sonicated. The homogenate was centrifuged at 12,000*g* for 20 min at 4°C to remove cell debris. The supernatant containing soluble protein was purified by immobilized metal affinity chromatography (IMAC). The clear supernatant was loaded by gravity onto a column containing pre-equilibrated Ni–NTA affinity resin (Qiagen, Hilden, Germany). The column was washed twice with wash buffer (50 mM phosphate buffer, pH 8.0; 0.3 M NaCl, and 50 mM imidazole). The bound MuSI protein was eluted with a buffer comprised of 50 mM phosphate buffer, pH 8.0; 0.3 M NaCl, and 0.25 M imidazole.

Stress tolerance in ectopic *MuSI*-expressing *E. coli*

Mid-log phase *E. coli* cells (*A*600 = 0.6) grown at 37°C were streaked onto LB agar supplemented with 1.0 mM or 1.5 mM $CdCl_2$ and incubated for 24–36 h at 37°C.

Plant material and Cd treatment

MuSI transgenic line 6 previously described by Seo et al. (2010) was used for this study. Transgenic T_3 seeds were germinated and grown for 4 weeks on MS medium containing 50 mg/L hygromycin to select transgenic plants. Hygomycin-resistant plants were transferred to the Yamazaki solution (Yamazaki 1982) and acclimatized for 5 days in a greenhouse prior to exposure to Cd. After acclimation, the nutrient solutions in each container were replaced by nutrient solutions containing no Cd, 100 μM Cd, or 200 μM Cd. The seedlings were grown for 3 weeks; the nutrient solutions were replenished every 3 days. The solution was maintained at pH 5.8 and EC 1.2 dS m^{-1} during the experiment.

Scanning electron microscopy

The tobacco T_3 plants leaf surface was examined using scanning electron microscopy (SEM) as previously described (Kim 2008). Pieces of the leaves were immersed overnight in modified Karnovsky's fixative (Karnovsky 1965) consisting of 2% (v/v) glutaraldehyde and 2% (v/v) paraformaldehyde in 0.05 M sodium cacodylate buffer (pH 7.2) at 4°C and washed with the same buffer three times for 10 min each. The specimens were fixed in a second fixative solution containing 1% (w/v) osmium tetroxide in 0.05 M sodium cacodylate buffer (pH 7.2) at 4°C for 2 h. The pieces of leaf were briefly washed twice with distilled water and dehydrated in a graded ethanol series (once in 30, 50, 70, 80, and 95% ethanol for 10 min, and three times in 100% ethanol) at room temperature. The specimens were treated twice with isoamyl acetate for 10 min each, and dried in a CPD 030 critical point drier (BAL-TEC, Balzers, Liechtenstein). Tobacco leaves specimens were individually mounted on metal stubs, sputter-coated with gold, and observed by SEM (JSM-5510; JEOL, Tokyo, Japan).

Determination of relative damage content and Cd contents

After 3 weeks of Cd treatment, whole tobacco plants were harvested and separated into leaves, shoots, and roots. All plant parts were then rinsed three times with deionized water and the excess water was removed from the surface of the tissues using ash-less paper. Weight of the fresh tissues and relative damage content (RDC) were determined prior to drying in a fan-forced oven at 65°C for 3 days to perform dry weight measurement (Steyn 1959; Westerman 1990). The roots, shoots, and leaves were digested in a solution of concentrated H_2SO_4 and 60% $HClO_4$ (1:10, v:v) (Cresser and Parsons 1979). The levels of Cd in the plant parts were then determined by atomic absorption spectrophotometry (AAS-6800; Shimadzu, Japan). The RDC was calculated using the following equation:

RDC (%)=(NumberofdamagedleavesNumberoftotalleaves)×100.RDC (%)=(NumberofdamagedleavesNumberoftotalleaves)×100.

RESULTS AND DISCUSSION

In our previous study, expression of the *MuSI* gene isolated from sweet potato was shown to be highly induced by various stresses including dehydration, high salt, heavy metals, oxidation, and plant hormones (Seo et al. 2010). Sequence analysis showed that the full-length cDNA has a length of 998 bp with a 717-bp open reading frame and encodes 238 amino acids. The function(s) of the putative protein(s) are unknown. Stress tolerance experiments using transgenic plants over-expressing the *MuSI* gene showed that all independent transgenic tobacco lines (lines 1, 3, and 6) have enhanced tolerance to high temperature and drought stress (Seo et al. 2010). Based on this previous study, *MuSI* seems to be involved in cellular responses to Cd stress in transgenic tobacco.

Enhanced tolerance to Cd stress in transgenic *E. coli*

To analyze the effect of *MuSI* over-expression on *E. coli*, the *MuSI* gene was introduced into a *E. coli* expression vector (pET28a(+)::*MuSI*), and each construct was transformed into *E. coli* (Figure 1). Ectopic over-expression of *MuSI* in *E. coli* was confirmed by immunoblotting analysis (Figure 1). No *MuSI* expression was detected in cells transformed with the empty vector. After confirming *MuSI* over-expression, cells transformed with *MuSI* and the empty vector was exposed to Cd stress. Ectopic over-expression of the *MuSI* gene enhanced Cd tolerance of the transgenic *E. coli* (Figure 1) compared to the control cells. These results suggest that *MuSI* may have a role in protection against Cd stress, at least in prokaryotic cells. Our findings in *E. coli* are similar to those from a previous study showing that *MuSI* is involved in defense mechanisms that act against heavy metal stress, wounding, and virus infection (Oh et al. 1999; Kim et al. 2010; Seo et al. 2010). Taken together, the results from our study and previous investigations imply that the *MuSI* gene may be responsible for Cd stress tolerance.

Cd tolerance in transgenic plants over-expressing *MuSI*

Tobacco plants can absorb Cd through roots and accumulate the metal in shoots and leaves even though Cd is not an essential nutrient for plant growth and development (Welch and Norvell 1999). However, plants exposed to high concentrations of Cd show decreased growth rates and develop chlorosis (Kahel 1993; Das et al. 1997). To test whether *MuSI* can improve heavy metal resistance in plants, we generated transgenic tobacco plants in which *MuSI* expression is driven by the cauliflower mosaic virus 35S promoter (Seo et al. 2010). Four-week-old transgenic T_3 and wild-type seedlings were acclimated in a hydroponic system and then treated with 100 or 200 μM $CdCl_2$ for 3 weeks. Under these conditions, we observed a Cd resistance phenotype in the 35S::*MuSI* plants compared to wild-type plants (Figure 2). Both the transgenic and wild-type plants showed visible injuries at the early stage of Cd treatment, but differences in injuries were later observed

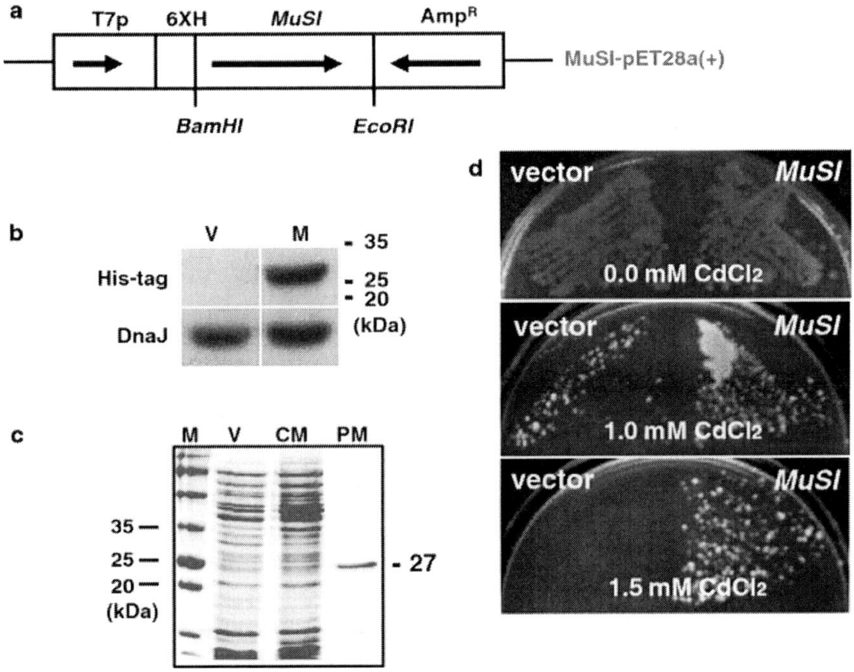

Figure 1. Construction of *MuSI* expression vector and expression in *E. coli* B21 cells. **A** Schematic diagram of the *MuSI* expression vector. *MuSI* gene with *Bam*HI and *Eco*RI restriction enzyme sites was subcloned downstream to the T7 promoter ($T7_P$), generating the recombinant *MuSI*::pET28a(+) expression vector; *6xH* six histidine-tagged resides; Amp^R ampicillin resistance gene. **B** To examine whether *MuSI* gene was expressed in *E. coli*, western blotting with anti-His tag antibody was performed. Anti-DnaJ antibody was used as a control. *V* Cells with the empty vector, *M* cells with the *MuSI*::pET28a(+) expression vector. **C** MuSI protein expression in crude protein extract and purified protein extract from *E. coli* cell expressing *MuSI*. *M* Protein marker, *V* crude protein extract from cells with the empty vector, *CM* crude protein extract from cells with the *MuSI* expression vector; *PM* purified *MuSI* protein form cells with the *MuSI* expression vector. **D** Stress tolerance of *E. coli* expressing *MuSI* in the presence of 1.0 mM (upper) or 1.5 mM (lower) $CdCl_2$. *Vector* Cells with the empty vector, *MuSI* cells with the *MuSI*::pET28a(+) expression vector

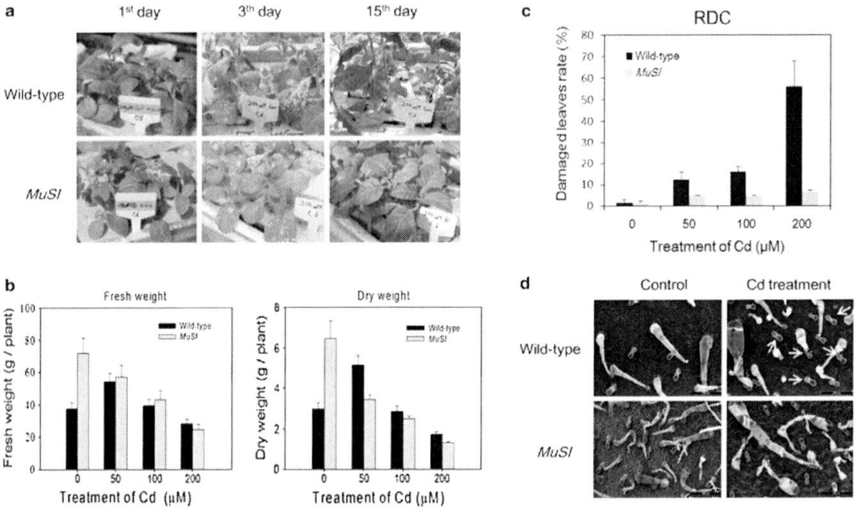

Figure 2. The tolerance of *MuSI* transgenic tobacco for Cd stress. **A** Effect of 200 μM Cd on the growth of wild-type tobacco and *MuSI* transgenic tobacco plants. Symptoms were visually observed at harvest. **B** Comparison of the fresh and dry weight of *MuSI* transgenic and wild-type tobacco plants treated with 0, 50, 100, or 200 μM Cd. Values are the average of six plants from each treatment. *Error bars* standard error. **C** Comparison of the growth of *MuSI* transgenic tobacco and wild-type tobacco plants in the presence of Cd. RDCs of the leaves from plants exposed to 50, 100, or 200 μM Cd are shown. Values are the average of six plants from each treatment. *Error bars* standard error. **D** Effect of Cd treatment on the formation of trichomes in the leaves from *MuSI* and wild-type tobacco plants as analyzed by SEM. *Arrowheads* indicate that short trichomes were sprung up by Cd treatment

(Figure 2a). While the wild-type plants displayed symptoms of severe toxicity at both concentrations of Cd, *MuSI* transgenic plants had attenuated symptoms. The leaves of the transgenic plants were in better condition than those of the wild-type plants that showed severe necrosis. Differences in plants growth rates were also noted between the transgenic and wild-type plants without Cd stress. *MuSI* transgenic plants showed higher growth rates than wild-type plants (Figure 2b). However, the growth rates of both transgenic and wild-type plants were decreased in the presence of increasing Cd concentrations (Figure 2b). Furthermore, although *MuSI* plants showed less damage than wild-type plants in the presence of 200 μM Cd, their growth rate was reduced as much as wild-type plants.

RDC is the number of damaged leaves relative to the total number of total leaves and indicates the degree of leaf damage. The RDC of wild-type tobacco plants exposed to progressively higher levels of Cd increased, especially at 200 μM Cd, indicating that the seedlings were severely damaged by the Cd (Figure 2c). In contrast, *MuSI* transgenic tobacco plants had a constant RDC at all tested Cd doses. Our previous study reported that three transgenic tobacco plant lines over-expressing *MuSI* are resistant to Cd and Cu in a germination experiment (Seo et al. 2010). The results of the present study also imply that the transgenic tobacco plants developed Cd tolerance through *MuSI* gene expression.

Authentic surface cell types can be distinguished by SEM. Specified portions of tobacco leaves were used for SEM analysis. The leaf cells were characterized by the presence of globular-tipped trichomes, stomata, and irregularly shaped cells. The trichomes of *MuSI* transgenic plants were altered in terms of size irregularities and increased density (Figure 2d). Tobacco trichomes have head cells coated with polysaccharides containing terpenoid (Choi et al. 2004) which transform into large crystals as Cd accumulates (Salt et al. 1995; Choi et al. 2001). This resistance mechanism involves the formation of complexes containing Cd thiol compounds and translocated metallothioneins and phytochelatins, and subsequent accumulation of these complexes in vacuoles, trichomes, leaves, and shoots (Choi et al. 2001; Song et al. 2003; Kawashima et al. 2004).

The SEM observations made in the present were consistent with this mechanistic scenario. Compared to Cd-free control plants, the number of short trichomes in the leaves (0.4 mm × 0.25 mm) doubled in plants treated with 200 μM Cd. On the other hand, the number of short trichomes was not increased by Cd treatment (Figure 2d) in the head cells of trichomes in *MuSI* transgenic tobacco plants that harbored modified (larger) crystals. These results suggest that the resistance mechanism of *MuSI* transgenic tobacco plants involves reduced Cd translocation from the roots and leaves. Put another way, the mechanism underlying the recovery of transgenic tobacco plants grown in a high concentrations of Cd is mainly due to impaired Cd uptake rather than sequestration of Cd. This is achieved by the formation of complexes containing thiol compounds in thricomes and/or Cd immobilization around the roots, resulting in decreased Cd translocation to the shoots. Longer exposure of *MuSI* transgenic tobacco plants to Cd could be used to more accurately measure Cd accumulation.

Cd contents of transgenic *MuSI* tobacco plants

Wild-type tobacco plants showed increased accumulation of Cd and suffered from more damage in a Cd concentration-dependent manner (Figure 3). In particular, chlorosis and reduced growth rates were observed in wild-type plants. In contrast, *MuSI* transgenic tobacco plants avoided Cd toxicity by blocking Cd absorption in the roots (Figure s. 2, 3). In the present study, *MuSI* transgenic tobacco plants accumulated less Cd than wild-type plants. When grown in the presence of 200 μM Cd, wild-type plants were found to contain 4.65 ± 0.32 mg of Cd per plant while and transgenic plants contained 2.37 ± 0.12 mg per plant. There was also a difference in the amount of Cd transported from the roots to leaves. In 200 μM Cd, the percent of Cd translocated to the leaves of wild-type plants was 81.6% while that of the *MuSI* transgenic plants was 37.1% (Figure 3). *MuSI* expression may repress Cd translocation from roots to shoots, leading to an overall decrease of Cd concentration in the leaves. These data suggest that *MuSI* may be playing unknown important role(s) in regulating Cd absorption in roots and translocation to the shoots.

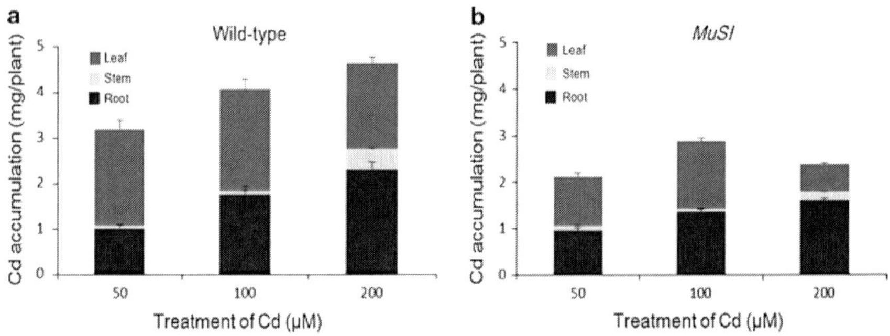

Figure 3. Cd accumulation in roots, shoots, and leaves of *MuSI* transgenic and wild-type tobacco plants treated with 50, 100, and 200 μM Cd. Values are the average of six plants from each treatment. *Error bars* standard error

Several genes such as *hMTll, YCF1*, and cysteine synthase, confer resistance to Cd toxicity in transgenic tobacco or *Arabidopsis* plants (Maiti et al. 1989; Elmayan and Tepfer 1994; Song et al. 2003; Kawashima et al. 2004). In previous studies, tobacco seedlings expressing a MT gene encoding the metal chelating protein metallothionein showed a modified distribution of Cd in various plant parts (Maiti et al. 1989; Elmayan and Tepfer 1994). Elmayan and Tepfer (1994) reported that Cd accumulation in the leaves of transgenic seedling was reduced and Cd translocation rate of transgenic plants (20%) to the shoots also decreased comparing with that of control plants (50%). These results are similar to the ones from the present study, and indicate that Cd resistance is closely related to regulating Cd distribution in the plant. Our findings demonstrate that Cd tolerance of *MuSI* transgenic tobacco plants may be associated with decreased Cd absorption and transport to shoots compared to wild-type plants.

In conclusion, the results from our study show that *MuSI* gene expression confers tolerance to Cd stress. *MuSI* transgenic tobacco plants display less Cd-mediated damage than wild-type tobacco plants when grown in the presence of Cd. This observation seems to be related to decreased translocation of Cd from the roots to the leaves in the *MuSI* transgenic plant compared to their wild-type counterparts. Although *MuSI* transgenic plants did not accumulate Cd in their shoots, they might still be useful for phytoremediation processes, such as phytostabilization, in which plants reduce the mobility and bioavailability of contaminants in soil and water by immobilization (Cherian and Oliveira 2005). Thus, future studies should focus on Cd detoxification mechanisms in *MuSI* transgenic plants, such as ones involving metallothionein and phytochelatin, as well as the suitability of *MuSI* transgenic plants for phytoremediation of Cd-contaminated soil.

OPEN ACCESS

REFERENCES

1. Cherian S, Oliveira MM (2005) Transgenic plants in phytoremediation: recent advance and new possibilities. Environ Sci Technol 39(24):9377–9390
2. Choi YE, Harada E, Wada M, Tsuboi H, Morita Y, Kusano T, Sano H (2001) Detoxification of cadmium in tobacco plants: formation and active excretion of crystals containing cadmium and calcium through trichomes. Planta 213:45–50
3. Choi YE, Harada E, Kim GH, Yoon ES, Sano H (2004) Distribution of elements on tobacco trichomes and leaves under cadmium and sodium stresses. Plant Biol 47(2):75–82
4. Cobbett CS (2000) Phytochelatins and their roles in heavy metal detoxification. Plant Physiol 123:825–832

5. Cresser MS, Parsons JW (1979) Sulfuric-perchloric acid digestion of plant material for the determination of nitrogen, phosphorus, potassium, calcium and magnesium. Anal Chem Acta 109:431–436
6. Das P, Samantaray S, Rout GR (1997) Studies on cadmium toxicity in plants: a review. Environ Poll 98:29–36
7. Elmayan T, Tepfer M (1994) Synthesis of a bifunctional metallothionein/β-glucuronidase fusion protein in transgenic tobacco plants as a means of reducing leaf cadmium levels. Plant J 6:433–440
8. Espen S, D'Souza SF (2005) Prospects of genetic engineering of plants for phytoremediation of toxic metals. Biotechnol Adv 23:97–114
9. Kahel H (1993) Response of roots of trees to heavy metals. Environ Exp Bot 33:99–119
10. Karnovsky MJ (1965) A formaldehyde-glutaraldehyde fixative of high osmolality for use in electron microscopy. J Cell Biol 27:137–138
11. Kawashima CG, Noji M, Nakamura M, Ogra Y, Suzuki KT, Saito K (2004) Heavy metal tolerance of transgenic tobacco plants over-expressing cysteine synthase. Biotechnol Lett 26:153–157
12. Kim KW (2008) Visualization of micromorphology of leaf epicuticular waxes of the rubber tree *Ficus elastic* by electron microscopy. Micron 39:976–984
13. Kim SH, Song WK, Kim YH, Kwon SY, Lee HS, Lee IC, Kwak SS (2009) Characterization of full-length enriched expressed sequence tags of dehydration-treated white fibrous roots of sweetpotato. J Biochem Mol Biol 42(5):271–276
14. Kim EY, Seo YS, Lee H, Kim WT (2010) Constitutive expression of CaSRP1, a hot pepper small rubber particle protein homolog, resulted in fast growth and improved drought tolerance in transgenic *Arabidopsis* plants. Planta 232:71–83
15. Lee JH, Bae HJ, Jeong JY, Lee JY, Yang YY, Hwang IH, Martinoia E, Lee YS (2003) Functional expression of a bacterial heavy metal transporter in Arabidopsis enhances resistance to and decreases uptake of heavy metals. Plant Physiol 133:589–596
16. Maiti IB, Wagner GJ, Yeargan R, Hunt AG (1989) Inheritance and expression of the mouse metallothionein gene in tobacco. Plant Physiol 91:1020–1024
17. Oh SK, Kang HS, Shin DH, Yang JM, Chow KS, Yeang HY, Wagner B, Breiteneder H, Han KH (1999) Isolation, characterization, and functional analysis of a novel cDNA clone encoding a small rubber particle protein from *Hevea brasiliensis*. J Biol Chem 274:17132–17138
18. Pan A, Tie F, Duau Z, Yang M, Wang Z, Li L, Chen Z, Ru B (1994) α-Domain of human metallothionein I_A can bind to metals in transgenic tobacco plants. Mol Gen Genet 242:666–674
19. Robinson BH, Mills TM, Petit D, Fung LE, Green SR, Clothier BE (2000) Natural and induced cadmium-accumulation in poplar and willow: implications for phytoremediation. Plant Soil 227:301–306
20. Salt DE, Prince RC, Pickering IJ, Raskin I (1995) Mechanisms of cadmium mobility and accumulation in Indian mustard. Plant Physiol 109:1427–1433
21. Seo SG, Kim JS, Yang YS, Jun BK, Kang SW, Lee GP, Kim W, Kim JB, Lee HU, Kim SH (2010) Cloning and characterization of the new multiple stress responsible gene I (*MuSI*) from sweet potato. Genes Genom 32:552–554
22. Song WY, Sohn EJ, Martinoia E, Lee YJ, Yang YY, Jasinski M, Forestier C, Hwang I, Lee YS (2003) Engineering tolerance and accumulation of lead and cadmium in transgenic plants. Nat Biotechnol 21:914–919
23. Steyn WJA (1959) Leaf analysis. Errors involved in the preparative phase. J Agric Food Chem 7:344–348
24. Welch RM, Norvell WA (1999) Mechanisms of cadmium uptake, translocation and deposition in plants. In: McLaughlin MJ, Singh BR (eds) Cadmium in soils and plant. Kluwer, Dordrecht, pp 125–150
25. Westerman RL (1990) Soil testing and plant analysis, 3rd edn. SSSA book series, no. 3, pp 389–427

26. Yamazaki K (1982) Nutrient solution culture (Japanese). Pak-kyo, Tokyo, p 251
27. Zenk MH (1996) Heavy metal detoxification in higher plants. Gene 179:21–30
28. Zhu YL, Pilon-Smits EAH, Jouanin L, Terry N (1999) Overexpression of glutathione synthetase in Indian mustard enhances cadmium accumulation and tolerance. Plant Physiol 119:73–79

CHAPTER 18

Cloning and Analysis of Expression Patterns and Transcriptional Regulation of *Rghbng* in Response to Plant Growth Regulators and Abiotic Stresses in *Rehmannia Glutinosa*

Yanqing Zhou[1,2]*, Yonghua Zhang[1]† , Jun Wei[1]† , Yu Zhang[1] , Jingyun Li[1] , Wanshen Wang[1] , Hongying Duan[1] and Juanjuan Chen[1]

[1] *College of Life Sciences, Henan Normal University, No.46 Jianshe Road, Xinxiang 453007, Henan, China*

[2]*Key Laboratory for Microorganisms and Functional Molecules, University of Henan Province, No.46 Jianshe Road, Xinxiang 453007, Henan, China*

ABSTRACT

RghBNG, a gene of unknown function, was cloned from *Rehmannia glutinosa* by reverse transcription PCR and rapid amplification of cDNA ends. The full-length cDNA of *RghBNG* was 548 bp with a282-bp open reading frame. It encoded a polypeptide of 93 amino acids with a predicted molecular weight of 10.5 kDa and a theoretical isoelectric point of 9.25. Bioinformatics analysis indicated that *RghBNG* had no homology to any known plant genes, whereas the RghBNG polypeptide was highly similar to other plant proteins and possessed one conserved B12D protein family functional domain. Phylogenetic analysis revealed that *RghBNG* encoded for a dicot protein. *RghBNG* spatial and temporal expression patterns and responses to abiotic stresses and plant growth regulators were investigated by qRT-PCR. RghBNG transcripts were detected in roots, stems, leaves, petals, receptacles, stamens and pistils with the highest and lowest levels respectively observed in petals and leaves of mature plants. Additionally, RghBNG transcripts were detected at three developmental stages of roots, stems and leaves; the highest levels were observed in roots at seedling stage; Transcript levels changed to varying degrees in different tissues and stages; We also studied the effects of abiotic stress and plant growth regulators in roots and leaves. *RghBNG* expression was significantly increased ($p < 0.01$) by chromium, gibberellic acid and NaCl, with the highest levels induced by chromium stress; In contrast, 6-benzyladenine reduced expression. These results strongly suggest that *RghBNG* is involved in *R. glutinosa* growth, development and response to plant growth regulators and abiotic stresses.

Keywords: *Rehmannia glutinosa RghBNG* gene Cloning and expression; qRT-PCR; Abiotic stress; Plant growth regulator

INTRODUCTION

Abiotic stresses reduce average yields of most crops by more than 50% (Wang et al. 2003; Bhatnagar-Mathur et al. 2008). Drought, salinity and extreme temperature are among the major abiotic stresses, hampering plant growth and productivity and frequently cause a series of morphological, physiological and biochemical changes (Mehrotra et al. 2014). In addition, heavy metal contamination of soil and water is a global problem giving rise to crop yield losses and having hazardous effects on human health when these metals enter the food chain (Vernay et al. 2007). Because plant vulnerability to such abiotic stresses is a serious threat to the entire plant ecosystem, molecular and genetic mechanisms of abiotic stress tolerance in plants are of great scientific interest (Alexandra et al. 2013). Plants respond and adapt to stress conditions with an array of biochemical and physiological changes, many of which are regulated by stress-responsive gene expression. The regulation of these abiotic stress responses to generate stress resistant plants is of great importance to plant growers (Sarika and Aaron 2014). One important strategy related to this goal is the development of stress-tolerant transgenic plants (Tuli et al. 2010). To develop such plants, stress-related plant genes must be identified, characterized and tested in plant systems for their potential use in commercially important crops (Kumar et al. 2013; Quaggiotti et al. 2007). *Rehmannia glutinosa* (*Scrophulariaceae*) is a well-known medicinal plant widely cultivated in Asian countries such as China, Korea, Japan, Vietnam and others. The tuberous roots are used to treat fever, nervous conditions, diabetes and hypertension, to strengthen liver function, and for hematopoietic, immuno-enhancing, and tonic purposes (Kim et al. 2012; Sun et al. 2010). Because root yield and quality are limited by various stresses. the incorporation of some stress-tolerant genes is needed to improve *R. glutinosa* resistance. The effects of several abiotic stresses, such as temperature, water, NaCl, Paraquat and choline chloride have been previously reported for *R. glutinosa* (Chung et al. 2006; Jia et al. 2009; Zhao et al. 2007; Piao et al. 2008). The NaCl- and waterlogging stress-intolerant *Aux/IAA* gene from *R. glutinosa* has also been recently described (Wang et al. 2013). Nevertheless, the effects of many other abiotic stresses and the functions of other stress-related genes remain unknown in *R. glutinosa*. In the present study, we therefore isolated the *RghBGN* gene from *R. glutinosa*. To elucidate the role of *RghBGN* in *R. glutinosa* growth, development, and stress responses, we investigated its expression patterns in the presence and absence of abiotic stresses and plant growth regulators.

RESULTS

Cloning and sequencing of *RghBNG*

One 221 bp cDNA fragment (middle fragment) was amplified by RT-PCR with the primer pair P1, cloned and sequenced. Its 3′-end product and 5′-end product were generated by 3′-RACE and 5′-RACE, respectively. After both were sequenced, three fragments were aligned linked together to give a 548 bp full-length cDNA known as *RghBNG* (GenBank Accession No.JX290370) including a 5′-untranslated region of 68 bp and 3′-untranslated region of 199 bp. Its 282-bp open reading frame (ORF) was searched by ORF Finder (Figure 1). *RghBNG* gene was amplified by RT-PCR with primer pair P4.

```
1   gaaaatcgacctaatttctcattcattcattcattcaacaaatctcgttccgcgtcaactgtgaata
68  ATGGCT TCA TCC ACC ACT CTC AAG CGA TGG CTG AGG CCC GAGGTGTATCCTTTGTTCGCT
1     M   A   S   S   T   T   L   K   R   W   L   R   P   E   V   Y   P   L   F   A   20
128 GCC GTT GGT GTG GCC GTC GGG ATT TGC GGA ATG CAA TTG ATC CGTAATATTACTATTAAC
21    A   V   G   V   A   V   G   I   C   G   M   Q   L   I   R   N   I   T   I   N   40
188 CCT GAA GTC AGG GTG ACT AAG GAA AAC AGA GCT GCTGGTGTGTTGGATGACTTTGCAGAA
41    P   E   V   R   V   T   K   E   N   R   A   A   G   V   L   D   D   F   A   E   60
248 GGG GAG AAG TAT GCT GAA CAT GCT CTT AGG AAG TAC GTTCGCAACAGGTCTCCTGAAATC
61    G   E   K   Y   A   E   H   A   L   R   K   Y   V   R   N   R   S   P   E   I   80
308 ATGCCGTCCCTTAATAGCTTCTTCACCGATCCTAAGTAT[TGA]ttatgcaagattatgattgtct
81    M   P   S   L   N   S   F   F   T   D   P   K   Y                               93
373 ggtgatacactgattattgaaaatacatttttatggagatcaaatactcaaataagtttgatagaa
440 ctttattggaagattggtatggaatttccaacaccaattcatggatatttacgattttattccccaa
507 gaatgggaagaatgagatgaatgctttactgaaaaaaaaaaa
```

Figure 1. NUCLEOTIDE SEQUENCE OF THE FULL-LENGTH CDNA AND DEDUCED AMINO ACID SEQUENCE OF ***RghBNG*****.** Numbers on the left and right correspond to nucleotide and amino acid sequences, respectively. In the nucleotide sequence, the stop codon is boxed and the polyadenylated tail is indicated in bold italic letters. In the amino acid sequence, the B12D superfamily motif is shaded and in bold.

Bioinfomatics analyses of *RghBNG* gene

Bioinfomatics analyses indicated that *RghBNG* has no homology to any known gene in NCBI at the cDNA level. However, its open reading frame may encode a protein of 93 amino acid residues in length (Figure 1).The protein is composed of 20 amino acids, of which alanine accounts for 9.7%, both arginine and valine for 8.6%, each of the rest for 1.1%-7.5%, alkaline amino acids for14.0% and acidic amino acids for10.8%, in which hydrophilic amino acids and hydrophobic amino acids are evenly distributed, and there are no signal peptide but four phosphorylation sites including 1 Ser site, 1 Thr site and 2 Tyr sites. Its secondary structure is composed of four parts such as α-helixes (56.99%), random coils (27.96%), β-sheets (9.68%) and β-turns (5.38%). Its theoretical isoelectric point (pI), molecular weight and molar extinction coefficient are, 9.25, 10.52 and 56.74, respectively. Furthermore, it has one conserved functional domain of B12D superfamily (Figure 1). Moreover, similarity analyses by the NCBI-blastp showed that the RghBNG protein had a high degree of similarity (58%--86%) to some unknown functional proteins from 14 known species (Figure 2). Meanwhile, it also has another high degree of similarity (>72%) to B12D proteins such as 80% to *Beta vulgaris* (CAK22419.1), 79% to *Camellia sinensis var. assamica* (Mast.) Kitam (AEC10990.1), 79% to *Arabidopsis thaliala* (NP_190397.1), 74% to *Ipomoea batatas* (AAD22104.1), 73% to *Castanea sativa* (AAL17696.1) and 72% to *Wolffia arrhiza* (ADB08699.1).

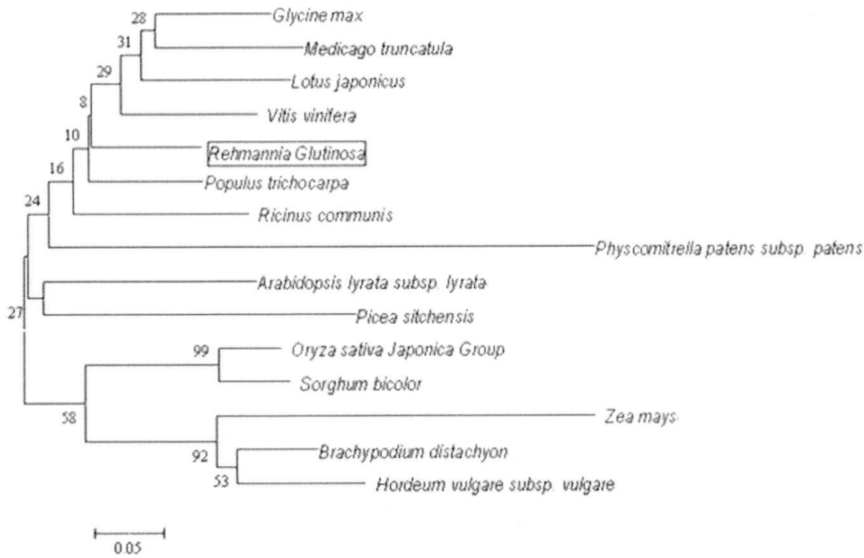

Figure 2. A MOLECULAR PHYLOGENETIC TREE OF RGHBNG AND RELATED PROTEINS GENERATED BY NEIGHBOR-JOINING METHOD USING MEGA 4.0.The unrooted phylogenetic tree was generated based on the alignment of amino acid sequences from 14 plant species: *Populus trichocarpa* (XP_002318014.1; 86% similarity to the RghBNG), *Glycine max* (XP_003527039.1; 84%), *Ricinus communis* (XP_002515002.1; 82%), *Lotus japonicus* (AFK42697.1; 81%), *Medicago truncatula* (ACJ84101.1; 79%), *Arabidopsis lyrata subsp* (XP_002875880.1; 79%), *Vitis vinifera* (XP_002283743.1; 78%), *Brachypodium distachyon* (XP_003557311.1; 72%), *Picea sitchensis* (ABK24106.1; 70%), *Oryza sativa Japonica Group* (NP_001057280.1; 70%), *Sorghum bicolor* (XP_002438146.1; 70%), *Hordeum vulgare subsp. Vulgare* (BAK01669.1; 70%), *Zea mays* (ACN31786.1; 59%) and *Physcomitrella patens subsp. Patens* (XP_001779363.1; 58%). Numbers above nodes are bootstrap support percentages based on 1,000 replicates. The scale bar indicates evolutionary distance of amino acid substitutions per position. RghBNG from *Rehmannia glutinosa* is boxed.

To investigate the phylogenetic relationship of RghBNG with other known homologous proteins, the amino acid sequences of available unknown functional proteins from 14 known species were used to construct the phylogenetic tree (Figure 2). This analysis showed that RghBNG clusters with *Glycine max*, *Medicago truncatula*, *Lotus japonicus* and *Vitis vinifera* with a support value of 29, and that RghBNG protein is attributed to dicots (Figure 2), which is in accordance with APG classification system.

Expression pattern of *RghBNG* in different *Rehmannia glutinosa* tissues and stages

To determine the spatial expression pattern, cDNA was synthesized from three tissues of the plants at seedling stage, seven ones at full blooming stage and three ones at mature stage of *Rehmannia glutinosa*. It was found that RghBNG mRNA was expressed in all the tissues tested, with the strongest signal in root and the significant difference between root and leaf at seedling stage ($\alpha = 0.05$); with the strongest signal in petal and the extremely significant one between petal and each of others, the one between receptacle or stamen and root or stem or leaf, the one between leaf and pistil at full blooming stage ($\alpha = 0.01$); with the strongest one in root, and the extremely significant one between root and stem or leaf ($\alpha = 0.01$), and the significant one between stem and leaf ($\alpha = 0.05$) at mature stage (Figure 3). All qRT-PCR products of RghBNG gene was confirmed by sequencing.

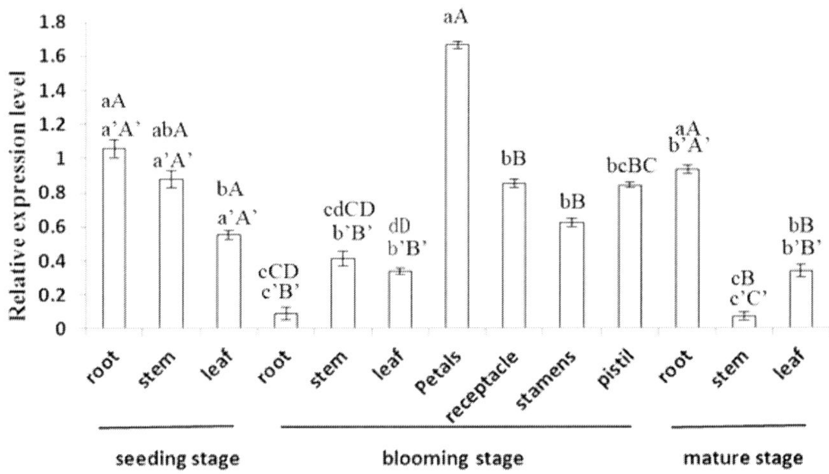

Figure 3. EXPRESSION PATTERN OF THE *Rehmannia glutinosa* *RghBNG* GENE IN DIFFERENT TISSUES AT THREE DIFFERENT DEVELOPMENTAL STAGES AS DETERMINED BY QUANTITATIVE REAL-TIME PCR. Expression levels are normalized relative to those of *TIP41*, which was used as an internal control. Data are means ± SE of three replicates. a-d and A-D in relative (spatial) expression level indicate the significance at 0.05 and 0.01, respectively; a'-c'and A'-C' in relative (temporal) expression level indicate the significance at 0.05 and 0.01, respectively.

The temporal expression profile of RghBNG RNAs during root, stem and leaf developments was determined by qRT-PCR. Their expression levels at three stages were summarized in Figure 3. It was found that RghBNG mRNA was expressed at all the stages tested, with the strongest signal at seedling stage, the extremely significant difference between blooming stage and seedling stage or mature stage ($\alpha = 0.01$) and the significant one between seedling stage and mature stage ($\alpha = 0.05$) for root; with the strongest signal at seedling stage and the extremely significant one between one and each of others among three ones ($\alpha = 0.01$) for stem; with the strongest one at seedling stage, and the extremely significant one between seedling stage and blooming one or mature one ($\alpha = 0.01$) for leaf. All qRT-PCR products of RghBNG gene was confirmed by sequencing.

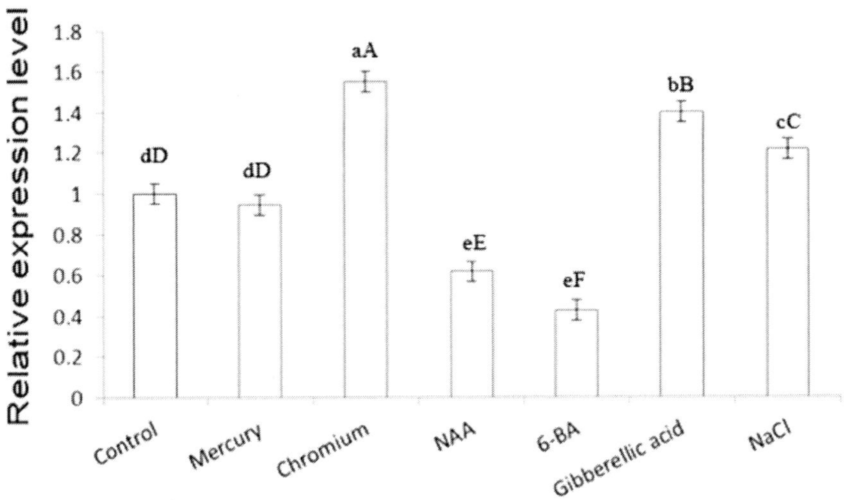

Figure 4. EXPRESSION OF THE *Rehmannia glutinosa RghBNG* GENE IN ROOTS OF PLANTS EXPOSED TO CHROMIUM, MERCURY, NACL, GIBBERELLIC ACID, 1-NAPTHALENE ACETIC ACID (NAA) OR 6-BENZYLADENINE (6-BA). Following treatments, total RNAs were isolated, and RghBNG expression levels were monitored by quantitative real-time PCR. Error bars indicate standard deviation values of three independent analyses. a-c and A-C in relative expression level indicate the significance at 0.05 and 0.01, respectively.

Gene transcription regulation of the *RghBNG* gene

In order to investigate if the expression of the *RghBNG* gene is associated with the abiotic stress and plant growth regulators-induced defense mechanism in *Rehmannia glutinosa*, the expression pattern of the *RghBNG* was examined in young *Rehmannia glutinosa* plants subjected to six factors such as chromium, mercury, NaCl, Gibberellin, NAA and 6-BA treatments by qRT-PCR (Figure ures 4 and 5). For root as shown in Figure 4, compared to control, Chromium, exogenous Gibberellin and NaCl resulted in different increases in *RghBNG* transcript levels with the highest one induced by chromium, whose ones were extremely significant ($\alpha = 0.01$), and that NAA and 6-BA down-regulated *RghBNG* transcript levels, whose ones were extremely significant ($\alpha = 0.01$). However, mercury did not significantly decrease the expression level ($\alpha = 0.05$). For leaf as shown in Figure 5, Chromium, NAA, Gibberellin and NaCl also resulted in different increases in *RghBNG* transcript levels with the highest one under chromium stress, whose ones were extremely significant ($\alpha = 0.01$), while mercury and 6-BA decreased *RghBNG* transcript levels, whose ones was also extremely significant ($\alpha = 0.01$) compared to control.

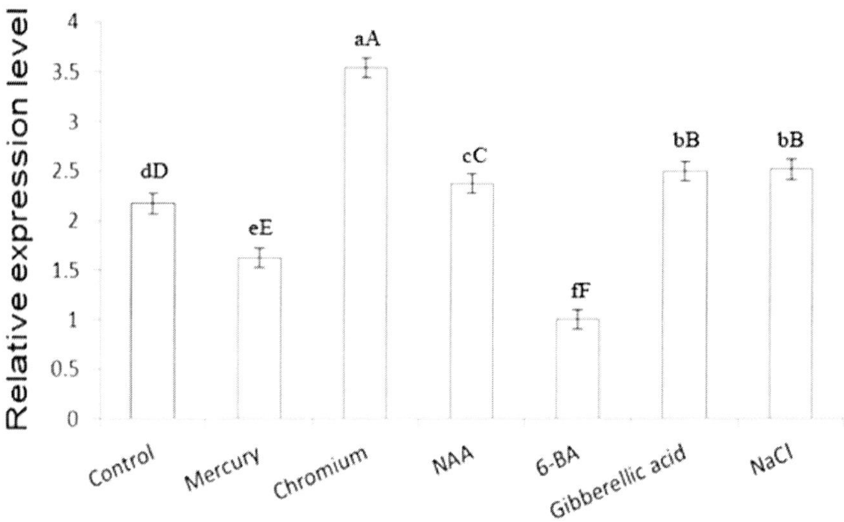

Figure 5. Expression of the Rehmannia glutinosa RghBNG gene in leaves of plants exposed to chromium, mercury, NaCl, gibberellic acid, 1-naphthalene acetic acid (NAA) or 6-benzyladenine (6-BA). Following treatments, total RNAs were isolated, and RghBNG expression levels were monitored by quantitative real-time PCR. Error bars indicate standard deviation values of three independent analyses. a-c and A-C in relative expression level indicate the significance at 0.05 and 0.01, respectively.

DISCUSSION

Cloning, sequencing, and structural analysis of *RghBNG*

In this study, we cloned the full-length sequence of the *RghBNG* gene, deduced its amino acid sequence, and analyzed the structure of the encoded protein. The *RghBNG* gene was obtained from *R. glutinosa* using RT-PCR and RACE technology. No homology was found between the *RghBNG* cDNA sequence and that of any other plant species gene, but considerable similarity was uncovered between the amino acid sequence of the deduced RghBNG protein and proteins from other species, which could be divided into two groups: B12D proteins, and proteins of unknown function. In addition, RghBNG protein possesses the conserved structure domain of B12D protein family, B12D superfamily,(Walker et al. 1992, Figure 1), belonging to one kind of subunits of the NADH: ubiquinone oxidoreductase (Complex I) in mitochondrial membrane (Gabaldón et al. 2005; Zhang et al. 2012).

The roles of other known B12D protein genes

The B12D gene was first isolated from barley (Aalen et al. 1994) and was initially characterized as a transcript related to barley aleurone and embryo development (Steinum et al. 1998). Several B12D protein-like protein genes have been subsequently cloned and their roles have been studied. For example, B12D protein has been shown to be localized in peroxisomes of Arabidopsis and rice (Kaur and Hu 2011), and GA3 up-regulates B12D protein gene expression in imbibed barley embryoless grains (Aalen et al. 1994). However, abscisic acid (ABA) suppresses B12D expression in imbibed barley embryoless grains (Steinum et al. 1998), while ABA and 20% mannitol inhibit its expression in barley embryos (Aalen et al. 1994). In another study, B12D protein was specifically expressed by the induction of ABA (Aalen and Steinum 2001). A B12D-like protein of unknown function was found to be constitutively expressed in Arabidopsis seedlings, roots, leaves, inflorescences, flowers and siliques at different developmental stages (Zhu et al. 2001). A soybean B12D protein gene (Glyma11g37370.1) has been observed to increase the female index of soybean cyst nematode (Matthews et al. 2013). The C832 (DT045059) gene, which has been found to be differentially expressed between wheat near-isogenic lines Chisholm-T and Chisholm-S under Al stress, shows similarity with known putative B12D protein genes in GenBank and TIGR databases (Guo et al. 2007). Plant-specific B12D proteins of unknown function are strongly down-regulated in root tissues of wheat genotypes with distinct responses to water deficit (Ergen et al. 2009). A B12D protein encoded by TDF41

(NP_001057280) is induced by sorbitol, and indirectly promotes plantlet regeneration frequency of rice calli (Feng et al. 2011). An abiotic stress-regulated cDNA clone (CD725202) from *Pennisetum* encodes proteins involved in cellular homeostasis, including a B12D-like protein that protects plants from stress-induced damages and facilitates the establishment of cellular homeostasis (Mishra et al. 2007). A cDNA sequence (CdR311) with identity to the mRNA sequence of a *Zea mays* B12D-like protein (AY104072) is up-regulated in roots of Bermuda grass growing under petroleum stress, but its biological activity remains unknown (Peña-Castro et al. 2006), B12D proteins may be involved in general functions such as signal transduction, disease resistance, cell growth, and differentiation (Vinod et al. 2010). A B12D protein cDNA (BP947551) showed 100% transcript variation in NaCl-treated leaves and roots of *Burma mangrove*, making it a perfect candidate as a stress-responsive protein (Miyama et al. 2006). Another B12D-like protein gene (BP945437) has been discovered to be highly repressed in salt-treated *Burma mangrove* roots (Miyama and Hanagata 2007). *Castanea sativa* B12D-like protein mRNA is differentially expressed upon wounding and infection of chestnut with chestnut blight fungus (Schafleitner and Wilhelm 2002). B12D proteins also have been inferred to participate in ATP synthesis and energy metabolism (Gabaldón et al. 2005; Zhang et al. 2012).

Spatial and temporal expression of *RghBNG*

The roles of *RghBNG* in plant development were examined by characterizing its spatial and temporal expression patterns during *R. glutinosa* growth and development by qRT-PCR. RghBNG mRNA transcripts were detected in all tested tissues and organs, but the transcript levels varied. The highest levels were clearly detected in petals at the full-blooming stage and roots at both seedling and mature stages (Figure 3), demonstrating that *RghBNG* may play an important role in *R. glutinosa* growth and development.

Transcriptional expression of *RghBNG* in response to abiotic stresses and plant growth regulators

Plants are frequently exposed to many forms of stress. Plant adaptation to stress has been suggested to be mediated by both preexisting and induced defense mechanisms. Plant signaling pathways are driven by plant growth regulators and reactive oxygen species. Phytohormones generate a signal transduction network that leads to a cascade of events responsible for plant adaptation to external conditions (Al-Momany and Abu-Romman 2014). To assign possible gene physiological functions, detailed expression analysis of the target gene is an important and necessary step (Pérez-Torres et al. 2009). In our study, *RghBNG* responses to abiotic stresses and plant growth regulators were investigated by qRT-PCR. The *RghBNG* gene was induced by GA3 and NaCl, in agreement with previous studies of B12D proteins (Steinum et al. 1998; Aalen et al. 1994; Aalen and Steinum 2001; Miyama et al. 2006; Zhang et al. 2012). Furthermore, our results revealed that the *RghBNG* gene significantly responded to abiotic stress induced by chromium and plant growth regulators such as NAA and 6-BA (Figure ures 4 and 5). These findings reveal that RghBNG proteins possessed some known roles for B12D proteins such as GA3 and NaCl induction and novel roles for B12D proteins (one group) or the proteins of unknown function (the other group) mentioned above, for example, responses to Cr, Hg, NAA and 6-BA.

CONCLUSIONS

A full-length cDNA of the *RghBNG* gene from *R. glutinosa* was cloned and characterized, and its spatial and temporal expression profiles and abiotic stress and plant growth regulator-responsive patterns were determined. This study has provided essential molecular data that can be used in follow-up studies to elucidate the biological roles of *RghBNG* in growth, development, and stress and plant growth regulator-responses of *R. glutinosa*.

MATERIALS AND METHODS

Plant materials and treatment conditions

Sterile 25-day-old plants of *R. glutinosa* f. *hueichingensis* 'Wen85-5' were grown in hormone-free Murashige-Skoog (MS) media or transplanted into flower pots in a greenhouse to obtain full-blooming plants and mature plants (Zhou et al. 2009). Leaf sample were collected for gene cloning. For expression pattern analysis, samples of fresh leaves, stems, roots, petals, receptacles, stamens and pistils were collected from *R. glutinosa* plants during corresponding vegetative growth, full-blooming and early senescence phases. To investigate gene expression regulation under abiotic and hormonal stresses, 1.5-cm-long shoots with two leaves from 25-day-old plants, five shoots per treatment, were inserted and maintained for 20 days in a 450-mL glass bottle containing MS medium either lacking (control) or supplemented with one of the following impact factors: 200 μM NAA,

500 μM 6-BA, 150 μM $K_2Cr_2O_7$ or 200 μM $HgCl_2$. At the end of the treatment period, five control plants each were sprayed every other hour for 48 h either with a solution of 100 μM GA3 or 250 mM NaCl, and leaf and root samples were then collected. The experiment was repeated three times in a greenhouse. All samples were immediately frozen in liquid nitrogen after collection and stored at −80°C until further analysis.

Total RNA extraction and first-strand cDNA preparation

Total RNA was extracted from leaf samples using an RNA kit (Lifefeng Biotechnology Co., Shanghai, China). RNA concentration and purity were estimated spectrophotometrically based on absorbance at 260 and 280 nm. First-strand cDNA was synthesized from 1 μg total RNA, which was reversely transcribed using M-MLV reverse transcriptase with oligo $(dT)_{15}$ as a primer in a 10-μL reaction volume (Takara, Japan).

Cloning of *RghBNG* cDNA

The middle region of the *RghBNG* gene was amplified using first-strand cDNA as a template with primer pair P1 (Table 1). The amplification reaction was performed in a total volume of 25 μL using a CodeDR011 PCR amplification kit (Takara). Amplification conditions were as follows: 94°C for 5 min followed by 30 cycles of 94°C for 30 s, 46°C for 1 min and 72°C for 1 min, with a final extension of 72°C for 10 min. *RghBNG* 5′ and 3′ ends were cloned with primer pairs P2-P2′ and P3′-P3 using 5′-and 3′-Full RACE kits (Takara) according to the manufacturer's instructions. The full-length cDNA of the *RghBNG* gene was obtained by aligning the three cDNA fragments using DNAMAN4.0 software. The ORF was determined with ORF Finder and cloned by RT-PCR with primer pair P4 (Table 1), which was designed based on the ORF nucleotide sequence. All amplicons were separated, excised, purified, ligated, transformed and sequenced following Al-Momany and Abu-Romman (2014), except that we replaced the T vector and host used in that study with pMD19-T vector and *Escherichia coli* DH5α.

Bioinformatics analyses

The *RghBNG* nucleotide sequence was aligned using BLAST on the NCBI website. The ExPASy proteomics server (http://www. expasy.ch /tools/protparam.html) was used to determine the amino acid sequence, molecular weight, isoelectric point and other physicochemical properties of the RghBNG protein. The Expert Protein Analysis System proteomics server (http://www.expasy.org/) and the Predict Protein server (http://www.Predictprotein.org/) were used to predict the structure and function of the encoded protein. Phylogenetic trees were constructed by the neighbor-joining method with 1,000 bootstrap replicates using MEGA 4.0 software.

Table 1. Primers used in this study

Primer pair	Primer sequence (5′ → 3′)	Description
P1	F: GGTCTAACRGCRTCWYTGTT	Gene cloning
	R: AATBGGRACYAAYARCARCATA	
P2	OuterF: GCTGGGCTGACTCCACCTACTCT	Gene cloning
	InnerF: CGCTTGCCCAATTGTTTGTGGCC	
P3	OuterR: GACTTCAGGGTTAATAGTAAT	Gene cloning
	InnerR: CGCAAATCCCGACGGCCACAAAC	
P2′	OuterR:TACCGTCGTTCCACTAGTGATTT	Gene specific primer for 3′ RACE
	InnerR: CGCGGATCCTCCACTAGTGATTTCACTATAGG	
P3′	OuterF: CATGGCTACATGCTGACAGCCTA	Gene specific primer for 5′ RACE
	InnerF: CGCGGATCCACAGCCTACTGATGATCAGTCGATG	
P4	F: CGCGGATCCAAAATCGACCTAATTTCTC	Gene-specific primer for ORF
	R: CCGCTCGAGCAGTAAAGCATTCATCTC	
P5	F: GTTGGTGTGGCCGTCGGGATTT	Gene-specific primers for real-time qPCR
	R: AGCATACTTCTCCCCTTCTGCAAAG	
P6	F: TGGCTCAGAGTTGATGGAGTGCT	Gene-specific primers for *TIP41*
	R: CTCTCCAGCAGCTTTCTCGGAGA	

Analyses of *RghBNG* gene expression and transcriptional regulation

We used qRT-PCR to monitor *RghBNG* transcript levels in tissues and organs of *R. glutinosa* plants at three developmental phases as well as leaves and roots subjected to abiotic and hormonal stress treatments. PCR experiments were carried out using the first-strand cDNAs as templates with primer pair p5 (Table 1) as described above. As a control, *TIP41* was also amplified by RT-PCR using primer pair P6 (Table 1). qRT-PCR was performed on an ABI 7500 Real-Time PCR instrument (Cwbiotech, Beijing, China) in 20-μL reaction volumes comprising 1 μL of cDNA template, 10 μL of 2× UltraSYBR mixture (with ROX), 1 μL of each primer (5 μM), and 8 μL ddH_2O. PCR conditions were as follows: 95°C for 10 min followed by 40 cycles of 95°C for 15 s, 60°C for 1 min and 72°C for 5 min. The fluorescent product was detected in the last step of each cycle. Amplicons were diluted 16-fold and used to generate a calibration curve for determination of the PCR amplification efficiency of each gene. After acquiring *Ct* values of *RghBNG* and *TIP41*, relative mRNA expression (relative quantity) of target genes was determined with the formula described in Liu et al. (2012). Results were presented as means ± standard error (SE) of data from triplicate replicates. Data were additionally analyzed by one-way analysis of variance and the least significant difference test in SPSS 13.0.

NOTES

Yonghua Zhang and Jun Wei contributed equally to this work.

ACKNOWLEDGMENTS

The work was supported by Personnel Training United Fund of NSFC-Henan province (U1304304), National Training Programs of Innovation and Entrepreneurship for Undergraduates (201310476102), Program for Innovative Research Team (in Science and Technology) in University of Henan Province (No. 13IRTSTHN009) and Key Research Program of Science and technology of Education department of Henan province (No.14B180028).

COMPETING INTERESTS

The authors declare that they have no competing interests.

AUTHORS' CONTRIBUTIONS

Yanqing Zhou: AB and FG. Yonghua Zhang: AB. Yu Zhang: JY and MT. Jingyun Li; ES. Wanshen Wang: ES. Hongying Duan: FG. Jun Wei: AB. Juanjuan Chen: MT. All authors read and approved the final manuscript.

REFERENCES

1. Aalen RB, Steinum SZ (2001) Stability of barley aleurone transcripts: Dependence on protein synthesis, influence of the starchy endosperm and destabilization by GA3. Physiol Plantarum 112:403–413
2. Aalen RB, Opsahl-Ferstad HG, Linnestad C, Olsen OA (1994) Transcripts encoding an oleosin and a dormancy related protein are present in both the aleurone layer and the embryo of developing barley (Plordeum vitlgare L.) seeds. Plant J 5:385–396

3. Alexandra SD, Konstantin VK, Valeriya SK (2013) Expression of calcium-dependent protein kinase (CDPK) genes under abiotic stress conditions in wild-growing grapevine Vitis amurensis. J Plant Physiol 170:1491–1500
4. Al-Momany B and Abu-Romman S (2014) Cloning and molecular characterization of a flavin-dependent oxidoreductase gene from barley. J Appl Genetics.
5. Bhatnagar-Mathur P, Vadez V, Sharma KK (2008) Transgenic approaches for abiotic stress tolerance in plants: retrospect and prospects. Plant Cell Rep 27:411–424
6. Chung IM, Kim JJ, Lim JD, Yu CY, Kim SH, Hahn SJ (2006) Comparison of resveratrol, SOD activity, phenolic compounds and free amino acids in Rehmannia glutinosa under temperature and water stress. Environ Exp Bot 56(1):44–53
7. Ergen NZ, Thimmapuram J, Bohnert HJ, Budak H (2009) Transcriptome pathways unique to dehydration tolerant relatives of modern wheat. Funct Integr Genomics 9:377–396
8. Feng XL, Zhao PK, Hao JQ, Hu JQ, Kang DM, Wang H (2011) Effects of sorbitol on expression of genes involved in regeneration of upland rice (Oryza sativa L.). Plant Cell Tissue Organ Cult 106:455–463
9. Gabaldón T, Rainey D, Huynen MA (2005) Tracing the evolution of a large protein complex in the eukaryotes, NADH: ubiquinone oxidoreductase (Complex I). J Mol Biol 348(4):857–870
10. Guo PG, Bai GH, Carver B, Li RH, Bernardo A, Baum M (2007) Transcriptional analysis between two wheat near-isogenic lines contrasting in aluminum tolerance under aluminum stress. Mol Genet Genomics 277:1–12
11. Jia YY, He YJ, Liang ZS, Dong F, Fang L (2009) Effect of different moisture treatments on photosynthetic characteristics of Rehmannia glutinosa Libosch. J Northwest Sci-Tech Univ Agri For (Nat Sci Ed) 8:182–186 (in Chinese)
12. Kaur N, Hu JP (2011) Defining the plant peroxisomal proteome: from Arabidopsis to rice. Front Plant Sci 2:1–20
13. Kim YS, Ryuk JA, Ko BS (2012) Discrimination of Korean Rehmannia glutinosa from Chinese Rehmannia glutinosa using sequence-characterized amplified region marker. J Korean Soc Appl Biol Chem 55:1–6
14. Kumar S, Asif MH, Chakrabarty D, Tripathi RD, Dubey RS, Trivedi PK (2013) Expression of a rice Lambda class of glutathione S-transferase, OsGSTL2, in Arabidopsis provides tolerance to heavy metal and other abiotic stresses. J Hazard Mater 248–249:228–237
15. Liu HH, Wang JW, Si JM, Jia J, Li L, Han CC et al (2012) Molecular cloning and in silico analysis of the duck (Anas platyrhynchos) MEF2A gene cDNA and its expression profile in muscle tissues during fetal development. Genet Mol Biol 35:182–190
16. Matthews BF, Beard H, Macdonald MH, Kabir S, Youssef RM, Hosseini P et al. (2013) Engineered resistance and hypersusceptibility through functional metabolic studies of 100 genes in soybean to its major pathogen, the soybean cyst nematode. Planta 237:1337–1357
17. Mehrotra R, Bhalothia P, Bansal P, Basantani MK, Bharti V, Mehrotra S (2014) Abscisicacid and abiotic stress tolerance – Different tiers of regulation. J Plant Physiol 171:486–496
18. Mishra RN, Reddy PS, Nair S, Markandeya G, Reddy RR, Sopory SK et al. (2007) Isolation and characterization of expressed sequence tags (ESTs) from subtracted cDNA libraries of Pennisetum glaucum seedlings. Plant Mol Biol 64:713–732
19. Miyama M, Hanagata N (2007) Microarray analysis of 7029 gene expression patterns in burma mangrove under high-salinity stress. Plant Sci 172:948–957
20. Miyama M, Shimizu H, Sugiyama M, Hanagata N (2006) Sequencing and analysis of 14,842 expressed sequence tags of burma mangrove, Bruguiera gymnorrhiza. Plant Sci 171:234–241
21. Peña-Castro JM, Barrera-Figure ueroa BE, Ferna'ndez-Linares L, Ruiz-Medrano R, Xoconostle-Ca'zares B (2006) Isolation and identification of up-regulated genes in bermudagrass roots (Cynodon dactylon L.) grown under petroleum hydrocarbon stress. Plant Sci 170:724–731

22. Pérez-Torres E, ParedesM PV, Becerra V (2009) Gene expression analysis: a way to study tolerance to abiotic stresses in crops species. Chil J Agri Res 69(2):260–269
23. Piao RZ, Zhao HY, Jin YJ, Lin H (2008) Influences of paraquat stress on membrane lipid peroxidation and antioxidative activity in Rehmannia glutinosa and piemarker. Jiangsu Agril Sci 3:271–274 (in Chinese)
24. Quaggiotti S, Barcaccia G, Schiavon M, Nicolé S, Galla G, Rossignolo V et al (2007) Phytoremediation of chromium using Salixs pecies: Cloning ESTs and candidate genes involved in the Cr response. Gene 402:68–80
25. Sarika GS, Aaron MR (2014) Expression patterns and regulation of SlCRF3and SlCRF5 in response to cytokinin and abiotic stresses in tomato (Solanum lycopersicum). J Plant Physi 171:349–358
26. Schafleitner R, Wilhelm E (2002) Isolation of wound-responsive genes from chestnut (Castanea sativa) microstems by mRNA display and their differential expression upon wounding and infection with the chestnut blight fungus (Cryphonectria parasitica)". Physiol Mol Plant Pathol 61:339–348
27. Steinum TM, Berner HS, Stacy RAP, Salehian Z, Aalen RB (1998) Differential regulation of the barley (Hordeum vulgare)transcripts B22E and B12D in mature aleurone layers. Physiologia plantarum 102:337–345
28. Sun P, Guo YH, Qi JJ, Zhou LL, Li XE (2010) Isolation and expression analysis of tuberous root development related genes in Rehmannia glutinosa. Mol Bio Rep 37:1069–1079
29. Tuli R, Chakrabarty D, Trivedi PK, Tripathi RD (2010) Recent advances in arsenic accumulation and metabolism in rice. Mol Breed 26:307–323
30. Vernay P, Gauthier-Moussard C, Hitmi A (2007) Interaction of bioaccumulation of heavy metal chromium with water relation, mineral nutrition and photosynthesis in developed leaves of Lolium perenne L. Chemosphere 68:1563–1575
31. Vinod MS, Sankararamasubramanian HM, Priyanka R, Ganesan G, Parida A (2010) Gene expression analysis of volatile-rich male flowers of dioecious Pandanus fascicularis using expressed sequence tags. J Plant Physiol 167:914–919
32. Walker JE, Arizmendi JM, Dupuis A, Ian MF, Moshe F, Simon MM et al (1992) Sequences of 20 subunits of NADH: ubiquinone oxidoreductase from bovine heart mitochondria. application of a novel strategy for sequencing proteins using the polymerase chain reaction. J Mol Biol 226:1051–1072
33. Wang W, Vinocur B, Altman A (2003) Plant responses to drought, salinity and extreme temperatures towards genetic engineering for stress tolerance. Planta 218:1–14
34. Wang FQ, Tian YH, Li MJ, Yang JF, Zhang B, Lin WX et al (2013) Molecular cloning and expression analysis of an Aux/IAA gene (RgIAA1) from Rehmannia glutinosa. China J C Mater Med 38(23):10–15 (in Chinese)
35. Zhang YY, Huang N, Xiao XH, Huang L, Su WH, Xu LP, Que YX (2012) Cloning and expression analysis of B12D gene in sugarcane. http://www.paper.edu.cn
36. Zhao HJ, Tan JF, Qi CM (2007) Photosynthesis of Rehmannia glutinosa subjected to drought stress is enhanced by choline chloride through alleviating lipid peroxidation and increasing proline accumulation. Plant Growth Regul 51(3):255–262
37. Zhou YQ, Duan HY, Zhou CE, Li JJ, Gu FP, Wang F et al (2009) Hairy root induction and plant regeneration of Rehmann glutinosa Libosch. f. hueichingensis Hsiao via Agrobacterium rhizogenes-mediated transformation. Russ J Plant Physl 56(2):224–231
38. Zhu T, Budworth P, Han B, Brown D, Chang HS, Zou GZ et al. (2001) Toward elucidating the global gene expression patterns of developing Arabidopsis: Parallel analysis of 8300 genes by a high-density oligonucleotide probe array Plant Physiol. Biochem 39:221–242

CHAPTER 19

RNA Interference: A Novel Source of Resistance to Combat Plant Parasitic Nematodes

Sagar Banerjee[1,2,3], Anamika Banerjee[1], Sarvajeet S. Gill[2], Om P. Gupta[3], Anil Dahuja[3], Pradeep K. Jain[4] and Anil Sirohi[1*]

[1]*Division of Nematology, Indian Agricultural Research Institute (ICAR), New Delhi, India*

[2]*Centre for Biotechnology, Maharshi Dayanand University, Rohtak, India*

[3]*Division of Biochemistry, Indian Agricultural Research Institute (ICAR), New Delhi, India*

[4]*National Research Centre on Plant Biotechnology (ICAR), New Delhi, India*

ABSTRACT

Plant parasitic nematodes cause severe damage and yield loss in major crops all over the world. Available control strategies include use of insecticides/nematicides but these have proved detrimental to the environment, while other strategies like crop rotation and resistant cultivars have serious limitations. This scenario provides an opportunity for the utilization of technological advances like RNA interference (RNAi) to engineer resistance against these devastating parasites. First demonstrated in the model free living nematode, *Caenorhabtidis elegans*; the phenomenon of RNAi has been successfully used to suppress essential genes of plant parasitic nematodes involved in parasitism, nematode development and mRNA metabolism. Synthetic neurotransmitants mixed with dsRNA solutions are used for *in vitro* RNAi in plant parasitic nematodes with significant success. However, host delivered *in planta* RNAi has proved to be a pioneering phenomenon to deliver dsRNAs to feeding nematodes and silence the target genes to achieve resistance. Highly enriched genomic databases are exploited to limit off target effects and ensure sequence specific silencing. Technological advances like gene stacking and use of nematode inducible and tissue specific promoters can further enhance the utility of RNAi based transgenics against plant parasitic nematodes.

Keywords: plant parasitic nematodes, host delivered RNAi, root-knot nematodes, cyst nematodes, dsRNA, siRNA

INTRODUCTION

Plant parasitic nematodes (PPNs) have emerged as a severe threat to crop production and are responsible for an estimated loss of US $173 billion annually to world agriculture (Elling, 2013). In addition to the direct damage caused to the plants, nematode infection facilitates subsequent attack by other plant pathogens such as bacteria and fungi. Sedentary endoparasites are the most significant and economically damaging PPNs which include the genera *Meloidogyne* [root-knot nematodes (RKNs)], *Heterodera* and *Globodera* [Cyst Nematodes (CNs)]. The RKNs have a broad host range and form characteristic root galls on the host roots, while CNs have a comparatively constricted host range. The life cycles of these parasites are characteristically different from each other involving complex interactions with their hosts (Williamson and Gleason, 2003). However, both CNs and RKNs inject their salivary secretions in plant cells and withdraw nutrients (Atkinson et al., 1988; Li et al., 2011) by forming multi-cellular feeding sites called syncytia and giant cells, respectively. The infected plants remain stunted, show wilting symptoms and become prone to enhanced susceptibility to other diseases (Fairbairn et al., 2007). Migratory endoparasitic nematodes such as *Radopholus* spp. and *Pratylenchus* spp. are also highly damaging as in contrast to sedentary endoparasites, the migratory endoparasites remain motile and migrate through the course of their development.

Various methods have been used for management of PPNs including development of resistant cultivars, chemical nematicides and cultural practices. Use of nematicides as means of chemical control has been most effective in managing PPNs. However, detrimental effects of the major nematicides on environment and human health have compelled various developed and developing countries to impose bans on their use. Nematode resistant genes are present in some cultivars of species like tomato, potato and soybean but they are only limited to specific pathotypes while many crops do not have the presence of resistance loci (Fairbairn et al., 2007). This situation leads us to utilize technological advancements like RNA interference (RNAi) for engineering resistance against important nematode pests in crop plants. RNAi refers to sequence specific and homology-dependent gene silencing through a complex mechanism in which double stranded RNA (dsRNA) is recognized which leads to a chain of events resulting in the degradation of both the dsRNA and homologous RNA. Since its first description in *Caenorhabtidis elegans* (Fire et al., 1998), this highly conserved mechanism of RNAi has been demonstrated in various organisms belonging to different species across the animal and plant kingdoms (Jones et al., 2011). After the onset of RNAi by exogenous dsRNA, the intestinal epithelial cells of the nematodes are generally involved in dsRNA uptake. In case, there is no expression of the target gene in the uptake cells, silencing of the target transcript is systematically spread to more distant cells (Whangbo and Hunter, 2008). The most imperative aspect of RNA interference is that, it is highly precise, remarkably powerful and the interference can be caused in cells and tissues far away from the site of introduction (Rosso et al., 2009; Tomoyasu et al., 2008).

RNAI IN PPNS BY SOAKING IN DSRNA SOLUTIONS

In vitro RNAi has been successfully demonstrated in CNs (Urwin et al., 2002), RKNs (Rosso et al., 2005) and even migratory nematodes (Haegeman et al., 2009; Soumi et al., 2012; Li et al., 2015) by feeding dsRNA solutions. Three different methods have been used in *C. elegans* for introduction of dsRNA which include feeding on bacteria expressing target gene dsRNA (Timmons and Fire, 1998; Kamath et al., 2001; Timmons et al., 2001), soaking of nematodes in dsRNA solution facilitating its oral uptake (Tabara et al., 1998) and microinjection (Fire et al., 1998; Mello and Conte, 2004). But, in case of PPNs, microinjection has not been effective because of the small size of the infective stages and their inability to ingest fluid without host plant infection. However, a variety of technical and chemical advances have successfully enhanced efficient dsRNA uptake from solutions by soaking. Urwin et al. (2002) first demonstrated *in vitro* RNAi in PPNs successfully by using a neuroactive compound, octopamine to facilitate dsRNA uptake by second stage juveniles (J2s) of CNs *H. glycines* and *G. palida*. In further studies, induction of dsRNA uptake by *Meloidogyne incognita* J2s was enhanced by using the same method (Bakhetia et al., 2005; Shingles et al., 2007). In other studies, resorcinol and serotonin were used for successful uptake of dsRNA in *M. incognita* (Rosso et al., 2005; Huang et al., 2006) and lipofectin was used in case of *Bursaphelenchus xylophilus* (Park et al., 2008). Fluoroscein isothiocyanate (FITC) have been used as a visual marker for monitoring dsRNA uptake and selection of individuals in various studies (Urwin et al., 2002; Rosso et al., 2005; Dutta et al., 2015b).

Different time periods of exposure of J2s to dsRNA in a range of 1 h to 7 days have been employed in different studies. For the knockdown of *Mi-gsts-1, M. incognita* J2s were incubated with *Mi-gsts-1* dsRNA for 1 h with 90% reduction in the transcript expression (Dubreuil et al., 2007). Bakhetia et al. (2005) demonstrated highly efficient *in vitro* dsRNA uptake by J2s of *M. incognita* with 4 h incubation period. However, further studies showed increase in transcript reduction resulting in desired phenotypic effects with increase in the incubation time (Kimber et al., 2007; Jones et al., 2011). Majority of the reports showing *in vitro* dsRNA soaking have not assessed the stability of the resulting gene knock down. However, the efficiency and duration of the silencing effect was assessed for *M. incognita* calreticulin (*Mi-crt*) and polygalacturonase (*Mi-pg-1*) by Rosso et al. (2005). The transcript repression was highest at 20 and 44 h after soaking into dsRNA solution for *Mi-crt* and *Mi-pg-1* respectively. However, the silencing effect was not detectable for both the genes 68 h after soaking. In *H. glycines*, the reduction in the transcript levels of β-1,4- endoglucanase was observed immediately after an incubation of 16 h with the corresponding dsRNA and after 15 days of dsRNA treatment, the transcript levels returned to normal (Bakhetia et al., 2007). These reports suggest that the silencing achieved due to soaking in dsRNA solutions is often transient and lack stability. In general, it can be concluded that the soaking method is an efficient tool for identification of gene function and expression. However, the obligate parasitic nature of plant parasitic nematodes and their exclusivity to feed on plant cells throughout their life cycle inside the host makes *in planta* RNAi technology a suitable approach to combat plant parasitic nematodes. The advantage of the host delivery strategy is that it provides a continuous availability of the dsRNA to the nematode, thereby making the chances of gene suppression reversal remote.

HOST GENERATED RNAI TO SILENCE NEMATODE SPECIFIC GENES

Host generated RNAi has proved to be a revolutionary approach for the delivery of dsRNAs or siRNAs into the feeding nematodes for the silencing of vital nematode specific genes. Purposely, those genes should be targeted whose expression is essential for the nematodes after the feeding starts to ensure a highly lethal phenotype. A dsRNA construct for the target gene is developed by cloning a part of the target gene cDNA in sense and antisense orientation separated by an intron or spacer region. A strong tissue specific or constitutive promoter may be used to drive the expression of the dsRNA. Transcription of the sense and antisense strands results in the formation of a self-complimentary hairpin structure with the removal of the intron by splicing (Smith et al., 2000). The dsRNAs so formed can either be directly ingested by the PPNs or can be processed by the host plant's own RNAi machinery and the resulting siRNAs can be subsequently ingested by the PPNs (Bakhetia et al., 2005; Dutta et al., 2015a; Figure 1).

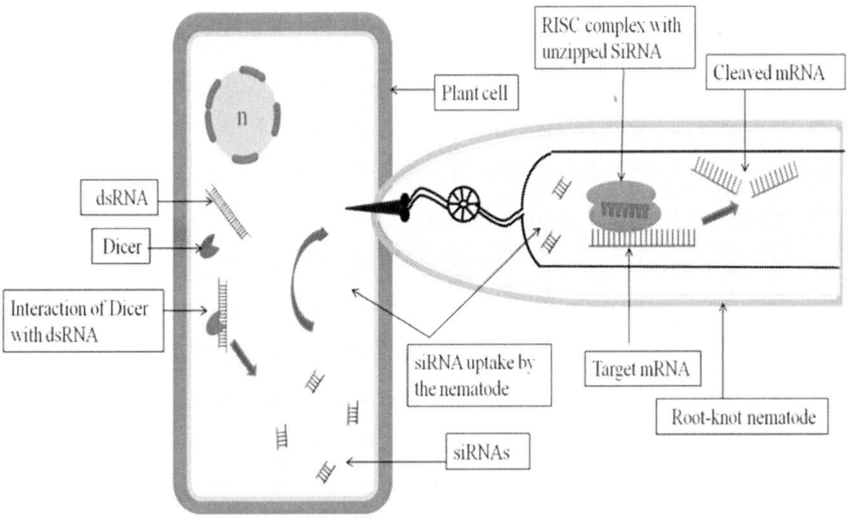

FIGURE 1. HOST GENERATED RNAI THROUGH INTERACTION BETWEEN HOST PLANT CELL AND ROOT-KNOT NEMATODE. The dsRNA introduced into the host plant is recognized by the cellular RNAse III type enzyme dicer, which cuts the dsRNA into shorter fragments of 20–25 nucleotides called siRNAs. During infection into the host, the nematode ingests the siRNAs through its stylet. These host derived siRNAs are then processed by the nematode RNAi machinery where the unzipped siRNAs bound to the RISC complex cleaves the target mRNA in a sequence specific manner and inhibits further translation of the target mRNA.

Host generated RNAi has been demonstrated by targeting different nematode genes which may be broadly classified under three categories: housekeeping genes, parasitism or effector genes and genes associated with nematode development. Yadav et al. (2006) first demonstrated host delivered RNAi in tobacco to silence two nematode specific housekeeping genes; splicing factor and integrase of *M. incognita*. They reported more than 90% reduction in established nematodes in both the cases of host generated RNAi. Since then, host generated RNAi has been effectively demonstrated against PPNs by various groups of scientists all over the world (Table 1). Recently, Kumar et al. (2017) reconfirmed the utility of splicing factor and integrase as lethal RNAi targets for *M. incognita* by demonstrating significant reduction in number of galls, females and egg masses by targeting these genes for host generated RNAi in *A. thaliana*. Silencing of some of the other housekeeping genes like ribosomal protein 3a, ribosomal protein 4, spliceosomal SR protein, *Mi-Rpn7, Prp-17* etc. have resulted in substantial reduction in number and parasitism of incursive PPNs (Klink et al., 2009; Li et al., 2010b; Niu et al., 2012). In an innovative approach toward gene stacking, host generated RNAi was used to obtain combined resistance against migratory root lesion nematode, *Pratylenchus vulnus* and crown gall in walnut through co-transformation (Walawage et al., 2013). They used *A. tumefaciens* carrying self-complementary *iaaM* and *ipt* transgenes and *A. rhizogenes* carrying *Pv010* gene from *P. vulnus* for co-transformation. Most of the reports involving nematode housekeeping genes indicate their effectiveness as efficient targets for host generated RNAi. However, silencing of a putative transcription factor of *M. javanica, MjTis11* through host generated RNAi in tobacco did not result in lethal phenotypic effect on the nematodes (Fairbairn et al., 2007). siRNA generation was detected in the transgenic plants confirming the successful processing of the delivered dsRNAs. Therefore, it can be concluded that either this gene is not a suitable candidate for host generated RNAi or the achieved downregulation levels of the transcript is not sufficient to generate lethal phenotypic effects. The unsuitability of genes as targets for host generated RNAi can possibly result from the genetic redundancy of such genes.

Both, *in vitro* and *in vivo* RNAi approaches were used to silence a parasitism gene, *16D10*, expressed in the subventral gland cells of *M. incognita* leading to substantial reduction in the number of galls in the range of 63–90% in *Arabidopsis* (Huang et al., 2006). In addition, the gall size also decreased leading to reduction in the total number of eggs as compared to control. *Mc16D10L* was also targeted for host delivered RNAi in potato and *Arabidopsis* against *M. chitwoodi* by Dinh et al. (2014a,b) leading to significant reduction in nematode numbers in terms of the number of eggs and egg masses. They also reported substantial reduction in target gene expression in the second generation *M. chitwoodi* eggs and J2s which demonstrates transmission of the RNAi effect into the progenies. *16D10* is conserved among the *Meloidogyne* spp. which further makes it a suitable target for engineering resistance against a broader range of PPNs.

Table 1. Host generated rnai in plant parasitic nematodes.

Target gene	Nematode species	Host plant	promoter	phenotype	Time point	References
Integrase	*M. incognita*	Tobacco	35S	>90% reduction in number of established nematodes	6–7 weeks after infection	Yadav et al., 2006
Splicing factor	*M. incognita*	Tobacco	35S	>90% reduction in number of established nematodes	6–7 weeks after infection	Yadav et al., 2006
Secreted peptide *16D10*	*M. incognita* *M. arenaria* *M. javanica* *M. hapla*	Arabidopsis	35S	63–90% reduction in number of galls and gall size	4 weeks after infection	Huang et al., 2006
Major sperm protein	*H. glycines*	Soybean	ACT2	Upto 68% reduction in number of eggs	8 weeks after infection	Steeves et al., 2006
Putative transcription factor	*M. javanica*	Tobacco	35S	None	6 weeks after infection	Fairbairn et al., 2007
Ribosomal protein 3a	*H. glycines*	Soybean	FMV-sgt	87% reduction in number of female cysts	8 days after infection	Klink et al., 2009
Ribosomal protein 4	*H. glycines*	Soybean	FMV-sgt	81% reduction in number of female cysts	8 days after infection	Klink et al., 2009
Spliceosomal SR protein	*H. glycines*	Soybean	FMV-sgt	88% reduction in number of female cysts	8 days after infection	Klink et al., 2009
4G06, Ubiquitin-like	*H. schachtii*	Arabidopsis	35S	23–64% reduction in number of developing females	14 days after infection	Sindhu et al., 2009
3B05, cellulose binding protein	*H. schachtii*	Arabidopsis	35S	12–47% reduction in number of developing females	14 days after infection	Sindhu et al., 2009
8H07, SKP1-like	*H. schachtii*	Arabidopsis	35S	>50% reduction in number of developing females	14 days after infection	Sindhu et al., 2009
10A06, Zinc finger protein	*H. schachtii*	Arabidopsis	35S	42% reduction in number of developing females	14 days after infection	Sindhu et al., 2009
Y25, beta subunit of COPI complex	*H. glycines*	Soybean	35S	81% reduction in number of nematode eggs	5 weeks after infection	Li et al., 2010a,b
Prp-17, Pre-mRNA splicing factor	*H. glycines*	Soybean	35S	79% reduction in number of nematode eggs	5 weeks after infection	Li et al., 2010b
Cpn-1	*H. glycines*	Soybean	35S	95% reduction in number of nematode eggs	5 weeks after infection	Li et al., 2010b
Rpn7	*M. incognita*	Tomato	35S	Reduction in motility and infectivity of J2s	40 days after infection	Niu et al., 2012
AF531170, parasitism gene	*M. incognita*	Tomato	35S	54–59% reduction in number of developing females	7, 21, 30 days after infection	Choudhary et al., 2012
8D05, Parasitism gene	*M. incognita*	Arabidopsis	35S	Reduction in number of galls	8 weeks after infection	Xue et al., 2013
flp-14, FMRF amide like peptide	*M. incognita*	Tobacco	35S	Reduction in parasitic ability from 67–68%	30 days after infection	Papolu et al., 2013
flp-18, FMRF amide like peptide	*M. incognita*	Tobacco	35S	Reduction in parasitic ability from 53–82%	30 days after infection	Papolu et al., 2013
Mi-ser-1, serine protease	*M. incognita*	Tobacco	35S	Reduction in number of eggs per gram of root. Reduction in egg hatching ratio	28 days after infection	Antonino de Souza Júnior et al., 2013
Mi-cpl-1, Cysteine protease	*M. incognita*	Tobacco	35S	Reduction in number of eggs per gram of root.	28 days after infection	Antonino de Souza Júnior et al., 2013
Mi-asp-1+Mi-ser-1+Mi-cpl-1 (fusion)	*M. incognita*	Tobacco	35S	Reduction in number of eggs per gram of root.	28 days after infection	Antonino de Souza Júnior et al., 2013
Pv010	*P. vulnus*	Walnut	35S	Reduction in number of nematodes	60 days after infection	Walawage et al., 2013
Mc16D10L	*M. chitwoodi*	Potato	35S	65–68% Reduction in the number of egg masses	35, 55 days after infection	Dinh et al., 2014b
Mc16D10L	*M. chitwoodi*	Arabidopsis	35S	57 and 67% reduction in number of egg masses and eggs respectively	35, 55 days after infection	Dinh et al., 2014a
Mi-cpl-1	*M. incognita*	Tomato	35S	60–80% reduction in infection and multiplication	35 days after infection	Dutta et al., 2015b
Pp-pat-10	*P. penetrans*	Soybean	35S	Upto 40% reduction in number of nematodes	3 months after infection	Vieira et al., 2015
Pp-unc-87	*P. penetrans*	Soybean	35S	Upto 50% reduction in number of nematodes	3 months after infection	Vieira et al., 2015
Rs-cb-1	*R. similis*	Tobacco	35S	Reduced reproduction and pathogenicity	75 days after infection	Li et al., 2015
HSP90, Heat Shock Protein	*M. incognita*	Tobacco	35S	Delayed gall formation and upto 46% reduction in the number of eggs		Lourenço-Tessutti et al., 2015
ICL, isocitrate lyase	*M. incognita*	Tobacco	35S	Upto 77% reduction in egg oviposition		Lourenço-Tessutti et al., 2015
Unc-15	*Ditylenchus destructor*	Sweet potato	35S	50% reduction in the infection area	45 days after infection	Fan et al., 2015
MeTCTP	*M. enterolobii*	Tomato	35S	Reduction in number of nematodes	15 and 30 days after infection	Zhuo et al., 2017
MiMSP40	*M. incognita*	Arabidopsis	35S	Reduction in the number of galls	7 weeks after infection	Niu et al., 2016

Parasitism of PPNs was successfully suppressed by targeting some other parasitism or effector genes like *4G06, 3B05, 8H07, 10A06, AF531170, 8D05, MeTCTP etc. through host generated RNAi* (Sindhu et al., 2009; Choudhary et al., 2012; Xue et al., 2013; Zhuo et al., 2017). Recently, *M. incognita* effector *MiMSP40* was reported to facilitate parasitism through manipulation of plant immunity (Niu et al., 2016). Overexpression of *MiMSP40* lead to increased susceptibility in *Arabidopsis*, while host generated RNAi of *MiMSP40* resulted in reduction in parasitism and reproductive potential of *M. incognita*. Similarly, overexpression of *M. enterolobii* effector *MeTCTP* imparted increased susceptibility. Conversely, *in planta* RNAi targeting *MeTCTP* caused attenuated parasitism in *Arabidopsis* (Zhuo et al., 2017). Therefore, recent reports indicate the suitability of nematode effector genes as targets for host generated RNAi to achieve resistance.

Genes involved in nematode development and reproduction are often targeted with considerable success in hindering the development and reproductive potential of PPNs (Steeves et al., 2006; Charlton et al.,

2010; Antonino de Souza Júnior et al., 2013; Papolu et al., 2013; Dutta et al., 2015b). Decreased number of eggs was phenotypically observed by targeting a major sperm protein from *H. glycines* for host generated silencing in soybean (Steeves et al., 2006). Silencing of the major sperm protein hampered the reproductive potential of *H. glycines* which continued to the next generation of the nematodes as the progeny also showed an impaired ability to reproduce successfully. FMRF amide like neuropeptides were targeted for silencing through *in vitro* and *in vivo* RNAi by Papolu et al. (2013). Transgenic tobacco lines expressing flp-*14* and *flp-18* dsRNAs showed significant reduction in the range of 50–80% in the infection and multiplication of *M. incognita*. This study proved that neuropeptides can be exploited as potential targets for host delivered RNAi considering the involvement of these genes in nematode physiology including locomotion, feeding, parasitism and reproduction. *In vitro* silencing of carrying cathepsin L cysteine proteinase (*Mi-cpl-1*), led to reduced attraction and penetration of *M. incognita* in tomato suggesting its role in nematode parasitism (Dutta et al., 2015b).

The development of dsRNA constructs for host delivered RNAi through conventional methods have been a time taking and tedious. Therefore, a much easier, quick and effective process of gateway cloning system has been employed by different groups for the development of RNAi constructs (Klink et al., 2009; Papolu et al., 2013; Dutta et al., 2015b). *A. rhizogenes* mediated hairy root method for transformation in crops like tomato (Remeeus et al., 1998), soybean (Klink et al., 2009; Li et al., 2010b) and sugar beet (Kifle et al., 1999; Cai et al., 2003) has been instrumental for rapid screening of the target genes. The genes found effective through hairy root transformation can further be utilized for the development of stable transgenic plants through *A. tumifeciens* mediated transformation. The studies on host delivered RNAi have showed that the success in obtaining significant degree of resistance depends on the role of the target gene and the degree of transcript knockdown. However, of all the strategies implied till date to achieve resistance against PPNs, host-delivered RNAi appears to provide more effective resistance when it targets nematode genes involved in essential cellular processes.

BIOSAFETY ASPECTS

Application of RNAi for management of PPNs needs thorough risk assessment and proper designing of the experiments to overcome the limitations. Avoiding off-target effects is an important consideration for all RNAi experiments. Off target effects can originate due to sequence identity between dsRNA and non target mRNA transcripts resulting in a compromised specificity of RNAi and false interpretation of the resulting phenotype. According to Rual et al. (2007); in *C. elegans*, a sequence of mRNA having more than 95% identity with the dsRNA for over 40 nucleotides results in off-target effects. Unexpected and unintended gene silencing in the plant may lead to harmful and deleterious effects on its phenotype and physiology and can cause serious environmental consequences. Xing and Zachgo (2007) explained the phenomenon of pollen lethality in *Arabidopsis* where surprising pleiotropic effects such as reduced pollen viability was observed due to RNAi, however other plant growth parameters were found to be normal. Apart from these off target effects, non-target organisms may encounter harmful non target effects of RNAi on being exposed to the plant parts or debris of genetically modified plants through RNAi. Therefore, development of strategies for preventing the off-target and non-target effects is a crucial biosafety consideration for a wider employment of RNAi as a novel tool in plant disease management.

Availability of suitable genomic databases can be utilized extensively for *in silico* homology searches for selection of target genes to avoid off-target effects (Banerjee et al., 2017). Genes showing high degree of sequence conservation among plant and animal kingdom should be avoided and use of species specific targets should be encouraged. The 5′-3′ untranslated regions (UTR) sequences can also be used as siRNA targets owing to their less degree of conservation as compared to coding regions. Thorat et al. (2016) used a nematode responsive and root specific promoter of *Arabidopsis* origin to transform tomato with a GUS reporter gene. A strong GUS activity was reported at nematode infection site starting from 10 days up to 21 days post infection. Further, this promoter was used to drive the expression of *M. incognita* splicing factor dsRNA and upon transformation in tomato; 50–70% reduction in nematode galls over the control plant was reported. This strategy of using root specific and/or nematode inducible promoters can avoid the expression of siRNAs in undesirable parts of the plants.

CONCLUSION AND FUTURE PROSPECTS

RNAi has emerged as a powerful strategy to control multiple pest and pathogens including nematodes especially as we are moving toward the goal of phasing out chemicals that are harmful to environments and ecosystems. Management of nematodes however, presents some unique challenges as these are obligatory parasites requiring living host for feeding. The use of host induced RNAi to combat plant pathogenic nematodes has so far been effective especially with respect to RKNs and CNs. The advancement in the area of functional genomics availability of genome sequence data and new bioinformatics tools have enabled design and engineering of effective dsRNA expression constructs addressing concerns of off-target silencing. Stacking of dsRNA sequences to target multiple genes has emerged as an attractive proposition for effective nematode control. Use of nematode induced and plant tissue specific promoters limiting dsRNA gene expressions to specific plant tissue/s in response to particular nematode can also mitigate biosafety concerns. The ability to precisely edit genomes is rapidly transforming the landscape of novel ways to target plant pathogens. CRISPR/Cas system is emerging as a powerful approach for loss of function analysis, insights into host–parasite and parasite–vector interactions, and the genetic basis of parasitism. A number of CRISPR/Cas9 genome editing protocols have been established in *C. elegans* (Friedland et al., 2013) opening new doors to studying the biology of closely related nematode parasites. Translation of CRISPR/Cas9 technology from *C. elegans* to *Strongyloides spp.*, *Ascaris suum, Brugia malayi* and *Haemonchus contortus* have been recently outlined (Ward, 2015; Britton et al., 2016; Zamanian and Andersen, 2016). Recent reports on topical application of dsRNA for resistance against viruses using layered double hydroxide clay nanosheets (Mitter et al., 2017) opens up possibilities to exploit such innovations for specific and combinatorial resistance against PPNs, insects and plant pathogenic fungi.

AUTHOR CONTRIBUTIONS

SB, AB, SG, OG, and AS drafted the manuscript. SB and AB collected background information. SG, PJ, AD, and AS critically revised the manuscript. All authors read and approved the final manuscript.

CONFLICT OF INTEREST STATEMENT

The authors declare that the research was conducted in the absence of any commercial or financial relationships that could be construed as a potential conflict of interest.

ACKNOWLEDGMENTS

The authors gratefully acknowledge the financial support from Indian Council of Agricultural Research (ICAR) through National Agricultural Innovative Project (NAIP/C4/C1092) and National Fund for Basic Strategic and Frontier Application Research in Agriculture (NFBSFARA/RNA-3022/2012-13). The authors apologize to colleagues whose work could not be cited due to space constraints.

REFERENCES

1. Antonino de Souza Júnior, J. D., Ramos Coelho, R., Tristan Lourenço, I., da Rocha Fragoso, R., Barbosa Viana, A. A., Lima Pepino de Macedo, L., et al. (2013). Knocking-down Meloidogyne incognita proteases by plant-delivered dsRNA has negative pleiotropic effect on nematode vigor. *PLoS ONE* 8:e85364.
2. Atkinson, H. J., Harris, P. D., Halk, E. J., Novitski, C., Leighton-Sands, J., Nolan, P., et al. (1988). Monoclonal antibodies to the soya bean cyst nematode, *Heterodera glycines*. *Ann. Appl. Biol.* 112, 459–469.
3. Bakhetia, M., Charlton, W., Atkinson, H. J., and McPherson, M. J. (2005). RNA interference of dual oxidase in the plant nematode Meloidogyne incognita. *Mol. Plant Microb. Interact.* 18, 1099–1106.
4. Bakhetia, M., Urwin, P. E., and Atkinson, H. J. (2007). qPCR analysis and RNAi define pharyngeal gland cell-expressed genes of Heterodera glycines required for initial interactions with the host. *Mol. Plant Microb. Interact.* 20, 306–312
5. Banerjee, S., Gill, S. S., Jain, P. K., and Sirohi, A. (2017). Isolation, cloning, and characterization of a cuticle collagen gene, *Mi-col-5*, in *Meloidogyne incognita*. *3 Biotech* 7:64.
6. Britton, C., Roberts, B., and Marks, N. (2016). Functional genomics tools for Haemonchus contortus and lessons From Other Helminths. *Adv. Parasitol.* 93, 599–623.
7. Cai, D., Thurau, T., Tian, Y., Lange, T., Yeh, K. W., and Jung, C. (2003). Sporamin-mediated resistance to beet cyst nematodes (*Heterodera schachtii* Schm.) is dependent on trypsin inhibitory activity in sugar beet (*Beta vulgaris* L.) hairy roots. *Plant. Mol. Biol.* 51, 839–849.
8. Charlton, W. L., Harel, H. Y. M., Bakhetia, M., Hibbard, J. K., Atkinson, H. J., and Mcpherson, M. J. (2010). Additive effects of plant expressed double-stranded RNAs on root-knot nematode development. *Int. J. Parasitol.* 40, 855–864.
9. Choudhary, D., Koulagi, R., Rohatagi, D., Kumar, A., Jain, P. K., and Sirohi, A. (2012). "Engineering resistance against root-knot nematode, Meloidogyne incognita, by host delivered RNAi," in *Abstracts of International Conference on Plant Biotechnology for food Security: New Frontiers* (New Delhi: National Agricultural Science Centre), 21–24.
10. Dinh, P. T. Y., Brown, C. R., and Elling, A. A. (2014a). RNA interference of effector gene Mc16D10L confers resistance against Meloidogyne chitwoodi in Arabidopsis and Potato. *Phytopathology* 104, 1098–1106.
11. Dinh, P. T. Y., Zhang, L., Brown, C. R., and Elling, A. A. (2014b). Plant-mediated RNA interference of effector gene Mc16D10L confers resistance against Meloidogyne chitwoodi in diverse genetic backgrounds of potato and reduces pathogenicity of nematode offspring. *Nematology* 6, 669–682.
12. Dubreuil, G., Magliano, M., Deleury, E., Abad, P., and Rosso, M. N. (2007). Transcriptome analysis of root knot nematode functions induced in early stage of parasitism. *New Phytol.* 176, 426–436.
13. Dutta, T. K., Bankar, P., and Rao, U. (2015a). The status of RNAi-based transgenic research in plant nematology. *Front. Microbiol.* 5:760.
14. Dutta, T. K., Papolu, P. K., Banakar, P., Choudhary, D., Sirohi, A., and Rao, U. (2015b). Tomato transgenic plants expressing hairpin construct of a nematode protease gene conferred enhanced resistance to root-knot nematodes. *Front. Microbiol.* 6:260.

15. Elling, A. A. (2013). Major emerging problems with minor Meloidogyne species. *Phytopathology* 103, 1092–1102.
16. Fairbairn, D. J., Cavallaro, A. S., Bernard, M., Mahalinga-Iyer, J., Graham, M. W., and Botella, J. R. (2007). Host-delivered RNAi: an effective strategy to silence genes in plant parasitic nematodes. *Planta* 226, 1525–1533
17. Fan, W., Wei, Z., Zhang, M., Ma, P., Liu, G., Zheng, J., et al. (2015). Resistance to Ditylenchus destructor infection in sweet potato by the expression of small interfering RNAs targeting unc-15, a movement-related gene. *Phytopahology* 105, 1458–1465.
18. Fire, A., Xu, S., Montgomery, M. K., Kostas, S. A., Driver, S. E., and Mello, C. C. (1998). Potent and specific genetic interference by double-stranded RNA in *Caenorhabditis elegans*. *Nature* 391, 806–811.
19. Friedland, A. E., Tzur, Y. B., Esvelt, K. M., Colai acovo, M. P., Church, G. M., and Calarco, J. A. (2013). Heritable genome editing in *C. elegans* via a CRISPR-Cas9 system. *Nat. Methods* 10, 741–743.
20. Haegeman, A., Vanholme, B., and Gheysen, G. (2009). Characterization of putative endoxyalanase in the migratory plant-parasitic nematode Rhodopholus similis. *Mol. Plant Pathol.* 10, 389–401.
21. Huang, G., Allen, R., Davis, E. L., Baum, T. J., and Hussey, R. S. (2006). Engineering broad root-knot resistance in transgenic plants by RNAi silencing of a conserved and essential root-knot nematode parasitism gene. *Proc. Natl. Acad. Sci. U.S.A.* 103, 14302–14306.
22. Jones, L. M., Giorgi, C. D., and Urwin, P. E. (2011). "C. elegans as a resource for studies on plant parasitic nematodes," in *Genomics and Molecular Genetics of Plant-nematode Interactions*, eds J. Jones, G. Gheysen and C. Fennoll (New York, NY: Springer), 175–220.
23. Kamath, R. S., Martinez-Campos, M., Zipperlen, P., Fraser, A. G., and Ahringer, J. (2001). Effectiveness of specific RNA-mediated interference through ingested double-stranded RNA in *Caenorhabditis elegans*. *Genome Biol.* 2, research 0002.1–2.10.
24. Kifle, S., Shao, M., Jung, C., and Cai, D. (1999). An improved transformation protocol for studying gene expression in hairy roots of sugar beet (*Beta vulgaris* L.). *Plant Cell Rep.* 18, 514–519.
25. Kimber, M. J., McKinney, S., McMaster, S., Day, T. A., Flemming, C. C., and Maule, A. G. (2007). Flp gene disruption in a parasitic nematode reveals motor disfunction and unusual neuronal sensitivity to RNA interference. *FASEB J.* 21, 1233–1243.
26. Klink, V. P., Kim, K. H., Martins, V., Macdonald, M. H., Beard, H. S., Alkharouf, N. W., et al. (2009). A correlation between host-mediated expression of parasite genes as tandem inverted repeats and abrogation of development of female *Heterodera glycines* cyst formation during infection of Glycine max. *Planta* 230, 53–71.
27. Kumar, A., Kakrana, A., Sirohi, A., Subramaniam, K., Srinivasan, R., Abdin, M. Z., et al. (2017). Host-delivered RNAi-mediated root-knot nematode resistance in Arabidopsis by targeting splicing factor and integrase genes. *J. Gen. Plant Pathol.* 83, 91–97.
28. Li, J., Todd, T. C., Lee, J., and Trick, H. N. (2011). Biotechnological application of functional genomics towards plant parasitic nematode control. *Plant Biotech. J.* 9, 936–944.
29. Li, J., Todd, T. C., Oakley, T. R., Lee, J. L., and Trick, H. N. (2010a). Host-derived suppression of nematode reproductive and fitness genes decreases fecundity of *Heterodera glycines* Ichinohe. *Planta* 232, 775–785.
30. Li, J., Todd, T. C., and Trick, H. N. (2010b). Rapid in planta evaluation of root expressed transgenes in chimeric soybean plants. *Plant Cell Rep.* 29, 113–123.
31. Li, Y., Wang, K., Xie, H., Wang, D. W., Xu, C. L., Huang, X., et al. (2015). Cathepsin B Cysteine Proteinase is Essential for the Development and Pathogenesis of the Plant Parasitic Nematode *Radopholus similis*. *Int. J. Biol. Sci.* 11, 1073–1087.

32. Lourenço-Tessutti, I. T., Souza Junior, J. D. A., Martins-de-Sa, D., Viana, A. A. B., Carneiro, R. M. D. G., Togawa, R. C., et al. (2015). Knock-down of heat-shock protein 90 and isocitrate lyase gene expression reduced root-knot nematode reproduction. *Phytopathology* 105, 628–637
33. Mello, C. C., and Conte, D. Jr. (2004). Revealing the world of RNA interference. *Nature* 431, 338–342.
34. Mitter, N., Worrall, E. A., Robinson, K. E., Li, P., Jain, R. G., Taochy, C., et al. (2017). Clay nanosheets for topical delivery of RNAi for sustained protection against plant viruses. *Nat. plants* 3:16207.
35. Niu, J. H., Jian, H., Xu, J., Chen, C., and Guo, Q. (2012). RNAi silencing of the Meloidogyne incognita Rpn7 gene reduces nematode parasitic success. *Euro. J. Plant Pathol.* 134, 131–144.
36. Niu, J., Liu, P., Liu, Q., Chen, C., Guo, Q., Yin, J., et al. (2016). Msp40 effector of root-knot nematode manipulates plant immunity to facilitate parasitism. *Sci. Rep.* 6:19443.
37. Papolu, P. K., Gantasala, N. P., Kamaraju, D., Banakar, P., Sreevathsa, R., and Rao, U. (2013). Utility of host delivered RNAi of two FMRF amide like peptides, flp-14 and flp-18, for the management of root knot nematode, Meloidogyne incognita. *PLoS ONE* 8:e80603.
38. Park, J. E., Lee, K. Y., Lee, S. J., Oh, W. S., Jeong, P. Y., Woo, T., et al. (2008). The efficiency of RNA interference in *Bursaphelenchus xylophilus*. *Mol. Cells* 26, 81–86.
39. Remeeus, P. M., Bezooijen, J. V., and Wijbrandi, J. (1998). "*In vitro* testing is a reliable way to screen the temperature sensitivity of resistant tomatoes against *Meloidogyne incognita*," in *Proceedings of 5th International Symposium on Crop Protection, Universiteit Gent Vol. 63*, (Ghent), 635–640.
40. Rosso, M. N., Dubrana, M. P., Cimbolini, N., Jaubert, S., and Abad, P. (2005). Application of RNA interference to root-knot nematode genes encoding esophageal gland proteins. *Mol. Plant Microb. Interact.* 18, 615–620.
41. Rosso, M. N., Jones, J. T., and Abad, P. (2009). RNAi and functional genomics in plant parasitic nematodes. *Annu. Rev. Phytopathol.* 47, 207–232.
42. Rual, J. F., Klitgord, N., and Achaz, G. (2007). Novel insights into RNAi off-target effects using *C. elegans* paralogs. *BMC Genomics* 8:106.
43. Shingles, J., Lilley, C. J., Atkinson, H. J., and Urwin, P. E. (2007). Meloidogyne incognita: molecular and biochemical characterization of a cathepsin L cysteine proteinase and the effect on parasitism following RNAi. *Exp. Parasitol.* 115, 114–120.
44. Sindhu, A., Maier, T. R., Mittchum, M. G., Hussey, R. S., Davis, L. R., and Baum, J. T. (2009). Effective and specific in planta RNAi in cyst nematodes: expression interference of four parasitism genes reduces parasitic success. *J. Exp. Bot.* 1, 315–324.
45. Smith, N. A., Singh, S. P., Wang, M. B., Stoutjesdijk, P. A., Green, A. G., and Waterhouse, P. M. (2000). Gene expression: total silencing by intron-spliced hairpin RNAs. *Nature* 407, 319–320.
46. Soumi, J., Gheysen, G., and Subramaniam, K. (2012). RNA interference in Pratylenchus coffeae: knockdown of Pc-pat-10 and Pc-unc-87 impedes migration. *Mol. Biochem. Parasitol.* 186, 51–59.
47. Steeves, R. M., Todd, T. C., Essig, J. S., and Trick, H. N. (2006). Transgenic soybeans expressing siRNAs specific to a major sperm protein gene suppress *Heterodera glycines* reproduction. *Func. Plant Biol.* 33, 991–999.
48. Tabara, H., Grishok, A., and Mello, C. C. (1998). RNAi in *C. elegans*: soaking in the genome sequence. *Science* 282, 430–431.
49. Thorat, Y., Kumar, A., Jain, P. K., and Sirohi, A. (2016). "Development of root-knot nematode, Meloidogyne incognita specific gene expression system in tomato, *Solanum lycopersicum* L.," in *Abstracts of 3rd International IUPAC Conference on Agrochemicals Protecting Crops, Health and Natural Environment: New Chemistries for Phytomedicines and Crop Protection Chemicals* (New Delhi).

50. Timmons, L., Court, D. L., and Fire, A. (2001). Ingestion of bacterially expressed dsRNAs can produce specific and potent genetic interference in *Caenorhabditis elegans*. *Gene* 263, 103–112.
51. Timmons, L., and Fire, A. (1998). Specific interference by ingested dsRNA. *Nature* 395:854.
52. Tomoyasu, Y., Miller, S., Tomita, S., Schoppmeier, M., Grossmann, D., and Bucher, G. (2008). Exploring systemic RNA interference in insects: a genome-wide survey for RNAi genes in Tribolium. *Genome Biol.* 9:R10.
53. Urwin, P. E., Lilley, C. J., and Atkinson, H. J. (2002). Ingestion of double-stranded RNA by pre-parasitic juvenile cyst nematodes leads to RNA interference. *Mol. Plant Microb. Interact.* 15, 747–752.
54. Vieira, P., Eves-van den Akker, S., Verma, R., Wantoch, S., Eisenback, J. D., and Kamo, K. (2015). The Pratylenchus penetrans transcriptome as a source for the development of alternative control strategies: mining for putative genes involved in parasitism and evaluation of in planta RNAi. *PLoS ONE* 10:e0144674.
55. Walawage, S. L., Britton, M. T., Leslie, C. A., Uratsu, S. L., Li, Y., and Dandekar, M. (2013). Stacking resistance to crown gall and nematodes in walnut rootstocks. *BMC Genomics* 14:668.
56. Ward, J. D. (2015). Rendering the intractable moretractable: tools from Caenorhabditis elegans ripe for import into parasitic nematodes. *Genetics* 201, 279–1294.71.
57. Whangbo, J. S., and Hunter, C. P. (2008). Environmental RNA interference. *Trends Gen.* 24, 297–305.
58. Williamson, V. M., and Gleason, C. A. (2003). Plant–nematode interactions. *Curr. Opin. Plant Biol.* 6, 327–333.
59. Xing, S., and Zachgo, S. (2007). Pollen lethality: a phenomenon in Arabidopsis RNA interference plants. *Plant Physiol.* 145, 330–333
60. Xue, B., Hamamouch, N., Li, C., Huang, G., Hussey, R. S., Baum, T. J., et al. (2013). The 8D05 parasitism gene of Meloidogyne incognita is required for successful infection of host roots. *Phytopathology* 103, 175–181.
61. Yadav, B. C., Veluthambi, K., and Subramaniam, K. (2006). Host generated double stranded RNA induces RNAi in plant parasitic nematodes and protects the host from infection. *Mol. Biochem. Parasitol.* 148, 219–222.
62. Zamanian, M., and Andersen, E. C. (2016). Prospects and challenges of CRISPR/Cas genome editing for the study and control of neglected vector-borne nematode diseases. *FEBS J.* 283, 3204–3221.
63. Zhuo, K., Chen, J., Lin, B., Wang, J., Sun, F., Hu, L., et al. (2017). A novel *Meloidogyne enterolobii* effector MeTCTP promotes parasitism by suppressing programmed cell death in host plants. *Mol. Plant Pathol.* 18, 45–54.

CHAPTER 20

Transgenic Strategies for Enhancement of Nematode Resistance in Plants

Muhammad A. Ali[1,2*], Farrukh Azeem[3], Amjad Abbas[1], Faiz A. Joyia[2], Hongjie Li[4] and Abdelfattah A. Dababat[5]

[1]*Department of Plant Pathology, University of Agriculture, Faisalabad, Pakistan*

[2]*Centre of Agricultural Biochemistry and Biotechnology, University of Agriculture, Faisalabad, Pakistan*

[3]*Department of Bioinformatics and Biotechnology, Government College University, Faisalabad, Pakistan*

[4]*National Key Facility for Crop Gene Resources and Genetic Improvement, Institute of Crop Science, Chinese Academy of Agricultural Sciences, Beijing, China*

[5]*International Maize and Wheat Improvement Center, Ankara, Turkey*

ABSTRACT

Plant parasitic nematodes (PPNs) are obligate biotrophic parasites causing serious damage and reduction in crop yields. Several economically important genera parasitize various crop plants. The root-knot, root lesion, and cyst nematodes are the three most economically damaging genera of PPNs on crops within the family Heteroderidae. It is very important to devise various management strategies against PPNs in economically important crop plants. Genetic engineering has proven a promising tool for the development of biotic and abiotic stress tolerance in crop plants. Additionally, the genetic engineering leading to transgenic plants harboring nematode resistance genes has demonstrated its significance in the field of plant nematology. Here, we have discussed the use of genetic engineering for the development of nematode resistance in plants. This review article also provides a detailed account of transgenic strategies for the resistance against PPNs. The strategies include natural resistance genes, cloning of proteinase inhibitor coding genes, anti-nematodal proteins and use of RNA interference to suppress nematode effectors. Furthermore, the manipulation of expression levels of genes induced and suppressed by nematodes has also been suggested as an innovative approach for inducing nematode resistance in plants. The information in this article will provide an array of possibilities to engineer resistance against PPNs in different crop plants.

Keywords: plant parasitic nematodes, *R* genes, protease inhibitors, RNAi, plant resistance

INTRODUCTION

The word 'nematode' comes from the Greek word 'nema,' which means thread. Nematodes are thread like, long, cylindrical, sometimes microscopic worms, which can be found in a variety of environments. They belong to a huge phylum of animals called 'Nematoda' that comprises of plant and animal parasites, as well as numerous free-living species. They are omnipresent in nature inhabiting in all types of environments and habitats (Ali et al., 2015). However, most of the nematodes are free-living and feed on bacteria, fungi or algae. Some of them invade and parasitize both vertebrates and invertebrates including human beings, thus causing serious health damage and even human death, i.e., guinea worm (*Dracunculus medinensis*) and intestinal worm (*Ascaris lumbricoides*) (Decraemer and Hunt, 2006). On the other hand, some of these worms are serious parasites of plants and result in huge crop losses. Phyto-parasitic nematodes count for 7% of total species of the phylum Nematoda from 197 genera and 4300 species that can infect a wide range of economically important crop plants including wheat (*Triticum aestivum*), potato (*Solanum tuberosum*), tomato (*Solanum lycopersicum*), maize (*Zea mays*), soybean (*Glycine max*), and sugar beets (*Beta vulgaris*) (Decraemer and Hunt, 2006).

Most of the nematodes are obligate biotrophic parasites of plants and their infection results in above-ground symptoms in plants like leaf necrosis, chlorosis, plant wilting, stunted growth, and enhanced susceptibility to other pathogens, which mostly occur in patches (Webster, 1995). Penetration of plant parasitic nematodes (PPNs) to the root systems results in the hindrance of translocation of water from roots toward areal plant parts and translocation of nutrients from leaves toward the plant roots. The developing nematodes feed on cell sap and hinder the food supply to the plant physiological systems, which results in stunted growth, yellowing and drooping of leaves. Generally, below-ground symptoms include gall and knot formation in infected roots, root lesions and necrosis and root abbreviations. Figure 1 demonstrates the above- and below-ground symptoms on different plant species. Moreover, in addition to direct damage to crops in the form of yield reduction, these worms serve as disease transmitting vectors for viruses, thereby resulting in crucial economic and social impacts worldwide (Ali et al., 2015).

The members of family Heteroderidae (Superfamily: Tylenchoidea) are divided into three economically important genera *Meloidogyne*, *Heterodera*, and *Globodera*, which parasitize a huge number of crop plants. These are obligate endo-parasites of host plant roots, enter the plant roots as second stage juvenile larva (J2 larva) and develop specific feeding cells in the plant roots. The members of the genus *Meloidogyne*, known as root-knot nematodes (RKNs), develop giant cells (Jones and Payne, 1978). On the other hand, the cyst nematodes (CNs) that belong to the genera *Heterodera* and *Globodera* induce a very specialized feeding cell called syncytium (plural: Syncytia) (Jones, 1981). Migratory endo-parasitic nematodes are another category that is economically important. These nematodes follow destructive mode of feeding by continuously moving through the cells of root tissues and resulting in enormous tissue necrosis (Moens and Perry, 2009). The important genera from this category of nematodes are *Pratylenchus*, *Radopholus*, and *Hirschmanniella*. Moreover, *Aphelenchoides*, *Ditylenchus*, and *Anguina* are the main genera that infect above-ground plant parts like leaves, stem, and grains, respectively.

In the last two decades, our understanding of plant–nematode interactions has increased significantly. The first genome sequences of two root-knot nematodes species, *M. incognita* (Abad et al., 2008) and *M. hapla* (Opperman et al., 2008), have been described, which were significantly different from genome of the free-living nematode *Caenorhabditis elegans*. Both *M. incognita* and *M. hapla* have definite set of proteins that determine the virulence in plant species. The secretomes (set of secreted proteins through the stylets) of different PPNs have demonstrated a number of effector proteins that are involved in compatible plant–nematode interactions (Huang et al., 2003; Bellafiore et al., 2008; Caillaud et al., 2008). In response to infection of various nematodes, plant's transcriptome resulted in increased metabolic activity in the feeding cells and suppression of defense mechanisms of the plants in

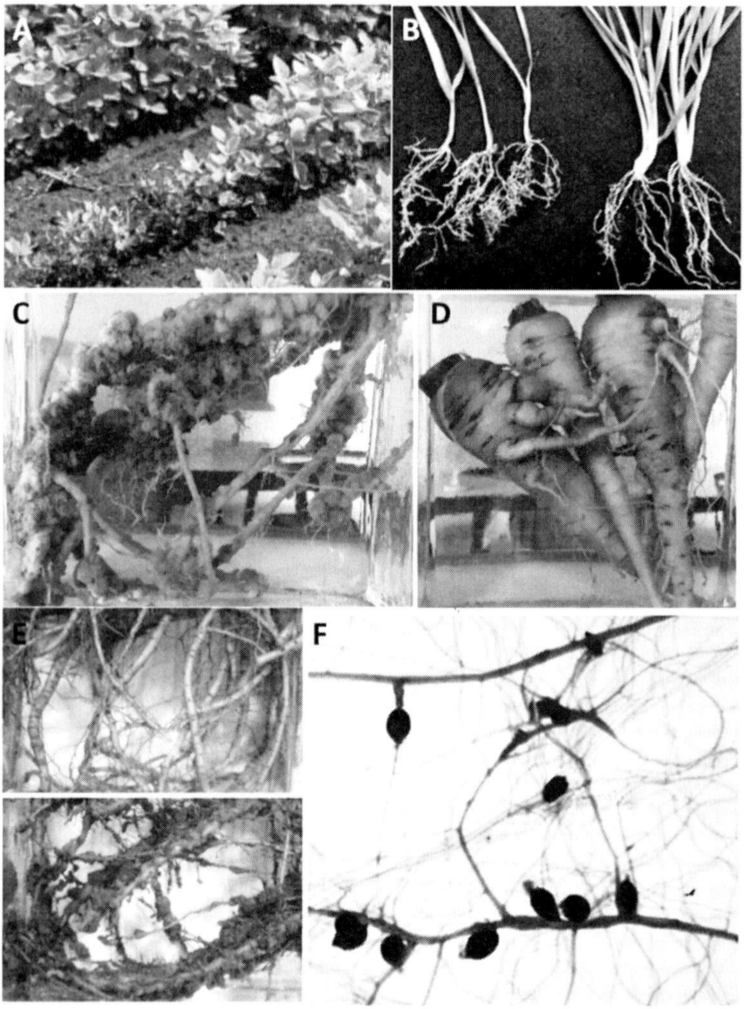

FIGURE 1. Above-ground and below-ground symptoms from different plant species in response to nematode infections. **(A)** Soybean plants infected with soybean cyst nematodes (*Heterodera glycines*) (http://www.extension.umn.edu, reproduced with the permission of Assoc. Prof. Malvick from UMN, US). **(B)** Infected and uninfected wheat plants with cereal cyst nematode *H. avenae* (Courtesy Prof. Honglian Li, China, reproduced with permission from Riley et al., 2009). **(C–E)** Roots of sponge gourd, carrots, and okra infected with root-knot nematode *Meloidogyne incognita*, respectively. **(F)** Arabidopsis roots showing development of cysts induced by beet cyst nematode *H. schachtii*.

most of the cases (Szakasits et al., 2009; Barcala et al., 2010; Kyndt et al., 2012; Ali et al., 2015). Most of these studies revealed considerable progress toward an understanding of plant–nematode interactions under natural conditions. On the other hand, many works have been published in the past two decades regarding the transgenic resistance in model plants, as well as the crop species using natural resistance (*R*) genes, proteinase inhibitors and RNA interference of nematode effector genes (Bakhetia et al., 2005; Fuller et al., 2008; Gheysen et al., 2010; Atkinson et al., 2012). In this article, we have focused on the most recent literature on PPNs with the emphasis on the use of different conventional and transgenic approaches to manage PPNs in different plant species of economic importance.

MANAGEMENT OF PPNS

The management of PPNs has been a big challenge for the agricultural scientists and farming community. The control of nematodes below the threshold level is very important for agricultural sustainability and food security. Several strategies have been tried for this purpose, which includes management through cultural, chemical, biological, and transgenic means. The cultural practices for nematode management include solarization (Katan, 1981), flooding (MacGuidwin, 1993), fallowing (Brodie and Murphy, 1975), crop rotation (Westphal, 2011; Dababat et al., 2015), cover crops (Kirkegaard and Sarwar, 1998), and a combination of some of these (Singh et al., 2009). Similarly, the utilization of natural host resistance from various crop species is a preferred approach for nematode management because it is environmentally safe and cost-saving option as compared to chemical control (Williamson and Kumar, 2006; Dababat et al., 2016). In chemical control, fumigants, organophosphates, and carbamates have been used for inhibiting nematode populations in the soil. Most of the nematicides are now being banned due to their high cost and health and environment issues (Sorribas et al., 2005). Similarly, currently available organophosphate and carbamate compounds, i.e., oxamyl, fosthiazate, and ethoprophos, are at risk of withdrawal by EU Directive 91/414/EEC due to their hazardous nature (Clayton et al., 2008).

A lot of work has been carried out and going on to find biological antagonistic microorganisms like fungi and bacteria to minimize the effect of pathogens (Kerry, 1980, 1988; Rodriguez-Kabana et al., 1987; Sayre and Walter, 1991; Stirling, 1991; Sikora, 1992; Dababat et al., 2015). Sayre (1986) listed nematode fungal parasites *Dactylella oviparasitica*, *Nematophthora gynophila*, and *Paecilomyces lilacinus* along with a bacterium *Pasteuria penetrans* as important candidates for management of nematodes. Similarly, a *Fusarium oxysporum* strain, *Fo*162, has been shown promising for controlling *M. incognita* in various studies (Dababat and Sikora, 2007a,b; Martinuz et al., 2012). However, it has been found difficult to develop a biological control agent that is effective worldwide for any plant parasitic nematode. Due to high cost and health hazards, nematicides are losing their value with the passage of time thus paving the way toward the use of nematode resistance crop varieties, biocontrol and transgenic strategies for nematode management.

ENGINEERING PLANTS FOR NEMATODE RESISTANCE

Recent advancements in biotechnological approaches have made it possible to incorporate and express indigenous and heterologous proteins from one organism to another. This has brought about new era of crop improvements after the advent of so-called "Green Revolution" in the 1960s. Genetic engineering of different crop plants has led to improvement of both quality and quantity of the produce in addition to enhancement of tolerance against various stresses. Several disease and pest resistance genes have been introduced into crop plants through genetic engineering. The important examples in this context are *Bt* cotton (*Gossypium hirsutum*) resistant against bollworms (*Helicoverpa armigera*), herbicide resistance for weed management, salinity tolerance, and rust resistance in wheat. Additionally, genetic engineering has been exceptional for enhancement of nematode resistance in plants. Here, we discuss different ways of engineering resistance genes in crop plants to suppress the nematode infection and populations in the soil below the threshold level.

In the past, several transgenic strategies have been used for enhancement of nematode resistance in plants. The resistance genes from natural resources have been cloned from numerous plant species and could be transferred to other plant species, for instance, *Mi* gene from tomato for resistance against *M. incognita*, $Hs1^{pro\text{-}1}$ from sugar beet (*Beta vulgaris*) against *H. schachtii*, *Gpa*-2 from potato against *Globodera pallida* and *Hero A* from tomato against *G. rostochiensis* (reviewed by Fuller et al., 2008). The overexpression of different protease inhibitors (PIs) such as cowpea trypsin inhibitor (CpTI), PIN2, cystatins, and serine proteases has been used for producing nematode resistant plants (Lilley et al., 1999). Another main strategy was the targeted suppression of important nematode effectors in plants using RNA interference (RNAi) approach. Unlike these strategies, some recent researches have suggested that nematode resistance could be enhanced in plants by modifying the expression of particular genes in syncytia (Klink and Matthews, 2009; Ali, 2012; Ali et al., 2013a,b). An overview of various transgenic strategies aimed at nematode resistance is provided in Table 1.

Table 1. Various transgenic strategies used for nematode resistance in plants.

Molecular strategy	Name of gene	Source	Trangenic plant	Effective against	Resistance response	Reference
Natural resistance genes	*Mi-1.2*	Wild tomato (*S. peruvianum*)	Tomato (*Solanum lycopersicum*)	*M. incognita*	Triggering of HR before significant establishment of giant cells on roots	Milligan et al., 1998; Williamson, 1998
	Hs1^{pro-1}	*wild beet (B. procumbens)*	Sugar beet (*B. vulgaris*)	*H. schachtii*	Abnormal syncytium development leading to starvation of nematodes beyond J2 stage	Cai et al., 1997
	Gpa-2	Potato	Potato (*Solanum tuberosum*)	*G. pallida*	Development of stagnated and translucent female nematodes on plant roots	van der Vossen et al., 2000
	Hero A	Tomato	Potato	*G. pallida* *G. rostochiensis*	Hypersensitive response after the initiation of syncytia which become abnormal and necrotic due to degeneration of surrounding cells	Sobczak et al., 2005
	Gro1-4	Potato	Potato	*G. rostochiensis* pathotype Ro1	Unknown	Paal et al., 2004
	Rhg1	Soybean (*Glycine max*)	Soybean	*H. glycines*	Hypersensitive response leading to abnormal syncytia and necrosis due to degeneration of cells surrounding syncytia	Kandoth et al., 2011
	Rhg4	Soybean	Soybean	*H. glycines*	Hypersensitive response following the initiation of syncytia, which become abnormal and necrotic due to degeneration of surrounding cells this HR breaks down with the passage of time	Liu et al., 2012; Matthews et al., 2013
	Cre loci	*Aegilops* spp.	Wheat (*Triticum aestivum*)	*H. avenae*	Unknown	Williamson, 1998; Safari et al., 2005
Proteinase/ protease inhibitors	*CpTI* *SpTI-1*	Cowpea (*Vigna unguiculata*) *Sweet potato (Ipomoea batatas)*	Potato Sugar beet	*G. pallida* *M. incognita* *H. schachtii*	Effect on the sexual fate of newly established *G. pallida* and reduce the fecundity of females of without inhibition of growth and development of female *H. schachtii*	Hepher and Atkinson, 1992; Cai et al., 2003
	PIN2	Potato	Wheat	*H. avenae*	Unknown	Vishnudasan et al., 2005
	Oc-IΔD86	Rice (*Oryza sativa*)	Potato	*G. pallida* *M. incognita*	Reduced reproductive success of PPNs	Urwin et al., 1995, 2003; Lilley et al., 2004
			Arabidopsis thaliana	*H. schachtii* *M. incognita* *R. reniformis*	-do-	Urwin et al., 1997, 2000
			Rice	*M. incognita*	-do-	Vain et al., 1998
			Cavendish dessert bananas (*Musa acuminata*)	*Radopholus similis*	-do-	Atkinson et al., 2004
			Lily (*Lilium longiflorum*)	*Pratylenchus penetrans*	-do-	Vieira et al., 2015

			Eggplant (*Solanum melongena*)	*M. incognita*	-do-	Papolu et al., 2016
	CeCPI	*Taro (Colocasia esculenta)*	Tomato	*M. incognita*	Interferes with nematode ability of sex determination and gall formation	Chan et al., 2010, 2015
	CCII	Maize (*Zea mays*)	Plantain (*Musa* spp.)	*R. similis*, *Helicotylenchus Multicinctus*, and *Meloidogyne* spp.	Anti-feedant, reduces the reproductive success of nematodes	Roderick et al., 2012; Tripathi et al., 2015
Lectins	lectin *GNA*	*Snowdrop (Galanthus nivalis)*	Arabidopsis, oilseed rape (*Brassica napus*), and potato	*G. pallida*, *Pratylenchus bolivianus*, *M. incognita*	Decrease in number of females and galls developed on plant roots	Burrows et al., 1998; Ripoll et al., 2003
Bt toxins	*Cry6A*, *Cry5B*	*B. thuringiensis*	Tomato	*M. incognita*	Significant reduction in nematode reproduction	Li et al., 2007, 2008
Anti-invasion peptides	ACHE-I-7.1	*Synthetic*	Potato	*G. pallida*	Inhibits nematode acetylcholinesterase (ACHE) leading to disorientation of invading J2s	Winter et al., 2002; Liu et al., 2006; Lilley et al., 2011
	LEV-I-7.1	*Synthetic*	Potato	*G. pallida*	Results in chemodisruption of J2s and avoids invasion	Winter et al., 2002; Liu et al., 2006
	nAChRbp	*Synthetic*	Plantain (*Musa* spp.)	*R. similis* *H. multicinctus* *Meloidogyne* spp.	Disrupts chemosensory function of invading J2s	Roderick et al., 2012; Tripathi et al., 2015
Dual resistance	*CpTi* + *Oc-IΔD96*	*Cowpea and rice*	*Arabidopsis thaliana*	*G. pallida* and *H. schachtii*	Abnormal sexual development PPNs	Urwin et al., 1998
	Oc-IΔD86 + nAChRbp	*Rice* + *Synthetic*	Potato	*G. pallida*	Reducedreproductive success of PPNs coupled with disruption of chemosensory function of invading J2s	Green et al., 2012
	CeCPI + *PjCHI-1*	*Taro and Paecilomyces javanicus fungus*	Tomato	*M. incognita*	Reduced chitin content and retardation in embryogenesis in the nematode eggs	Chan et al., 2015
	CCII + nAChRbp	Maize + *Synthetic*	Plantain (*Musa* spp.)	*R. similis* *H. multicinctus* *Meloidogyne* spp.	Anti-feedant and an anti-root invasion plants with reducedreproductive success of PPNs coupled with disruption of chemosensory function of invading J2s	Tripathi et al., 2016, 2017

PLANT NATURAL RESISTANCE GENES

Natural resistance genes could exist in both polygenic manner and single dominant nature. The resistance conferred by host plant single dominant resistance genes, the *R* genes from plants, interacts specifically with corresponding avirulence (*Avr*) genes in the nematode, resulting in a so-called 'gene-for-gene' interaction. This type of interaction initiates a cascade of defense responses in the plants. A short summary of natural nematode resistance genes is recently given by Fuller et al. (2008) to provide the basis for this kind of resistance in plants.

Several natural host resistance genes have been cloned from some plant species and could be transferred to other plant species. For instance, *Mi-1.2* from tomato against *M. incognita* (Milligan et al., 1998; Williamson,

1998), *Hs1*$^{pro-1}$ from *Beta procumbens* against beet cyst nematode *H. schachtii* (Cai et al., 1997), *Gpa*-2 from potato against potato cyst nematode (PCN, *Globodera pallida*) (van der Vossen et al., 2000) and *Hero A* from tomato against *G. pallida* and *G. rostochiensis* (Sobczak et al., 2005) and *Cre* loci from *Aegilops* spp. against cereal cyst nematodes in wheat (Williams et al., 1994; Safari et al., 2005) are some examples that could be used in future to develop cyst nematode resistance in crop plants. Transgenic expression of resistance proteins also induces the expression of PR (pathogenesis related) proteins to establish nematode resistance in plants. The potato roots expressing *Hero A* gene showed high levels of several salicylic acid (SA)-dependent PR genes in the incompatible interaction with PCN at 3 dpi (Uehara et al., 2010). They confirmed that SA inducible PR-1(P4) was a hallmark for the cultivar resistance conferred by *Hero A* against PCN and that nematode parasitism resulted in the inhibition of the SA signaling pathway in the susceptible cultivars. Similar effects were found in resistant line of hexaploid wheat carrying *Cre2* gene, which showed upregulation of ascorbate peroxidase coding gene in response to cereal cyst nematode (*H. avenae*) when compared with the expression in the susceptible lines (Simonetti et al., 2010).

Another example is *Gro1-4*, the constitutive expression of which has increased resistance in potato plants against *G. rostochiensis* pathotype Ro1 (Paal et al., 2004). *Rhg1* is another natural resistance gene identified in soybean against soybean cyst nematode (SCN), *H. glycines* (Kandoth et al., 2011). A recent study has shown that map-based cloning of a gene at the *Rhg4* locus, which is a major quantitative trait locus (QTL), contributing resistance against SCN (Liu et al., 2012). In that study, the *Rhg4* mutant was analyzed through transgenic complementation, which revealed that this gene confers resistance in soybean by encoding a serine hydroxymethyltransferase. This was further confirmed through overexpression of serine hydroxymethyltransferase in soybean roots that demonstrated 45% decrease in the number of mature cyst nematode (Matthews et al., 2013). In the same study, overexpression of nine other putative resistance genes (including short chain dehydrogenase, ascorbate peroxidase, lipase, β-1,4-endoglucanase, calmodulin, DREPP membrane protein, and three proteins with unknown function) resulted in more than 50% decrease in the number of adult females in soybean roots. Similarly, *Hero A* gene gives almost complete resistance (>95%) against root cyst nematode (*G. rostochiensis*), while it provides around 80% of resistance against *G. pallida* (Ernst et al., 2002). Transgenic expression of *Hs1*$^{pro-1}$ (resistance gene against *H. schachtii*) from *B. procumbens* into sugar beet led to nematode resistance, but unluckily was linked with the genes that were negatively correlated with beet yield (Panella and Lewellen, 2007). Moreover, the *R* genes are typically effective against one or limited range of nematode species/pathotypes. Another limitation with this strategy is the development of different nematode pathotypes, which have the effectors (*avr* genes) that would not be recognized by the *R* genes (Jung et al., 1998).

USE OF PROTEINASE INHIBITOR CODING GENES

Proteinase inhibitors/PIs are molecules, mostly protein in nature, which inhibit the function of proteinases/proteases released by the pathogens. After the attack of herbivores and wounding, a variety of proteinase inhibitors are produced into the plants. In case of PPNs, these PIs become active against all the four classes of proteinases from nematodes, i.e., serine, cysteine, metalloproteinases, and aspartic. The PIs used for nematode resistance studies are CpTI (Hepher and Atkinson, 1992), sweet potato (*Ipomoea batatas*) serine PI (sporamin or SpTI-1) (Cai et al., 2003), PIN2 (Vishnudasan et al., 2005), rice (*Oryza sativa*) cystatin (Oc-IΔD86) (Urwin et al., 1997, 1998), and some others cystatins from maize (*Zea mays*), taro (*Colocasia esculenta*), and sunflower (*Helianthus annuus*) (Fuller et al., 2008; Chan et al., 2015).

The anti-nematode potential of plant PIs was firstly described in transgenic potato expressing the serine PI, the cowpea (*Vigna unguiculata*) trypsin inhibitor (CpTI) against PCN (*G. pallida*) (Hepher and Atkinson, 1992). Similarly, the Arabidopsis plants overexpressing cystatin Oc-IΔD86 suppress both the growth and fecundity in *H. schachtii* and *M. incognita* (Urwin et al., 1997). Oc-IΔD86 was effective against different nematode species in various plants species (Urwin et al., 1995, 2000, 2003; Vain et al., 1998; Atkinson et al., 2004; Lilley et al., 2004).

Moreover, the combinations of different PIs could be helpful to couple specificity with wide range of resistance. Transgenic expression of two proteinase inhibitors, CpTI and Oc-I1Δ86, as a translational fusion protein in Arabidopsis resulted in an additive effect against *G. pallida* and *H. schachtii* (Hepher and Atkinson, 1992; Urwin et al., 1998). Other important proteinase inhibitors are sporamin (SpTI-1) (Cai et al., 2003) and PIN2 (Vishnudasan et al., 2005), which have shown good resistant response in plants.

As compared to other proteinase inhibitors, cystatins from different plant species have met the promise of enhancement of nematode resistance in a variety of crop plants (Urwin et al., 1997, 1998; Chan et al., 2010, 2015; Green et al., 2012; Tripathi et al., 2015; Papolu et al., 2016). Heterologous expression of a taro cystatin established considerable degree of resistance in tomato against *M. incognita* infection by interference with nematode ability of sex determination and gall formation (Chan et al.,

2010). Recently, a dual strategy has been used to control PCN without affecting the soil quality (Green et al., 2012). In the first approach, a peptide was precisely targeted under control of a root tip-specific promoter that disrupts chemoreception of nematodes and suppresses root invasion without a lethal effect in both containment as well as under field trial (Lilley et al., 2011). In addition to this chemoreception disruptive peptide, OcIΔD86 cystatin from rice was incorporated to control the invading larvae that are able to cross the barrier of chemodisruptive peptide. This approach establishes that a combination of these genes offers distinct bases for the transgenic plant resistance to *G. pallida* without harmful impact on the non-target nematode soil community (Green et al., 2012). This rice cystatin has also shown good control in lily (*Lilium longiflorum* cv. 'Nellie White') against lesion nematode (*P. penetrans*) (Vieira et al., 2015).

Recently, an anti-feedant maize cystatin and an anti-root invasion synthetic peptide were transformed into plantain (*Musa* spp., cv. Gonja manjaya), individually and in combination (Tripathi et al., 2015). The field trials of the best transgenic event containing the peptide only demonstrated 186% more yield in addition to 99% control against *R. similis*, *H. multicinctus*, and *Meloidogyne* spp. as compared to non-transgenic control. Moreover, transgenic expression of proteinase inhibitor from maize and synthetic chemodisruptive peptide resulted in enhance yield and nematode resistance in plantain. Similarly, dual overexpression a taro cysteine proteinase inhibitor (CeCPI) and a fungal chitinase (PjCHI-1) in tomato under the control of a synthetic promoter, pMSPOA, (with NOS-like and SP8a elements), had negative effects on reproduction of *M. incognita* (Chan et al., 2015). This study further revealed that dual gene transformation had more inhibition of nematodes than plants transformed with a single gene. This advocated that the use of gene pyramiding could be employed for developing and improving nematode resistance in plants (Chan et al., 2015; Tripathi et al., 2015, 2017).

Very recently, a modified rice cystatin (Oc-IΔD86) was expressed in the roots of eggplant (*S. melongena*) under the control of the root-specific promoter, TUB-1 (Papolu et al., 2016). Five putative transformants containing this cystatin exhibited detrimental effects on both the development and the reproduction of *M. incognita* in eggplant. In that study, a single copy transgenic event showed 78.3% reduction in reproductive success of *M. incognita*. This concludes that proteinase inhibitors are potential candidates for induction of nematodes resistance in a variety of crop plants to increase crop yield and minimize the damage caused by these parasitic worms.

NEMATICIDAL PROTEINS

These proteins could be characterized as anti-nematode proteins because these are directly involved in inhibiting the nematode development on the plants. Lectins, some antibodies, and *Bt* Cry proteins are some examples of these proteins. The toxicity of *lectins* is characterized by their ability to obstruct intestinal function of organisms exhibiting or ingesting them (Vasconcelos and Oliveira, 2004). The defense mechanism conditioned by the lectins is vital as several lectins bind with glycans (Peumans and van Damme, 1995). Overexpression of a snowdrop (*Galanthus nivalis*) lectin GNA driven by cauliflower mosaic virus promoter (CaMV35S) has been exploited to exhibit anti-nematode activity in several plants, i.e., Arabidopsis, oilseed rape (*Brassica napus*), and potato, in response to RKNs, CNs and lesion nematodes (Burrows et al., 1998; Ripoll et al., 2003).

Plantibodies are the antibodies expressed in plants and also potential candidates for the development of nematode resistance. These are important because the establishment of a compatible plant–nematode interaction engages a series of processes against which plantibodies may be directed. RKNs and CNs depend on secretions of their pharyngeal glands to mimic re-differentiation of plant cells into specialized nematode feeding sites like giant cells or syncytia. Direction of plantibodies opposite to the active proteins from these secretions could be attenuated to suppress the parasitic ability of the nematode. There are only a few reports available in the literature regarding the use of plantibodies for nematode resistance (Fioretti et al., 2002; Sharon et al., 2002).

In addition to lectins and plantibodies, different variants of *Bt toxins* (Cry proteins) derived from *Bacillus thuringiensis* have shown promise to induce plant resistance against nematodes. These variants of *Bt* toxins are, however, more frequently used against phytophagous insects. *Bt* toxin was firstly used as an anti-nematode protein by Marroquin et al. (2000), when *C. elegans* was exposed to Cry5B and Cry6A which resulted in the reduction in nematode fertility and viability. The PPNs use feeding tube at the stylet orifice while feeding on the plants roots. The feeding tube operates as molecular sieve, allowing the uptake of certain molecules and excluding the others. The ultrastructure of feeding tubes from root-knot and cyst nematode differ which is based on the observation that root-knot nematodes are able to ingest larger proteins compared with cyst nematodes (Hussey and Mims, 1990; Böckenhoff and Grundler, 1994; Urwin et al., 1998; Sobczak et al., 1999; Li et al., 2007). Transgenic expression of 54 kDa Cry6A and Cry5B proteins in tomato hairy roots affected the reproduction of root-knot nematode *M. incognita* (Li et al., 2007, 2008). Western blotting technique showed that the 54 kDa Cry6A protein was shown to be ingested by *M. incognita*. On the other hand, this large protein could not be ingested by cyst nematodes (i.e., *H. schachtii*) due to small orifice of the feeding tube having the size limit up to approximately 23 kDa (Urwin et al., 1998). This limitation severely restricts the agronomic application of these toxins against PPNs.

CHEMODISRUPTIVE PEPTIDES

Plant parasitic nematodes are highly dependent on their chemoreceptive neurons to sense distinct chemical stimuli for invasion into the plants. Nematodes use acetylcholinesterase (*AChE*) and/or nicotinic acetylcholine receptors for proper functioning of the nervous system. Chemodisruptive peptides are another important strategy to minimize the invasion of PPNs into the plant roots. Two peptides have been shown to bind with these receptors to inhibit their proper function (Winter et al., 2002). Both of these peptides disrupted nematode ability of chemoreception by blocking their reaction to chemical signal at very minute concentrations of up to 1 nm. Transgenic potato plants expressing a secreted peptide that inhibited nematode *AChE* leading to disorientation of invading nematode *G. pallida*, which resulted in a 52% decline in the number of female nematodes (Liu et al., 2005). The peptide is considered effective after its uptake from chemoreceptor sensillae through retrograde transport along nematode neurons to cholinergic synapses. Costa et al. (2009) has demonstrated that cyst nematode acetylcholinesterase gene (*AChE*) is expressed in chemo-and mechanosensory neurons of *C. elegans*, which further supports this hypothesis. Similarly, Wang et al. (2011) reported indirect evidence to support the mechanism by which such peptide disrupts chemosensory function in cyst nematodes. The peptide exhibits disulphide-constrained 7-mer with the amino acid sequence CTTMHPRLC that binds to nicotinic acetylcholine receptors. Incubation in the peptide solution or root-exudate from transgenic plants that secrete the peptide disrupted normal orientation of infective cyst nematodes to host root diffusate.

Moreover, chemosensory disruptive peptide that inhibits *AChE* has recently been expressed under the control of the constitutive CaMV35S promoter and the root tip-specific promoter in Arabidopsis and potato plants, where it confers resistance against *H. schachtii* and *G. pallida* (Lilley et al., 2011). This root tip-specific promoter from Arabidopsis gene (*MDK4-20*; *At5g54370*) directed expression of the nematode repellant peptide only at the sites of cyst nematode invasion and has shown strong level of resistance against PCN. This strategy has now been combined with the transgenic expression of a rice cystatin in potato to maintain high level of resistance against PCN without affecting soil quality (Green et al., 2012). By using the same technique, The International Institute of Tropical

Agriculture (IITA), in partnership with the University of Leeds, UK, developed transgenic plantain for nematode resistance using maize digestive protease inhibitor cystatin and synthetic nematode repellent peptides (Roderick et al., 2012; Tripathi et al., 2013). Furthermore, pyramiding of cystatins and chemodisruptive peptide into different crop plants has shown high degree of nematode resistance and enhanced crop yields against root-knot nematodes (Chan et al., 2015; Tripathi et al., 2015, 2017).

UTILIZATION OF RNA INTERFERENCE TO SUPPRESS NEMATODE EFFECTORS

RNA interference has emerged as a very useful tool for gene-silencing aimed at functional analysis of different genes by suppressing their expression in a wide variety of organisms including PPNs. In this strategy, the nematodes uptake double-stranded RNA (dsRNA) or short interfering RNAs (siRNAs) from the plants expressing these RNAs, which elicit a systemic RNAi response in nematodes. A schematic diagram elaborating the mechanism of *in planta* RNAi is shown in Figure 2. The transgenic expression of dsRNA targeting a specific nematode effector gene could be handful to suppress the expression of that effector gene, which is crucial for infection process. There are many review articles mushroomed out recently emphasizing the usefulness and application of RNAi technology to induce nematode resistance in plants by silencing the expression of nematode effectors mainly (Gheysen and Vanholme, 2007; Lilley et al., 2007; Fuller et al., 2008; Rosso et al., 2009; Maule et al., 2011; Tamilarasan and Rajam, 2012). In a recent review article most of the aspects of the RNAi application in nematode resistance are reviewed (Lilley et al., 2012). Lilley et al. (2012) have reviewed various features ranging from *in vitro* assays with *C. elegans* to delivering RNAi *in planta* as an important strategy for crop protection against cyst nematodes.

Recently, Youssef et al. (2013a) have tested this approach by silencing the *H. glycines* gene *HgALD*, encoding fructose-1,6-diphosphate aldolase important in the conversion of glucose into energy and actin-based motility during parasite invasion into its host. Transgenic soybean roots expressing an RNAi construct targeted to silence *HgALD* revealed 58% reduction of females formed by *H. glycines*. Very recently, Tripathi et al. (2017) have reviewed the application of RNAi for the enhancement of nematode resistance by suppression of important effector proteins. Nevertheless, RNAi has become an established experimental tool for the enhancement of resistance against PPNs and also offers the prospect of being developed into a novel control strategy when delivered from transgenic plants.

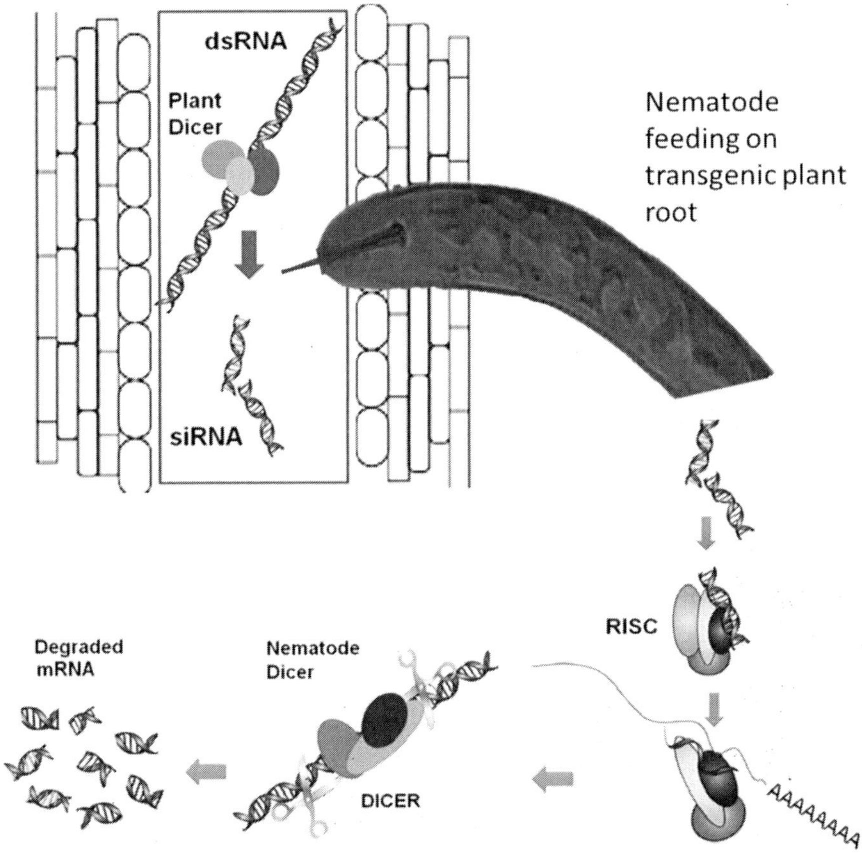

FIGURE 2. The mechanism of RNA interference (RNAi), the double-stranded RNA (dsRNA) is processed by the plant dicer enzyme **(A)** into **(B)**. Once the dsRNA is uptaken by the nematode from the plant cell while feeding **(C)**, the processing from dsRNA to short interfering RNA (siRNA) can be executed by the nematode dicer. Then, the siRNA is recognized by the RNA-induced silencing (RISC) complex of the nematode **(D)** and its unwinding into sense and antisense strands takes place. A proportion of the RISC complex loaded with the antisense strand interacts with the corresponding mRNA in the nematode **(E)** as a result the mRNA is cleaved by the RISC **(F)** and subsequently degraded **(G)**. Moreover, the targeted mRNA can be made double-stranded after binding of the siRNA, and this dsRNA is then processed to produce additional siRNAs, intensifying the initial silencing signal (Gheysen and Vanholme, 2007).

OTHER STRATEGIES FOR NEMATODE RESISTANCE IN PLANTS

The manipulation of expression of genes from the plants could also be useful for induction of resistance against cyst nematodes. Transgenic plants that overexpress genes correlated with resistance or have silenced genes which are important to syncytium development and nematode success are two obvious areas to explore (Klink and Matthews, 2009; Ali, 2012). This could be achieved by using precise delivery of transgene into the feeding sites as the constitutive overexpression or suppression of a particular gene could have detrimental effects on the growth and development of the plants (Ali et al., 2013b). For this purpose, promoters that are specifically expressed in the feeding sites of PPNs, i.e., *Pdf2.1* or *MIOX5*, could be used (Siddique et al., 2009, 2011). Although constitutive promoters can deliver the gene of interest or suppress the genes which are important for compatible interaction (Ali and Abbas, 2016), however, it could be lethal for the plants to silence the genes which are also important for other physiological processes in addition to establishment of the nematodes. The promoters of the genes like *Pdf2.1* or *MIOX5* are specially and highly expressed in syncytia and several studies have shown the utility of these promoters to overexpress and silence the genes of interest specially into the feeding sites (Siddique et al., 2009, 2011; Ali et al., 2012, 2013b, 2014). In addition to syncytia specific promoters, root and root tip-specific promoters have also been used to drive site specific expression of proteinase inhibitors and nematode chemodistruptive peptides in several plant species (Lilley et al., 2011; Atkinson et al., 2012; Green et al., 2012).

The transcriptomes of various plant species infected with different nematode species demonstrated the upregulation of genes important for development of nematode feeding structures in the plant roots. The knockout mutants of two endo-1,4-β-glucanases, which were highly upregulated in syncytia, revealed less susceptibility in Arabidopsis in response to beet cyst nematode (Wieczorek et al., 2008). Likewise, an ATPase gene from Arabidopsis (*At1g64110*) was reported to be induced in syncytia caused by *H. schachtii* (Ali et al., 2013b). The knocking down of this gene using syncytia specific promoters (prom.AtPDF2.1 and prom.AtMIOX5) supported less number of nematodes.

Conversely, the nematodes are smart enough to suppress the defense mechanisms of the plants as most of the genes involved in defense related pathways were downregulated in the feeding sites revealed by plant transcriptomes in response to nematode infection (Szakasits et al., 2009; Barcala et al., 2010; Kyndt et al., 2012; Ali et al., 2015). One strongly downregulated group comprised peroxidase gene family, as out of top 100 differentially expressed genes with the strongest decrease in expression, 14 were peroxidases (Szakasits et al., 2009). Similarly, ethylene responsive transcription factor from Arabidopsis, AtRAP2.6, was one of the strongly downregulated transcripts in syncytia. This gene was driven through the constitutive promoter CaMV35S to overexpress in Arabidopsis at the global level, which resulted in reduced susceptibility in overexpression lines (Ali et al., 2013a). This also supports the debate of use of CaMV35S promoter to expression gene of interest in the syncytia induced by beet cyst nematodes in Arabidopsis roots (Ali and Abbas, 2016).

Another strategy is the expression of the genes involved in the defense pathways like camalexin and callose synthesis in plants (Hofmann et al., 2010; Millet et al., 2010; Mao et al., 2011; Birkenbihl et al., 2012). Recently, expression of *AtPAD4* under the control of FMV-sgt promoter has resulted in enhanced resistance against soybean cyst and root-knot nematodes in soybean (Youssef et al., 2013b). Expression of *AtPAD4* in soybean roots decreased the number of mature *H. glycines* females and *M. incognita* galls up to 68 and 77%, respectively. We have recently dissected the pathway of camalexin synthesis in plant–nematode interactions based on infection assays of *AtWRKY33* and *AtPAD3* mutants and overexpression lines (Ali et al., 2014). In this report, the syncytia specific overexpression of *WRKY33* resulted in the suppression of susceptibility in Arabidopsis. Similarly, the overexpression of a soybean salicylic acid methyltransferase (*GmSAMT1*) gene is found to confer resistance to SCN (Lin et al., 2013).

CONCLUSION REMARKS

As a consequence of enormous yield losses in crop plants imposed by the PPNs, the understanding of plant–nematode interaction is becoming of utmost importance. The nematode effectors include, for instance, cell wall degrading enzymes, the genes involved in molecular mimicry of plant genes for both compatible and incompatible plant–nematode interactions. Targeted silencing of known nematode effector proteins through *in planta* RNAi technology holds a great potential for plant resistance against various species of nematodes (Gheysen and Vanholme, 2007). A recently developed virus-induced gene silencing (VIGS) method provides a new tool to identify genes involved in soybean–nematode interactions (Kandoth et al., 2013). Similarly, the application of bioinformatics in the form of approaches like OrthoMCL could be very important for computational identification and analysis of effector proteins from PPNs (Williams et al., 1994). The development of population specific markers along with character compatibility method for diagnosis and phylogenetic inference of inter and intra specific populations of nematodes could also be done (Phiri et al., 2013). During the compatible plant–nematode interaction, nematodes are somehow able to suppress the defense related genes, the overexpression of which has led to enhanced resistance (Ali et al., 2013a). In addition to CaMV-35S promoter, several syncytia specific promoters could be used to overproduce the defense related genes in the feeding sites of the nematode to enhance resistance (Ali et al., 2013a, 2014; Ali and Abbas, 2016). This could be the interesting starting point for further studies to explicate how nematodes are able to suppress systemic plant defense mechanisms. It is concluded that the use of different transgenic strategies has shown good promise for nematode resistance. These have been helpful for reduction of nematode population on the plants on individual basis; however, by stacking all these molecular strategies together in the one plant will result in additive resistance, almost near to immunity against nematodes in crop plants.

AUTHOR CONTRIBUTIONS

MA conceived, designed and mainly develop the article; FA, AA, and FJ contributed toward the write up in main body of article; HL and AD reviewed and finalize the article.

FUNDING

This study is partially supported by Higher Education Commission of Pakistan.

CONFLICT OF INTEREST STATEMENT

The authors declare that the research was conducted in the absence of any commercial or financial relationships that could be construed as a potential conflict of interest.

ACKNOWLEDGMENTS

The authors declare that all the appropriate permissions have been obtained from the copyright holders of any work that has been reproduced in the manuscript. We also thank Higher Education Commission of Pakistan for partial support.

REFERENCES

1. Abad, P., Gouzy, J., Aury, J. M., Castagnone-Sereno, P., Danchin, E. G. J., Deleury, E., et al. (2008). Genome sequence of the metazoan plant-parasitic nematode *Meloidogyne incognita*. *Nat. Biotechnol.* 26, 909–915.
2. Ali, M. A. (2012). *Induction of Resistance Against Cyst Nematodes by Modifying Gene Expression in Syncytia*. Ph.D. thesis, University of Natural Resources and Life Sciences, Vienna.
3. Ali, M. A., and Abbas, A. (2016). Analysis of reporter proteins GUS and DsRed driven under the control of CaMV35S promoter in syncytia induced by beet cyst nematode *Heterodera schachtii* in Arabidopsis roots.*Adv. Life Sci.* 3, 89–96.
4. Ali, M. A., Abbas, A., Azeem, F., Javed, N., and Bohlmann, H. (2015). Plant-nematode interactions: from genomics to metabolomics. *Int. J. Agric. Biol.* 17, 1071–1082.
5. Ali, M. A., Abbas, A., Kreil, D. P., and Bohlmann, H. (2013a). Overexpression of the transcription factor RAP2.6 leads to enhanced callose deposition in syncytia and enhanced resistance against the beet cyst nematode *Heterodera schachtii* in Arabidopsis roots. *BMC Plant Biol.* 13:47.
6. Ali, M. A., Plattner, S., Radakovic, Z., Wieczorek, K., Elashry, A., Grundler, F. M., et al. (2013b). An Arabidopsis ATPase gene involved in nematode-induced syncytium development and abiotic stress responses. *Plant J.* 74, 852–866.
7. Ali, M. A., Shah, K. H., and Bohlmann, H. (2012). pMAA-Red: a new pPZP-derived vector for fast visual screening of transgenic Arabidopsis plants at the seed stage. *BMC Biotechnol.* 12, 37.
8. Ali, M. A., Wieczorek, K., Kreil, D. P., and Bohlmann, H. (2014). The beet cyst nematode *Heterodera schachtii* modulates the expression of WRKY transcription factors in syncytia to favour its development in Arabidopsis roots. *PLoS ONE* 9:e102360.
9. Atkinson, H. J., Grimwood, S., Johnston, K., and Green, J. (2004). Prototype demonstration of transgenic resistance to the nematode *Radopholus similis* conferred on banana by a cystatin. *Transgenic Res.* 13, 135–142.
10. Atkinson, H. J., Lilley, C. J., and Urwin, P. E. (2012). Strategies for transgenic nematode control in developed and developing world crops. *Curr. Opin. Biotechnol.* 23, 251–256.
11. Bakhetia, M., Charlton, W. L., Urwin, P. E., Mcpherson, M. J., and Atkinson, H. J. (2005). RNA interference and plant parasitic nematodes. *Trends Plant Sci.* 10, 362–367.
12. Barcala, M., Garcia, A., Cabrera, J., Casson, S., Lindsey, K., Favery, B., et al. (2010). Early transcriptomic events in microdissected Arabidopsis nematode-induced giant cells. *Plant J.* 61, 698–712.
13. Bellafiore, S., Shen, Z. X., Rosso, M. N., Abad, P., Shih, P., and Briggs, S. P. (2008). Direct identification of the *Meloidogyne incognita* secretome reveals proteins with host cell reprogramming potential. *PLoS Pathog.* 4:e1000192.
14. Birkenbihl, R. P., Diezel, C., and Somssich, I. E. (2012). Arabidopsis WRKY33 is a key transcriptional regulator of hormonal and metabolic responses toward *Botrytis cinerea* infection. *Plant Physiol.* 159, 266–285.
15. Böckenhoff, A., and Grundler, F. M. W. (1994). Studies on the nutrient-uptake by the beet cyst-nematode *Heterodera schachtii* by *in situ* microinjection of fluorescent probes into the feeding structures in *Arabidopsis thaliana*. *Parasitology* 109, 249–255.

16. Brodie, B. B., and Murphy, W. S. (1975). Population dynamics of plant nematodes as affected by combinations of fallow and cropping sequence. *J. Nematol.* 7, 91–92.
17. Burrows, P. R., Barker, A. D. P., Newell, C. A., and Hamilton, W. D. O. (1998). Plant-derived enzyme inhibitors and lectins for resistance against plant-parasitic nematodes in transgenic crops. *Pestic. Sci.* 52, 176–183.
18. Cai, D., Thurau, T., Tian, Y., Lange, T., Yeh, K. W., and Jung, C. (2003). Sporamin-mediated resistance to beet cyst nematodes (*Heterodera schachtii* Schm.) is dependent on trypsin inhibitory activity in sugar beet (*Beta vulgaris* L.) hairy roots. *Plant Mol. Biol.* 51, 839–849.
19. Cai, D. G., Kleine, M., Kifle, S., Harloff, H. J., Sandal, N. N., Marcker, K. A., et al. (1997). Positional cloning of a gene for nematode resistance in sugar beet. *Science* 275, 832–834.
20. Caillaud, M. C., Dubreuil, G., Quentin, M., Perfus-Barbeoch, L., Lecomte, P., De Almeida Engler, J., et al. (2008). Root-knot nematodes manipulate plant cell functions during a compatible interaction. *J. Plant Physiol.* 165, 104–113.
21. Chan, Y.-L., He, Y., Hsiao, T.-T., Wang, C.-J., Tian, Z., and Yeh, K.-W. (2015). Pyramiding taro cystatin and fungal chitinase genes driven by a synthetic promoter enhances resistance in tomato to root-knot nematode *Meloidogyne incognita*. *Plant Sci.* 231, 74–81.
22. Chan, Y.-L., Yang, A.-H., Chen, J.-T., Yeh, K.-W., and Chan, M.-T. (2010). Heterologous expression of taro cystatin protects transgenic tomato against *Meloidogyne incognita* infection by means of interfering sex determination and suppressing gall formation. *Plant Cell Rep.* 29, 231–238.
23. Clayton, R., Storey, M., Parker, B., Ballingall, M., and Davies, K. (2008). *Impact of Reduced Pesticide Availability on Control of Potato Cyst Nematodes and Weeds in Potato Crops.* Oxford: Potato Council Ltd.
24. Costa, J. C., Lilley, C. J., Atkinson, H. J., and Urwin, P. E. (2009). Functional characterisation of a cyst nematode acetylcholinesterase gene using *Caenorhabditis elegans* as a heterologous system. *Int. J. Parasitol.* 39, 849–858.
25. Dababat, A. A., Ferney, G.-B. H., Erginbas-Orakci, G., Dreisigacker, S., Imren, M., Toktay, H., et al. (2016). Association analysis of resistance to cereal cyst nematodes (*Heterodera avenae*) and root lesion nematodes (*Pratylenchus neglectus* and *P. thornei*) in CIMMYT advanced spring wheat lines for semi-arid conditions. *Breed. Sci.* 66, 692–702.
26. Dababat, A., Imren, M., Erginbas-Orakci, G., Ashrafi, S., Yavuzaslanoglu, E., Toktay, H., et al. (2015). The importance and management strategies of cereal cyst nematodes, *Heterodera* spp., in Turkey. *Euphytica* 202, 173–188.
27. Dababat, A., and Sikora, R. A. (2007a). Importance of application time and inoculum density of *Fusarium oxysporum* 162 for biological control of *Meloidogyne incognita* on tomato. *Nematropica* 37, 267–275.
28. Dababat, A. A., and Sikora, R. A. (2007b). Induced resistance by the mutualistic endophyte, *Fusarium oxysporum* strain 162, toward *Meloidogyne incognita* on tomato. *Biocontrol Sci. Technol.* 17, 969–975.
29. Decraemer, W., and Hunt, D. J. (2006). "Structure and classification," in *Plant Nematology*, eds R. N. Perry and M. Moens (Wallingford: CABI Publishing).
30. Ernst, K., Kumar, A., Kriseleit, D., Kloos, D. U., Phillips, M. S., and Ganal, M. W. (2002). The broad-spectrum potato cyst nematode resistance gene (*Hero*) from tomato is the only member of a large gene family of NBS-LRR genes with an unusual amino acid repeat in the LRR region. *Plant J.* 31, 127–136.
31. Fioretti, L., Porter, A., Haydock, P. J., and Curtis, R. (2002). Monoclonal antibodies reactive with secreted-excreted products from the amphids and the cuticle surface of *Globodera pallida* affect nematode movement and delay invasion of potato roots. *Int. J. Parasitol.* 32, 1709–1718.
32. Fuller, V. L., Lilley, C. J., and Urwin, P. E. (2008). Nematode resistance. *New Phytol.* 180, 27–44.

33. Gheysen, G., Kyndt, T., Haegeman, A., Joseph, S., Remy, S., and Swennen, R. (2010). RNAi for research and applications in plant-nematode interactions. *In Vitro Cell. Dev. Biol. Anim.* 46, S28–S29.
34. Gheysen, G., and Vanholme, B. (2007). RNAi from plants to nematodes. *Trends Biotechnol.* 25, 89–92.
35. Green, J., Wang, D., Lilley, C. J., Urwin, P. E., and Atkinson, H. J. (2012). Transgenic potatoes for potato cyst nematode control can replace pesticide use without impact on soil quality. *PLoS ONE* 7:e30973.
36. Hepher, A., and Atkinson, H. J. (1992). Nematode Control with Proteinase Inhibitors. EP 0502730 B1.
37. Hofmann, J., Youssef-Banora, M., De Almeida-Engler, J., and Grundler, F. M. (2010). The role of callose deposition along plasmodesmata in nematode feeding sites. *Mol. Plant Microbe Interact.* 23, 549–557.
38. Huang, G. Z., Gao, B. L., Maier, T., Allen, R., Davis, E. L., Baum, T. J., et al. (2003). A profile of putative parasitism genes expressed in the esophageal gland cells of the root-knot nematode *Meloidogyne incognita*. *Mol. Plant Microbe Interact.* 16, 376–381.
39. Hussey, R. S., and Mims, C. W. (1990). Ultrastructure of esophageal glands and their secretory granules in the root-knot nematode *Meloidogyne incognita*. *Protoplasma* 156, 9–18.
40. Jones, M. G. K. (1981). Host cell responses to endoparasitic nematode attack: structure and function of giant cells and syncytia. *Ann. Appl. Biol.* 97, 353–372.
41. Jones, M. G. K., and Payne, H. L. (1978). Early stage of nematode-induced giant-cell formation in roots of *Impatiens balsamina*. *J. Nematol.* 10, 70–84.
42. Jung, C., Cai, D. G., and Kleine, M. (1998). Engineering nematode resistance in crop species. *Trends Plant Sci.* 3, 266–271.
43. Kandoth, P. K., Heinz, R., Yeckel, G., Gross, N. W., Juvale, P. S., Hill, J., et al. (2013). A virus-induced gene silencing method to study soybean cyst nematode parasitism in *Glycine max*. *BMC Res. Notes* 6:255.
44. Kandoth, P. K., Ithal, N., Recknor, J., Maier, T., Nettleton, D., Baum, T. J., et al. (2011). The soybean *Rhg1* locus for resistance to the soybean cyst nematode *Heterodera glycines* regulates the expression of a large number of stress- and defense-related genes in degenerating feeding cells. *Plant Physiol.* 155,1960–1975.
45. Katan, J. (1981). Solar heating (solarization) of soil for control of soilborne pests. *Annu. Rev. Phytopathol.* 19, 211–236.
46. Kerry, B. (1980). Biocontrol: fungal parasites of female cyst nematodes. *J. Nematol.* 12, 253–259.
47. Kerry, B. (1988). Fungal parasites of cyst nematodes. *Agric. Ecosyst. Environ.* 24, 293–305.
48. Kirkegaard, J. A., and Sarwar, M. (1998). Biofumigation potential of Brassicas. I. Variation in glucosinolate profiles of diverse field-grown Brassicas. *Plant Soil* 201, 71–89.
49. Klink, V. P., and Matthews, B. F. (2009). Emerging approaches to broaden resistance of soybean to soybean cyst nematode as supported by gene expression studies. *Plant Physiol.* 151, 1017–1022.
50. Kyndt, T., Nahar, K., Haegeman, A., De Vleesschauwer, D., Hofte, M., and Gheysen, G. (2012). Comparing systemic defence-related gene expression changes upon migratory and sedentary nematode attack in rice. *Plant Biol.* 14, 73–82.
51. Li, X.-Q., Tan, A., Voegtline, M., Bekele, S., Chen, C.-S., and Aroian, R. V. (2008). Expression of Cry5B protein from *Bacillus thuringiensis* in plant roots confers resistance to root-knot nematode. *Biol. Control* 47, 97–102.
52. Li, X. Q., Wei, J. Z., Tan, A., and Aroian, R. V. (2007). Resistance to root-knot nematode in tomato roots expressing a nematicidal *Bacillus thuringiensis* crystal protein. *Plant Biotechnol. J.* 5, 455–464.

53. Lilley, C. J., Bakhetia, M., Charlton, W. L., and Urwin, P. E. (2007). Recent progress in the development of RNA interference for plant parasitic nematodes. *Mol. Plant Pathol.* 8, 701–711.
54. Lilley, C. J., Davies, L. J., and Urwin, P. E. (2012). RNA interference in plant parasitic nematodes: a summary of the current status. *Parasitology* 139, 630–640.
55. Lilley, C. J., Devlin, F., Urwin, P. E., and Atkinson, H. J. (1999). Parasitic nematodes, proteinases and transgenic plants. *Parasitol. Today* 15, 414–417.
56. Lilley, C. J., Urwin, P. E., Johnston, K. A., and Atkinson, H. J. (2004). Preferential expression of a plant cystatin at nematode feeding sites confers resistance to *Meloidogyne incognita* and *Globodera pallida*. *Plant Biotechnol. J.* 2, 3–12.
57. Lilley, C. J., Wang, D., Atkinson, H. J., and Urwin, P. E. (2011). Effective delivery of a nematode-repellent peptide using a root-cap-specific promoter. *Plant Biotechnol. J.* 9, 151–161.
58. Lin, J., Mazarei, M., Zhao, N., Zhu, J. J., Zhuang, X., Liu, W., et al. (2013). Overexpression of a soybean salicylic acid methyltransferase gene confers resistance to soybean cyst nematode. *Plant Biotechnol. J.* 11, 1135–1145.
59. Liu, B., Hibbard, J. K., Urwin, P. E., and Atkinson, H. J. (2005). The production of synthetic chemodisruptive peptides in planta disrupts the establishment of cyst nematodes. *Plant Biotechnol. J.* 3, 487–496.
60. Liu, S., Kandoth, P. K., Warren, S. D., Yeckel, G., Heinz, R., Alden, J., et al. (2012). A soybean cyst nematode resistance gene points to a new mechanism of plant resistance to pathogens. *Nature* 492, 256–260.
61. MacGuldwin, A. E. (1993). "Management of nematodes," in *Potato Health Management*, ed. R. C. Rowe (St. Paul, MN: APS Press).
62. Mao, G., Meng, X., Liu, Y., Zheng, Z., Chen, Z., and Zhang, S. (2011). Phosphorylation of a WRKY transcription factor by two pathogen-responsive MAPKs drives phytoalexin biosynthesis in *Arabidopsis*. *Plant Cell* 23, 1639–1653.
63. Marroquin, L. D., Elyassnia, D., Griffitts, J. S., Feitelson, J. S., and Aroian, R. V. (2000). *Bacillus thuringiensis* (*Bt*) toxin susceptibility and isolation of resistance mutants in the nematode *Caenorhabditis elegans*. *Genetics* 155, 1693–1699.
64. Martinuz, A., Schouten, A., and Sikora, R. A. (2012). Post-infection development of *Meloidogyne incognita* on tomato treated with the endophytes *Fusarium oxysporum* strain Fo162 and *Rhizobium etli* strain G12. *Biocontrol* 58, 95–104.
65. Matthews, B. F., Beard, H., Macdonald, M. H., Kabir, S., Youssef, R. M., Hosseini, P., et al. (2013). Engineered resistance and hypersusceptibility through functional metabolic studies of 100 genes in soybean to its major pathogen, the soybean cyst nematode. *Planta* 237, 1337–1357.
66. Maule, A. G., Mcveigh, P., Dalzell, J. J., Atkinson, L., Mousley, A., and Marks, N. J. (2011). An eye on RNAi in nematode parasites. *Trends Parasitol.* 27, 505–513.
67. Millet, Y. A., Danna, C. H., Clay, N. K., Songnuan, W., Simon, M. D., Werck-Reichhart, D., et al. (2010). Innate immune responses activated in *Arabidopsis* roots by microbe-associated molecular patterns. *Plant Cell* 22, 973–990.
68. Milligan, S. B., Bodeau, J., Yaghoobi, J., Kaloshian, I., Zabel, P., and Williamson, V. M. (1998). The root knot nematode resistance gene *Mi* from tomato is a member of the leucine zipper, nucleotide binding, leucine-rich repeat family of plant genes. *Plant Cell* 10, 1307–1319.
69. Moens, M., and Perry, R. N. (2009). Migratory plant endoparasitic nematodes: a group rich in contrasts and divergence. *Annu. Rev. Phytopathol.* 47, 313–332.
70. Opperman, C. H., Bird, D. M., Williamson, V. M., Rokhsar, D. S., Burke, M., Cohn, J., et al. (2008). Sequence and genetic map of *Meloidogyne hapla*: a compact nematode genome for plant parasitism. *Proc. Natl. Acad. Sci. U.S.A.* 105, 14802–14807.
71. Paal, J., Henselewski, H., Muth, J., Meksem, K., Menendez, C. M., Salamini, F., et al. (2004). Molecular cloning of the potato *Gro1-4* gene conferring resistance to pathotype Ro1 of the root cyst nematode *Globodera rostochiensis*, based on a candidate gene approach. *Plant J.* 38, 285–297.

72. Panella, L., and Lewellen, R. T. (2007). Broadening the genetic base of sugar beet: introgression from wild relatives. *Euphytica* 154, 383–400.
73. Papolu, P. K., Dutta, T. K., Tyagi, N., Urwin, P. E., Lilley, C. J., and Rao, U. (2016). Expression of a cystatin transgene in eggplant provides resistance to root-knot nematode, *Meloidogyne incognita. Front. Plant Sci.* 7:1122.
74. Peumans, W. J., and van Damme, E. J. M. (1995). The role of lectins in plant defence. *Histochem. J.* 27, 253–271.
75. Phiri, A. M., De Pomerai, D., Buttle, D. J., and Behnke, J. M. B. (2013). Developing a rapid throughput screen for detection of nematicidal activity of plant cysteine proteinases: the role of *Caenorhabditis elegans* cystatins. *Parasitology* 141, 164–180.
76. Riley, I. T., Nicol, J. M., and Dababat, A. A. (2009). *Cereal Cyst Nematodes: Status, Research and Outlook.* Ankara: CIMMYT.
77. Ripoll, C., Favery, B., Lecomte, P., Van Damme, E., Peumans, W., Abad, P., et al. (2003). Evaluation of the ability of lectin from snowdrop (*Galanthus nivalis*) to protect plants against root-knot nematodes. *Plant Sci.* 164, 517–523.
78. Roderick, H., Tripathi, L., Babirye, A., Wang, D., Tripathi, J., Urwin, P. E., et al. (2012). Generation of transgenic plantain (*Musa* spp.) with resistance to plant pathogenic nematodes. *Mol. Plant Pathol.* 13, 842–851.
79. Rodriguez-Kabana, R., Morgan-Jones, G., and Chet, I. (1987). Biological control of nematodes: soil amendments and microbial antagonists. *Plant Soil* 100, 237–247.
80. Rosso, M. N., Jones, J. T., and Abad, P. (2009). RNAi and functional genomics in plant parasitic nematodes. *Annu. Rev. Phytopathol.* 47, 207–232.
81. Safari, E., Gororo, N. N., Eastwood, R. F., Lewis, J., Eagles, H. A., and Ogbonnaya, F. C. (2005). Impact of *Cre1*, *Cre8* and *Cre3* genes on cereal cyst nematode resistance in wheat. *Theor. Appl. Genet.* 110, 567–572.
82. Sayre, R. M. (1986). Pathogens for biological control of nematodes. *Crop Prot.* 5, 268–276.
83. Sayre, R. M., and Walter, D. E. (1991). Factors affecting the efficacy of natural enemies of nematodes. *Annu. Rev. Phytopathol.* 29, 149–166.
84. Sharon, E., Spiegel, Y., Salomon, R., and Curtis, R. H. (2002). Characterization of *Meloidogyne javanica* surface coat with antibodies and their effect on nematode behaviour. *Parasitology* 125, 177–185.
85. Siddique, S., Endres, S., Atkins, J. M., Szakasits, D., Wieczorek, K., Hofmann, J., et al. (2009). Myo-inositol oxygenase genes are involved in the development of syncytia induced by *Heterodera schachtii* in Arabidopsis roots. *New Phytol.* 184, 457–472.
86. Siddique, S., Wieczorek, K., Szakasits, D., Kreil, D. P., and Bohlmann, H. (2011). The promoter of a plant defensin gene directs specific expression in nematode-induced syncytia in Arabidopsis roots. *Plant Physiol. Biochem.* 49, 1100–1107.
87. Sikora, R. A. (1992). Management of the antagonistic potential in agricultural ecosystems for the biological control of plant parasitic nematodes. *Annu. Rev. Phytopathol.* 30, 245–270.
88. Simonetti, E., Alba, E., Montes, M. J., Delibes, A., and Lopez-Brana, I. (2010). Analysis of ascorbate peroxidase genes expressed in resistant and susceptible wheat lines infected by the cereal cyst nematode, *Heterodera avenae. Plant Cell Rep.* 29, 1169–1178.
89. Singh, A. K., Sharma, A. K., and Shoran, J. (2009). "*Heterodera avenae* and its management on wheat in India," in *Cereal Cyst Nematodes: Status, Research and Outlook*, eds I. T. Riley, J. M. Nicol, and A. Dababat (Ankara: CIMMYT).
90. Sobczak, M., Avrova, A., Jupowicz, J., Phillips, M. S., Ernst, K., and Kumar, A. (2005). Characterization of susceptibility and resistance responses to potato cyst nematode (*Globodera* spp.) infection of tomato lines in the absence and presence of the broad-spectrum nematode resistance *Hero* gene. *Mol. Plant Microbe Interact.* 18, 158–168.
91. Sobczak, M., Golinowski, W. A., and Grundler, F. M. W. (1999). Ultrastructure of feeding plugs and feeding tubes formed by *Heterodera schachtii. Nematology* 1, 363–374.

92. Sorribas, F., Ornat, C., Verdejo-Lucas, S., Galeano, M., and Valero, J. (2005). Effectiveness and profitability of the *Mi*-resistant tomatoes to control root-knot nematodes. *Eur. J. Plant Pathol.* 111, 29–38.
93. Stirling, G. R. (1991). *Biological Control of Plant-Parasitic Nematodes.* Wallingford: CAB International.
94. Szakasits, D., Heinen, P., Wieczorek, K., Hofmann, J., Wagner, F., Kreil, D. P., et al. (2009). The transcriptome of syncytia induced by the cyst nematode *Heterodera schachtii* in Arabidopsis roots. *Plant J.* 57, 771–784.
95. Tamilarasan, S., and Rajam, M. V. (2012). Engineering crop plants for nematode resistance through host-derived RNA interference. *Cell Dev. Biol.* 2, 114.
96. Tripathi, L., Atkinson, H., Roderick, H., Kubiriba, J., and Tripathi, J. N. (2017). Genetically engineered bananas resistant to *Xanthomonas* wilt disease and nematodes. *Food and Energy Security*
97. Tripathi, L., Babirye, A., Roderick, H., Tripathi, J. N., Changa, C., Urwin, P. E., et al. (2015). Field resistance of transgenic plantain to nematodes has potential for future African food security. *Sci. Rep.* 5:8127.
98. Tripathi, L., Tripathi, J. N., Roderick, H., and Atkinson, H. J. (2013). Engineering nematode resistant plantains for sub-Saharan Africa. *Acta Hortic.* 974, 99–107.
99. Uehara, T., Sugiyama, S., Matsuura, H., Arie, T., and Masuta, C. (2010). Resistant and susceptible responses in tomato to cyst nematode are differentially regulated by salicylic acid. *Plant Cell Physiol.* 51, 1524–1536.
100. Urwin, P. E., Atkinson, H. J., Waller, D. A., and Mcpherson, M. J. (1995). Engineered Oryzacystatin-I expressed in transgenic hairy roots confers resistance to *Globodera pallida. Plant J.* 8, 121–131.
101. Urwin, P. E., Green, J., and Atkinson, H. J. (2003). Expression of a plant cystatin confers partial resistance to *Globodera*, full resistance is achieved by pyramiding a cystatin with natural resistance. *Mol. Breed.* 12, 263–269.
102. Urwin, P. E., Levesley, A., Mcpherson, M. J., and Atkinson, H. J. (2000). Transgenic resistance to the nematode *Rotylenchulus reniformis* conferred by *Arabidopsis thaliana* plants expressing proteinase inhibitors. *Mol. Breed.* 6, 257–264.
103. Urwin, P. E., Lilley, C. J., Mcpherson, M. J., and Atkinson, H. J. (1997). Resistance to both cyst and root-knot nematodes conferred by transgenic *Arabidopsis* expressing a modified plant cystatin. *Plant J.* 12, 455–461.
104. Urwin, P. E., Mcpherson, M. J., and Atkinson, H. J. (1998). Enhanced transgenic plant resistance to nematodes by dual proteinase inhibitor constructs. *Planta* 204, 472–479.
105. Vain, P., Worland, B., Clarke, M. C., Richard, G., Beavis, M., Liu, H., et al. (1998). Expression of an engineered cysteine proteinase inhibitor (Oryzacystatin-IΔD86) for nematode resistance in transgenic rice plants. *Theor. Appl. Genet.* 96, 266–271.
106. van der Vossen, E. A., Van Der Voort, J. N., Kanyuka, K., Bendahmane, A., Sandbrink, H., Baulcombe, D. C., et al. (2000). Homologues of a single resistance-gene cluster in potato confer resistance to distinct pathogens: a virus and a nematode. *Plant J.* 23, 567–576.
107. Vasconcelos, I. M., and Oliveira, J. T. (2004). Antinutritional properties of plant lectins. *Toxicon* 44, 385–403.
108. Vieira, P., Wantoch, S., Lilley, C. J., Chitwood, D. J., Atkinson, H. J., and Kamo, K. (2015). Expression of a cystatin transgene can confer resistance to root lesion nematodes in *Lilium longiflorum* cv. 'Nellie White'. *Transgenic Res.* 24, 421–432.
109. Vishnudasan, D., Tripathi, M. N., Rao, U., and Khurana, P. (2005). Assessment of nematode resistance in wheat transgenic plants expressing potato proteinase inhibitor (*PIN2*) gene. *Transgenic Res.* 14, 665–675.
110. Wang, D., Jones, L. M., Urwin, P. E., and Atkinson, H. J. (2011). A synthetic peptide shows retro- and anterograde neuronal transport before disrupting the chemosensation of plant-pathogenic nematodes. *PLoS ONE* 6:e17475.

111. Webster, J. M. (1995). "The host-parasite relationships of plant-parasitic nematodes," in *Advances in Parasitology*, ed. B. Dawes (Cambridge, MA: Academic Press Inc).

112. Westphal, A. (2011). Sustainable approaches to the management of plant-parasitic nematodes and disease complexes. *J. Nematol.* 43, 122–125.

113. Wieczorek, K., Hofmann, J., Blochl, A., Szakasits, D., Bohlmann, H., and Grundler, F. M. W. (2008). Arabidopsis endo-1,4-beta-glucanases are involved in the formation of root syncytia induced by *Heterodera schachtii*. *Plant J.* 53, 336–351.

114. Williams, K. J., Fisher, J. M., and Langridge, P. (1994). Identification of RFLP markers linked to the cereal cyst nematode resistance gene (*Cre*) in wheat. *Theor. Appl. Genet.* 89, 927–930.

115. Williamson, V. M. (1998). Root-knot nematode resistance genes in tomato and their potential for future use. *Annu. Rev. Phytopathol.* 36, 277–293.

116. Williamson, V. M., and Kumar, A. (2006). Nematode resistance in plants: the battle underground. *Trends Genet.* 22, 396–403.

117. Winter, M. D., Mcpherson, M. J., and Atkinson, H. J. (2002). Neuronal uptake of pesticides disrupts chemosensory cells of nematodes. *Parasitology* 125, 561–565.

118. Youssef, R. M., Kim, K. H., Haroon, S. A., and Matthews, B. F. (2013a). Post-transcriptional gene silencing of the gene encoding aldolase from soybean cyst nematode by transformed soybean roots. *Exp. Parasitol.* 134, 266–274.

119. Youssef, R. M., Macdonald, M. H., Brewer, E. P., Bauchan, G. R., Kim, K. H., and Matthews, B. F. (2013b). Ectopic expression of AtPAD4 broadens resistance of soybean to soybean cyst and root-knot nematodes. *BMC Plant Biol.* 13:67.

Index